Agricultural
Systems
Modeling and
Simulation

BOOKS IN SOILS, PLANTS, AND THE ENVIRONMENT

Modern Soil Microbiology, edited by J. D. van Elsas, J. T. Trevors, and E. M. H. Wellington

Growth and Mineral Nutrition of Field Crops: Second Edition, N. K. Fageria, V. C. Baligar, and Charles Allan Jones

Fungal Pathogenesis in Plants and Crops: Molecular Biology and Host Defense Mechanisms, P. Vidhyasekaran

Plant Pathogen Detection and Disease Diagnosis, P. Narayanasamy

Agricultural Systems Modeling and Simulation, edited by Robert M. Peart and R. Bruce Curry

Agricultural Biotechnology, edited by Arie Altman

Additional Volumes in Preparation

Plant–Microbe Interactions and Biological Control, edited by Gregory J. Boland and L. David Kuykendall

Handbook of Soil Conditioners, edited by Arthur Wallace and Richard E. Terry

Environmental Chemistry of Selenium, edited by William T. Frankenberger Jr. and Richard A. Engberg

Sulfur in the Environment, edited by Doug Maynard

Soil-Machine Interactions, edited by Jie Shen and Radhey Lal Kushwaha

Principles of Soil Chemistry: Third Edition, Revised and Expanded, Kim H. Tan

Agricultural Systems Modeling and Simulation

EDITED BY

ROBERT M. PEART

R. BRUCE CURRY
The University of Florida
Gainesville, Florida

CRC Press
Taylor & Francis Group
Boca Raton London New York

CRC Press is an imprint of the
Taylor & Francis Group, an **informa** business

CRC Press
Taylor & Francis Group
6000 Broken Sound Parkway NW, Suite 300
Boca Raton, FL 33487-2742

First issued in paperback 2019

© 1998 by Taylor & Francis Group, LLC
CRC Press is an imprint of Taylor & Francis Group, an Informa business

No claim to original U.S. Government works

ISBN-13: 978-0-367-40085-9

Visit the Taylor & Francis Web site at
http://www.taylorandfrancis.com

and the CRC Press Web site at
http://www.crcpress.com

Preface

Modeling has become a well-developed discipline, and modeling research on agricultural processes and operations is being conducted in many locations throughout the world. One of the most productive aspects of this work is that it makes it possible for workers at distant locations to cooperate by obtaining a copy of a program, and applying it to their own agricultural environment. The speed and inexpensiveness of the internet have also enhanced communication. This book is designed to further such interaction by presenting the recent work of many of the top agricultural systems modeling researchers in the world.

Models and simulation are mentioned throughout this book, so brief definitions are necessary. We think of the model of a system as the set of equations and rules that quantitatively describe the operation of the system through time. Simulation is the process of solving these equations within the rules with changing time, that is, simulating or mimicking the performance of the system over time by calculating values of the variables at each series of time steps.

The process of modeling and simulation in agriculture is unique. First, most agricultural problems require simulation of continuous processes, as well as discrete events. Second, agricultural problems face the possibility of discontinuities, such as those caused by a killing frost or a severe drought; changes in the process function, such as shifting from daytime to dark respiration; and phenology changes, such as the plant's "switch" from leaf expansion to seed production.

In addition, agricultural models usually seek to mimic real-life situations, so the effect of economics is often a part of the problem. High-capacity field equipment helps to service agronomic crops in a timely manner, to avoid yield losses due to late operations. However, the fixed cost of the equipment might be more than the savings from yield losses. Models that aid in selection and design of field equipment for crop agriculture are presented in chapters on selection of machinery (Chapter 14) and on simulation of a whole-farm system of specific equipment, crops, and operations that must be carried out in a timely fashion (Chapter 15). As irrigation water becomes more and more expensive, decisions are necessary, not only about timing of irrigation scheduling, but also about the amount of irrigation water to apply. Chapter 9 addresses these questions. The simulation of evaporation of water from the soil and from the leaves of plants has been the subject of a great deal of research, and a thorough review is presented in Chapter 7.

The amazing price/performance facts of today's personal computers, as well as the continuing decrease in this ratio, make the equipment cost factor for modeling a trivial question. Today, equipment costs are minor compared to software development costs. Software engineers are helping to ease this cost problem, by developing various new software tools. Object-oriented programming systems (OOPS) are now favored for beginning a new simulation project. Such systems are generally thought of as a formalized extension of the concept of modular programming, through the use of a series of subprograms and a relatively brief main or executive program. This method is covered in Chapter 17.

Most of the models described here are one-dimensional in the sense that time is the main variable. However, in many situations, the space or areal dimension is also important, as in the growth of the amount of space occupied by an insect, or changes in the energy use over time, and in particular areas. This latter example is covered in Chapter 4, in which data from India are presented.

Many models of agricultural systems programmed by graduate students have been subsequently laid to rest on the thesis shelf when students graduated and went on to new pursuits. A smaller number of models have been written and rewritten, and are continually being revised and released in newer versions, as with any commercial software. These are models that are used by a number of researchers and teachers (including extension specialists), often in several countries throughout the world. This book encourages the development of more useful models, to be shared with other scientists and engineers working in agriculture. Examples of models that have been used by a substantial number of researchers are CROPGRO (Chapters 2 and 18) and GRAZE (Chapter 10).

Many people in agriculture wonder why more of these models are not used directly by growers on the farm; implying that if they are not used on the farm, then the cost of the development was not justified. However, this overlooks the fact that historically most agricultural research involved finding out how a particular plant, animal, or machine system behaves under varied conditions, and models were developed from that research. When questions came up about the possible effect of climate change on agricultural crops, models of the major grain crops were already available. With rather simple modifications to account for the effects of carbon dioxide concentration, the models were the quickest and least expensive method of estimating the answers to these questions.

In addition to the previously mentioned reasons for the value of developing crop models even before they have been applied to a particular farm, the second answer is that some crop models are being used on the farm today. GOSSYM/COMAX, the cotton growth and management model described in Chapter 8, has been used by many farmers for about 10 years, and the California program called CALEX, while not strictly a simulation program, has also been used by farmers and their technical advisers for several years. The growth of the profession of crop and pest management consulting is an indication that the use of crop models by this group of experts will probably be more widespread than the direct use of models by farmers themselves. In addition, many of these models will be modified and adapted for easier use with new software methods such as the various "Visual" tools for developing Graphical User Interfaces with particular programs. Chapter 1 reviews important principles by which to adapt these research models to farm use, according to what is important to the farm manager at the time. We predict growing use of agricultural models on the farm with private or public extension workers involved in a great many of these uses.

Other specialized methods of modeling are covered in this book, including the older linear programming method (Chapter 5), and the newer neural network theory (Chapter 16). Linear programming is a unique method of finding specific solutions to problems that can be formulated to meet the criteria of the technique (briefly, linear constraints, costs, and non-negativity). It is used routinely in many farm and agribusiness applications. Neural network theory is much newer, and has been applied only in the past few years. Linear programming works when the mathematical relationships are linear and known. Neural network methods work when there is a great amount of data on inputs to and outputs from the system but the mathematical relationships are not known. Another newer technique—expert systems, or decision support systems—is presented here as a method of "fine-tuning" the parameter values of a crop model which includes the process of moisture movement in the soil (Chapter 6). Fundamental mathematical methods applied to biological processes, such as moisture and gas movement through the stomata of a leaf, are covered (Chapter 3) to round out the mathematical and computational methods.

Modeling of livestock systems probably does not have as long a history as that of crop systems, because modeling livestock on pasture involves modeling the pasture as well. Four very different approaches are presented. First, a fundamental method of modeling, using differential equations and classical solutions for those equations, is well documented in Chapter 11. Then, a dairy cow is simulated with the usual time steps and physiological processes, but the feed supply does not include growth modeling of a crop to grazing (Chapter 12). Chapter 13 shows how the detailed process of the animal chewing on and consuming the forage can be modeled and simulated as part of the animal–pasture interaction. Chapter 10 uses the physiological or process approach, with numerical integration, and models the growth of the pasture and harvesting of the pasture by the animal (or stopping this grazing for a time) as interacting processes.

As the population of the world continues to grow, and with current grain surpluses becoming almost nonexistent, agricultural production has become very important. Modeling and simulation of food production systems will be increasingly used in the future, to help meet the global need for food. We hope this book will help to spread the knowledge of agricultural systems modeling and simulation to meet this task.

Robert M. Peart
R. Bruce Curry

Contents

Contributors

John R. Barrett Agricultural and Biological Engineering Department, Agricultural Research Service, U.S. Department of Agriculture, Purdue University, West Lafayette, Indiana.

William D. Batchelor Agricultural and Biosystems Engineering, Iowa State University, Ames, Iowa

John Bolte Bioresource Engineering Department, Oregon State University, Corvallis, Oregon

Kenneth J. Boote Agronomy Department, University of Florida, Gainesville, Florida

Susan M. Bridges Department of Computer Science, Mississippi State University, Mississippi State, Mississippi

J. Robert Cooke Department of Agricultural and Biological Engineering, Cornell University, Ithaca, New York

R. Bruce Curry Agricultural and Biological Engineering Department, University of Florida, Gainesville, Florida

J. B. Dent Institute of Ecology and Resource Management, University of Edinburgh, Edinburgh, Scotland

R. H. Fawcett   Institute of Ecology and Resource Management, University of Edinburgh, Edinburgh, Scotland

M. Herrero   Institute of Ecology and Resource Management, University of Edinburgh, Edinburgh, Scotland

Harry F. Hodges   Department of Plant and Soil Sciences, Mississippi State University, Mississippi State, Mississippi

Gerrit Hoogenboom   Department of Biological and Agricultural Engineering, University of Georgia, Griffin, Georgia

James W. Jones   Agricultural and Biological Engineering Department, University of Florida, Gainesville, Florida

Harbans Lal   Pacer Infotec Inc., Portland, Oregon

Neil L. Lecler   Department of Agricultural Engineering, University of Natal, Pietermaritzburg, South Africa

Otto J. Loewer, Jr.   College of Engineering, University of Arkansas, Fayetteville, Arkansas

Joep C. Luyten   Agricultural and Biological Engineering Department, University of Florida, Gainesville, Florida

James M. McKinion   Crop Simulation Research Unit, Agricultural Research Service, U.S. Department of Agriculture, Mississippi State, Mississippi

John L. Monteith   Institute of Terrestrial Ecology, Penicuik, Midlothian, Scotland

Mark A. Nearing   Agricultural and Biological Engineering Department, Agricultural Research Service, U.S. Department of Agriculture, Purdue University, West Lafayette, Indiana

B. S. Panesar   School of Energy Studies for Agriculture, Punjab Agricultural University, Ludhiana, Punjab, India

Robert M. Peart   Agricultural and Biological Engineering Department, University of Florida, Gainesville, Florida

K. Raja Reddy   Department of Plant and Soil Sciences, Mississippi State University, Mississippi State, Mississippi

John C. Siemens   Agricultural Engineering Department, University of Illinois at Urbana–Champaign, Urbana, Illinois

Jan Tind Sørensen   Department of Animal Health and Welfare, Danish Institute of Animal Science, Tjele, Denmark

Robert S. Sowell Agricultural and Biological Engineering Department, North Carolina State University, Raleigh, North Carolina

Steven J. Thomson U.S. Cotton Ginning Laboratory, Agricultural Research Service, U.S. Department of Agriculture, Stoneville, Mississippi

R. C. Ward Colorado State University, Fort Collins, Colorado

Frank D. Whisler Department of Plant and Soil Sciences, Mississippi State University, Mississippi State, Mississippi

Simon J. R. Woodward AgResearch Ruakura, Hamilton, New Zealand

1

Humanization of Decision Support Using Information from Simulations

John R. Barrett and Mark A. Nearing

Agricultural Research Service, U. S. Department of Agriculture, Purdue University, West Lafayette, Indiana

I. INTRODUCTION

Agricultural researchers have tried from the turn of the 20th century to develop data and information bases that when interpreted can be used to improve the management of agricultural systems. Early on, researchers recorded and published descriptions that, for the most part, were observations of situations involving abnormal and unusual occurrences. With the invention of adding machines and calculators, scientists began to define and determine relationships. With the advent of computers, engineers began to consider continuous and discrete happenings with respect to time. Beginning in the late 1950s, descriptive and mathematical modeling of processes evolved. These mimics were called simulations. Figure 1 is suggested, somewhat with tongue in cheek.

In the mid-1960s, the concept of systems dynamics emerged, facilitating time-related representations of process flows and interactions. During the 1970s, systems dynamics became formalized. Re-

1

FIGURE 1   Simulating is stimulating!

finements continued through the 1980s, primarily in computer programming techniques, in verification, validation, and evaluation. Simulation of dynamic agricultural and industrial systems has become an integral part of agricultural science. Increased understanding of ecosystem interactions, as influenced by the environment and management practices, has greatly expanded the potential for decision support systems. In the mid-1990s we experienced the emergence of the era of information technologies.

Bridging the gap between simulations that have been developed to mimic and describe agricultural processes and procedures and the use of these simulations to supply information that supports decisions being made by managers is difficult and challenging. Supplying information to managers involves interfacing between computer output and the people who are the information users (see Fig. 2).

II. DECISION-MAKING PROCESS AS IT OCCURS WITHIN HUMANS

A first point is that all decisions are made in a person's mind. Some human makes each, any, and all decisions, either directly or indirectly. Further, decisions are always based on whether to continue doing things as they are being done or to change to a different procedure. And, people who control the resources have the authority to make the decisions. The icon of Fig. 3 emphasizes all this.

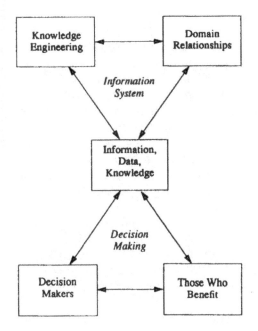

FIGURE 2 Information system and decision making.

FIGURE 3 Where decisions are made.

A. Thinking Process and Human Memory

The mind, including the brain and sensory mechanisms for perception, functions to process and store symbolic expressions. An individual feels, perceives, thinks, remembers, and reasons in an adaptive conscious and nonconscious manner. One way to think of this is that the conscious mind is the interface between the source of symbols that have been sensed and nonconscious memory, where they are stored. Contained in a person's memory is everything that has been observed and learned from all experiences, right or wrong, good or bad. Certainly, some things have become buried so deeply that they are obscure. Stored in memory may be data, information, and/or knowledge. Words are stored symbolically.

In resolving a problem, the contents of memory are reviewed, perhaps intuitively, and ordered according to importance, and alternatives are identified and automatically ranked. Risks are considered. Short- and long-term consequences are explored, and choices are made. Several matters may be thought through interactively and simultaneously.

When an individual is confronted with a problem to be solved, several things happen. Some considerations occur almost instantly, and some are more protracted. What cognitive actions occur, the order in which they occur, and whether some of these actions recur are a function of experience and knowledge of the domain where a decision is to be made. Most important, they are not a function of how others see the problem, but of how the individual with the problem sees it.

Mostly, decisions are choices to continue a course of action or to take a divergent course. Many preparatory dilemmas are resolved over time before a decision is made to take action. Typically, when an individual is confronted with a new situation or one where he or she wants the action taken to be more effective, an attempt will be made to review the situation. Thus, a fresh look may be taken to reevaluate and reduce risks.

The blackboarding process that has been developed for computer information searching and problem solving is a mimic of what in complex fashion goes on in a human's mind. That is, several problem areas may be addressed simultaneously. The processing goes on night and day, asleep or awake, consciously and nonconsciously, until the equations are complete or the situation is resolved to the point that risk is adequately reduced and it is felt that action can be safely taken. Then the resolution of the dilemma becomes evident. This may

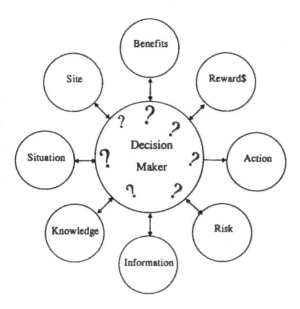

FIGURE 4 Decision maker.

occur in the middle of the night as if out of a dream or at some other time.

All of this is important to an understanding of how simulations are potentially excellent sources of important information that can support decision making. Symbolically, the interaction of high impact areas is portrayed in Fig. 4.

III. DECISION SUPPORT

So what does all this mean to decision support? Foremost, the job is easier than may have been thought. Recognizing that decisions are always made in a human's mind, support of the decision-making process can best be accomplished by supplying missing or incomplete data or correcting inaccurate information about interactive critical areas in logic and achieving wording that is understandable and acceptable to the decision maker. This should point out the methodology that potentially can be used in any decision support process.

Simple? Not necessarily, because we are instilled through training and experience to view problems from scientific perspectives, not from the applications-oriented viewpoints of people who manage businesses. Scientists use the scientific method, which gives verified and validated information with defined error, while others seek the interpretation of information and data in changing situations. Scientists think in terms of probability, and decision-making managers look to possibility.

A. Example Sources of Information

Excellent farmers who produce grain said that the following sources are most used for needed information (Barrett and Jacobson, 1995):

> Networking with other experts.
> Suppliers of chemicals, equipment, seed, etc.
> Extension and university specialists, trusted, with no profit ties.
> Consultants, although rarely used.
> Manufacturers' technical departments.
> Publications, trade and other.
> Specification statements and sheets.
> Computers for grain price and weather information.

Note that this list does not include social contacts.

B. Managerial Time

These same excellent grain farmers explained "managerial time" by pointing out that they are more receptive to information during certain periods of the year. Assume that a season begins with the end of harvest, targeted in the Midwest to be about the first of November. Following this are

1. The postharvest period, which lasts about 2 months, or until the first of the new calendar year. Postharvest is when decision support can best occur, as during this time all plans are finalized for the next season, considering prior experiences, especially the immediate past growing season.
2. Preplanting or the 3+ month period used for getting everything ready for planting.
3. Planting, targeted to be 2+ weeks.
4. The growing period, which lasts about 4+ months.
5. Harvest, which occupies the remaining 1+ month.

To support decisions with maximal effect, information should be presented at times when information users are most receptive. In the case of grain farmers, this would be from late harvest until the end of the year for strategic planning, and/or as they walk out the door to accomplish an operation such as planting, tilling, or irrigating. This latter is tactical and short-run operational.

C. Computer Programs

Computer programs including spreadsheets, expert systems, decision support systems, simulations, and similar items can be excellent sources of information for decision support. For more information on knowledge engineering in agriculture, see Barrett and Jones (1989). Unfortunately, there has been limited acceptance of computer programs by information users. Why? Perhaps it is partly because the information as presented is not what is needed. Of course there are other reasons. Computer output must be in wording and logic that can be understood by the information users.

The founders of artificial intelligence—Minsky, Feigenbaum, Englemore, and others—stated emphatically, in personal communications about 1983, that energies should be spent in defining problems, that representative clients should be included on research teams, and that programming, language development, information storage and retrieval, machine learning, and other mechanistic computer-related developments should be left to specialists. The greatest challenge would be to develop software programs that would actually be used. Advice was to specifically:

1. *Make needs assessments.* Determine from discussions with the proposed clients that there is a real and economically viable need for program development and that there is a chance of meeting that need. Many programs are produced only because there is a body of knowledge that can be expressed.
2. *Involve users.* Both users of programs and users of the information generated should be included on developmental teams from the beginning, not simply to follow progress and to evaluate output.
3. *Clearly define important problems.* Program logic needs to be expressed from the perspective of the people who have the problem. Problem definition must be from an information user's viewpoint and not be limited by the domain specialist.
4. *Not worry about software and hardware.* Developers were advised to leave software writing to specialists and consider program needs, economics, and availability when selecting hardware.

5. *Address the needs of potential users.* Users may be farmers, extension service specialists, scientists, agribusiness people, or other clients. They seek information and interpretations. To be understood, outputs should be in the language of the information users, not that of the domain experts.

6. *Allow for maintenance.* The 80/20 rule characterizes the development process. It takes 20% of the effort to develop a system or program, and the other 80% for validation, delivery, and maintenance.

7. *Know the benefactors.* Keep in mind the people who are funding the work.

IV. SIMULATION

Simulation is the process of using a model, or models, dynamically to follow changes in a system over time. This is dynamic mimicry. Models may be descriptive and/or mathematical. Technically, models are equations and rules defining and describing a system. Simulation involves calculating values over time in dynamic fashion. For thorough understanding of modeling and simulation, look to Peart and Barrett (1979) and Barrett and Peart (1982).

Almost all phenomena related to agriculture have been simulated, some accurately. Everything—for example, biochemical pathways, erosion, water runoff, crop drying, pesticide leaching, crop growth and development, harvest, transport, residue decomposition, the aspects of cash flow and economics, cost effectiveness, weed stress, pest (insect, weed, disease) management, machinery selection and management, animal production, dairy herd improvement, hydrology, pollution, irrigation, biomass burning, global warming, weather—has been simulated. Most of these simulations came out of a scientist-researcher's desire to express his or her understanding of system interactions. Nevertheless, simulations are resources that can be used to supply data, information, and knowledge to decision makers in agriculture (Fig. 5). For example, a crop growth and development simulation can be used to predict the amount of dry matter in a plant or crop at the time of harvest, which can be related to residue left on the surface after harvest. Considering dispersal and decomposition, the residue cover at planting can be predicted and related to the potential for erosion or weed suppression. All of these relationships can be simulated to help a farm manager make strategic and procedural decisions. The difficulty lies in expressing relationships of critical factors in a way the information user can understand.

FIGURE 5　Data, information, and knowledge are resources.

V.　GRAIN PRODUCTION EXAMPLE

Research carried out to support decisions in grain production was used as an example by Barrett and Jacobson (1995). Excellent farmers were observed over a 2 year cycle (rotation of corn and soybeans) to determine the factors critical to their success (Rockart, 1979; Bullen and Rockart, 1981). These factors were related to the best management practice areas. A computer process/program was developed that emphasizes the most important management practices where optimization can occur. These areas are listed below in the expert's priority to show where simulations can be used to supply information.

> *Cash flow.*　This involves timing and the dynamics of selling the crop, buying resources, attaining maximum return in dollars and yield, achieving sustainability in farming, making major payments, taxes, interest, etc. From a producer's viewpoint, economics per se is a subset of cash flow.
>
> *Moisture.*　Conditions are basic to all thinking and decisions. Management must always be aware of conditions, local, regional, and international, including rain; the effect on scheduling; surface conditions; subsoil levels; if and when to irrigate. Managing with respect to surface, soil, and grain moisture is paramount to success.
>
> *Equipment.*　Here "equipment" includes maintenance, manner of operation, repair, replacement and change, purchase, labor required for use, innovations.

Information. Information must be continuously searched for with respect to prices, yield, policies, international affairs, tillage, variety performance, new technologies, etc. Networking is useful here.

Crop effects. Status of growth and development, local and regional; performance considering moisture and soil status; vigor after emergence; stress factors; stability at harvest; shatter; size; stamina; etc.

Weed suppression. Existing and potential pressures related to tillage; effectiveness of chemicals, spray scheduling, etc. The producers were observed to think of conventional tillage, reduced tillage, and no-till as variations of seedbed preparation directly affecting potential weed pressures.

Planting. Varieties; scheduled over 2+ week period; when to start with respect to harvest; tillage; labor; use of planter or drill; etc.

Fertility. Maintenance, soil testing, scheduling, etc.

Harvest. Schedule over 1+ month period; preparation; storage; labor; timing relative to planting schedule; etc.

Erosion. Environmental stewardship, sustainable agriculture, sustainable production, conservation, compliance, etc.

Diseases, insects. Scouting; chemical selection; spraying; etc.

Other enterprises. All observed expert farmers maintain an associated non-production-agriculture business enterprise where resources are used interactively.

Labor, personal preferences, other crops, other factors.

Any information from simulations about interactions of the above areas can potentially improve decision making.

A. Structure of Program to Supply Information

To continue the example being discussed, after considering the management practice areas that have been identified, a prototype computer decision support system of programs was assembled that includes an interface between the computer user and the physiologically based soybean and corn crop growth and development simulations, environmental quality simulations, hydrology simulations, and weed and herbicide effect models. There is also an additional program structure to compute profitability. For detailed discussion, see Jacobson et al. (1995).

The important problem to be solved is to complete the interface between the computer as a source of data and information and the information user, who is usually not the keyboard user. Relating the outputs of these programs with the informational needs of a manager of production is not a straightforward task. The information being sought relative to the critical areas must be extracted and phrased in words and logic appropriate to the decision maker.

VI. EROSION EXAMPLES

Early erosion simulations were empirical equations. Later, process-based models were developed. Information on rates of erosion related to land management on a specified site continues to be objectively sought.

A. Universal Soil Loss Equation

The first widely used erosion simulation was the Universal Soil Loss Equation (USLE) (Wischmeier and Smith, 1978). This statistically derived regression model was developed to estimate average annual soil loss from a hillslope. USLE does not address sediment deposition on the toe slope of the hill, the amount of sediment leaving a field, or the composition of the sediment eroded.

USLE has been the basis for the development and implementation of conservation practices throughout the world. It has been applied to nearly every acre of cropland in the United States. The primary reason for its success is its simplicity. The mathematics can be done on a calculator, on the back of an envelope, or in the head. The answer is also simple, being a single average annual soil loss value for a hillslope. USLE is easily implemented and easily integrated with large and complex simulations and GIS applications.

B. Revised Universal Soil Loss Equation

The Revised Universal Soil Loss Equation (RUSLE) (Renard et al., 1996) is a computer version of USLE. RUSLE includes additional process descriptions such as temporal changes in soil susceptibility to erosion and the time decay of surface crop residue.

Using RUSLE involves a computer, as it requires extensive data input, including access to files for plants, climate, and management systems. It makes simulations on a daily to biweekly time step for various processes that influence erosion, such as residue decay, tillage, seasonal soil erodibility, and plant biomass production. The results of RUSLE are allegedly better than those of USLE because time-variable factors are considered that USLE does not include and because RUSLE was designed to better predict erosion on rangelands with minimum tillage conditions. Yet overall, RUSLE provides the same level of information as USLE, i.e., average annual soil loss on a hillslope.

C. Water Erosion Prediction Project

The Water Erosion Prediction Project (WEPP) model (Nearing et al., 1989; Flanagan and Nearing, 1995) is composed of a complex set of process-based models. It provides information on soil loss, deposition, sediment leaving the field, and the characteristics of this sediment on a daily, monthly, annual, and average annual basis. This is important in terms of off-site environmental impact. It provides information on soil water content over time and surface water runoff rates and amounts. It provides information on the size distribution of sediment that leaves the field. Characterization of sediment is important for quantifying chemical transport.

The WEPP model predicts crop yields. WEPP also gives spatial and temporal distribution information on erosion. This should be of benefit in meeting the needs of the precision farmer. On the other hand, WEPP is data-intensive, both for input and for output. Interpretation of the WEPP output information requires an erosion expert.

It will be interesting over the next few years to observe how these simulations are used. The challenge will be to make the models effective in supplying information that supports decisions.

D. Uses of Erosion Simulations

Erosion simulations are used for conservation planning, engineering design, erosion inventories, and government regulations. USLE, and now RUSLE, are the worldwide standards for evaluating conservation practices. Farmers in the United States, with the aid of soil conservationists, use these in estimating soil losses from their fields as they choose among strategic and tactical alternatives.

Erosion simulations are used extensively in the design of engineering structures such as terraces, diversion ditches, sediment catch basins, reservoirs, and drop structure plunge pools. They are also used as the basis for local, state, and national soil resource inventories, such as the USDA Natural Resource Inventory, which is conducted on a five-year basis.

The USLE was mentioned in the language of the 1985 USDA Farm Bill to define highly erodible lands. These lands were then tied to farm price support and loan program regulations. The 1985 farm bill was instrumental in both identifying, on the broadest scale, the field areas with high erosion potential and encouraging compliance with conservation guidelines on erosion-susceptible lands. Basically, fields that were identified as highly erodible had to be placed into approved conservation plans in order for the producer to be eligible for federal farm programs.

E. Futuring

Simulations are critically important for addressing all problems associated with erosion, from issues related to sustainable productivity and sustainable agriculture to water quality and associated environmental impact. Environmental stewardship can be enhanced by use of simulations.

The information on soil loss on the field and sediment delivery from the field and their respective impact on productivity losses and the downstream environment needs to be delivered when the decision maker is receptive, at the level of precision the decision maker can use, and in a way he or she understands, along with associated information on related issues such as chemical use, yields, and environmental impact.

Implementing soil conservation practices involves deciding whether or not to change from the present way of doing things, and if so, in what way. This is especially true when new conservation tillage practices are considered. Traditional practices include extensive tillage to manage weeds, insects, and diseases. There has been a gradual shift in production methodology from crop rotations, often including meadow crops, to alternative row cropping of corn and soybeans. This has been accompanied by an increase in the susceptibility of cropland to soil erosion. With the advancement of disease and pest control technology it became more feasible to leave crop residue on the soil surface and thus make farming more viable in terms of long-term soil conservation. The transition from moldboards and discing tillage to reduced, or minimum, tillage has been difficult. Four basic factors came into play that have driven the move to reduced tillage: economics, research, regulation, and conservation ethics.

VII. PRESCRIPTION (PRECISE/PRECISION) FARMING

The effectiveness of more precise (prescription) farming can be evaluated through the use of simulations and economic analysis. Benefits will vary among farms, with the technology used, and with the levels of management abilities. The underlying ideas of precision farming are solid and consistent with the concepts of good management. One farmer of the grain production example said that he had been practicing prescription farming all his life. His only questions were: How precise should he be, what size areas should he consider, and whether and when he should change! He said that practicing precise, site-specific farming was what made excellent management.

A.   Weed Suppression Example

Weed control, or suppression, is complicated when viewed from a grain producer's perspective. To show how farmers think about weed suppression, consider that weed control is a continuous and ongoing problem area. The potential for excessive competition is constantly being evaluated. Critical actions affect sustaining production and sustaining agriculture and environmental stewardship. Fall tillage, the purchase of herbicides, preplanting tillage and herbicide application, planting time tillage and herbicide application, postplanting tillage or herbicide application, labor utilization, safety, equipment purchase, repair and management, continuous and crisis control, etc., all interact with each other.

Weed identification, herbicide selection, and cost effectiveness are rather straightforward. Their interaction is where simulations can fit together to potentially supply critical information. Information users want to know how the resolution of each dilemma over time will affect yield and profitability and if they can sustain production over their lifetime and for generations to come. Simply stated, the question is "If I use more or less, here or there, will it pay?" where "pay" does not necessarily refer to dollar amounts.

B.   Fertilization Example

Relative to prescription (precise) farming, farmers in the Midwest are asking for help in determining whether changes in fertilization application rates, fertilizer types, or application methodologies will improve their profitability. If they are good producers, they monitor and maintain fertility levels on an ongoing, long-term basis according to characteristics of the fields—slopes, soil types, and so on. Should their methods be more refined? Simulations can supply the answers to their site- and situation-specific questions.

C. Dairy Management

In similar thinking, dairy production can be improved by using simulations to evaluate the effect of changes in herds, both numbers and quality of animals. This combination of cow and herd genetics, resource management, economics including sales methods, government programs, and quality control is basic to good business management.

D. State of Preciseness

The application of site- and situation-specific production concepts is destined to bring a whole new type of management to agriculture, one based on data and information technologies. To effectively implement more precise (prescription) production, the variability associated with small units (areas, cows) must be determined so that production yields and profitability can be predicted along with the potential for beneficial responses to input resources. This has the potential of extending into all facets of agriculture.

VIII. PROCESS OF DECISION SUPPORT

From an information user's viewpoint, there are several somewhat simplistic considerations that will help improve decision support. After the computer has been introduced as a source of help (not trivial), it becomes important to establish a believable base. A simulated base must be shown to replicate reality—a situation that the information user knows to be true. Then it becomes a matter of showing the results of each proposed change. If a ranked group of alternatives are presented as program output, each alternative will be viewed as information that when added to all the other information in memory supports change from the procedure being presently used.

Acceptance can be improved by leading information users to see that their personal data have been properly and accurately input, including weather, and that the yield, profit, and corresponding results predicted are what they observed to have happened. Then the person who is making the decision will be ready to accept as accurate the effects of changed inputs. Simulations come into play as they mimic procedures and show the potential results of changing resource use.

IX. CHALLENGES AND OPPORTUNITIES

It is a great challenge to develop simulations and use existing ones objectively to produce information that supports decisions. Otherwise, in most cases it is still necessary to determine what information, known and unknown, is critical to the resolution of problems and the improvement of decision making management. It is necessary to determine what available simulations are relevant and how to get the required information out of them. It is necessary to interface between simulations and other software, to interface output with information users, to determine where data and information bases can be found, to determine what can be easily input in user-specific manner, to limit the items that must be left to the information user to supply, and how to play believable what-if games.

X. PROGNOSIS

Much political, social, and scientific emphasis is being placed on global warming, environmental pollution, chemical leaching, pesticide and herbicide runoff erosion, conservation, and other quality-related matters. Correspondingly, we now have the ability and the opportunity to bring the processes of decision support together with simulation and the emerging methods of information technologies to better bridge the gap between research and practical applications. It is exciting to catch glimpses of the future and see that a farmer's sustainable production can be balanced with the environmentalist's sustainable agriculture so that we can be better environmental stewards while enjoying a high quality of life.

The future is unlimited! Grab the brass ring!

REFERENCES

Barrett, J. R., and B. M. Jacobson. 1995. Humanization of decision support for managing U.S. grain (soybean and corn) production. In: Proc. 2nd IFAC/IFIP/EurAgEng Workshop on AI in Agriculture. Wageningen, The Netherlands, 29–31 May, pp. 3–13.

Barrett, J. R., and D. D. Jones. 1989. *Knowledge Engineering in Agriculture.* Monograph No. 8. St. Joseph, MI: ASAE.

Barrett, J. R., and R. M. Peart. 1982. Systems simulation in U.S. agriculture. In: *Progress in Modelling Simulation.* F. E. Collier (Ed.), London: Academic Press, pp. 39–59.

Bullen, C. V., and J. F. Rockart. 1981. *A Primer on Critical Success Factors.* Center for Information Research No. 69. Sloan School of Management WP No. 1220-81, Cambridge, MA: Massachusetts Institute of Technology.

Flanagan, D. C., and M. A. Nearing. 1995. USDA-Water Erosion Prediction Project: Hillslope Profile and Watershed Model Documentation. NSERL

Report No. 10. West Lafayette, IN: USDA-ARS, National Soil Erosion Research Laboratory.

Jacobson, B. M., J. W. Jones, and S. M. Welch. 1995. Decision support system to assess agronomic, economic, and environmental impacts of soybean and corn management. ASAE Paper No. 95-2696. St. Joseph, MI: ASAE.

Nearing, M. A., G. R. Foster, L. J. Lane, and S. C. Finkner. 1989. A process-based soil erosion model for USDA-Water Erosion Prediction Project technology. *Trans. ASAE* 32:1587–1593.

Peart, R. M., and J. R. Barrett. 1979. The role of simulation. In: *Modification of the Aerial Environment of Plants.* Monograph No. 2. St. Joseph, MI: ASAE, pp. 467–480.

Renard, K. G., G. R. Foster, G. A. Weesies, D. K. McCool, and D. C. Yoder. 1996. *Predicting Soil Erosion by Water: A Guide to Conservation Planning with the Revised Universal Soil Loss Equation (RUSLE).* USDA Handbook No. 703. Washington, DC: Government Printing Office.

Rockart, J. R. 1979. Chief executives define their own data needs. *Harvard Business Review.* March/April, p. 81.

Wischmeier, W. H., and D. D. Smith. 1978. *Predicting Rainfall Erosion Losses: A Guide to Conservation Planning.* USDA Handbook No. 537. Washington, DC: Government Printing Office.

2

Simulation of Biological Processes

James W. Jones and Joep C. Luyten
University of Florida, Gainesville, Florida

I. INTRODUCTION

Biological systems are made up of interacting chemical and physical processes. Living systems are composed of many subsystems and components, each having its own unique characteristics and behavior while contributing to the overall form and function of an entire system. These systems are highly complex; many components interact simultaneously, and their interactions are highly nonlinear or chaotic in nature. These interactions and nonlinearities must be taken into account when attempts are made to understand or predict system behavior. Our understanding of these interactions is incomplete and often guided only by empirical evidence of overall system behavior instead of empirical data on processes that lead to overall system behavior. Because of these complexities, the classical mathematical methods used to study nonliving physical or chemical systems have been inadequate for living systems.

Simulation based on quantitative models of biological processes and their interactions can provide considerable insight into the behavior of living systems and into ways of managing these systems to achieve specific goals. During the last several decades, computer simulation has proved to be a powerful tool in basic and applied biological sciences. The incredible growth and acceptance of personal computers during this same time period has made it possible for simulation to become an integral part of biological research in many basic and applied fields. The purpose of this chapter is to present an introduction to the concepts and techniques used in the simulation of agricultural and biological systems with a few examples that demonstrate the approach.

Because of our own interest and experience in the simulation of agricultural production systems (particularly crop systems), we focus on the simulation of crop systems in this chapter and include examples related to plants, soil, and insects. Additional material on crop simulation can be found in a number of references, including Thornley and Johnson (1990), Leffelaar (1993), Penning de Vries et al. (1989), and Goudriaan and van Laar (1994). The concepts and methodology could have been presented equally well with examples on animal, disease, or other biological systems (e.g., see Curry and Feldman, 1987; Keen and Spain, 1990; France and Thornley, 1984; Odum, 1973, and Dent and Blackie, 1979).

II. TERMINOLOGY

A. System

A system is a collection of components and their interrelationships that have been grouped together for the purpose of studying some part of the real world. The selection of the components to include in a system depends on the objectives of the study and actually represents our simplified view of reality. Systems can be described as collections of mutually interacting objects that are affected by outside forces (Pritsker, 1995; Rabbinge et al., 1994). For example, typical crop models define the crop and soil root zone as components that interact in complex ways and are also affected by weather conditions and management practices. Other models may define a leaf or a cell as the system for which models are to be developed.

One of the complicating features of biological systems is that they are hierarchically organized and can be studied at a number of levels. Figure 1 shows hierarchical levels for crop systems. At different hierarchical levels, models and simulation analysis are being performed by scientists, engineers, and economists. Note the parallel between research and models at each of these levels. The fundamental goals and objectives of a model should guide one to determine which hierarchical level to use.

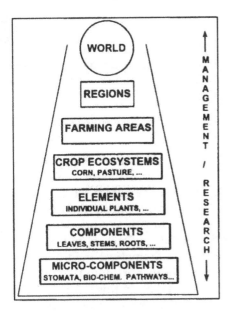

FIGURE 1   Hierarchy of models used in agriculture.

B.   Environment and System Boundary

To consider the scope of a system, one must prescribe its boundaries and its contents. The environment of a system includes everything except the components of the system. A system boundary is an abstraction of the limits of the system components, separating them from the environment. The boundary may be physical; however, it is better to think of it in terms of cause and effect. The environment may affect the system in a number of ways, but the system does not affect the environment. For example, there may be a flow of mass and energy from the environment to the system that affects the system's behavior. The model of the system must take into account how these flows affect the changes in system components. In contrast, the environment is not affected by the system, and the environment does not have to be modeled. For example, in a crop model, the crop is affected by the temperature and radiation of the environment but does not itself influence these environmental conditions. Of course, crop processes do affect the microclimate and, in a very small way, other meteorological conditions, but these are usually assumed to be negligible for a number of purposes.

C. Model

A model is defined as a mathematical representation of a system, and modeling is the process of developing that representation. Conceptually, the process of developing a model of a system is different from and a prerequisite to computer simulation.

D. Computer Simulation

Computer simulation includes the processes necessary for operationalizing or solving a model to mimic real system behavior. Developing computer logic and flow diagrams, writing the computer code, and implementing the code on a computer to produce desired outputs are necessary tasks in the simulation process. In practice, modeling and computer simulation are usually interrelated. One should have in mind techniques for implementing the mathematical model as it is being conceptualized and developed. Some authors define simulation to include the modeling process (Pritsker, 1995). In this chapter, "simulation of biological systems" refers to the overall procedures outlined below, including mathematical model development and computer simulation to produce numerical model results for analysis.

E. Inputs and Outputs

Inputs to the system are factors in the environment that influence the behavior of the system but are not influenced by the system. Inputs are also referred to as exogenous variables, driving variables, or forcing functions. Inputs may vary with time, and numerical values for inputs are needed for all values of time for which the model is to be simulated by prescribed values, tables, or equations. Rainfall, temperature, and light are examples of inputs to crop systems. These environmental variables affect the soil water and crop growth dynamics in a field but are not themselves affected by the crop system. The choice of components and inputs for various models may differ depending on model objectives and the availability of data. For example, if one chooses to continuously monitor soil water conditions, then soil water could be an input to a crop model having no soil component. However, soil components are usually added because of the difficulty in monitoring soil water at any site where one wishes to apply the model. This is in contrast to the relative ease with which rainfall can be monitored and used as input to a crop model with a soil component.

Outputs from the system represent the characteristic behavior of the system that is of interest to the modeler. For example, outputs from a crop system model would include biomass of the crop and water content of the soil component.

F. Parameters and Constants

Parameters and constants are characteristics of the components of the model that are constant throughout simulated time. Usually, constants are quantities with reliable and accurate values that remain the same when experimental conditions are varied or when the model is applied to different genotypes or organisms (France and Thornley, 1984). Examples of constants are the molecular weight of glucose, the gravitational constant, and the number of seconds in a day. Parameters, on the other hand, are quantities whose values are less certain but are assumed to be the same throughout a simulation. For example, parameters may define the functional response of photosynthesis to light, the resistance to soil water flow, respiration loss for tissue synthesis, and the number of degree-days from crop emergence to flowering.

G. State Variables

State variables are quantities that describe the conditions of system components. These state variables change with time in dynamic models as system components interact with each other and with the environment. For example, soil water content and crop biomass are two state variables that change with time in most crop models. A model may have one state variable or it may have many to describe the various characteristics of a system that change with time. State variables are of critical importance because these are the dynamic characteristics of the crop or other system that are of interest to researchers.

H. Process Models

The interrelationships between components in a system, and therefore between state variables in the system, exist because of various processes. We sometimes use the term "process-oriented" to describe models that describe the flow and storage of mass, energy, or other substances. For example, the crop biomass state variable changes as a result of photosynthesis and respiration processes, and the soil water state variable changes as a result of rainfall, runoff, percolation, and evapotranspiration processes. A crop model is the set of mathematical relationships that describe the changes in state variables as a result of the various processes that occur.

Continuous models are characterized by state variables that can change smoothly over small time intervals and are not restricted to integer values. *Discrete* models, on the other hand, are those in which the variables describing the system states take on integer values. Crop models are usually classified as continuous models. An example of a discrete biological model is one that predicts the birth and death of individual insects. The number of insects alive at any point in time is a whole number. Discrete models usually require information on the times required to complete activities, such as the time required for an insect egg to hatch. In contrast, continuous models require information on processes such as the flow rate of material or energy between components and between components and the environment.

Continuous models are usually represented by a set of differential or difference equations derived from the structure of the system and the interrelationships among components. Some systems can be modeled as either continuous or discrete, depending on the purpose of the model. For example, the fate (birth, death) of individual insects may be viewed as a series of discrete events. The population number at any time would be a count of the total number of individuals, each considered separately. Obviously, if the population is large, a lot of computer time is required to keep up with each insect. One may also view the insect population density (number of insects per unit area) as a continuous state variable for large numbers and when population processes (births, deaths) occur smoothly over time, and thus use differential equations to model their dynamics.

I. Verification

Verification involves the evaluation of the accuracy with which the computer code represents the mathematical model and the programmer's intentions. Verification also involves the careful checking of mathematical manipulations, units and their conversions, and programming logic and code to ensure that neither the mathematical model nor its translation into one or more computer programs has errors in it.

J. Calibration

Calibration consists of making adjustments to model parameters to give the best fit between simulated results and results obtained from measurements on the real system. In other words, calibration involves adjusting certain model parameters by systematically comparing simulated results with observations of state variables. Model structure remains the same, and parameters are adjusted to more closely describe observed behavior. For example, suppose that partitioning of new crop growth to leaves varies with variety, all else being equal. An experiment could be conducted in which the total crop and leaf dry matter are measured and a leaf partitioning parameter estimated for each variety by fitting simulated results to match observed data. Calibration should be conducted only within the confines of a given data set. In many cases, calibration is the only practical way to estimate some parameter values that are used in biological models.

K. Validation

Validation is the process of comparing simulated results to real system data not previously used in any calibration or parameter estimation process. The purpose of validation is to determine if the model is sufficiently accurate for its application as defined by objectives of the simulation study. Simulated state variables are compared with measured values of state variables. Usually, in crop simulation studies only a few state variables out of many possibilities are measured, and thus a complete comparison is usually not possible. Validation involves subjective judgment. First, the areas for comparison must be selected. Then a measure of "accuracy" or "closeness of fit" must be established, such as the final crop yield. The choice of criteria and important state variables for comparison should be based on model objectives. Validation efforts are essential in the application of crop models.

III. SIMULATION APPROACH

Simulation involves the development of a model and its use in gaining insight about the system itself or its management. The simulation approach usually entails the use of a number of steps, which are summarized below. Following this summary, techniques for biological model development and implementation are presented along with an example of a simplified crop model.

A. Statement of Objectives

A clear statement of the reasons for undertaking the simulation study is essential. All remaining steps in the simulation study depend on this initial step. The problem to be addressed and the information to be derived from the model should be explicitly identified. Although this is one of the most critical steps in a simulation project, it is often overlooked or is not clearly communicated. As a part of the statement of objectives, there should be a clear definition of the intended end product and the intended users of the models that will be developed within a project. Without such documentation, project goals may remain poorly defined and reduce the effectiveness of contributors to the project. This step is not as easy as it may seem. The objective of the simulation effort should be used to determine the level of detail needed in the model, the type of experiment that is needed, the type of data that should be collected, and every other step that follows.

Broadly speaking, there are two fundamental objectives for biological simulation models. First, scientists may wish to obtain a better understanding of the behavior of a system, the interactions that occur, and cause-and-effect relationships in a system. Thus, simulation of biological systems can help evaluate one's understanding by testing hypotheses about its behavior, using models and well-defined experiments (Boote et al., 1996; Sinclair, 1990; Monteith, 1990; Goudriaan and Monteith, 1990; Goudriaan, 1988). For example, a scientist may be interested in developing a quantitative model of crop canopy photosynthesis under varying temperature and carbon dioxide conditions to gain insight into possible crop responses to changing climatic conditions. A model would serve as the scientist's hypothesis concerning crop process and overall response, and experiments would be designed specifically for determining the adequacy or inadequacy of the hypothesis. The end result of such a study could be a model with improved capabilities for prediction, or it could lead to a new research when a model and current hypotheses are shown to be inadequate. This fundamental objective for modeling could be referred to as a scientific goal since the major expected result is an increase in knowledge.

The second fundamental objective is more problem-oriented or applied and could be referred to as an engineering goal. In this case, the major goal relates to better prediction of system behavior for use in improving control or management of a system. For example, a research team may wish to develop a model for use in a computerized irrigation system controller or to be used by water resource policy board members to help them decide how to allocate water during periods of drought. The end result of this type of project is a software product or combination hardware and software product designed for a specific application.

It can be argued that scientific models can also be applied and that there is really no distinction between these two goals. However, for problem-oriented or engineering objectives, it is highly important that the model predict the system behavior with an acceptable level of accuracy. The type of model is not important as long as it is reliable and performs as needed. In contrast, scientific models may not predict the system's behavior well at all, but the study itself could be very valuable. It could demonstrate where knowledge is inadequate and where additional research is needed. In this case, the form of the model is important; it should be structured to represent current scientific understanding of the processes and used to help design experiments and data collection methods for testing the understanding. Both scientific and engineering objectives are relevant. General arguments about which is the most appropriate approach serve no useful purpose. Applied models of biological systems evolve as understanding is improved; scientific modeling is a highly effective method for helping researchers improve their understanding of biological systems.

One can think of biological simulation efforts as being "product-oriented." The end product may be increased knowledge and an improved understanding of crop behavior, or the end product may be a tool that is designed for application to specific problems. This characterization of objectives sets the stage for all activities in a simulation project.

B. Definition of the System

The components to be included in the study and the system boundary should be identified. Inputs and outputs should be described. As additional information is gathered, it may later be necessary to redefine the system. Or, for example, lack of available data may force a change in the system definition.

C. Literature Review and Data Analysis

Prior to model development, a review of information availability is needed in order to evaluate the possibility of meeting objectives as stated. In this step, the essential features of the system are established and a conceptual model that will depend on the data available and the ability to quantify processes in the model is developed.

D. Model Development

In the model development step, diagrams may be used to represent the components in the system and to summarize their interrelationships. The mathematical representation of the system should be developed, including specific functions and relationships to be used in the model. Experiments may be needed to develop functions and estimate parameters for the model. The model is translated into computer code for simulating the behavior of the real system. Computer flowcharts are usually helpful in translating the model to computer code and documenting the relationships between the model and the code. Techniques for model development are presented in Section IV.

E. Model Accuracy Evaluation

The question of whether a model is accurate must be followed by "for what objectives?" and "as compared to what?" In the final analysis, one must ascertain whether the cost of obtaining an additional unit of accuracy is greater than the benefits to be gained in terms of the study objectives. Accuracy is usually defined in terms of verification, calibration, and validation.

Dent and Blackie (1979) provide an excellent perspective on model evaluation:

> When we are confident that the behavior of the model is satisfactory, the formal validation process is over. Although a formal process of validation is always recommended, the process of gaining confidence in the model is generally a slowly emerging one over the period of model construction, through formal validation, to application of the model. During this time, assessment and modification will proceed, essentially two sets of judgments being made on the way:
> a.   That the model is not different from the real existing system to a degree that will detract from the value of the model for the purposes for which it was designed;
> b.   That if the model is accepted as being adequate then the decisions made with its assistance will not be measurably less correct than those made without the benefit of the model. Such an exact assessment is extremely difficult and cannot be other than subjective in nature. Subjective judgments in this regard are, however, not without value and this aspect of validation may often be the most relevant.

It is because of the difficulties associated with these two sets of judgments that validation remains an elusive issue in the simulation procedure. Statistical tests are available to determine whether the model behavior is different from real-system behavior to some stated level of significance. The precision of such tests should not overshadow the conceptual problems in applying them. There is no mechanical procedure that permits the conditions for acceptable validation by a single (or even a series of) statistical comparison(s) of the model-performance against some recorded or measured data from an existing system which the model represents. Any statistical tests that may be carried out and which are positive in favor of the model add to the model-builder's confidence that his creation is functioning well. The acceptable level of confidence is achieved normally by a series of assessments and subsequent modifications until such time as the model is to be applied in support of decision making.

F.   Sensitivity Analysis

Sensitivity analysis involves exploring the behavior of the model for different values of parameters. This is done to determine how much a change in the value of a parameter influences the important outputs from the model. Techniques for performing sensitivity analysis are presented in Section VI.

G.   Model Application

The model application may relate to management of a system. In this case, the model may be viewed as "complete," and its users may not have participated in its development. Documentation of the model and the computer code is essential. The original objective may be related to research and the use of systems modeling to assist in a more thorough understanding of the system. In this case, the use of sensitivity analysis in guiding research efforts may be the intended application.

Simulation steps should not be viewed as a strict sequential process. Rather, in most systems modeling and simulation studies, these steps are repeated in an interactive fashion as information is gained and progress is made.

IV.   MODELING

Continuous system modeling is a process-oriented approach for describing the behavior of a system. Processes fall into three categories: transport (or flow), transformation, and storage (Smerage, 1977).

These processes are described by two classes of variables: *extensive* or through variables and *intensive* or across variables. Extensive variables are characterized by flow-through quantities such as mass, volume, electric charge, force, and heat flows. Intensive variables are measures of energy intensities or potentials across system components. Intensive variables represent the driving force for the extensive variables. Examples are pressure, temperature, voltage, and velocity.

Extensive and intensive variables are well-defined for many physical and chemical systems and have been widely used. In such systems, a component process is described by relationships between its extensive and intensive variables. For example, heat flow through a conductor is described by its heat conduction property, and the difference in temperature is the intensive variable. Water flow through soil is described by soil properties and the pressure gradient in the soil. The flow of charge (or current) through a resistor is described by the voltage drop across the resistor and its resistive or energy-dissipative property. Water flow through soil–plant–atmosphere systems is sometimes represented by resistances and water potential gradients. Extensive variables are measured at a point, and intensive variables are measured across an object of interest.

By proper identification of intensive and extensive variables and system components, a system diagram can be drawn that can be used to derive the mathematical model of the system. Because of this commonality of extensive and intensive variables among systems, similar methodologies can be used to model the results of analogous mathematical models from different systems. Koenig et al. (1967) and Martens and Allen (1969) describe in detail the use of these procedures for physical systems. Smerage (1977) extended these concepts to a broader class of agricultural, biological, and ecological systems.

One difficulty in modeling agricultural and biological systems using this methodology stems from the fact that we do not always know the intensive variable for a particular extensive or flow process, or there may be several mechanisms causing flow. In this case, the inclusion of the details of all intensive variables causing flow on a diagram may unduly complicate the model and detract from its purposes. In other cases, our objectives for modeling the system and the resultant scope and hierarchical detail may make it impractical to express flow processes mechanistically as related to intensive variables. For example, we know that the rate of flow of water through a plant is regulated by water potential differences between the leaf and soil and by resistances to flow along this path. However, if our objective is to predict transpiration water use of a crop for application to irrigation management, it is neither necessary nor desirable to include these mechanisms in the model. Including these details would require model descriptions of water potentials of soil and plant and associated details of plant and soil parameters that might be useful for scientific purposes but not practical for these purposes.

A. Compartment Models

So-called compartment modeling is a useful tool in conceptualizing systems in which the primary emphasis is on the flow and storage of system variables. Compartment models have been widely used to schematically represent various agricultural, biological, and ecological systems (Dent and Blackie, 1979; Patten, 1973; Odum, 1973; Penning de Vries and van Laar, 1982; Keen and Spain, 1992; Leffelaar, 1993). This approach is particularly useful for conceptualizing continuous systems. A mathematical model in the form of a set of first-order differential equations describing the system structure can be obtained directly from the compartment model diagram. Although there are several variations in the symbols used to represent components in compartment models (Forrester, 1971; Patten, 1973), we will use the Forrester (1971) notation because of its widespread acceptance in agricultural and biological literature. Forrester (1961) originally developed the compartment diagrams to provide a pictorial representation of systems of equations for industrial dynamics models.

A *compartment* is defined to be a quantity or level of a state variable. Compartments represent the state variables in a system or transformations of state variables. Examples of these state variables are mass, volume, electric charge, and population. Many applications in agricultural and biological systems relate to mass flow among various compartments in the system. A conceptual model of this type consists of a network of compartments representing all relevant components of the system. The compartments are connected by lines that represent the flow rates of the quantities between compartments. Flow may also occur between the environment and compartments.

Figure 2 shows an abbreviated set of symbols from Forrester (1961). The level represents the state variable at a point in time and is shown by a rectangle. Sources and sinks represent the environment of the system in that flow can occur from a source in the environment into the system without affecting the environment. Likewise, flow

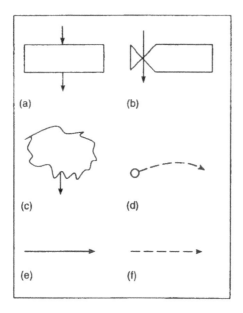

FIGURE 2 Compartment model symbols adapted from Forrester (1961). (a) Level; (b) Rate; (c) Source; (d) Auxiliary Variable; (e) Pathway for Material Flow; (f) Pathway for Information Flow.

from the system into a sink in the environment has no effect on the environment. The rate symbol represents flow from one compartment to another, from a source to a compartment, or from a compartment to a sink. Auxiliary variables are shown by circles and represent factors (inputs and parameters) that influence the rates. For example, temperature may affect the rate of biomass accumulation in a crop growth model and would be classified in this diagram as an auxiliary variable. Solid lines represent pathways for material flow whereas dashed lines represent information flow.

After the compartment model is constructed, one can write the mathematical model representing the system using structural constraints and component descriptions. The Appendix lists variables used in this chapter along with their definitions and units. From the diagram itself the structural constraint or continuity can be applied to each compartment level, or

$$\frac{dx_i}{dt} = \sum_j I_{i,j} - \sum_k O_{i,k} \qquad (1)$$

where

$$x_i = \text{level of } i\text{th variable}$$
$$dx_i/dt = \text{rate of change in the level of the } i\text{th variable}$$
$$I_{i,j} = \text{rate of flow into level } i \text{ from source } j$$
$$O_{i,k} = \text{rate of flow out of level } i \text{ to source } k$$

By applying this condition to each compartment in the system, a set of first-order differential equations is derived.

B. Water Tank Example

For example, consider the simple water flow system in Fig. 3a. The volumes of water in each tank are the integrated extensive variables of interest in the system. Equations describing this system are

$$\frac{dV_1}{dt} = i(t) - f_{1,2}(t), \qquad \frac{dV_2}{dt} = f_{1,2}(t) - O(t) \tag{2}$$

Now we must describe the component flow processes. Since flow from each tank is controlled by an orifice, we know that flow rate is proportional to the square root of the height of the water surface above the orifice multiplied by 2 times the gravitational constant. Thus, we can further describe $f_{1,2}(t)$ and $O(t)$ as follows.

Component descriptions:

$$f_{1,2}(t) = C_1(2gH_1)^{1/2}, \qquad O(t) = C_2(2gH_2)^{1/2} \tag{3}$$

Since H_1 and H_2 can be expressed in terms of V_1 and V_2, we can write the complete model for this two-compartment system as (compare with Fig. 3b):

$$\frac{dV_1}{dt} = i(t) - C_1\left(\frac{2gV_1}{A_1}\right)^{1/2}, \qquad \frac{dV_2}{dt} = C_1\left(\frac{2gV_1}{A_1}\right)^{1/2} - C_2\left(\frac{2gV_2}{A_2}\right)^{1/2} \tag{4}$$

When developing the structural and component equations as in Eqs. (1)–(4), care should be taken to ensure the dimensional consistency of the relationships. The model developers must always make sure that all terms have the same dimensions and that the same measurement system is used for each parameter, input, variable, and process. Otherwise, the resulting model results will not be meaningful. Figure 3c shows an example of simulated results obtained by programming the model and simulating its behavior for 100 h, for the

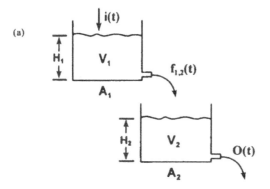

FIGURE 3   (a) Schematic of a two-tank water flow system; (b) the corresponding compartment model using Forrester (1961) notation; and (c) example results.

parameters shown in the figure. When the input to the first tank
was set to 5 m³/h for the first 5 h of simulation, then set to 0.0 for
the remaining time, the volume of water in the first tank first increased to 16.9 m³, then dropped to 0.0 after 19.1 h. Volume in
the second tank lagged behind that in the first tank, peaking at only
7.1 m³.

This example demonstrates the simplicity with which this approach can be used to obtain a first-order differential equation model
of a system. Agricultural examples of compartment modeling are insect population dynamics and growth of crops. In some such systems,
the mechanisms causing flow are not well enough understood or are
so complex that research may be required to determine relationships
between flow and other system characteristics and inputs.

In general, a model developed in this way will have state variables defined by levels $x_1, x_2, \ldots, x_n$. The mathematical model will
be of the form

$$\frac{dx_1}{dt} = f_1(x_1, \ldots, x_n) + b_1(t)$$

$$\vdots \tag{5}$$

$$\frac{dx_n}{dt} = f_n(x_1, \ldots, x_n) + b_n(t)$$

(b)

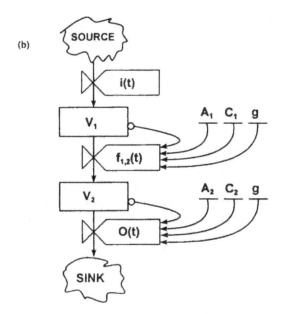

(c)

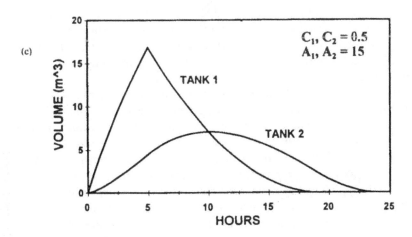

FIGURE 3   Continued

where $f_1(x_1, x_2, \ldots, x_n)$ represents the rates as affected by any of the variables $x_1, x_2, \ldots, x_n$, and $b_i(t)$ are constant or time-varying inputs to the ith storage compartment. For special cases when the relationships are linear, Eq. (5) can be abbreviated as

$$\frac{dx}{dt} = Ax + b \tag{6}$$

where x and b are vectors and A is a matrix of coefficients.

Gold (1982) describes another useful diagrammatic approach called a signal-flow graph. In this approach, variables for the system, including state variables, flow variables, inputs, and parameters, are defined. The variables are connected by directional arrows in a diagram to denote functional dependency. For example, $A \rightarrow B \leftarrow C$ would indicate a functional dependence of A and C on B, i.e., $B = f(A, C)$. This approach is very useful during the early model formulation stages to obtain an overview of the needed mathematical relationships, such as $f(A, C)$ in the above example, and to identify important feedback loops that may occur in a system.

C. Distributed System Modeling

A distributed system is one in which the state variables and flow processes vary over the spatial region of interest. In other words, the state of the system may change through space as well as through time. The mathematical model that results from an analysis of a distributed system is a partial differential equation.

It is useful to break distributed systems into a set of connected, homogeneous elements for modeling purposes. For example, we may wish to study the flow of water and its state in a soil profile. At any point in time, soil water content and its flow may vary with depth. If we considered the entire profile to be a single compartment, it would not adequately represent the vertical variations in water content with time. Therefore, consider the soil profile to be divided into small, discrete layers, with each layer having a volume of water w_i stored in it. Each layer is connected to layers above and below by flow of water; $q_{i,j}$ is the volume flow rate of water from layer i to layer j. Figure 4 shows a schematic of the system and corresponding Forrester diagram representation. From this, we can write a series of first-order differential equations describing the structure of the system.

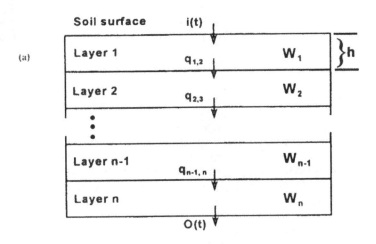

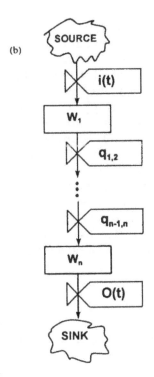

FIGURE 4 (a) Schematic of a distributed soil system (one-dimensional) for studying soil water processes and (b) the corresponding compartment model using Forrester (1961) notation.

$$\frac{dw_1}{dt} = i(t) - q_{1,2}(t)$$

$$\frac{dw_2}{dt} = q_{1,2}(t) - q_{2,3}(t)$$

$$\vdots$$

$$\frac{dw_n}{dt} = q_{n-1,n}(t) - O(t)$$

(7)

Ignoring gravity effects, flow q can be represented by the diffusivity property of the soil, D, and the gradient of volumetric water content, θ (Hillel, 1971). Thus,

$$q_{i-1,i} = \frac{D_{i-1,i}(\theta_i - \theta_{i-1})A}{h_i}$$

(8)

where $\theta_i = w_i/V = w_i/Ah_i$ and h_i is the thickness of layer i.

This system of equations can be simulated if initial conditions $w_i(0)$ and boundary conditions $i(t)$, $O(t)$ are known for $i = 1, 2, \ldots,$ n and $t > 0$. Initial conditions refer to the values of all state variables at the start of the simulation, usually at time $t = 0$; they are required for simulating the behavior of a model. As an example, if a system has two state variables $x(t)$ and $y(t)$, then $x(0)$ and $y(0)$ must be specified in order to simulate system behavior for $t > 0$. Boundary conditions refer to values of variables at the boundaries of a system. Boundary conditions may refer to flows across the system boundary or to known values of intensive variables at the boundary. Values of boundary conditions must be specified for all time to be simulated.

Now we show how this set of equations is related to a partial differential equation. Let h_i and dt be approximated by Δx and Δt, respectively. Since $w_i = Ah_i\theta_i$, we can approximate dw_i/dt in difference form as

$$\frac{dw_i}{dt} \approx Ah_i \frac{d\theta_i}{dt} \approx A \frac{\Delta x(\theta_{i,t} - \theta_{i,t-\Delta t})}{\Delta t}$$

(9)

We can equate this expression with the right-hand side of Eq. (7) and expand it into

$$A \frac{\Delta x(\theta_{i,t} - \theta_{i,t-\Delta t})}{\Delta t} = A\left(\frac{-D_{i-1,i}(\theta_i - \theta_{i-1})}{\Delta x} - \frac{-D_{i,i+1}(\theta_{i+1} - \theta_i)}{\Delta x}\right)$$

(10)

Assuming D is variable, we can write

$$\frac{\theta_{i,t} - \theta_{i,t-\Delta t}}{\Delta t} = -\frac{1}{\Delta x}\left(\frac{D_{i-1,i}(\theta_i - \theta_{i-1})}{\Delta x} - \frac{D_{i,i+1}(\theta_{i+1} - \theta_i)}{\Delta x}\right) \tag{11}$$

which can be written

$$\frac{\Delta\theta_i}{\Delta t} = \frac{\Delta(D_i\,\Delta\theta_i/\Delta x)}{\Delta x} \tag{12}$$

Taking the limit as $\Delta t \to 0$ and as $\Delta x \to 0$ results in

$$\frac{\partial\theta}{\partial t} = \frac{\partial}{\partial x} D \frac{\partial\theta}{\partial x} \tag{13}$$

where θ is a function of both x and t. Equation (13) is the partial differential equation for water flow in one dimension through a soil of unit area, the same equation presented by Gardner (1959). The first-order differential equation system [Eq. (7)] is an approximate "lumped" representation of the distributed (one-dimension) soil system more accurately represented by the partial differential equation, Eq. (13). When approximate solutions are adequate, the lumped representation provides a relatively quick and simple method for simulating the behavior of distributed systems.

The same procedure has been used to lump age structures in population dynamics models. Figure 5 shows the Forrester diagram for an insect model with the population divided into age classes to simulate age structure. The levels are population numbers for individuals in specific age classes. An age class may be defined as a range of ages, for example 0–10, 10–20, 20–30, and >30 represent four age classes of insects. Flow between age classes occurs as insects age. The partial differential equation describing this system is

$$\frac{\partial N(a,\,t)}{\partial t} + \frac{\partial N(a,\,t)}{\partial a} = -u(a,\,t)N(a,\,t) \tag{14}$$

where

$N(a,\,t)$ = population of insects age a at time t, dimensionless
$u(a,\,t)$ = net flux of insects age a at time t across the system boundary, s^{-1}

$u(a,\,t)$ represents migration and mortality factors.

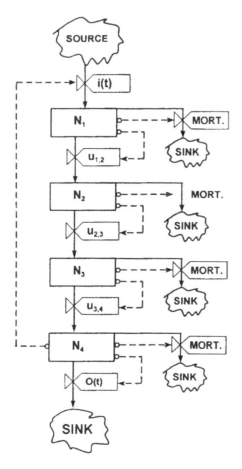

FIGURE 5   Compartment model of insect population, with insects distributed
into four age classes.

V.   COMPUTER SIMULATION

In continuous systems, computer simulation is performed to "solve"
the set of system equations to result in a time series behavior of state
variables of the system for a prescribed set of inputs. This involves
one of several procedures for (1) advancing time in small steps and
(2) calculating the state of the system for the new point in time. This
can easily be shown for one time step using a simple model $dx/dt =
ax + b$. Letting $dx \cong (x_{t+\Delta t} - x_t)$ and $dt \cong \Delta t$, we can write

$$\frac{dx}{dt} \cong \frac{x_{t+\Delta t} - x_t}{\Delta t} = ax_t + b \tag{15}$$

and

$$x_{t+\Delta t} = x_t + (ax_t + b)\Delta t \tag{16}$$

If x_t is known, we have all the information needed to calculate $x_{t+\Delta t}$ or x at time $t + \Delta t$. It is helpful to think of this in terms of a rate, expressed as

$$x_{t+\Delta t} = x_t + \text{Rate}_t \, \Delta t \tag{17}$$

where Rate$_t$ is any expression for the rate of change of the state variable x at time t. In the example above, Rate$_t = ax_t + b$.

Equation (16) is a finite-difference expression of the differential equation. In general, Eq. (16) is also referred to as the rectangular or Euler method of integration of a first-order differential equation. The value of x is known at time t and possibly prior to time t. The model expresses the instantaneous rate of change of the variable x. Although we do not know the value of x at $t + \Delta t$, we can choose a small Δt and predict a value of x at time $t + \Delta t$ based on the current rate of change of x. This procedure is repeated as long as desired. For example, by continuing with x at $t + \Delta t$ and projecting forward by one more Δt we can estimate x for time $t + 2\Delta t$.

The result will be a time series behavior of x defined by the model. If the system has more than one first-order differential equation, the finite difference procedure can be applied to each. The rate expression for any one differential equation may be functionally dependent on any of the state variables as well as on inputs to the system. This method is dependent on the system model being described by first-order differential equations (or difference equations as described earlier). Second-order differential equations can be simulated by transformation to two first-order equations. For example, the second-order equation

$$\frac{d^2y}{dt^2} - a\frac{dy}{dt} + y = u(t) \tag{18}$$

can be transformed into two first-order equations by letting $x_1 = y$ and $x_2 = dy/dt$. It follows that $dx_2/dt = d^2y/dt^2$. Then the two equivalent first-order differential equations are

$$\frac{dx_1}{dt} = x_2 \quad \text{and} \quad \frac{dx_2}{dt} = ax_2 - x_1 + u(t) \tag{19}$$

Other numerical solution techniques are more accurate than the Euler method for the same size Δt. For example, the trapezoid integration method can be used to develop a "predictor-corrector" integration technique. In this case,

$$y_{t+\Delta t} = y_t + \frac{1}{2} (\text{Rate}_t + \text{Rate}_{t+\Delta t}^*)\Delta t \qquad (20)$$

is the formula for calculating $y_{t+\Delta t}$. Note that Rate_{t+t}^* is on the right-hand side of Eq. (20). We have to "predict" $\text{Rate}_{t+\Delta t}^*$ since we do not know y or Rate at time $t + \Delta t$. The Euler method is used to predict $y_{t+\Delta t}^*$ as a first approximation. Then that $y_{t+\Delta t}^*$ is used to compute $\text{Rate}_{t+\Delta t}^*$ using the first-order differential equation model. We then have all that is needed to compute $y_{t+\Delta t}$ above. The Euler and trapezoid methods are *explicit* methods; all terms on the right-hand side of the equation are known or can be computed from what is known.

In general, a Taylor's series expansion about y_t can be used to express $y_{t+\Delta t}^*$. This expression is

$$y_{t+\Delta t} = y_t + \Delta t\, y_t^{(1)} + \frac{(\Delta t)^2}{2!} y_t^{(2)} + \frac{(\Delta t)^3}{3!} y_t^{(3)} + \cdots + \frac{(\Delta t)^n}{n!} y_t^{(n)} \quad (21)$$

where $y_t^{(n)}$ represents the nth derivative of y with respect to t and $n!$ represents factorial numbers (i.e., $n! = 1 \times 2 \times 3 \times 4 \times 5 \times \cdots \times n$). If we approximate y_t by using the first three terms of Eq. (21), then

$$y_{t+\Delta t} = y_t + \Delta t\, y_t^{(1)} + \frac{(\Delta t)^2}{2!} y_t^{(2)} + \epsilon_t \qquad (22)$$

where ϵ_t is the total truncation error, or the error caused by not including all the terms in the Taylor expansion.

Since $y_t^{(2)}$ can be approximated by $(y_{t+\Delta t}^{(1)} - y_t^{(1)})/\Delta t$, Eq. (22) can be written

$$y_{t+\Delta t} = y_t + \Delta t\, y_t^{(1)} + \frac{(\Delta t)^2}{2} \frac{y_{t+\Delta t}^{(1)} - y_t^{(1)}}{\Delta t} + \epsilon_t \qquad (23)$$

Equation (23) is equivalent to the trapezoid predictor-corrector technique with the added error term. The Euler technique is equivalent to the Taylor series expansion with only the first derivative term included explicitly. Thus, the trapezoid method is a more accurate method for simulating first-order differential equations. However, the Euler method is used for most models because of its simplicity and ease of use.

Because that is a numerical approximation, it is subject to numerical errors. The local truncation error using the Euler method is $[(\Delta t)^2/2][d^2y(\xi)/dt^2]$, where ξ is some point in the interval $[t, t + \Delta t]$. The truncation error is the error with each integration step and can thus accumulate. Truncation error for the trapezoid predictor-corrector method is $[(\Delta t)^3/6][d^3y(\xi)/dt^3]$, which, for a small Δt, is less than that of Euler method (Ortega and Poole, 1981). Generally, the higher order methods are more stable than lower order methods.

VI. SENSITIVITY ANALYSIS

The purpose of sensitivity analysis is to study the behavior of the model. Sensitivity analysis can also be structured to determine important subsystems, relationships, and inputs. Sensitivity analysis should be designed with respect to the objectives of the study. A sensitivity analysis is the process by which parameters or inputs are evaluated with regard to their effects on simulated results. For example, rainfall would be an environmental input to a crop balance model for estimating crop irrigation requirements. Soil characteristics such as soil water-holding capacity or root zone depth are examples of model parameters. One may be interested in the sensitivity of estimated irrigation requirements to changes in rainfall, changes in soil characteristics, or both. A sensitivity analysis also provides a mechanism for testing the simulation in the extremes; that is, using the extreme values of parameters will rigorously test the model in terms of mathematical logic and stability.

Computer graphics are highly useful in sensitivity analysis. Graphical display of simulated results for a number of parameter values or inputs provides a visual image of model behavior over a range of parameter values. A mathematical approach to sensitivity analysis may also be a useful way to organize and present model behavior. This approach compares the change in one or more simulated outputs relative to the change in one or more parameters by approximating partial derivatives using numerical results. Absolute sensitivity, $\sigma(y|k)$, of some model output y to a variable k (a parameter of the model or input to the system) is defined by

$$\sigma(y|k) = \frac{\partial y}{\partial k} \approx \frac{y(k + \Delta k/2) - y(k - \Delta k/2)}{\Delta k} \tag{24}$$

Relative sensitivity, $\sigma_r(y|\kappa)$, is often used to provide a normalized measure for comparing the sensitivity of a model to several variables. Relative sensitivity is defined by

$$\sigma_r(y|k) = \frac{\partial y/y}{\partial k/k} = \sigma(y|k)\frac{k}{y} \tag{25}$$

For example, if $\sigma_r(y|k) = 2.0$ and k is changed by 30% ($\Delta k/k = 0.03$), we can expect y to change by 6%. Also, if $\sigma_r(y|k_1) = 0.5$ and $\sigma_r(y|k_2) = 1.5$, then y is more sensitive to a percentage change in k_2 than the same percentage change in k_1 (3 times as sensitive).

In simulation studies Δk is used to represent a small change in k so that changes in y, Δy, can be simulated. Then Eqs. (24 and (25) can be used to approximate $\sigma(y|k)$ and $\sigma_r(y|k)$ using the discrete changes in k and y, or $\partial y/\partial k \cong \Delta y/\Delta k$. By varying parameters through their expected ranges of extremes, one can use this approach to compare the sensitivity of model results.

Sensitivity analysis usually begins with the selection of model output results that are considered to be crucial to the study. A set of "base" conditions are then established; these usually comprise the set of the best estimates of each parameter and input. Base results are the simulation outputs obtained when base condition values are used. A range of values are then selected representing the extreme conditions associated with each parameter to be evaluated. Simulation runs are made using each value while holding all other base conditions constant. A comparison between changes in the base condition values and changes in the base results provides an indication of the relative importance of the variable. Results can be compared in graphs or tables or by computing the sensitivity variable values. Note that the values of σ and σ_r and graphical results depend on the "base" conditions selected.

Regression analysis or response surface techniques are also useful techniques in sensitivity analysis. Computer experiments are conducted by varying parameters over a range of interest followed by regressing simulated results against one or more parameters to determine if outputs are affected by changes in the parameter, and if so, to what extent.

VII. COMPUTER IMPLEMENTATION

There are many options for computer implementation of biological models today. Because there are many reasons for developing biological models, some consideration should be given to which option to choose for a particular model. Some biological models are developed by an individual or small team of scientists to test a hypothesis, and there is no attempt to make these models available to others except through the scientific literature. Other models are developed by teams of researchers for use in a number of applications and for distribution to others who may want to use them for their own purposes. Biological models are also being used in formal university classrooms and in specialized short courses and training programs. Programs should be constructed using modules or objects designed for describing well-defined components of the system being modeled and for ease of documentation, revisions, replacement, and maintenance. Van Kraalingen (1995) describes a structure for programming in Fortran that provides these characteristics. In many applications, biological models need to be linked with other software, such as geographical information systems (e.g., Engel et al., 1996; Fraisse et al., 1994) and graphical user interface programs (e.g., Jacobson and Jones, 1996). Standardized formats for data storage and exchange are needed to facilitate model evaluation and to link models with such applications. Some progress has been made on standardizing data for crop models and their applications (Jones et al., 1994); this effort is continuing at an international scale (Ritchie, 1995). However, considerably more effort is needed by biological science communities on standard protocols for biological model module design and data storage and exchange.

There are many choices of programming languages and software to help those who are developing biological models. The choice of language is not a major issue today because of the ability to link together modules developed in different languages, particularly if the models use modular structure and adhere to well-defined data standards. However, each has its own advantages and disadvantages. Procedural languages such as Fortran, Basic, and Pascal are widely used for biological models. Most existing crop models, for example, are programmed in Fortran (Jones and Ritchie, 1990; Hoogenboom et al., 1994; Bouman et al., 1996). Procedural languages provide excellent capabilities for scientific computations, they can easily handle nonlinearities and other complexities common to many biological systems, and they can be used on many computer platforms with little or no modification. Because of limitations in these languages with respect to handling modern user interface programming, new applications of these models are being developed in which the model is treated as a module in an overall software package and the user interface and data manipulations are handled by one of the modern visual programming tools such as Visual Basic (Microsoft, 1995). Some biological models are now being developed using object-oriented programming languages such as SMALLTALK and C++.

These new languages have potential for helping improve the modularity of biological models.

There are a number of specialized computer simulation languages that allow model developers with little or on computer programming skills to implement models. Continuous simulation languages solve systems of continuous differential equations. For example, the SLAM system (Pritsker, 1995) and FSE (van Kraalingen, 1995) packages provide modules that perform most of the required computer simulation tasks. Users only have to "program" their specific model equations and provide parameters and other inputs in order to obtain simulated behavior of a system. Probably the simplest and easiest to use simulation software allows users to build conceptual models on computer screens using icons, such as Forrester symbols presented earlier in this chapter. Two widely used packages are Stella (High Performance Systems, 1990) and VisSim (Visual Solutions, 1990), both available for use on personal computers. These packages guide users in linking together different icons to depict a system with its storage compartments and flow paths, then ask for needed parameters and other inputs to run the model. They will then simulate the system and graph or print results for users; no programming is needed. These packages have the distinct advantage of rapid evaluation of variations in model structure to test different hypotheses about the system. Each computer language has its advantages and disadvantages. The best choice for a particular model will depend on many factors such as model complexity, intended users of the model, how it will be used, the need to integrate it with other software and databases, and the need to maintain it.

VIII. DEVELOPING A CROP SYSTEM MODEL

Dynamic growth and yield models have been developed for a wide range of crops, including cotton, grain sorghum, wheat, corn, soybeans, alfalfa, and others (Jones and Ritchie, 1990). Researchers have used these models for a number of purposes, such as irrigation management (Boggess et al., 1981; Swaney et al., 1983), nutrient management (Keating et al., 1993; Singh et al. 1993), pest management (Pinnschmidt et al., 1990; Batchelor et al., 1993), land use planning (Beinroth et al., 1997), crop sequencing (Thornton et al., 1995), climate change assessment (Curry et al., 1995; Rosenzweig et al., 1995; Penning de Vries, 1993), and yield forecasting (Swaney et al., 1983). These applications have demonstrated conclusively the value of simulation applied to cropping systems. However, these models have limitations because they do not include all factors that occur in reality and contain empiricism that may require calibration and testing for site-specific applications. Research directed to improving these shortcomings and limitations is needed to enhance the applicability of the crop models.

A useful characterization of crop models for application at the field scale was attributed to C. T. de Wit (Bouman et al., 1996; Penning de Vries and van Laar, 1982). Initially, four levels were defined, but these were later simplified to three levels (Lovenstein et al., 1993). First is the potential yield level; crop growth is dependent on weather factors, such as temperature, solar radiation, CO_2 concentration, and day length, and on genetic factors of the crop. This level of model includes basic crop growth processes such as photosynthesis, respiration, tissue growth, and development. Its main use is to gain an understanding of how these factors affect potential production of a crop, assuming that water and fertilizers are adequate and that no pest damage occurs. A second level of detail is characterized as attainable yield; at this level, the model has components that describe water and nitrogen fertilizer limitations to production. Attainable yield thus depends on water and nitrogen supplied by management as well as by rainfall and soil organic matter. Many existing crop models have the capability to describe crop growth and yield with these assumptions (Jones, 1993). The third level of detail is actual yield level, which attempts to explain all the factors that influence growth in a field, including pest damage and mismanagement. This characterization of levels of detail has been useful in providing both a practical pathway for model development and a conceptual framework for model applications. It has also served to determine data requirements for model development, taking into consideration the need for more comprehensive data sets as the level of model detail increases.

There are many choices of state variables for crop models. Simple models may have only one or two state variables, such as leaf area and dry weight, whereas others may have hundreds of state variables describing leaf area, leaf weight, number of leaves, stem weight, number of fruit, and weight of fruit for various age classes (Jones et al., 1991). Crop models with many state variables provide more details about crop behavior, including lags in fruit growth and size distributions, but require much more time to develop and test and more time to solve in a computer. For some problems, models with two to four state variables provide adequate predictions of crop responses. In this chapter, a simple crop model at the potential yield

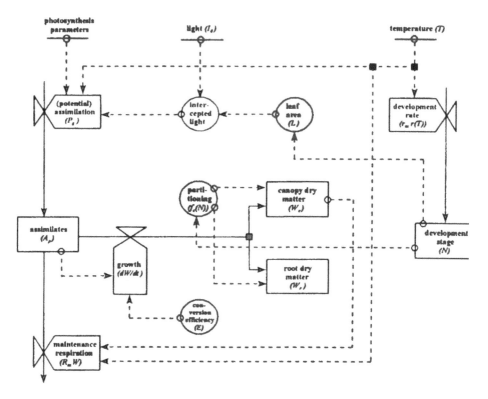

FIGURE 6   Forrester diagram for the crop model with three state variables at the potential production level of detail.

level of detail is presented along with its sensitivity to temperature and light.

Figure 6 shows a Forrester diagram of a simple crop model with three state variables: stage of development (N, number of vegetative nodes), canopy biomass (W_c, g/m^2), and root biomass (W_r, g/m^2). The diagram represents a potential yield level crop model as described by Lovenstein et al. (1993). The figure also shows a compartment for assimilates (A_p, g CH$_2$O/m). However, in the mathematical model, it is assumed that $dA_p/dt = 0.0$, and thus the assimilates do not accumulate but are transformed directly into plant tissue.

A. Crop Development

"Development" refers to the progression of the crop through its life stages, which may be measured by various vegetative and reproductive plant characteristics such as number of leaves on the plant or the appearance of the first flower or first fruit. Development rate is usually expressed as the inverse of time between events, such as the inverse of hours between successive leaves on the main stem or of days from plant emergence to flowering. These rates of development can thus be calculated from experiments in which stages or numbers of leaves are recorded. Development rates in plants are very sensitive to temperature and sometimes to day length but are usually insensitive to light, CO_2, and other factors such as water and nutrient stresses unless the stresses are severe. Crop development is usually insensitive to rate of dry matter production.

There have been many models for predicting crop development; most have related development rate to temperature and some, to day length. For example, Wolf et al. (1986) developed a model for scheduling field operations that predicted emergence, flowering, and harvest dates for field tomatoes within 3–5 days of actual dates. Kiker et al. (1991) described a system for predicting development stages for cucumbers, eggplant, and other crops growing in greenhouses. These models are mostly of the form $r = r(T, \eta)$, where r is the rate of development (h^{-1}), T is temperature, and η is day length. In models for predicting flowering data, r is the rate of progression toward flowering, and the state variable in these models is the development unit, which is 0.0 when the development starts and 1.0 at flowering.

For the three-state variable model, N is the state variable for leaf number for a vegetative plant or for node number for plants like tomatoes that alternately produce leaves and fruiting trusses on the main axis. Assuming that leaf appearance rate, dN/dt, depends only on temperature in this example, it can be computed as

$$\frac{dN}{dt} = r_m r(T) \tag{26}$$

where r_m is the maximum rate of leaf appearance (h^{-1}) and $r(T)$ is a function of temperature. Various functional forms have been used for $r(T)$. In the model presented here, a piecewise-linear form was used for $r(T)$, based on the tomato model (Jones et al., 1991), where $r(T)$ is 0 when T is below 8°C or above 50°C, and has the values of 0.55, 1.0, and 1.0 at temperatures of 12, 30, and 35°C, respectively. Between 30 and 35°C for this example, leaf appearance rate is maximum at r_m leaves per hour.

B. Dry Weight Growth

The rate of dry matter increase in crops is of major importance in determining yield of fruit and vegetative tissue, and it depends directly on photosynthesis. Photosynthesis converts carbon dioxide from the air into sucrose (CH_2O) in plant leaves. Some of the carbon in this CH_2O is combined with other elements (N, P, K, S, ...) and retained in plant tissue, and some is used to provide energy for synthesizing tissue, releasing CO_2 in the process. Penning de Vries et al. (1989) describe a sound quantitative basis for computing the rates of tissue synthesis based on the photosynthetic rate and on the composition of tissue being synthesized. They computed the net product weights of protein, carbohydrate, lipid, lignin, organic acid, or mineral if 1 g of CH_2O was used to provide both the carbon in the final product and the energy for product synthesis. Thus, depending on the composition of plant tissue, a conversion efficiency, E, can be computed and used to convert photosynthesis rate into dry matter accumulation rate. For vegetative tissue, this conversion efficiency is in the range of 0.65–0.75 (g tissue)/(g CH_2O).

Respiration is the loss of CO_2 from plants in growth and maintenance processes. Growth respiration is accounted for in the conversion efficiency described above. Maintenance respiration is the loss of CO_2 due to the breakdown and resynthesis of existing tissue and depends on temperature. Gent and Enoch (1983) expressed maintenance respiration rate as

$$R_m = k_m \exp[0.0693(T - 25)] \tag{27}$$

where

R_m = maintenance respiration rate, (g CH_2O)/[(g tissue)·h]
T = temperature, °C
k_m = respiration rate at 25°C, (g CH_2O)/[(g tissue)·h]

The equation for crop dry weight growth rate is

$$\frac{dW}{dt} = E(P_g - R_m W) \tag{28}$$

where

dW/dt = rate of dry weight growth of crop, (g tissue)/(m²·h)
W = total plant dry weight, g/m²
R_m = maintenance respiration rate, (g CH_2O)/[(g tissue)·h]
E = conversion efficiency of CH_2O to plant tissue, (g tissue)/(g CH_2O)
P_g = canopy gross photosynthesis rate, (g CH_2O)/(m²·h)

Total biomass growth rate is partitioned into canopy and root biomass with the function $f_c(N)$, or $dW_c/dt = (dW/dt)f_c(N)$, and $dW_r/dt = (dW/dt)[1.0 - f_c(N)]$. This partitioning of new growth between canopy and root varies with stage of development; however, in this example, we will assume that $f_c(N)$ is constant for all N.

An expression is needed to predict canopy gross photosynthesis rate, P_g, as affected by light, CO_2, temperature, and plant size. There are many models that describe leaf and canopy photosynthesis; Charles-Edwards (1981) gives the mathematics of photosynthesis processes in detail. The model of Acock et al. (1978) was found to adequately describe tomato canopy photosynthesis rates (Jones et al., 1991). This equation, modified to include temperature effects, is

$$P_g = D \frac{\tau C p(T)}{K} \ln \left[\frac{\alpha K I_0 + (1 - m)\tau C}{\alpha K I_0 \exp(-KL) + (1 - m)\tau C} \right] \qquad (29)$$

where

> D = coefficient to convert photosynthesis calculations from $(\mu mol\ CO_2)/(m^2 \cdot s)$ to $(g\ CH_2O)/(m^2 \cdot h)$
>
> τ = leaf conductance to CO_2, $(\mu mol\ CO_2)/[(m^2\ leaf) \cdot s]$
>
> C = CO_2 concentration of the air, $(\mu mol\ CO_2)/(mol\ air)$
>
> $p(T)$ = photosynthesis reduction factor, dimensionless
>
> α = leaf light utilization efficiency, $(\mu mol\ CO_2)/(\mu mol\ photons)$
>
> K = canopy light extinction coefficient, dimensionless
>
> I_0 = light flux density at the top of the canopy, $(\mu mol\ photons)/[(m^2\ ground) \cdot s]$
>
> m = light transmission coefficient of leaves, dimensionless
>
> L = canopy leaf area index, $(m^2\ leaf)/(m^2\ ground)$

The function $p(T)$ expresses the effect of temperature on the maximum rate of photosynthesis for a single leaf, expressed as a quadratic equation with T:

$$p(T) = [1 - ((\phi_h - T)/(\phi_h - \phi_l))^2] \qquad (30)$$

where ϕ_h is the temperature at which leaf photosynthesis is maximum and ϕ_l is the temperature below which leaf photosynthesis is zero. Hourly values of I_0 were computed using the method of Goudriaan (1986), assuming 12 h day lengths (i.e., $6 < t_h < 18$):

$$I_0 = I_m \sin\{2\pi[(t_h - 6)/24]\} \qquad (31)$$

where I_m is the maximum light flux density (at noon) and t_h is the solar time in hours.

Equation (29) contains environmental variables (T, C, I_0), various parameters (τ, α, K, m, D), and the canopy leaf area, L, which depends on time. Leaf area ratio (square meters of leaf area per gram of plant) and specific leaf area, (m² leaf)/(g leaf), vary considerably with environmental conditions. Under cool temperatures, leaves appear more slowly [Eq. (26)], accumulate more dry weight during the extended growth period, but expand to about the same final area over a practical temperature range, as shown by Jones et al. (1991) for tomato. Under low light, leaves will tend to be thinner because of lower rates of photosynthesis per unit of development. The assumption that L is a function of N provides a model that mimics observed leaf area ratio responses to light, temperature, and CO_2. The expolinear equation (Goudriaan and Monteith, 1990) was used to fit leaf area vs. node number using data reported by Jones et al. (1991) for tomatoes grown at two CO_2 levels and three temperatures. The equation is

$$L = \rho(\delta/\beta)\ln\{1 + \exp[\beta(N - n_b)]\} \qquad (32)$$

where

L = leaf area index, (m² leaf)/(m² ground)
ρ = plant density, number/m²
N = leaf number

and δ, β, and n_b are empirical coefficients for the expolinear equation.

Therefore, the model predicts the rate of leaf appearance, which primarily depends on temperature, and calculates leaf area. The final three-state variable model can be expressed as

$$\frac{dN}{dt} = r_m r(T)$$

$$\frac{dW_c}{dt} = E(P_g - R_m W)f_c(N), \qquad \frac{dW_r}{dt} = E(P_g - R_m W)[1 - f_c(N)] \qquad (33)$$

with Eqs. (27) and (29)–(32) required to compute R_m, P_g, $p(T)$, I_0, and L, respectively. These equations could be substituted into Eq. (33), but this is not necessary because computer simulation is used to solve the equations. For this study, a time step (Δt) of 1 h was used, and the model was simulated for 80 days. For each simulation, environmental inputs (C, T, and I_0) were held constant for each run, and a number of combinations were simulated to demonstrate model behavior. Table 1 shows the values for coefficients used in this model, most of which were taken from Jones et al. (1991).

Table 1 Values of Coefficients Used in the Crop Model Example

$\alpha = 0.056$	$\tau = 0.0664$	$r_m = 0.021$	$I_m = 1200$
$\beta = 0.38$	$\rho = 4.0$	$C = 350$	$K = 0.58$
$\delta = 0.074$	$k_m = 0.0006$	$D = 0.108$	
$\phi_h = 30.0$	$m = 0.10$	$E = 0.70$	
$\phi_l = 5.0$	$n_b = 13.3$	$f_c(N) = 0.85$	

There are limitations in the use of this model. It does not account for senescence or picking of plant material, nor does it partition dry matter into different plant components, such as fruit. The model could be extended to describe the growth of fruit and other organs, thereby creating more state variables, equations, and a need for more coefficients and relationships.

C. Model Behavior

Figures 7a and 7b show the effects of temperature, light, and CO_2 concentration on the rates of photosynthesis predicted by Eq. (29) and the effects of temperature on respiration for a canopy with a leaf area index (L) of 4.0. These results demonstrate the importance of weather on the rate of dry matter accumulation. Figure 7c shows the effect of temperature on leaf area expansion over time. Low temperature reduces the rate of node formation [Eq. (26)], which results in delays in leaf area development. When plants are young, rapid leaf area expansion is important to create a full canopy for capturing light for photosynthesis.

The three-state variable model, Eq. (33), was simulated for 80 days under assumed constant environmental conditions to demonstrate the integrated and cumulative effects of growth and development rates on yield. For each set of computations, all variables but one were held at reference values and one variable was changed for successive runs. Reference conditions were 30/20°C day/night temperatures, 12 h days, 1200 (μmol photons)/($m^2 \cdot s$) maximum light flux density, and 350 (μmol CO_2)/(mol air).

Figure 7d demonstrates a major effect of temperature on dry weight over the 80 day simulations. Temperature affects both development and photosynthesis, but its major effect is on development. For the 18/8°C temperature case, 59 days were required to reach node 10 in contrast at 23 days for the 30/20°C temperature case. Photosynthesis for the 18/8°C case was reduced by only about 30%. For this

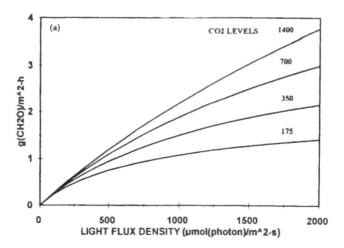

FIGURE 7   Simulated results from the three state variable crop model. (a) Photosynthesis vs. maximum light flux (I_m) at four CO_2 levels at temperature $T = 30°C$; (b) photosynthesis vs temperature at four maximum light flux levels for a canopy with leaf area index $L = 4.0$; (c) leaf area development vs time for four temperatures; (d) total dry weight vs. time for four day/night temperatures; and (e) total dry weight vs. time for four light levels at 30/20°C day/night temperatures.

coldest case, dry matter at 40 days was lower by a factor of 15. Note that growth was lower at 38/28°C than at 30/20°C day/night temperatures.

As light was varied from 400 to 1600 (μmol photons)/(m²·s), dry weight yield increased by a factor of 2.5 at 80 days (Fig. 7e). Numbers of leaves and leaf area were not affected by light. Increasing CO_2 levels from an ambient level of 350 (μmol CO_2)/(mol air) to 700 resulted in a 28% increase in dry weight yield at 80 days (data not shown).

APPENDIX.   VARIABLES USED IN THIS CHAPTER

Water Tank, Soil Layers, and Insect Population Examples

A_i area of water tank i, m²
C_i constant, m² (water tank case)
D diffusivity of the soil, m/s

(*Appendix continues p. 57*)

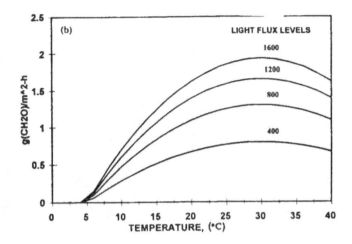

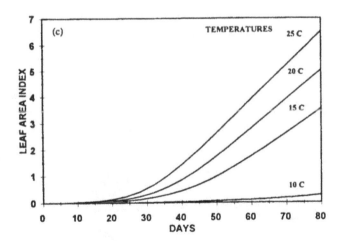

FIGURE 7 Continued

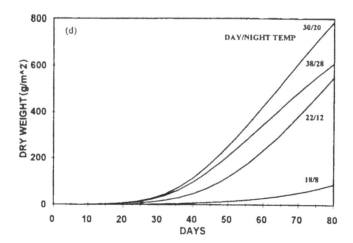

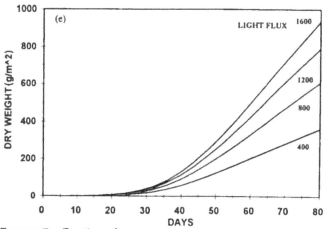

FIGURE 7   Continued

$f_{i,j}$ flow rate from tank i to tank j, m^3/s
g gravity constant, m/s^2
h_i thickness of soil layer i, m
H_i height of water in tank i, m
$I_{i,j}$ rate of flow into level i from source
$N(a, t)$ population of insects age a at time t, number
$O_{i,k}$ rate of flow out of level i to source
$q_{i,j}$ flow rate of water from layer i to layer j, m^3/s
$u(a, t)$ net flux of insects age a at time t across system boundary, s^{-1}
V_i volume in tank i, m^3
w_i volume of water stored in layer i, m^3
x_i level of ith state variable, various units
θ volumetric water content, m^3/m^3
$\sigma(y, k)$ absolute sensitivity of output y to input k, dimensionless
$\sigma_r(y, k)$ relative sensitivity of output y to input k, dimensionless

Crop Model Example

A_p assimilates, $(g\ CH_2O)/m$
C CO_2 concentration of the air, $(\mu mol\ CO_2)/(mol\ air)$
D photosynthesis conversion coefficient, $(g\ CH_2O)/(\mu mol\ CO_2)$
E conversion efficiency at CH_2O to plant tissue, (g tissue)/(g CH_2O)
$f_c(N)$ fraction of total crop growth partitioned to canopy, dimensionless
I_0 actual light flux density at top of canopy, $(\mu mol\ photons)/(m^2 \cdot s)$
I_m maximum light flux density at top of canopy, $(\mu mol\ photons)/(m^2 \cdot s)$
k_m respiration rate at 25°C, $(g\ CH_2O)/[(g\ tissue) \cdot h]$
K canopy light extinction coefficient, dimensionless
L canopy leaf area index, $(m^2\ leaf)/(m^2\ ground)$
m light transmission coefficient of leaves, dimensionless
n_b empirical coefficient for expolinear equation, dimensionless
N leaf number or node number, dimensionless
$p(T)$ photosynthesis reduction factor, dependent on T, dimensionless
P_g canopy gross photosynthesis rate, $(g\ CH_2O)/(m^2 \cdot h)$
r rate of development, h^{-1}
r_m maximum rate of leaf appearance, h^{-1}

R_m	maintenance respiration rate, $(g\ CH_2O)/[(g\ tissue)\cdot h]$
T	temperature, °C
t_h	solar time, h
W	total plant dry matter weight, g/m^2
W_c	canopy dry matter weight, g/m^2
W_r	root dry matter weight, g/m^2
α	light utilization efficiency, $(\mu mol\ CO_2)/(\mu mol\ photons)$
β	empirical coefficient for expolinear equation, dimensionless
δ	empirical coefficient for expolinear equation, dimensionless
η	day length, h
ρ	plant density, m^{-2}
τ	leaf conductance to CO_2, $(\mu mol\ CO_2)/[(m^2\ leaf)\cdot s]$
ϕ_h	temperature at which leaf photosynthesis is maximal, °C
ϕ_l	temperature at which leaf photosynthesis is zero, °C

REFERENCES

Acock, B., D. A. Charles-Edwards, D. J. Fitter, D. W. Hand, L. J. Ludwig, J. Warren Wilson, and A. C. Withers. 1978. The contribution of leaves from different levels within a tomato crop to canopy net photosynthesis: An experimental examination of two canopy models. *J. Exp. Bot.* 29:815–827.

Batchelor, W. D., J. W. Jones, K. J. Boote, and H. Pinnschmidt. 1993. Extending the use of crop models to study pest damage. *Trans. ASAE* 36(2):551–558.

Beinroth, F. H., Jones, J. W., E. B. Knapp, P. Papajorgji, and J. C. Luyten. 1997. Application of DSSAT to the evaluation of land resources. In: *Understanding Options for Agricultural Production*. G. Y. Tsuji, G. Hoogenboom, and P. K. Thornton (Eds.). Dordrecht: Kluwer Academic. (in press).

Boggess, W. G., J. W. Jones, D. P. Swaney, and G. D. Lynne. 1981. Evaluating irrigation strategies in soybeans: A simulation approach. In: Proceedings of the ASAE irrigation Scheduling Conference, Dec. 14–15, 1981. St. Joseph, MI: ASAE, pp. 45–53.

Boote, K. J., J. W. Jones, and N. B. Pickering. 1996. Potential uses and limitations of crop models. *Agron. J.* 88(5):704–716.

Bouman, B. A. M., H. van Keulen, H. H. van Laar, and R. Rabbinge. 1996. The "School of de Wit" crop growth simulation models: A pedigree and historical overview. *Agric. Syst.* 52(2–3):171–198.

Charles-Edwards, D. A. 1981. *The Mathematics of Photosynthesis and Productivity*. New York: Academic Press.

Curry, G. L., and R. M. Feldman. 1987. *Mathematical Foundations of Population Dynamics*. TEES Monograph Series, No. 4. College Station, TX: Texas A&M Univ. Press.

Curry, R. B., J. W. Jones, K. J. Boote, R. M. Peart, L. H. Allen, Jr., and N. B. Pickering. 1995. Response of soybean to predicted climate change in the USA. In: *Climate Change and Agriculture: Analysis of Potential International*

Impacts. C. Rosenzweig, J. W. Jones, G. Y. Tsuji, and P. Hildebrand (Eds.). ASA Special Publication No. 59. Madison, WI: American Society of Agronomy, pp. 163–182.

Dent, J. B., and M. J. Blackie. 1979. *Systems Simulation in Agriculture*. London: Applied Science.

Engel, T., J. W. Jones, and G. Hoogenboom. 1996. AEGIS/WIN—A powerful tool for the visualization and comparison of crop simulation results. In: *Proceedings of the Sixth International Conference on Computers in Agriculture*. F. S. Zazueta (Ed.). Cancun, Mexico, June 1–14, 1996. ASAE Pub. 701P0396. St. Joseph, MI: ASAE, pp. 605–612.

Forrester, J. W. 1961. *Industrial Dynamics*. New York: Wiley.

Forrester, J. W. 1971. *Principles of Systems*. Cambridge, MA: Wright-Allen.

Fraisse, C. W., K. L.Campbell, J. W. Jones, W. G. Boggess, and B. Negahban. 1994. Integration of GIS and hydrologic models for nutrient management planning. Proceedings of the National Conference on Environmental Problem Solving with Geographical Information Systems. Cincinnati, OH. Sept. 21–23, 1994. EPA/625/R-95/004, pp. 283–291.

France, J., and J. H. M. Thornley. 1984. *Mathematical Models in Agriculture*. London: Butterworths.

Gardner, W. R. 1959. Solutions of the flow equation for the drying of soil and other porous media. *Soil Sci. Soc. Proc.* 23:183–187.

Gent, M. P. N., and H. Z. Enoch. 1983. Temperature dependence of vegetative growth and dark respiration: a mathematical model. *Plant Physiol.* 71: 562–567.

Gold, H. J. 1982. *Mathematical Modeling of Biological Systems: An Introductory Guidebook*. New York: Wiley.

Goudriaan, J. 1986. A simple and fast numerical method for the computation of daily totals of crop photosynthesis. *Agric. Forest Meteorol.* 38:251–255.

Goudriaan, J. 1988. The bare bones of leaf-angle distribution in radiation models for canopy photosynthesis and energy exchange. *Agric. Forest Meteorol.* 43:155–169.

Goudriaan, J., and H. H. van Laar. 1994. *Modelling Potential Crop Growth Processes: Textbook with Exercises*. Current Issues in Production Ecology, Vol. 2. Dordrecht: Kluwer.

Goudriaan, J., and J. L. Monteith. 1990. Mathematical function for crop growth based on light interception and leaf area expansion. *Ann. Bot.* 66: 695–701.

High Performance Systems. 1990. *Stella II User's Guide*. Hanover, NH: High Performance Systems.

Hillel, D. 1971. *Soil and Water Physical Principles and Processes*. New York: Academic Press.

Hoogenboom, G., J. W. Jones, P. W. Wilkens, W. D. Batchelor, W. T. Bowen, L. A. Hunt, N. B. Pickering, U. Singh, D. C. Godwin, B. Bear, K. J. Boote, J. T. Ritchie, and J. W. White. 1994. Crop models. In: *DSSAT v3*. Vol. 2. G. Y. Tsuji, G. Uehara, and S. Balas (Eds.). Honolulu: University of Hawaii, pp. 95–244.

Jacobson, B. M., and J. W. Jones, 1996. Designing a decision support system for soybean management. In: *Proceedings of the Sixth International Conference on Computers in Agriculture*. F. S. Zazueta (Ed.). Cancun, Mexico, June 1–14, 1996. ASAE Pub. 701P0396. St. Joseph, MI: ASAE, pp. 394–403.

Jones, J. W. 1993. Decision support systems for agricultural development. In: *Systems Approaches for Agricultural Development*. F. Penning de Vries, P. Teng, and K. Metselaar (Eds.). Dordrecht: Kluwer Academic, pp. 459–471.

Jones, J. W., and J. T. Ritchie. 1990. Crop growth models. In: *Management of Farm Irrigation Systems*. G. J. Hoffman, T. A. Howell, and K. H. Solomon (Eds.). St. Joseph, MI: Am Soc. Agric. Eng., pp. 61–89

Jones, J. W., E. Dayan, L. H. Allen, H. van Keulen, and H. Challa. 1991. A dynamic tomato growth and yield model (TOMGRO). *Trans. ASAE 34*: 663–672.

Jones, J. W., L. A. Hunt, G. Hoogenboom, D. C. Godwin, U. Singh, G. Y. Tsuji, N. B. Pickering, P. K. Thornton, W. T. Bowen, K. J. Boote, and J. T. Ritchie. 1994. Input and output files. In: *DSSAT v3*. Vol. 2. G. Y. Tsuji, G. Uehara, and S. Balas (Eds.). Honolulu: University of Hawaii, pp. 1–94.

Keating, B. A., R. L. McCown, and B. M. Wafula. 1993. Adjustment of nitrogen inputs in response to a seasonal forecast in a region of high climatic risk. In: *Systems Approaches for Agricultural Development*. F. Penning de Vries, P. Teng, and K. Metselaar (Eds.). Dordrecht: Kluwer Academic, pp. 233–252.

Keen, R. E., and J. D. Spain. 1992. *Computer Simulation in Biology: A Basic Introduction*. New York, Wiley-Liss.

Kiker, G. A., J. W. Jones, and E. R. Muller. 1991. A model and database for predicting phenomenological stages. ASAE Paper No. 91-7035. St. Joseph, MI: ASAE.

Koenig, H. E., Y. Tokad, and D. R. Allen. 1967. *Analysis of Discrete Physical Systems*. New York: McGraw-Hill.

Leffelaar, P. A. (Ed.). 1993. *On Systems Analysis and Simulation of Ecological Processes, with Examples in CSMP*. Current Issues in Production Ecology. Vol. 1. Dordrecht: Kluwer Academic.

Lovenstein, H., E. A. Lantinga, R. Rabbinge, and H. van Keulen. 1993. *Principles of Theoretical Production Ecology*. Course book. Department of Theoretical Production Ecology. Wageningen, The Netherlands: Wageningen Agricultural University.

Martens, H. R., and D. R. Allen. 1969. *Introduction to Systems Theory*. Columbus, OH: Merrill.

Microsoft. 1995. *Microsoft Visual Basic, Programmer's Guide*. Redmond, WA: Microsoft Corporation.

Monteith, J. L. 1990. Conservative behaviour in response of crops to water and light. In: *Theoretical Production Ecology: Reflections and Prospects*. R. Rabbinge, J. Goudriaan, H. van Keulen, F. W. T. Penning de Vries and H. H. van Laar (Eds.). Simulation Monographs 34. Wageningen, The Netherlands: PUDOC, pp. 3–16.

Odum, H. T. 1973. An energy circuit language for ecological and social systems. In *Systems Analysis and Simulation in Ecology.* Vol. II. B. C. Patten
(Ed.). New York: Academic Press, pp. 140–211.

Ortega, J. M., and W. G. Poole, Jr. 1981. *An Introduction to Numerical Methods
for Differential Equations.* Marshfield, MA: Pitman.

Patten, B. C. 1973. *Systems Analysis and Simulation in Ecology.* Vol. II. New
York: Academic Press.

Penning de Vries, F. W. T. 1993. Rice production and climate change. In:
Systems Approaches for Agricultural Development. F. Penning de Vries, P.
Teng, and K. Metselaar (Eds.). Dordrecht, The Netherlands: Kluwer Aca-
demic, pp. 175–189.

Penning de Vries, F. W. T., and H. H. van Laar (Eds.). 1982. *Simulation of
Plant Growth and Crop Production.* Wageningen, The Netherlands: PUDOC.

Penning de Vries, F. W. T., D. M. Jansen, H. F. M. ten Berge, and A. Bakema.
1989. *Simulation of Ecophysiological Processes of Growth in Several Annual
Crops.* Simulation Monographs 29. Wageningen, The Netherlands: PUDOC.

Pinnschmidt, H., P. S. Teng, and J. E. Yuen. 1990. Pest effects on crop growth
and yield. In *Proceedings of the Workshop on Modeling Pest–Crop Interactions.*
P. Teng and H. Yuen (Eds.). Honolulu: University of Hawaii.

Pritsker, A. A. B. 1995. *Introduction to Simulation and SLAM II.* 4th ed. New
York: Wiley.

Rabbinge, R., P. A. Leffelaar, and H. C. van Latesteijn. 1994. The role of
systems analysis as an instrument in policy making and resource management. In *Opportunities, Use, and Transfer of Systems Research Methods in Agriculture to Developing Countries.* P. R. Goldsworthy and F. W. T. Penning
de Vries (Eds.). Dordrecht, The Netherlands: Kluwer Academic, pp. 67–
79.

Ritchie, J. T. 1995. International Consortium for Agricultural Systems Applications (ICASA): Establishment and purpose. *Agric. Syst.* 49:329–335.

Rosenzweig, C., J. W. Jones, G. Y. Tsuji, and P. Hildebrand (Eds.). *Climate
Change and Agriculture: Analysis of Potential International Impacts.* ASA Spec.
Publ. No. 59. Madison, WI: American Society of Agronomy.

Sinclair, T. R. 1990. Nitrogen influence on the physiology of crop yield. In:
Theoretical Production Ecology: Reflections and Prospects. R. Rabbinge, J.
Goudriaan, H. van Keulen, F. W. T. Penning de Vries, and H. H. van Laar
(Eds.). Simulation Monographs 34. Wageningen, The Netherlands: PU-
DOC, pp. 41–55.

Singh, U., P. K. Thornton, A. R. Saka, and J. B. Dent. 1993. Maize modeling
in Malawi: A tool for soil fertility research and development. In: *Systems
Approaches for Agricultural Development.* F. Penning de Vries, P. Teng, and
K. Metselaar (Eds.). Dordrecht, The Netherlands: Kluwer Academic, pp.
253–273.

Smerage, G. H. 1977. On the extension of complementary variables modeling
theory to biology and ecology. In: *Problem Analysis in Science and Engineering.* F. H. Branin, Jr. and K. Huseyin (Eds.). New York: Academic Press,
pp. 427–462.

Swaney, D. P., J. W. Jones, W. G. Boggess, G. G. Wilkerson, and J. W. Mishoe, 1983. A crop simulation model for evaluation of within season irrigation decisions. *Trans. ASAE 26*(2):362–568.

Thornley, J. H. M., and I. R. Johnson. 1990. *Plant and Crop Modeling.* New York: Oxford Univ. Press.

Thornton, P. K., G. Hoogenboom, P. W. Wilkens, and W. T. Bowen. 1995. A computer program to analyze multi-season crop model outputs. *Agron. J. 87*:131–136.

van Kraalingen, D. W. G. 1995. *The FSE System for Crop Simulation, Version 2.1. Quantitative Approaches in Systems Analysis,* No. 1. July 1995. Wageningen, The Netherlands: DLO Research Institute for Agrobiology and Soil Fertility and C. T. de Wit Graduate School for Production Ecology.

Visual Solutions. 1990. *VisSim. User's Manual.* Westford, MA: Visual Solutions Inc.

Wolf, S., J. Rudich, A. Marani, and Y. Rekah. 1986. Predicting harvesting date of processing tomatoes by a simulation model. *J. Am. Soc. Hort. Sci. 111*(1): 11–16.

3

Using Mathematics as a Problem-Solving Tool

J. Robert Cooke
Cornell University, Ithaca, New York

I. INTRODUCTION

Crop modeling's considerable popularity and success as a research and planning tool stems from its focus on the behavior of agronomic crops as a system. In addition to the inherent biological complexity of plant growth and development, there are numerous environmental conditions and insults such as drought and attacks by pests and diseases that must be considered in a truly comprehensive characterization of crop production processes and yield.

Two tools have come to occupy a central role in dealing with this complexity: mathematics and computers. The first provides a powerful language and framework for the conceptualization of the relationships, and the second provides a powerful technique for exploring the relationships encoded mathematically. This chapter explores the process of representing physical reality using mathematics and the process of extracting insights from these representations. In other words, I explore the process of using mathematics as a working tool to gain insight into a problem.

This discussion consists of

1. A brief review of the process scientists use to conceptualize the regularities of nature for predictive purposes
2. A brief discussion of the role of mathematical models in connecting the conceptual and experimental tasks
3. An extended discussion of the practical steps engineers have found useful in formulating, validating, and interpreting mathematical models
4. A consideration of a representative problem to illustrate and to make the discussion of the traditional modeling process more concrete

II. THE SCIENTIFIC METHODOLOGY FOR DESCRIBING REALITY

The process of creating new knowledge involves the exploration of a focus question or puzzlement. Figure 1 depicts this process (Novak and Gowan, 1984). Guided by the focus question phenomena of interest are examined. Beginning at the base of Gowin's "Knowledge Vee" certain events or objects are observed. Meaning is attached to these events or objects by an intimate interplay of conceptual (or thinking) and methodological (or doing) considerations, as depicted by the two sides of the "Vee."

Consider first the methodological considerations (on the right side). Records of the events or objects are made. The facts are then ordered or otherwise organized (transformations). The results are then presented as data using tables, graphs, etc. This is followed by interpretations, explanations and generalizations. These lead to knowledge claims about what was found. The final step consists of value claims about the meaning or significance of the knowledge claims.

On the conceptual side of the process, (beginning at the top-left of the "Vee"); going from the most general, i.e., World View (that nature is orderly and knowable) to Philosophies, to Theories (logically related concepts) to Principles (derived from prior knowledge claims) to Constructs and Conceptual Structures to Concepts (signifying regularities). During the creation of new knowledge there is an active interplay between the conceptual and methodological sides of

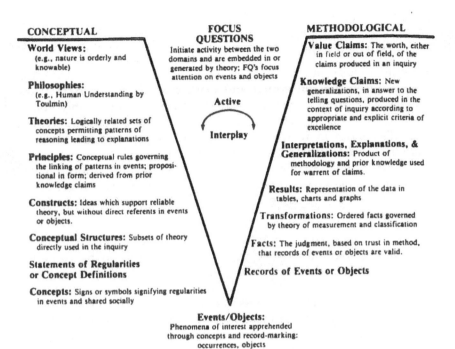

CONCEPTUAL

World Views: (e.g., nature is orderly and knowable)

Philosophies: (e.g., Human Understanding by Toulmin)

Theories: Logically related sets of concepts permitting patterns of reasoning leading to explanations

Principles: Conceptual rules governing the linking of patterns in events; propositional in form; derived from prior knowledge claims

Constructs: Ideas which support reliable theory, but without direct referents in events or objects.

Conceptual Structures: Subsets of theory directly used in the inquiry

Statements of Regularities or Concept Definitions

Concepts: Signs or symbols signifying regularities in events and shared socially

FOCUS QUESTIONS
Initiate activity between the two domains and are embedded in or generated by theory; FQ's focus attention on events and objects

Active

Interplay

METHODOLOGICAL

Value Claims: The worth, either in field or out of field, of the claims produced in an inquiry

Knowledge Claims: New generalizations, in answer to the telling questions, produced in the context of inquiry according to appropriate and explicit criteria of excellence

Interpretations, Explanations, & Generalizations: Product of methodology and prior knowledge used for warrent of claims.

Results: Representation of the data in tables, charts and graphs

Transformations: Ordered facts governed by theory of measurement and classification

Facts: The judgment, based on trust in method, that records of events or objects are valid.

Records of Events or Objects

Events/Objects:
Phenomena of interest apprehended through concepts and record-marking: occurrences, objects

FIGURE 1   The vee heuristic with descriptions of the interacting elements involved in the construction or analysis of knowledge in any discipline. Although all elements are involved in any coherent research program, the major sources of difficulty in individual inquiry usually begin at the bottom of the vee, where concepts, events/objects, and records must be scrutinized. (From Novak and Gowin, 1984, Reprinted with permission of Cambridge University Press, New York, New York.)

the "Vee." Novak and Gowin (1984) provide numerous specific illustrative examples from a range of subjects.

In a nutshell, this is the process we use to create new knowledge. The process is so familiar that we often take it for granted. However, being conscious of these steps makes the modeling process clearer; it is fundamentally an artistic process in which there is no single "right" model. It is fundamentally a matter of successive approximations deeply affected by matters of taste.

III. CHOOSE YOUR TOOLS

Scientific inquiry progresses more rapidly when there is a rich interplay between the conceptual and experimental at each step in the process of creating knowledge rather than relying upon only one or the other of these legs of the "vee." However, the literature is replete with disciplines in which the theoretical and experimental aspects have matured at dramatically different rates.

Often, however, we as individual scientists rely more heavily upon either one or the other, depending upon our training. For example, Fry (1941) contrasted the habits of thought of mathematicians and engineers:

> [1] Greater confidence in: The typical mathematician feels greater confidence in a conclusion reached by careful reasoning and is not convinced to the same degree by experimental evidence.
>
> An engineer has less interest in "pure" forms which have no apparent connection to the world of physical reality, but may have great interest in such useful information as a table of hardness which may be totally unrelated to any theory and which the typical mathematician would find quite boring.
>
> [2] Severe criticism of details: A mathematician is highly critical towards the details of a demonstration. For almost any other class of people, an argument may be good enough, even though some minor question remains open. For a mathematician an argument is either perfect in every detail, in form as well as substance, or else it is wrong. The mathematician calls this rigorous thinking and it is necessary for work to have permanent value; the engineer calls it hair-splitting and says if indulged in would never get anything done.
>
> [3] Idealization: The mathematician tends to idealize any situation—gases are 'ideal,' conductors, gases 'perfect,' surfaces 'smooth.' What the mathematician calls 'getting down to essentials,' the engineer is likely to dub somewhat contemptuously as 'ignoring the facts.'
>
> [4] Generality: When confronted with solving
>
> $$x^3 - 1 = 0$$

that simple question is likely to be turned into solving

$$x^n - 1 = 0.$$

The mathematician calls this "conserving energy." The engineer is likely to regard this as "wasting time."

One approach is not superior to the other. They are different and complementary.

One of the truly great contributions that engineering can make to crop modeling is the transfer of some of the concepts and techniques that have proven to be so fruitful for engineering purposes. Far greater reliance has been placed upon the use of mathematical techniques in the physical sciences than has been the case in the biological and agricultural sciences. Nevertheless, the use of empirical models has gained widespread acceptance in crop modeling in the past 30 years.

While a major portion of the body of knowledge associated with engineering in general and in crop modeling in particular is heavily empirical in nature, fairly extensive use has nevertheless been made of concepts and theories that can be most conveniently expressed in mathematical form. The following discussion applies equally for modeling, which relies almost exclusively on the use of computers and empirical relationships, and the more traditional approach, which relies on computers and first principles. The latter approach, when enough is known to make the approach feasible, often provides deeper insight and usually can be more readily generalized to a broader context.

IV.   THE NATURE OF MATHEMATICAL MODELS

At the interface region between engineering and biology there persists the lingering and mistaken belief that biologically based problems are inherently so complex that they simply are not amenable to mathematical treatment. One might imagine that this same attitude could have been applied to quantum mechanics, were it not for the astonishing success achieved using mathematics as a central tool.

Experimentalists, especially those who ordinarily work with biological problems, are sometimes inclined to be suspicious of mathematical work. Although a healthy skepticism is highly desirable, the role of mathematical modeling has many times been unfairly criticized (but certainly not in all cases). A particularly common difficulty is the failure to note the necessity and desirability of making assumptions—in both mathematical and experimental studies. In both instances attention is directed to a limited problem and the conclusions reached are valid only within the constraints of "the experimental assumptions." Mathematical models (Noble, 1967) can be used to cause attention to be directed to the most important questions in an experimental study. If data agree with a theory, the level of confidence in the predictive power of the model is enhanced. In other instances, mathematics can be used to reduce or eliminate certain experimentation.

Nahikian (1964) graphically depicted the successive approximations associated with a mathematical model (see Fig. 2). Beginning with a clear statement of a "problem," the most important properties of a physical or biological problem to be studied are abstracted and isolated by making judicious use of assumptions (left side of Fig. 2). A mathematical problem, using basic physical or chemical principles, is developed to embody the "essence" of the problem. The "mathematical crank" (bottom) is turned and checked. The implications of the model are articulated. These implications are then related to the original problem through the observed phenomena. Discrepancies are noted, and the cycle begins again. However, notice that the arrows are bidirectional throughout, indicating that the logical progression may reverse at any point.

The advent of the computer is having a great liberating effect on the use of mathematics. There is no longer the overriding need to formulate problems to conform to the limited tools of classical anal-

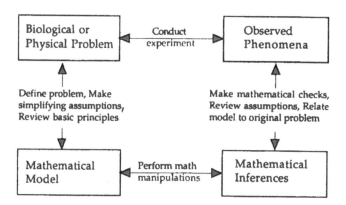

FIGURE 2   Development of mathematical models. [Adapted from Nahikian (1964).]

ysis. On the other hand, when using a computer we still must abstract and idealize a given engineering problem. We still must break the original problem into its elements, decide what is significant, specify the data we must start from, what the steps of the analysis are, and what end results are required. These steps require the habit of thought of the mathematician.

V.   FORMULATING A MATHEMATICAL MODEL

The process of forming a model (left side of Fig. 2) is both a science and an art form. The following discussion concentrates on the well-established techniques used in creating and validating a model and gaining insight from it. The more artistic, but just as essential, aspect of modeling is mentioned only in passing; nonetheless, that can be the most important and decisive step. It is mainly here that a reasonable balance is struck between complexity and simplicity. A model can be so inclusive of details that it becomes intractable or difficult to understand. A model might be endowed with so many empirical coefficients, discernible only from experimental measurements for a particular setting, that it has limited utility in a setting that is only slightly different because of a loss of predictive power and insight into the underlying mechanisms. On the other hand, a model can be made so simple that the essence of the situation is lost. For example, certain phenomena might not be encompassed if represented using a system of linear ordinary differential equations rather than a system of nonlinear ordinary differential equations. A linear model, for example, does not encompass stable limit cycle behavior. And a deterministic model might be unsuited for situations in which probabilistic models might be successful.

Suppose you are interested in examining the temperature variations within the walls of an internal combustion engine during the combustion cycle. How might you represent the problem as a mathematically tractable problem, yet achieve a good approximation to experimental values?

You probably would formulate this problem as one of heat conduction in a solid. Next you would probably realize that you must define an appropriate geometric configuration. Dealing with the actual geometry would probably require that you resort to a numerical approach such as the use of finite-difference or finite-element techniques. Suppose you wish to make a simpler first approximation using solutions from the existing literature?

Several possibilities come to mind. You might formulate this as (a) a thick cylinder subjected to a periodic temperature fluctuation on the inner wall of the cylinder or (b) a cylindrical hole in an infinite space. Case a will lead to the use of Hankel functions, whereas case b will lead to the use of the simpler Bessel functions. Depending upon the frequency of the periodic fluctuation at the inner cylinder, the depth of penetration might be approximately the same in both cases. If so, the simpler model might be adequate.

On the other hand, if the curvature of the inner cylinder is not a primary consideration, you might formulate this as (c) a finite-thickness slab or (d) a semi-infinite solid (a half-space having a flat edge). For case c, the solution will involve hyperbolic functions, while case d will involve the even simpler trigonometric functions (or the complex exponential function). As it turns out, the ratio of periodic surface temperature to the periodic temperature at any depth in the wall is roughly the same for all four cases when the frequency is in the range typical of an engine, but the mathematical burden becomes *very* much less for the simpler case d.

A crop modeler probably would recognize an analogy with the daily or annual periodic temperature fluctuations in the soil and how

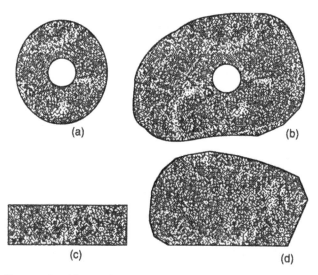

FIGURE 3 Alternative geometries. (a) A thick cylinder subjected to a periodic temperature fluctuation on the inner wall of the cylinder. (b) A cylindrical hole in an infinite space. (c) A finite-thickness slab. (d) A semi-finite solid.

the temperature varies with distance into the soil. Recognizing analogies can greatly strengthen one's confidence in a solution.

The greater the insight into the underlying processes you can muster at any stage of development, the more likely an acceptable balance between simplicity and complexity can be achieved. The crucial point to remember is that you are using a mathematical model to gain insight and that no single representation or model serves all purposes. A novice modeler is tempted to beg for a universal representation that fits all situations, but alas, if that is your goal, you are likely to trade an obscuring complexity for the clarity that often can be achieved more readily through simplicity.

Bowen (1966) described the basic issues to be considered in formulating a problem (Fig. 4). Before a problem can be properly defined, the decision maker must be identified. Usually the decision maker is both the person or group whose objectives are not being met and the one who controls the resources needed to effect a solution. Next the decision maker's objective and resources must be clarified and weighted. Usually objectives conflict and cannot be simultaneously satisfied. Furthermore, the environment with its associated constraints must be specified and understood. *A "solution" that ignores*

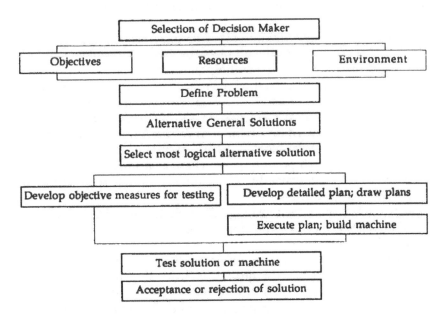

FIGURE 4 Steps in formulating a problem.

these constraints is not a solution. Only after all three of these issues (objectives, resources, and environment) are understood clearly can the problem be properly posed or defined with sufficient precision and clarity. How you define the problem fundamentally shapes the rest of the process!

To illustrate how a model fundamentally depends upon how the problem is framed, consider the approaches taken in solving the problem of mechanically harvesting tree fruits. The problem has been manifest in the literature in at least two different forms. An earlier view held that the criterion for fruit removal was the tensile forces developed from inertial reaction of the fruit stem while the tree is being vibrated. This inevitably led researchers to center on ways to produce vigorous shaking of the tree, e.g., at one of the natural frequencies of its *limbs*. On the other hand, if one defines the problem in terms of the bending stresses in the stem, at the fruit or at the branch, then the argument follows a substantially different pattern. The fruit and stem assembly can be regarded as a double physical pendulum, and one might then study the frequencies that produce the largest motions of the *fruit–stem system* rather than of the supporting tree structure. One should not be surprised to learn that the inferences concerning harvesting are quite different for the two statements of the problem and result in fundamentally different levels of understanding.

Only after a clear definition has been achieved—and this is often the most crucial and difficult phase in solving a problem—should you begin the process of generating alternative solutions. This stage requires imagination as well as a capacity to tap all your previous professional experience and knowledge. Novel associations and analogies often play an important role. In short, creativity plays the crucial role in this step. Many techniques can be used to stimulate this process. Often a mental block must be overcome if the problem is to be recast from a fresh point of view. There is a powerful tendency to view the problem in ways that are familiar even when objectives are no longer being satisfied.

The selection of the most logical alternative usually requires that a series of very specific problems be formulated and analyzed. This process is explored next.

As you develop plans to implement the alternative selected, you simultaneously articulate a quantifiable, or at least unambiguous, measure of performance by which you will judge the proposed solution. You do this before your plan is implemented and before you must make a rational assessment of whether you have succeeded.

After your solution has been implemented, there is a responsibility to determine whether your solution works. This usually involves some form of experimentation and an evaluation of whether you have met the objectives previously articulated.

VI. PRACTICAL STEPS IN FORMULATING MATHEMATICAL MODELS

Although numerous texts on mathematical modeling have been published, a very old text by Ver Planck and Teare (1954) serves well as the focal point for this discussion. Table 1 summarizes the approach presented in their book, but I've also made some adaptation. These are the time-honored steps that engineers routinely use in problem solving. Because the steps are so familiar, their significance is often overlooked.

Explicitly identifying these steps makes the process more comprehensible and more than just an art form. The steps are (A) define the problem, (B) plan its treatment, (C) execute the steps, (D) check your work, and (E) learn and generalize from your analysis. Let's consider each of the steps in this process.

A. Define the Problem

In dealing with non-textbook problems, great importance must be placed upon undertaking a heuristic first consideration of the problem before delving into the intricacies of a particular approach. In other words, consciously and explicitly decide among alternative approaches before you commit your time to any specific one.

By stating the anticipated results quite early in the study of a problem, you usually should be better able to judge the validity of your analysis. This will provide a basis for guiding the analysis as well as a background against which the analysis can be appraised. The appropriate point at which to review the relevant literature may well depend upon the nature of the problem. To avoid "reinventing the wheel," the literature might come very early, but on the other hand there is the possibility of prematurely channeling one's thinking to current ideas. There is also a need here for moderation, balance, and judgment.

How you state the problem to be solved, i.e., frame the puzzlement, will have a profound impact on the entire process. If you solve the wrong problem, your time may be wasted. If the problem

Table 1 Aspects of Using Mathematics in Engineering Problem Solving

A. Define the problem.
   1. Divide the problem into a series of specific questions.
   2. Discuss the problem in a qualitative fashion and present a heuristic motivation before plunging into equations.
   3. Consciously decide among alternative approaches.
   4. State what results you anticipate. This is equivalent to a hypothesis.
   5. Review the relevant literature.
B. Plan its treatment.
   1. Review the relevant physical or chemical principles. Express them verbally and then mathematically.
   2. Identify the assumptions that may be desirable and/or necessary.
   3. Formulate the problem mathematically.
      a. Use sketches freely.
      b. Define all symbols and give the associated units.
      c. Show all coordinates with positive directions identified.
      d. Identify all assumptions as they are embedded in the model.
      e. Cite all references consulted.
C. Execute the plan.
   1. Use more than one approach when possible.
   2. When carrying out the mathematical steps, number all equations.
   3. Use frequent intermediate checks.
   4. Provide an account of the motivation for each step, and indicate the logical connection between steps.
   5. If an impasse is reached, simplify the problem by altering the assumptions and/or the problem definition.
D. Check thoroughly.
   1. Check dimensions and units.
   2. Check limiting cases.
   3. Check symmetry.
   4. Check reasonableness.
   5. Check range of variables.
   6. Reexamine all assumptions for probable effect and importance.
   7. Use alternative derivation.
   8. Use analogies.
E. Learn and generalize from your analysis.
   1. Summarize the important findings, including limitations.
   2. Interpret the equations in terms of the original problem.
   3. Have important factors been overlooked?
   4. What is important by its absence?
   5. Use dimensionless numbers in the graphical presentation.
   6. Verbally state the meaning of the final mathematical equations.
   7. Can the proposed scheme be improved or altered so that it will not work?
   8. Check numerical cases for clarity and feasibility. This is a check of the magnitude of the effect and the required parameter values.
   9. How much parameter variation can be tolerated?
   10. How many significant figures should be used?
   11. Is the graphical presentation fully clarified, including clearly labeled axes, and is an interpretation of the information presented?
   12. What uncertainties still exist?
   13. How does the model relate to the original problem?
   14. Can and should the problem now be simplified or generalized?

Source: Adapted from Ver Planck and Teare (1954).

is stated carelessly, you may needlessly complicate your task. The clearer your grasp of the problem, the easier the task of defining the problem becomes. A poorly defined problem will likely increase the complexity of the mathematical formulation.

B. Plan Its Treatment

A verbal discussion of the problem (recorded in your notebook) is frequently important in refreshing one's grasp of the applicable physical principles, because the mathematical statement of the physical principles, especially if presented in mathematical form, may be terse and will usually contain many assumptions that are not immediately obvious. The actual formulation of the mathematical model should be accompanied by a very careful statement and review of the working assumptions that you must make in order to form a tractable mathematical model. Be continually aware of the trade-off between simplicity and complexity. Quite often it is advisable to formulate and solve a simplistic model before adding the level of detail that you believe will eventually be desirable. Whenever feasible, use multiple approaches. If the same result can be obtained by alternative methods, e.g., conservation of energy, and through equations of motion, you will increase your confidence in the results.

Furthermore, the validity and usefulness of your analysis usually suffer far more from an important unacknowledged assumption than from important acknowledged assumptions. Remember that modeling is an iterative process, i.e., not every detail is or should be included at the outset. In the spirit of Occam's razor, a successful simpler explanation is preferred to a successful but complex explanation. Formulating and solving one or more highly simplified models as a first approximation is usually prudent but is often overlooked.

Often a complex problem can be solved as a sequence of independent subproblems that, considered together, describe the more complex situation. Often several simpler problems will allow you to bracket the solution sufficiently to make a more sophisticated analysis unnecessary.

C. Execute the Plan

When you first introduce a variable, immediately define it in words and give its associated units. This will facilitate comprehension and subsequent checking. All equations must be dimensionally homogeneous, i.e., when the units replace the variables in the equation, the units on both sides of the equation must agree. Be wary of using numerical constants that have associated units; it is better to assign a parameter name. The parameter name is then treated on the same basis as the variables, and the dimensional homogeneity check is less error-prone.

Use sketches freely, number them, and provide captions. Even a crude sketch can often provide the clarity that would otherwise require many words. Label the axes, the positive directions, and all the components of the figure. Captions will be useful when you return to your work after any substantial time lapse. Likewise, for later recall and for use in the formal report writing stemming from your analysis, you should cite all the references you consulted. A methodical approach encourages careful thinking.

To facilitate the derivation and discussion of the equations, number every equation, even though you may choose to leave some unnumbered in your formal report. Remember that your analysis will progress more smoothly if you capture your thoughts as it proceeds. This not only helps you think through the problem, but also is an invaluable aid when you return to your analysis later. As the process proceeds, record your thoughts about the motivation of the analysis. Always provide a verbal account of the logical connections and the details in the analysis. You will not be well-served by dismissing the details with an assertion that "it can be shown." Also, your notes serve an archival purpose but should be treated as distinct from a formal report, which is used to present your ideas in an economical form for the benefit of others. Your notes are for your own use.

If you reach an impasse in the analysis, simplify the problem by making additional assumptions or by modifying the problem definition. Once you understand a simplified version, even if it is too simple to be included in the final report, it may provide just the insight you required to solve the more complicated version.

D.   Check Thoroughly

This section draws heavily upon two primary, out-of print sources: Ver Planck and Teare (1954, chapter 6) and Johnson (1944, chapter 8).

"Engineering work not thoroughly checked has little value." (Ver Planck and Teare, 1954, p. 229). Frequent use of intermediate checks is highly desirable. Your analysis must be free from essential errors—from using the incorrect methodology to getting the decimal point wrong. Your professional reputation depends upon your reliability so you must devote serious attention to checking and validating your work. In a college course perhaps the greatest harm resulting from your errors is to your course grade, but in professional practice, the safety and health of other persons may be at risk. In contrast with the classroom situation, reliability is vastly more important than speed. Also, when you publish a paper, an error remains for all the world to see and perhaps your errors may mislead others. Because most engineering problems are solved over a period of days rather than minutes, it is important to develop the habit of providing the appropriate step-by-step documentation and checking each step in your analysis. Engineering projects often involve team effort, but individuals retain responsibility for assuring the accuracy of their work. Team confidence in you depends fundamentally upon your ability to produce carefully checked work.

"By far the most effective way to check an analysis is by means of a properly designed experiment." (Ver Planck and Teare, 1954, p. 229). Depending upon one's perspective and experience, as the earlier Fry quote suggests, one will be more easily convinced of the validity of an analysis by experiment or by careful reasoning. The scientific method holds that a properly performed experiment serves as the final court of appeal when analysis and experiment differ. However, this does not mean that you must immediately and uniformly resort to experimentation—neglecting your mathematical tools. In the first place, experiments often require that you perform some nontrivial analysis to properly construct the experiment. Experiments are often expensive and time-consuming. Furthermore, the task may be inherently more amenable to either experimental or mathematical modeling. Your relative level of expertise as a modeler or as an experimentalist, independent of the particular project, may influence which path is less costly to pursue.

"Good professional method implies not only checking an end result in every reasonable way but also checking frequently as the work progresses so that the analysis is accurate at every stage of its development." (Ver Planck and Teare, 1954, p. 229–230). Checking is so important that it should become instinctive and habitual. Undoubtedly much of your checking efforts will occur after you have found results that look promising. However, checking must be performed at every stage of your analysis; not just after the derivation has been completed. Errors, even simple and easily detected ones, can propagate through your analysis and necessitate lost time retracing your steps.

In this chapter we are specifically interested in techniques that make your use of mathematical tools more effective (for validation and to help you gain insight), so let's consider some specific techniques, including some that can be applied rather easily as your work progresses.

Kinds of Checking

Two other kinds of checking are especially useful: (1) overall checks. The latter often merge with learning and generalizing, and (2) checks of the manipulations (arithmetic, dimensional, etc.).

Checking by Experience *Comparing your results with your previous experiences and common sense constitute the most important, but often overlooked form of checking.* If a calculation produces a result that is off by an order of magnitude (e.g., a car weighs twenty tons), your internal alarm should warn you. As you develop more experience in a particular specialty, your sense of scale becomes a powerful asset. Sometimes conversion of numbers into a more familiar set of units can facilitate an order of magnitude check with your experience. Checking by experience is extremely valuable, but of course, not infallible.

Dimension Checking To represent a physically meaningful relationship, an equation must be dimensionally homogeneous. For example, each term in an equation consisting of a sum must have the same units. This is a necessary, but not a sufficient condition. If the units are not consistent, an "equation" cannot be valid, but a dimensionally homogeneous equation may still be incorrect. That is, failure to meet this test is positive proof of incorrectness, but meeting the test does not establish its correctness. For example, missing terms will not be detected, nor can you be sure that dimensionless parameters have the correct magnitude or sign.

Names and Units

Ver Planck and Teare (1954, p. 231) illustrate the need to use the same units with the following:

> An equation is written with the symbol for each quantity followed by the name of the units in which that quantity is expressed.
>
>   5 apples + 6 apples = 11 apples
>
> is dimensionally homogeneous, but
>
>   5 apples + 1 cow = 6 apples
>
> is nonsense.

A more representative example:

$$a \left(\frac{\text{apple}}{\text{basket}} \right) \times A \text{ (baskets)}$$

$$+ b \left(\frac{\text{apples}}{\text{tree}} \right) \times B \text{ (trees)} = C \text{ (apples)}$$

The units for each term are treated algebraically. Contemporary symbolic mathematical software can be used to automate this task, as well as handle the mundane but error-prone task of conversion of units.

There are a few special situations to be considered. Numerical constants in an equation may represent pure numbers (e.g., π), but may also be dependent upon a particular set of units. For this reason, symbolic parameters are usually preferred because their units can be checked more easily if treated on a basis symmetrical with the variables. For example, if the variable x in the quadratic equation has the units of meters, then clearly the parameters a, b, and c must have units too.

The arguments of certain special functions, such as the trigonometric functions, exponential functions, and Bessel functions, are pure numbers and must be dimensionless. Beware that using units of measure such as degrees rather than radians can create an apparent discrepancy. Similarly, because the logarithm of a product can be separated, an apparent discrepancy may exist. Sometimes a physical law must be introduced to complete the conversion process to demonstrate dimensional homogeneity.

Dimensions

This is similar to units check, but each factor in the equaiton is replaced by its dimensional formula in terms of *primary* quantities such as mass (M), length (L), and time (T). Unlike the units check, this does not verify consistency of system units. Ver Planck and Teare (p. 231) illustrate the procedure:

> ... density is mass per unit volume, or mass divided by length cubed, hence [M L ^{3}] where M represents the dimensions of mass and L the dimension of length. Similarly, it might be desirable to have a dimensional formula for electrical resistance in terms of the dimensions voltage [V], charge

[Q], and time [T]; it is [VG^{-1}T], as may be deduced from Ohm's Law and the definition of current as a charge passing through a circuit per unit time. . . . If too many primary quantities happen to be chosen in a particular case, it will be necessary to eliminate one or more of them by using known physical relationships. For instance, in a problem in dynamics if all four of the dimensions, force [F], mass [M], length [L], and time [T] are used as primary quantities, it will be found necessary to eliminate one of them. This is done very easily by using Newton's law $f = ma$; it shows that force [F] has the dimensional formula [MLT^{-2}], or, if you prefer, that mass [M] has the formula [FL ^{1}T^2].

Limiting Case Checks

Limiting case checks are a particularly useful technique. Often your experience will suggest whether an expression should increase or decrease as each parameter is increased or decreased. Should the result increase or decrease without bound as a parameter or variable is changed, or should there be an asymptotic bound? Or should there be a relative maximum or minimum?

Checking a special value, such as 0, 1, e, multiples of π, or infinity, is often especially easy and revealing. Checking these special values often permit you to compare the expression with your intuitive sense of what should happen.

Identifying a steady state behavior often can be identified. An independent derivation of a simpler limiting case may be possible; such independent confirmation can greatly strengthen your confidence in the more general derivation.

Ver Planck and Teare (p. 46) illustrate this technique using a simple exponential result from heat transfer. The temperature T approaches the initial value $T = 0$ as time approaches zero and T approaches the steady state $q/(hA)$ as time increases without bound.

$$T = \frac{q}{hA}\left[1 - \exp\left(-\frac{ht}{C\rho L}\right)\right]$$

As $t \to 0$, $T \to 0$

As $t \to \infty$, $T \to \frac{q}{hA}$

Limiting case checks also plays an important role in helping you grasp the underlying physical significance of your model and is often useful in the modeling process when you are interpreting the physical significance of your result.

Symmetry as a Check

Symmetry also provides a powerful means for checking your derivations. Checking a result for even and odd functions is easy and often helpful. For example, if a variable has a positive direction, what result is expected if the sign of that variable is reversed? Will the sign of the expression be the same or reversed?

Is $F(x) = F(-x)$ (even) or

$F(x) = -F(-x)$ (odd)?

Ver Planck and Teare (p. 242) provide two examples of using symmetry to identify potential errors.

$$T = \frac{q}{hA}\left[1 - \exp\left(-\frac{ht}{C\rho L}\right)\right]$$
As $t \to 0$, $T \to 0$

As $t \to \infty$, $T \to \frac{q}{hA}$

(Ver Planck and Teare, 1954, p. 233.)

Symmetry as a Check

The equivalent resistance for the following circuit (see diagram) is known to be

$$R_T = R_1 + R_2 + \frac{R_3R_4}{R_3 + R_4}$$

Suppose your equation were

$$R_T = R_1 + R_3 + \frac{R_2R_4}{R_2 + R_4} \qquad \text{(error)}$$

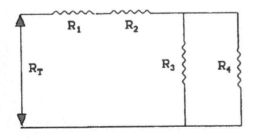

FIGURE 5   Circuit diagram.

Because R_3 and R_4 are similarly placed, they, of course, should be similarly placed in the equation (Ver Planck and Teare, 1954, p. 243).

Notice that the error below could not be detected this way.

$$R_T = R_3 + R_4 + \frac{R_1 R_2}{R_1 + R_2} \qquad \text{(error)}$$

Consider the following spring.

Suppose the relationship derived for the deflection of a leaf spring were

$$\frac{P}{\delta} = \frac{Ebnh^3 l_1}{6 l_2^2 l_3^2}$$

where

P = load (lb)
δ = deflection at the load (in.)
E = Young's modulus for the leaf material (lb in.$^{-2}$)
b = width of leaves (in.)
n = number of leaves
h = thickness of each leaf (in.)
l_1, l_2, l_3 = lengths (in.) as shown

E, b, n, and h should be in the numerator, for an increase would tend to increase the stiffness of the spring (Ver Planck and Teare, 1954, p. 244).

Placement of l's is less clear. Having l_3 in the denominator is less clear. But l_1 and l_2 should enter the formula symmetrically. Discover that subscripts 1 and 3 were swapped (Ver Planck and Teare, 1954, p. 244).

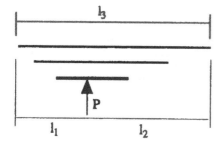

FIGURE 6   Spring schematic.

Let $f = l_1/l_3$ and $1 - f = l_2/l_3$. Then the corrected formula would read

$$\frac{P}{\delta} = \frac{Ebnh^3}{6f^2(1 - f)^2 l_3^3}$$

Here f and $(1 - f)$ enter symmetrically (Ver Planck and Teare, 1954, p. 245).

E. Learn and Generalize from Your Analysis

An analysis should always be concluded with a summary of important findings, with particular emphasis on the limitations imposed by the assumptions. Of course, all important assumptions should be made explicit rather than implicit. Mathematical equations are not useful to the engineer until they have been interpreted in terms of the original problem. You must verbally state the meaning of the appropriate mathematical equations. If you cannot state their meaning, quite likely you do not yet understand them.

The analysis should usually, but not always, involve numerical testing of the results. This will clarify the meaning and the feasibility of the particular problem. This provides a sense of scale, alerting you to gross blunders or possibly to your lack of success in finding a solution. Finally, the identifiable uncertainties that still exist should be clearly stated. Mistakes follow if you carelessly exceed the conditions on which your analysis is based. The analysis should not be concluded until the possibility of simplifying and/or generalizing the analysis has been considered, explicitly and individually.

Consider now a specific analysis in order to make the foregoing discussion more concrete. This also illustrates how a simple analysis can make an expensive experiment unnecessary.

VII. AN EXAMPLE: THE TEMPERATURE INCREASE DUE TO A GERMINATING SEED

In an effort to produce a more rational approach to the design of field machines and operations affecting the planting of agricultural crops, various factors must be investigated. Soil temperature is known to be one of the important factors affecting seed germination. It is also known that once germination has been initiated, the rate of internal heat generation due to respiration increases rapidly. Can this thermal effect be measured in situ with thermocouples?

A. Statement of the Problem

Does the heat of respiration of a germinating seed appreciably alter the soil temperature and internal seed temperature? And can this be measured in situ?

B. Assumptions

A1. If the other factors affecting germination are not limiting, e.g., available soil moisture, suitable soil temperature, adequate oxygen supply, acceptable soil physical impedance, no inhibiting chemicals (natural and artificial), previous temperature history, and activation of phytochrome by proper wavelength of light, the seed is viable.

A2. The heat transfer due to external sources (e.g., diurnal fluctuations) and the heat of vaporization and heat of wetting are neglected in this first analysis.

A3. The soil is homogeneous and isotropic with respect to thermal properties.

A4. Only the steady-state magnitudes are to be explored (at first). In other words, the seed and soil will be assumed to be in a (dynamic) thermal equilibrium independent of time, with the temperature never near freezing (to avoid latent heat difficulties).

A5. The thermal constants are independent of time, temperature, and position (for seed and for soil).

A6. The seed is taken to be (initially) spherical of radius $r = a$ (which would be valid for certain seeds), to be producing heat internally at the uniform, constant rate A_0 per unit volume per unit time, and to have thermal conductivity K_0.

A7. No heat is produced in the soil, which has conductivity K.

A8. There is imperfect thermal contact between the seed and soil such that there is a film temperature drop at the seed surface. Specifically, the thermal energy conducted to the seed surface passes into the soil at a rate proportional to the difference in temperature between the seed and soil.

A9. The seed is sufficiently far below the surface of the soil as to be relatively unaffected by the soil-to-air discontinuity.

A10. Heat transfer within the seed and within the soil is by conduction only, i.e., heat transfer due to moisture migration will be assumed negligible.

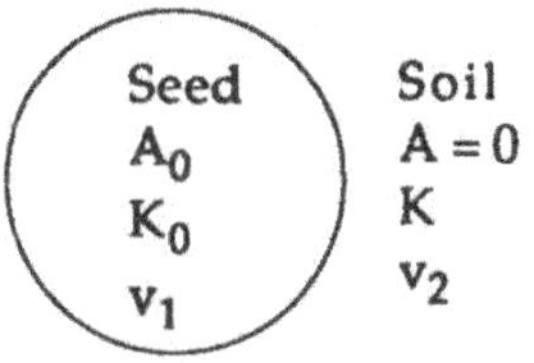

Seed	Soil
A_0	$A = 0$
K_0	K
v_1	v_2

FIGURE 7   Seed–soil schematic.

These assumptions are shown schematically in Fig. 7.

C. Mathematical Formulation

Since the soil temperature at large distances from the center of the seed will not be affected appreciably by the heat generated by the seed within the soil (which would be at uniform temperature if the seed were not present), that temperature may be taken as a reference. In other words, all other temperatures may be calculated with

$$\lim_{r \to \infty} v_2 = 0$$

as the zero reference.

From Assumption 10, we will use the heat conduction equation, with heat generation, as given by Carslaw and Jaeger (1959, p. 10).

$$\nabla^2 v - \frac{1}{\kappa}\frac{\partial v}{\partial t} = -\frac{A(x, y, z, t)}{K} \tag{1}$$

where

v = scaled temperature, C° (cgs units)
κ = diffusivity, cm²/s
A = heat generation rate, cal/(s·cm³)
K = thermal conductivity, cal/[s·cm²(°C/cm)]

and

$$\kappa = K/\rho c$$

where ρ is density (bulk) (g/cm³) and c is specific heat (at constant pressure approximately).

From Assumption 4, we may set

$$\frac{\partial v}{\partial t} \equiv 0 \tag{2}$$

In spherical polar coordinates (due to A9),

$$\nabla^2 v \equiv \frac{1}{r^2} \left[\frac{\partial}{\partial r} \left(r^2 \frac{\partial v}{\partial r} \right) \frac{1}{\sin \theta} \frac{\partial}{\partial \theta} \left(\sin \theta \frac{\partial v}{\partial \theta} \right) \frac{1}{\sin^2 \theta} \frac{\partial^2 v}{\partial \phi^2} \right] \tag{3}$$

Due to the symmetry in θ and ϕ, Assumptions 6 and 7, Eq. (1) may be applied to the seed and soil separately:

$$\frac{1}{r^2} \left[\frac{\partial}{\partial r} \left(r^2 \frac{\partial v_1}{\partial r} \right) \right] = -\frac{A_0}{K_0}, \quad 0 \le r \le a \tag{4}$$

$$\frac{1}{r^2} \left[\frac{\partial}{\partial r} \left(r^2 \frac{\partial v_2}{\partial r} \right) \right] = 0, \quad r > a \tag{5}$$

The conditions required for solution of Eqs. (4) and (5) are as follows. The temperature at great distances from the seed must approach the reference value of zero.

$$\lim_{r \to \infty} v_2(r) = 0 \tag{6}$$

where v_2 is not a temperature but a temperature difference—Celsius degrees from the reference temperature.

In order for the boundary between the seed and soil not to increase in temperature without bound, the continuity relation must be imposed. The flow of heat per unit area per unit time in the direction of increasing variable is given by

$$F = -K\nabla v \tag{7}$$

where

$$\nabla v = \mathbf{a}_r \frac{\partial v}{\partial r} + \frac{\mathbf{a}_\theta}{r} \frac{\partial v}{\partial \theta} + \frac{\mathbf{a}_\phi}{r \sin \theta} \frac{\partial v}{\partial \phi} \tag{8}$$

By symmetry, the temperature does not change in the θ and ϕ directions for a surface $r = a$, so

$$\frac{\partial v}{\partial \theta} = \frac{\partial v}{\partial \phi} = 0 \tag{9}$$

At the interface ($r = a$),

$$F_1 = F_2 \tag{10}$$

so

$$-K_0 \frac{\partial v_1}{\partial r} = -K \frac{\partial v_2}{\partial r} \quad \text{at} \quad r = a \tag{11}$$

The minus sign appears in Eqs. (7) and (11) because the flow occurs in the direction of decreasing temperature, i.e., $\partial v / \partial r$ will contain an inherent negative sign to compensate for the minus sign in Eq. (7).

From Assumption 8, we have, in place of a requirement for the continuity of temperature, the following:

$$-K_0 \frac{\partial v_1}{\partial r} = H(v_1 - v_2), \qquad r = a \tag{12}$$

where $H > 0$ is a proportionality constant that is called the coefficient of surface heat transfer (Carslaw and Jaeger, 1959, p. 19). (The H of Carslaw and Jaeger corresponds to the h more widely used in the literature on convective processes.) The surface thermal resistance per unit area may also be defined as

$$R \equiv \frac{1}{H} \tag{13}$$

This H should not be confused with the one in the expression

$$h = \frac{H}{K} \tag{14}$$

Equations (4) and (5) may be solved by integration and using the boundary conditions (6), (11), and (12).

Carslaw and Jaeger (1959, p. 232) give (without proof) the following solution:

$$v_1 = A_0[a^2 - r^2 + 2RaK_0 + 2a^2(K_0/K)]/6K_0, \qquad 0 \le r \le a \tag{15}$$

$$v_2 = A_0 a^3/3Kr, \qquad r > a \tag{16}$$

Equations (15) and (16) satisfy the differential equations (4) and (5) and the associated boundary conditions.

Since solutions to Poisson's equation are known to be unique, we may be confident that the actual solution has been found.

D. Checking and Discussion

Now that the temperature has been found at all points, a check should be made.

Limit Check Suppose heat generation were zero. Then

$$\lim_{A_0 \to 0} v_1 = 0 \qquad \text{OK} \tag{17}$$

$$\lim_{A_0 \to 0} v_2 = 0 \qquad \text{OK} \tag{18}$$

Suppose the size of the seed becomes vanishingly small. Then

$$\lim_{a \to 0} v_1 = 0 \qquad \text{OK for } r < a \tag{19}$$

$$\lim_{a \to 0} v_2 = 0 \qquad \text{OK for } r > a \qquad (20)$$

If $R \to 0$, then continuity in temperature at the interface should exist.

$$\lim_{r \to a^-} [\lim_{R \to 0} v_1] = \lim_{r \to a^+} v_2 \qquad \text{OK} \qquad (21)$$

If the thermal conductivity of the soil is very great (i.e., a very good conductor), then the heat would be conducted away and the soil temperature should not be affected, since the mass of the soil is assumed to be infinite. The seed temperature should be finite.

$$\lim_{K \to \infty} v_2 = 0 \qquad (22)$$

$$\lim_{K \to 0} v_1 = A_0(a^2 - r^2 + 2RaK_0)/6K_0 = \text{finite} \qquad (23)$$

On the other hand, if the soil behaves as a perfect insulator, then the soil temperature would become very great for finite r, and the seed temperature would increase without bound.

$$\lim_{K \to 0} v_2 = \infty, \qquad \text{all finite } r \qquad (24)$$

$$\lim_{K \to 0} v_1 = \infty, \qquad \text{all } r < a \qquad (25)$$

Other limiting checks may be performed, but a sense of understanding and confidence has begun to develop.

Let's embed this problem in a more general context to clarify what has been learned.

In order to obtain the results in terms of the rate of heat production Q per seed, the flux over the surface $r = a$ may be obtained by integration.

$$\lim_{r \to a^+} \frac{\partial v_2}{\partial r} = \frac{-A_0 a}{3K} \qquad (26)$$

$$Q = -\int_S K \frac{\partial v_2}{\partial r} dA \qquad (27)$$

$$Q = -4\pi a^2 K \left(-\frac{A_0 a}{3K} \right) \qquad (28)$$

$$Q = (4/3)\pi a^3 A_0 \qquad (29)$$

Using equation (16),

$$\lim_{r \to a^+} v_2 = \frac{A_0 a^2}{3K} \equiv v_0 \qquad (30)$$

and equation (29), then

$$v_0 = \frac{Q}{4\pi K a} \tag{31}$$

This is the same temperature that would have resulted from an isolated point source in a medium of conductivity K at a distance $r = a$.

Numerical Check   It will now be instructive to check the results numerically to establish reasonable bounds on the temperatures to be expected. If we now assume intimate seed–soil thermal communication ($R = 0$), i.e., seed coat temperature equals adjacent soil temperature, then the temperature of the center of the seed is found to be

$$v_c = \frac{A_0 a^2}{6K_0} + \frac{A_0 a^2}{3K}, \qquad \text{seed center} \tag{32}$$

The seed surface temperature is

$$v_s = \frac{A_0 a^2}{3K}, \qquad \text{seed surface} \tag{33}$$

Then,

$$v_c = \frac{A_0 a^2}{6K_0} + v_s \tag{34}$$

The soil temperature at the surface may be obtained from either Eq. (33) or Eq. (27). R may assume any value in Eq. (27), so we shall use that result. (The value of R will affect the seed temperature but not the soil temperature.)

v_s = soil temperature at $r = a$, C°
Q = rate of heat production, cal/(s·seed)
a = radius of seed, cm
K = soil thermal conductivity, cal/[s·cm²(°C/cm)]

Thermal Conductivity: Carslaw and Jaeger give for "average" soil

$K = 0.0023$ cal/[s·cm²(°C/cm)]

Kernel Size: A 12/64 in. round hole screen is used by the U.S. Department of Agriculture (USDA) as a basis for determining corn breakage. If we take twice that size, say radius of kernel

$$a = \left(\frac{12}{64} \text{ in.}\right)\left(\frac{2.54 \text{ cm}}{\text{in.}}\right) = 0.475 \text{ cm}$$

Heat Production of Kernel (Corn):   From Prat (1952, p. 391) the graphical and tabular data on respiration rates (cal/h) vs. time are given. Varieties of corn with peak respiration rates of 3.0, 2.1, and 2.3 cal/(h·g) with four seeds were given.

$$Q = [3.0 \text{ cal}/(h \cdot g)] \left(\frac{1 \text{ h}}{3600 \text{ s}}\right) \left(\frac{1 \text{ g}}{4 \text{ seeds}}\right) = 0.000208 \text{ cal}/(s \cdot \text{seed})$$

From Eq. (31),

$$
\begin{aligned}
v_s &= \frac{0.000208 \text{ cal}/(s \cdot \text{seed})}{(12.56)(0.475 \text{ cm})(0.0023 \text{ cal}/[s \cdot cm^2(^\circ C/cm)])} \\
&= 0.0151^\circ C
\end{aligned}
\tag{35}
$$

The units check.

However, this is far below the level that can be measured with a thermocouple!! Perhaps a series arrangement of thermocouples should be considered.

Conclusion The experimental arrangement proposed will probably not succeed.

Further Remarks The temperature rise at the center of the seed can also be computed from Eq. (34).

Let $R = 0$. K for corn can be taken as 1.22 Btu·in./(ft²·hr·°F) (ASAE, 1967, p. 285). This must be converted to cgs units, where the factor 0.00413 was obtained from Carslaw and Jaeger (1959, p. 3).

$$1.22 \frac{\text{Btu} \cdot \text{in.}}{\text{ft}^2 \cdot \text{hr} \cdot ^\circ F} \left(\frac{1 \text{ ft}}{12 \text{ in.}}\right) (0.00413) = 0.00042$$

Since

$$v_s = \frac{A_0 a^2}{3K} \tag{36}$$

$$v_c = \frac{K}{2K_0} v_s + v_s = \left[\frac{0.0023}{2(0.00042)} + 1\right] v_s = 0.056^\circ C$$

This is the maximum temperature rise above the undisturbed soil temperature. The temperature difference $v_c - v_s = 0.041^\circ C$ is also below the limits for a thermocouple. Under the most careful laboratory conditions, a thermocouple can be used to detect differences no smaller than $(0.05^\circ F) (5/9) = 0.03^\circ C$.

E. Discussion of Results, Assumptions, and Generalizations

In view of thermal conduction along a thermocouple and the difficulty of correctly placing the probe in the seed, this appears to be a futile experimental approach to the problem. The gradients that occur normally in the soil will be vastly greater than that produced by the heat of respiration.

However, if the contact resistance of the surface were great, then the internal temperature might be affected enough to alter the biochemical reactions within the seed. Therefore, the problem may be redefined, if desired, to explore this new problem suggested by the analysis.

Seed and soil temperatures will frequently be below those of the calorimeter tests with a consequently lower respiration rate. The water absorbed by the seed (imbibition) will, of course, alter the thermal energy status of the seed. The thermal conductivity of soil changes radically with moisture content. Also, the corn kernel is not spherical. However, the above computations may be taken as bounds on the actual values.

If the seed were very near the surface, the method of superposition of heat source and sink could be used. The other extreme case of a solid spherical seed producing heat at the constant rate A_0 per unit time per unit volume in air at constant temperature can be examined.

$$v = \frac{A_0}{6hK} [h(a^2 - r^2) + 2a], \qquad 0 \le r < a \qquad (37)$$

[See Carslaw and Jaeger (1959, p. 232).]

Finally, the heat of respiration is probably not produced uniformly throughout the seed. Specifically, the embryo temperature would be expected to be higher than that of the endosperm. This could have important consequences for behavior of the seed but is not likely to alter the soil temperature appreciably.

Heat of respiration experiments must necessarily be conducted in microcalorimeters as is the current practice of plant scientists. The air temperature change within an adiabatic finite chamber due to respiring seeds might be usefully examined.

VIII. CONCLUSION

The process of applying mathematics to the solution of engineering problems is an old art form that is enjoying increased attention at the interface between engineering and biology. The now ubiquitous presence of computational power unimagined only a few decades ago has massively expanded the importance of and scope of interest in mathematical tools and modeling. This chapter discusses the process we use to connect this form of conceptual thinking to problem solving. Elucidating the process—even those aspects that are obvious to the practitioner—makes the process more usable and comfortable. With greater attention to the process of marshaling more effectively all those mathematical tools built into our educational programs, we can increase the value of mathematics in our professional practice.

The example presented in Section VII is used to make the more abstract discussion concrete. Take time to compare its form to the steps listed in Table 1. Notice how the disciplined use of these steps facilitates the process of discovery.

ACKNOWLEDGMENT

I wish to express my appreciation to Henry D. Bowen, who stimulated my interest in this subject and helped me appreciate the magic inherent in modeling.

REFERENCES

ASAE. 1967. Yearbook.

Bowen, H. D. 1966. Developing tractable problems from abstract problem situations. ASAE Paper 66-510. St. Joseph, MI: ASAE.

Carslaw, H. C., and J. C. Jaeger. 1959. *Conduction of Heat in Solids*, 2nd ed. London: Oxford Univ. Press.

Fry, T. C. 1941. Industrial mathematics. *Am. Math. Monthly 48*, 1–38, Part II of the June–July Supplement.

Johnson, W. C. 1944. *Mathematical and Physical Principles of Engineering Analysis*. New York: McGraw-Hill, pp. 206–217.

Nahikian, H. M. 1964. *A Modern Algebra for Biologists*. Chicago, IL: Univ. Chicago Press, p. 2.

Noble, B. 1967. *Applications of Undergraduate Mathematics in Engineering*. New York: The Mathematical Association of America and The Macmillan Company.

Novak, J. D., and D. B. Gowin. 1984. *Learning How to Learn*. New York: Cambridge University Press.

Prat, H. 1952. *Can. J. Bot. 30*: 39.

Ver Planck, D. W., and B. R. Teare. 1954. *Engineering Analysis: An Introduction to Professional Method*. New York: Wiley, pp. 229–250.

4

Integrating Spatial and Temporal Models: An Energy Example

B. S. Panesar

Punjab Agricultural University, Ludhiana, Punjab, India

I. INTRODUCTION

Modeling has attracted the attention of many system analysts because it helps to predict the behavior of real-world systems without disturbing actual systems. Many models developed are either spatial or temporal owing to objective constraints or some limitations faced by the analysts (Chen, 1983; Chen and Hashimoto, 1980; Doorenbos and Pruitt, 1977; Driessen, 1986; Fluck et al., 1992a, 1992b; Hashimoto, 1984; Hill, 1983; Hitch, 1977, Hoffman, 1979; Hutber, 1973; Jensen et al., 1971, 1990; Jones et al., 1991; Lal et al., 1991, 1993; Negahban et al., 1993; Panesar et al., 1989; Papajorgji et al., 1993; Parikh, 1985; Parikh and Kromer, 1985; Searl, 1973; Smith, 1992). But the behavior of the systems predicted by these spatial/temporal models has both spatial and temporal characters. The predicted behavior of the model does not duplicate the actual behavior, and one of the major reasons listed for nonduplication of the actual system behavior is the lack of

93

either spatial or temporal character. Thus, these models have limited application.

Analysts and researchers have always wanted to enhance the capability of these spatial/temporal models so that the actual behavior of the system could be predicted with greater reality. The main aim of this chapter is to familiarize students with methodology that can be used to add the missing character in the spatial/temporal models. The chapter discusses the spatial and temporal characteristics of a system, the certainty of model output, time–space functions and their utility in developing methodology for adding the missing character, and integration methodology and gives an example of an energy model where the integration methodology has been successfully used.

II. SPATIAL FEATURES

The features that vary and/or are of concern for regional planning on the micro or macro levels are defined as spatial features. Spatial features are taken as attributes for database management of a region. These features help in identification and classification of homogeneous subregions in a region of interest. The spatial features are the core of spatial models, and their accurate identification by use of a set methodology becomes important to eliminate any errors. Both the variability and the equality among subregions have been used in the literature for the identification of spatial features (Bennett and Tan, 1981; Nelson, 1987; Negahban et al., 1993; Panesar et al., 1989; Parikh, 1985; Parikh and Kromer, 1985; Smith, 1992). The two approaches are dual, i.e., equality is the dual of variability and vice versa.

It is a known fact that spatial variability exists for almost all levels of systems, for example, in agricultural production, transport, industrial production, and the environment at the macro level and in nutrient transport in soils and nutrient uptake by plants and in biogas digesters at the micro level. Spatial variability is of concern to planners and modelers because they want equal performance from or equal distribution of benefits to spatially variable subregions. This leads to the concept of equality among subregions. Hence, spatial variability and equality concepts can be used to delineate homogeneous subregions or areas however small these subregions might be. Spatial variability will help in identifying some characteristics of the region, and equality might help to identify some other spatially variable characteristics of the region. For example, in allocating intergovernmental grant resources, the equality concept will help in identification of the tax base as a characteristic for a region that otherwise might be skipped.

III.   TEMPORAL FEATURES

All natural and man-made systems are dynamic. The dynamism in the system adds the time dimension as the behavior of a system is different at different times. System features that are due to the time dimension are referred to here as temporal features. The behavior of rice production or a processing system differs with time and hence has temporal features. The two most widely discussed and talked about temporal features are stability and dynamism. These are discussed next.

A.   Stability

A system has stability if it exhibits stable behavior from one period of time to another. If rice production in the United States is the same in 1991, 1992, 1993, and 1994, then the rice production system is said to have exhibited stable behavior. It is an important feature from the point of view of planners and forecasters that they do not have to worry about the complexity of a stable system. The stability feature of the system is good for short-term planning, but this should be used very carefully for long-term planning as it may lead to faulty decisions and actions.

B.   Dynamism

Temporal models are dynamic and hence require that past and future behavior of the system be considered in any analysis. The past is of utmost significance for biological systems, for these systems have memory. The future has great importance for man-made nonbiological systems such as industries, transport, roads, electricity supply, sewerage, and water supply. Man-made systems using biological systems as components require that analysts take into account both the future and past behavior of the system. The future behavior would determine how the system would meet the future requirements of the society for which it was made. All these factors add dynamism to both natural and man-made systems, which complicates the task of system analysts and modelers.

IV. CERTAINTY

The certainty in model output means that once the model output is predicted for an event it (the output) does not change when the event actually occurs. Thus model outputs should have a minimum of uncertainty. Uncertainty, the dual of certainty, is due mainly to four factors: (1) the stochastic nature of model inputs, (2) the stochastic determination of model parameters that are otherwise deterministic, (3) the deterioration of system behavior over time, and (4) uncertainty about model structure. The stochastic nature of model inputs is self-explanatory. The model parameters are deterministic, but the methods (experiments, instruments, etc.) used to determine them make them stochastic. For example, the spring constant of a spring is a deterministic feature of the spring, but the technique used to determine it makes it stochastic. The spring constant is determined by measuring force and displacements with the help of instruments (which have errors); plotting force vs. displacement, and finally determining slope with the use of regression analysis makes the spring constant a stochastic parameter.

Deterioration in a system's performance, i.e., changes in the system's behavior, with time is not uncommon. For example, the field capacity of a soil at a given location is a constant parameter. But the soil structure may change due to such activities as puddling and inundation of water, which are part of the rice production system. Thus, the field capacity of soil changes with multiple use of the rice production system, which therefore results in uncertainty in model output.

The model structure of a system is based on knowledge available at the time the system was modeled. Advances in scientific knowledge are being made at a rapid pace. These scientific advances might change the model structure, thereby creating uncertainty in model outputs at a later date. The certainty feature of temporal models is of utmost importance to planners and forecasters.

V. TIME–SPACE FUNCTIONS

Spatial and temporal features can be derived from time–space functions. The following discussion indicates how time–space functions can lead to spatial/temporal models and how missing temporal/spatial characteristics can be added to spatial/temporal models.

A function of the following form is called a time–space function:

$$Z = f(X, P, t) \tag{1}$$

where

Z = output set for a time–space model
X = space set (x, y, z for three dimensions and x, y for two-dimensions)
P = set of model parameters
t = time

Let $t = T$ where T is a specific time. Then, from Eq. (1), the spatial model output (Z_s) given by

$$Z_s = f(X, P, T) = f_s(X, P') \tag{2}$$

is a spatial model with Z_s as its outputs and P' as the model parameters. The parameter P' includes the effect of time period T. Thus, predicting system behavior from a spatial model over different time periods will add a temporal character to a spatial model.

Let $X = \underline{X}$, a specific space. Then, from Eq. (1), the temporal model output (Z_t) given by

$$Z_t = f(\underline{X}, P, t) = f_t(P'', t) \tag{3}$$

is a temporal model with Z_t as its output and P'' as the model parameters. The parameter P'' includes the effect of specific space, $\underline{X}$. Thus, predicting system behavior in different homogeneous subregions of a region with the help of a temporal model will add spatial character to the temporal models.

Comparing P, P' and P'', it is inferred that some or all model parameters of a spatial model would change with T whereas some or all model parameters of a temporal model would change with $\underline{X}$. Thus P' accumulates the effect of a time period while P'' accounts for the effect of spatial variability in a region.

Equation (3) illustrates how temporal models lose spatial variability and hence are not fit for regional equality. Equation (2) illustrates how spatial models lose dynamism and stability, and hence are not good for predicting the future behavior of the system.

In the literature, mathematical forms of spatial and temporal models have three basic characters: (1) model output, (2) model inputs, and (3) model parameters. Model coefficient terms have also been used in place of model parameters in econometric models. However, we use model parameters in this chapter. In mathematical form, the model output is written either as a function of model inputs, parameters, and outputs or in the form of a differential equation involving time and/or space derivatives of the model outputs along with the model inputs, parameters, and outputs. Either the model input or model parameters or both are functions of time or space or of both time and space. Thus, in temporal models, some or all of the model inputs are functions of time, and in some cases time may also appear as a model input, but model parameters are not a function of time. However, both parameters and inputs of temporal models might be functions of space. In spatial models also, similar to temporal models, some or all model inputs are functions of space variables, but model parameters are not functions of space. However, both parameters and inputs of spatial models might be functions of time.

If we compare Eqs. (2) and (3) in light of the above discussion, it can be inferred that P' and P'' consist of two distinct subsets of model inputs and parameters. The members of the subset of P' or P'' that are functions of space and time, respectively, are referred to as model inputs; and the members of the subset of P' or P'' that are not functions of space and time, respectively, are referred to as model parameters in more general terms.

Example   The modified expression for crop water requirements without water stress (CW_m) to include spatial character is (Panesar, 1993):

$$ET_0 = K\, ET_0 + I \tag{4}$$

where ET_0 is the reference crop evapotranspiration, m/crop season; K is the crop factor; and I is the total infiltration loss, m/crop season. CW_m is the model output.

The first factor in the estimation of CW_m is the crop factor K, which depends on the crop, season of planting or sowing, growth stage, and weather conditions such as wind speed and relative humidity. Doorenbos and Pruitt (1977) gave values of K for several crops during four stages of crop development, i.e., initial stage (approximately 10% ground cover), crop development stage (ground cover increases from 10% to about 70–80%), midseason stage (crop cover remains nearly complete), and late season stage (crop cover decreases from full to harvest or maturity crop cover). They outlined a procedure to determine crop factor K for each day after planting. Smith (1992) considered water demand during land preparation and nursery raising but only for rice. Water demand during the two operations is important for other crops also. Thus, factor K has spatial and temporal variations. Hence, K will form part of the model inputs.

The second factor in the estimation of CW_m is the reference crop evapotranspiration (ET_0). Several methods, such as pan evaporation, temperature, radiation, and combination methods (Jensen et al., 1990), have been used to estimate ET_0. The Penman–Monteith approach, a combination method, is most widely used as this approach represents a basic general description of the process of evaporation from vegetation and considers surface roughness and canopy rather than any specific crop. Smith (1992) gave the following equation based on the Penman–Monteith approach and as recommended in the FAO Expert Consultation held in May 1990 in Rome.

$$\frac{d}{dt}(ET_0) = \frac{\dfrac{\Delta(R_n - G)}{\lambda} + \gamma\,\dfrac{900}{T + 273}\,U(e_a)}{1000[\Delta + \gamma(1 + 0.34U)]} \tag{5}$$

where ET_0 is the reference crop evapotranspiration, m/crop season; R_n is the net radiation at crop surface, MJ/(m^2·day); G is the heat flux from the ground, MJ/(m^2·day); λ is the latent heat of vaporization of water, MJ/kg; T is the mean daily air temperature, °C; U is the average daily wind speed measured at 2 m above the ground; e_a is the saturation vapor pressure of water vapor at air temperature, kPa; e_d is the saturation vapor pressure at the dew point of the air, kPa; Δ is the slope of vapor pressure curve for water, kPa/K; γ is the psychrometric constant, kPa/K; t is time in days; and 900 is a constant in kg·K/kJ. The factors R_n, G, T, U, e_a, Δ, and e_d have spatial and temporal variations. However, the factors λ, γ, and 900 are not functions of time and space. Hence, the factors R_n, G, T, U, e_a, e_d, Δ, and t will be model inputs and the factors λ, γ, and 900 will be model parameters.

The third factor in estimation of CW_m is the infiltration loss I for wet cultivated crops. Soil is thoroughly worked and its structure destroyed during wetland preparation for planting. This operation is commonly known as puddling. Two main aims of the puddling operation are to loosen the soil to facilitate transplanting and to reduce the infiltration rate. The infiltration loss was estimated on a daily basis using the equation

$$I = I_0/k_i \tag{6}$$

where I is the loss of water due to infiltration, m/crop season; I_0 is the loss of water due to infiltration from saturated undisturbed soil, m/crop season; and k_i is the soil reduction factor (>1) for the infiltration rate. Both I_0 and k_i have spatial variations and therefore are categorized as model inputs.

The model for crop water requirements without water stress has

Output CW_m
Inputs K, R_n, G, T, U, e_a, e_d, Δ, t, I_0, k_i
Parameters λ, γ, 900

The inputs are functions of time and space as discussed earlier. Therefore output (CW_m) is also a function of time and space in addition to the parameters listed above. Thus, the model output (CW_m) has a form similar to Eq. (1). The exact mathematical functional form for inputs with time and space is not known. However, only an implicit functional relationship in the form of time–space tables for each input exists. Therefore the model has to be implemented in the form given by Eqs. (4)–(6). Panesar (1993) has developed the model as a temporal model by defining CW_m, ET_0, I, and I_0 in m/day and removing the time derivative from Eq. (5). Hence, the Panesar model does not provide spatial crop water requirements with water stress.

VI. INTEGRATION METHODOLOGY

Most models found in the literature are classified as either spatial or temporal. These models have limited application as they do not reflect the system behavior in both time and space. Researchers have mostly developed either spatial or temporal models due to certain limitations and hence have always felt the need to expand the scope of both types of models to predict both the spatial and temporal behavior of the system. The method of predicting the spatial and temporal behavior of a system from either a spatial or temporal model is called here an integration methodology. Figures 1 and 2 illustrate the integration methodology for temporal and spatial models, respectively, to predict the combined spatial and temporal behavior of a system. The integration methodology requires the design of input and output modules. The input module segregates the spatial and temporal data so that inputs and coefficients of the model with both spatial and temporal variations are made available to the model. The model will give outputs that will be stored and integrated by the output module to predict both the spatial and temporal behavior of the modeled system. The model is run a number of times for different time periods to add temporal characteristics to spatial models and for different homogeneous regions to add spatial characteristics to temporal models. Complete knowledge about the system, the model, and assumptions; the type of data formats for model inputs, parameters, and outputs; and language for computer implementation of the

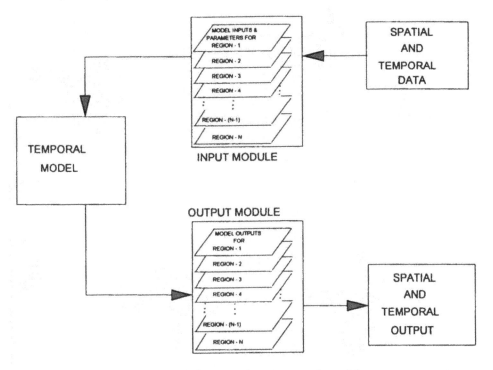

FIGURE 1   Integrating spatial aspects into temporal models.

model are prerequisites for the design and implementation of input and output modules.

The integration methodology is a multistep process. The sequential steps are

1. Revisit the system that has been modeled.
2. Study the modeling methodology used.
3. Understand the assumptions and the model structure.
4. Note model inputs, outputs, and parameters.
5. Study the units and dimensions of model inputs, outputs, and parameters.
6. Find values of the model parameters used for computer implementation of the model.
7. Study the language used for the implementation of the model on the computer.
8. Find computer data formats for inputs, outputs, and parameters of the model used for its implementation.

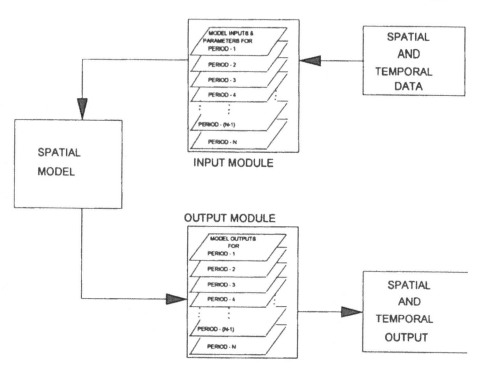

FIGURE 2   Integrating temporal aspects into spatial models.

9. Design input and output modules, and implement them on the computer in conjunction with the model.
10. Present model results in such a way as to facilitate interpretation.

Only the design and implementation of input and output modules need further discussion, as the other steps are self-explanatory. The design of the input/output module is model-specific, but the technique, as described below, is quite standard.

A.   Input Module

Design of the input module consists of spatial and temporal data manipulation for use by the spatial/temporal models. Two common types or formats of model outputs are maps and tables. The design of the input module must also consider these desired types or formats of the model output for better interpretation of the modeled results. Implementation of the input module on a computer comprises a structured program for the task and testing and debugging of the structured program modules are needed to eliminate programming errors. The spatial input module adapts spatial and temporal data for the spatial/temporal model such that the model processes all data. The adapted data of the system will be in the form of several files or subdirectories, each representing either a specific period or a specific homogeneous region. The file or subdirectory with model inputs and parameters is processed by the spatial/temporal model. Hence, the input module carries out two tasks: (1) data adaptation of inputs and parameters into files or subdirectories and (2) processing of adapted data from the file or subdirectory.

B. Output Module

Design of the output module consists of manipulation of the output from spatial/temporal models so that the missing character is added to the system behavior and on to the model output. As discussed earlier, two common types or formats of the model outputs are maps and tables. The design of the output module must therefore consider the desired type or format of the model output to provide for better interpretation of modeled results. The implementation of the output module comprises structured programming for the task and testing and debugging of the structured program modules to eliminate any programming errors. The output module should provide the capability of storing the model output for a specific period or homogeneous region in a file or subdirectory. The processing of these output files or subdirectories is done in such a way that the missing character is added to the predicted behavior of the system by the spatial/temporal models. Hence, the output module has two tasks: (1) to store the model output of a specific period or homogeneous region in a file or subdirectory, and (2) to integrate the output files or subdirectories in order to predict both the spatial and temporal behavior of the system in a desired fashion.

VII. ENERGY EXAMPLE

A spatial energy model for agricultural production (SENMAP) was developed by Panesar (1993). The model is depicted in Fig. 3. The model predicts requirements in terms of water, diesel fuel, electricity, labor, and total energy for crop cultivation over a period of 1 year. The model is driven by a large number of huge databases for Punjab (a state of the Union of India). These databases are on crops, imple-

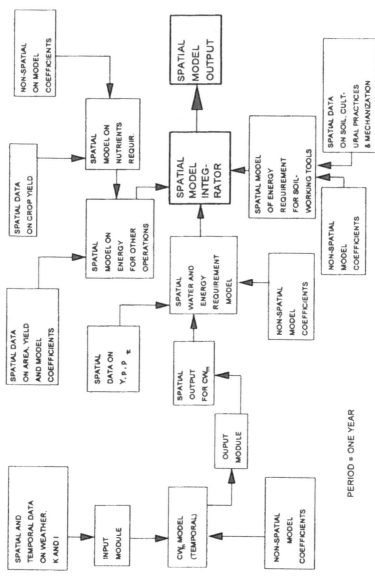

FIGURE 3   SENMAP model. (From Panesar, 1993.)

ments, soils, energy coefficients for other operations, mechanization, and irrigation. Panesar used the concepts of integration for both temporal and spatial models. SENMAP consists of three spatial component models: Water and Energy Requirements, Energy Requirements for Soil Working Tools, and Energy Requirements for Other Operations. The spatial model for water and energy requirements is driven by another temporal model on crop water requirements without water stress (CW_m). The output of the CW_m model was used by the spatial water and energy requirements model. But the CW_m model is a temporal model that requires weather data and rate of infiltration I in addition to crop coefficients K as discussed earlier, and these have both spatial and temporal features. The input in terms of CW_m needed for the water and energy requirements model must have a spatial character, but the output from the CW_m model (which is the input to that model) does not have a spatial character; it has a temporal character as discussed earlier. The CW_m model was implemented in Fortran. Panesar (1993) designed input and output modules as illustrated in Fig. 1 to adapt inputs, parameters, and outputs of the temporal model. The spatial and temporal data on weather, K and I, were in several ASCII files such that each file had temporal data of a homogeneous region. Panesar designed an input module and implemented it in Fortran to process data for different crops during their cropping period and for different homogeneous regions. He designed the output module to store model output in ASCII files so that each file represented a subregion and to process these files to create a CW_m database for the water and energy requirements model. He implemented the output module in Fortran and dBASE so that a spatial database for water requirements without water stress is created for different crops to be used by the spatial water and energy requirements model.

The SENMAP model gives spatial output for a period of 1 year in the form of maps and tables (Panesar, 1993). The sample format of the model output in the form of a table is given in Table 1, where each row is for a subregion and each column is for a model output. Similar tables formatted for average, maximum, and minimum values of the coefficients for diesel, electricity, water, labor, and total energy requirements were also presented by the model. Maps showing regions of high, medium, and low requirements of diesel, electricity, water, labor, and total energy also formed part of the presentation of modeled results for quicker appraisal. A method similar to Fig. 2 was developed by Panesar to add temporal character to the model output. He designed an input module for adapting spatial and temporal data

Table 1 Modeled Total Energy and Resource Requirements for Punjab State in 1991

District	Diesel (kL)	Electricity (GWh)	Labor 1000s of man-hr	Water (ha·m)	Energy (GJ)
Amritsar	46,809.68	186,508.60	396,871.2	638,287	11,373,274
Bathinda	34,738.44	27,325.54	248,950.8	733,073	7,435,317
Faridkot	49,619.08	42,922.50	409,184.4	471,335	11,019,501
Ferozpore	58,852.33	114,302.70	467,543.5	714,324	13,000,204
Gurdaspore	31,102.42	109,266.20	317,750.7	266,147	6,368,922
Hoshiarpore	17,866.85	96,769.38	184,199.6	215,237	4,442,695
Jallandar	35,997.36	181,792.30	417,121.5	361,110	9,485,957
Kapurthala	20,632.91	107,014.30	140,911.0	235,944	4,652,270
Ludhiana	48,323.08	248,312.10	439,159.9	513,610	13,220,959
Patiala	41,174.13	190,691.10	372,796.1	384,488	12,339,237
Rupnagar	12,458.31	40,538.72	152,361.5	106,591	2,545,513
Sangrur	49,350.39	202,080.50	210,476.7	575,977	14,459,249
State totals	446,925.10	1,547,524.00	3,757,328.0	5,216,123	110,343,072

Source: From Panesar, 1993.

such as weather, area and yield of different crops, cultural practices, irrigated area, area irrigated by tubewells, and mechanization (tractors, draft animals, diesel-operated tubewells, electricity-operated tubewells, etc.) for use by SENMAP. He implemented it in Fortran and dBASE such that the temporal character could be added to the outputs from SENMAP. The need for two languages/environments arose from the fact that a complex calculation task can be easily and time-efficiently handled by Fortran and database manipulations can be easily and efficiently done in dBASE. Thus the key in the use of computer language for implementation of the input module is ease and time-efficiency. Use of proper and suitable language makes the implementation easy, efficient, and productive.

Panesar (1993) did not design a complex output module for computer manipulation of spatial and temporal results. The output module carried out only one task—it stored the model output. It did not process spatial and temporal output. Panesar used computer manipulation of the spatial/temporal outputs from the model outside the model, and hence the manipulation was restricted to representing model output in a tabular form to display spatial and temporal characteristics. Table 2 shows the modeled results with temporal character

Table 2 Relative Total Energy and Resource Requirements for
Punjab State During the Period 1980–1991

(1980 = 1.000)

District	Year	Diesel (kL)	Electricity (GWh)	Labor (man·h)	Water (ha·m)	Energy (GJ)
Amritsar	1980	1.000	1.000	1.000	1.000	1.000
	1985	1.331	1.370	1.299	1.370	1.876
	1991	1.445	1.617	1.408	1.617	2.070
Bathinda	1980	1.000	1.000	1.000	1.000	1.000
	1985	1.259	1.659	1.182	1.660	1.696
	1991	1.339	1.849	1.195	1.850	2.002
Faridkot	1980	1.000	1.000	1.000	1.000	1.000
	1985	1.266	1.678	1.287	1.677	1.608
	1991	1.350	1.693	1.374	1.692	1.824
Ferozpore	1980	1.000	1.000	1.000	1.000	1.000
	1985	1.312	1.355	1.387	1.355	1.649
	1991	1.472	1.531	1.617	1.531	1.922
Gurdaspore	1980	1.000	1.000	1.000	1.000	1.000
	1985	1.530	1.278	1.462	1.279	1.951
	1991	1.521	1.392	1.424	1.392	1.915
Hoshiarpore	1980	1.000	1.000	1.000	1.000	1.000
	1985	0.848	1.068	0.850	1.068	1.035
	1991	1.028	1.547	1.021	1.547	1.579
Jallandar	1980	1.000	1.000	1.000	1.000	1.000
	1985	1.380	1.551	1.200	1.551	1.722
	1991	1.540	1.911	1.421	1.912	1.922
Kapurthala	1980	1.000	1.000	1.000	1.000	1.000
	1985	1.220	1.237	1.145	1.237	1.241
	1991	1.408	1.461	1.302	1.461	1.463
Ludhiana	1980	1.000	1.000	1.000	1.000	1.000
	1985	1.283	1.756	1.086	1.755	1.476
	1991	1.398	2.093	1.068	2.092	1.583
Patiala	1980	1.000	1.000	1.000	1.000	1.000
	1985	1.329	1.345	1.224	1.345	1.622
	1991	1.415	1.517	1.223	1.518	1.762
Rupnagar	1980	1.000	1.000	1.000	1.000	1.000
	1985	1.289	1.333	1.142	1.334	1.677
	1991	1.347	1.544	1.111	1.546	1.606
Sangrur	1980	1.000	1.000	1.000	1.000	1.000
	1985	1.406	1.953	0.990	1.955	1.689
	1991	1.641	2.556	0.882	2.559	2.070
State totals	1980	1.000	1.000	1.000	1.000	1.000
	1985	1.300	1.460	1.199	1.491	1.619
	1991	1.424	1.740	1.263	1.733	1.833

Source: From Panesar, 1993.

added to the model output. The results show relative diesel, electricity, water, labor, and total energy requirements in relation to 1980, yet the table effectively displays the spatial and temporal behavior of the agricultural production system.

The spatial output from the SENMAP model is in the form of either maps or tables. Thus the output module could also be designed and implemented in Fortran, dBASE, and ARC-INFO such that the spatial changes over different time periods could be displayed in the form of maps (results of one period overlaid over the results of another period to pinpoint differences) or in the form of tables with columns covering different periods and rows representing different regions. The Fortran language would be very useful for complex calculations, such as finding totals, computing results for presentation in tabular form, and getting hard copy of result tables. The dBASE language/environment helps in transferring model output to attributes of various homogeneous regions. ARC-INFO helps in developing maps for the model output, overlaying maps of different time periods for visual display of changes, and getting hard copy of these maps. Hence, implementation of the output module in a combination of several small modules formulated in a suitable computer language (which may differ among modules) may be helpful and easy for better presentation of modeled results.

VIII. SUMMARY

Researchers and system analysts have developed either spatial or temporal models for the system that actually have both spatial and temporal characters such as spatial variability or equality and temporal stability or dynamism. Uncertainty in model output limits the application of the model, and uncertainty of model structure is one of the reasons for the behavior. Lack of either spatial or temporal character in temporal or spatial models adds uncertainty to model structure. Thus, researchers have always felt the need to integrate the missing character into both spatial and temporal models.

Spatial and temporal models are derived from the time–space functional relationships within the system. Temporal models are derived from time–space functions for a specific region, whereas spatial models are derived from the same time–space functions for a specific time period. The concept of time–space functions helps in building a methodology for integrating the missing character into spatial and temporal models.

The integrating methodology is a sequential step method comprising revisiting the system that has been modeled; studying the modeling methodology used; understanding the assumptions and model structure; noting model inputs, outputs, and parameters; studying units and dimensions of model inputs, outputs, and parameters; finding values of the model parameters used for computer implementation of the model; studying the language used for the implementation of the model on the computer; finding computer data formats for inputs, outputs, and parameters of the model used for its implementation; designing input and output modules; implementing input and output modules on the computer in conjunction with the model; and presenting model results to facilitate interpretation. The input module is designed and implemented for two tasks: data adaptation of inputs and parameters into files or subdirectories for use by the model and processing of adapted data from the file or subdirectory. Similarly, the output module is designed and implemented for two tasks: storing the model output of a specific period or homogeneous region into a file or subdirectory and integrating the output files or subdirectories to predict spatial as well as temporal behavior of the system in a desired fashion.

The integration methodology has been demonstrated by taking an example from an agricultural production system that has both spatial and temporal character.

REFERENCES

Bennett, R. J., and K. C. Tan. 1981. Space–time models of financial resource allocation. In: *Dynamics Spatial Models*. D. A. Griffith and R. D. MacKinnow (Eds.). Proceedings of the NATO Advanced Study Institute on Dynamic Spatial Models, Rockville, MD: Sijthoff and Noordhoff.

Chen, Y. R. 1983. Kinetic analysis of anaerobic digestion of pig manure and its design implication. *Agric. Wastes* 8:65–81.

Chen, Y. R., and A. G. Hashimoto. 1980. Substrate utilization kinetic model for biological treatment processes. *Biotechnol. Bioeng.* 12:2081–2095.

Doorenbos, J., and W. O. Pruitt. 1977. *Crop Water Requirements*. FAO Irrigation and Drainage Paper No. 24. Rome: FAO of the United Nations.

Driessen, P. M. 1986. Nutrient demand and fertilizer requirements. In: *Modelling of Agricultural Production—Weather, Soils and Crops*. H. van Keulen and J. Wolf (Eds.). Wageningen: PUDOC, pp. 182–200.

Fluck, R. C. 1992a. Energy analysis of agricultural systems. In: *Energy in Farm Production. Energy in World Agriculture*, Vol. 6. R. C. Fluck (Ed.). Amsterdam, The Netherlands: Elsevier Science, pp. 45–52.

Fluck, R. C. 1992b. Energy conservation in agricultural transport. In: *Energy in Farm Production*. Energy in World Agriculture, Vol. 6. R. C. Fluck (Ed.). Amsterdam, The Netherlands: Elsevier Science, pp. 31–36.

Fluck, R. C., C. Fonyo, and E. Flaig. 1992a. Land-use-based phosphorus balances for Lake Okeechobee subbasins. *Appl. Eng. Agric.* 8(6):813–820.

Fluck, R. C., B. S. Panesar, and C. D. Baird. 1992b. Florida Agricultural Energy Consumption Model (FAECM). Unpublished report. Gainesville, FL: Agricultural Engineering Department, Institute of Food and Agricultural Sciences, University of Florida.

Hashimoto, A. G. 1984. Methane from swine manure: Effect of temperature and influent substrate concentration on kinetic parameter (K). *Agric. Wastes* 9:299–308.

Hill, T. H. 1983. Simplified Monod kinetics of methane formation of animal wastes. *Agric. Wastes* 5:1–16.

Hitch, C. J. 1977. *Modeling Energy-Economy Interactions: Five Approaches*. Washington, DC: Resources for the Future.

Hoffman, L. 1979. Energy demand in developing countries: Approaches to estimation and projection. In: *Proceedings of the Workshop on Energy Data of Developing Countries*. Vol. I. Paris: International Energy Agency, Organization for Economic Cooperation and Development, pp. 109–118.

Hutber, F. W. 1973. Modelling energy supply and demand. In: *Energy Modelling*. Special Energy Policy Publication. Guildford, UK: IPC Science and Technology Press, pp. 4–32.

Jensen, M. E., J. L. Wright, and B. J. Pruitt. 1971. Estimating soil moisture depletion from climate, crop and soil data. *Trans. ASAE* 14:954–959.

Jensen, M. E., R. D. Burman, and R. G. Allen (Eds.). 1990. *Evapotranspiration and Irrigation Water Requirements*. New York: American Society of Civil Engineers.

Jones, J. W., E. Dayan, L. H. Allen, H. V. Keulen, and H. Challa. 1991. A dynamic tomato growth and yield model (TOMGRO). *Trans. ASAE* 34(2): 663–672.

Lal, H., C. Fonyo, B. Negahban, J. W. Jones, W. G. Boggess, G. A. Kiker, and K. L. Campbell. 1991. Lake Okeechobee agricultural decision support system (LOADSS). ASAE Paper No. 91-2623. St. Joseph, MI: American Society of Agricultural Engineers.

Lal, H., G. Hoogenboom, J. P. Calixte, J. W. Jones, and F. H. Beinroth. 1993. Using crop simulation models and GIS for regional productivity analysis. *Trans. ASAE* 36(1):175–184.

Negahban, B., C. Fonyo, W. Boggess, J. Jones, K. Campbell, G. Kiker, E. Hamouda, E. Flaig, and H. Lal. 1993. LOADS: a GIS-based decision support system for regional environmental planning. Proceedings of the Conference on *Application of Advanced Information Technologies: Effective Management of Natural Resources*, St. Joseph, MI: American Society of Agricultural Engineers.

Nelson, R. 1987. State-space modelling of residential, commercial and peak demands. *J. Forecast* 6:97–115.

Panesar, B. S. 1993. *Spatial Energy Model of Agricultural Production.* A Ph.D. Dissertation. Ann Arbor, MI: U.M.I. Dissertation Services, A Belt and Howell Company.

Panesar, B. S., A. P. Bhatnagar, J. K. Parikh, and J. P. Painully. 1989. *Modelling of Energy and Agriculture Interactions: A Case Study of Punjab State.* Ludhiana, Punjab, India: School of Energy Studies for Agriculture, Punjab Agricultural University.

Papajorgji, P., J. W. Jones, J. P. Calixte, and G. Hoogenboom. 1993. A generic geographic decision support system for policy making in agriculture. In: Proceedings of the 13–14 Dec. 1993 Conf. on Integrated Resource Management and Landscape Modifications for Environmental Protection. St. Joseph, MI: ASAE, pp. 340–348.

Parikh, J. 1985. Modeling energy and agriculture interactions. I: A rural energy interaction model. *Energy* 10(7):793–804.

Parikh, J., and R. Kromer. 1985. Modelling energy and agriculture interactions. II: Fuel–fodder–food–fertilizer relationship for biomass in Bangladesh. *Energy* 10(7):805–817.

Searl, M. F. 1973. *Energy Modeling.* Washington, DC: Resources for the Future.

Smith, M. 1992. *CROPWAT—A Computer Program for Irrigation Planning and Management.* FAO Irrig. & Drain. Paper 46. Rome: Food and Agriculture Organization of the United Nations.

5

Modeling Processes and Operations with Linear Programming

Robert S. Sowell

North Carolina State University, Raleigh, North Carolina

R. C. Ward

Colorado State University, Fort Collins, Colorado

I. INTRODUCTION

Linear programming is a modeling tool that can assist in the solution of many problems in agriculture. In particular, linear programming is useful in selecting the best alternative from a number of available courses of action.

For example, consider the situation where a farmer is planning for the upcoming growing season. She has three crops she is considering planting on her 200 hectares (ha) of land. Considering all factors, the farmer estimates that 300 h of labor time can be devoted to these three crops. Assuming that machines and capital are not limiting, the farmer wants to know how many hectares she should plant in each crop in order to maximize her profits. She receives $50 profit per hectare for crop 1, $35 for crop 2, and $40 for crop 3.

This example illustrates the type of problem linear programming can assist in solving. The farmer has to decide how much land to allocate to each crop. She wants to maximize profits but has limited land and labor. There are many combinations of cropping plans she can choose from, but she wants to know which will yield the most profit.

To begin to model this problem into a linear programming format, several definitions are needed. A linear programming problem consists of three major components:

1. Decision variables
2. Objective function
3. Constraints

The decision variables relate to the decision that must be made and are expressed with algebraic symbols. In the above example, the decision variables (x_j, $j = 1, 2, 3$) can be defined as

x_j = number of hectares of land that should be planted in

crop j if profits are to be maximized

The objective function is an expression of the measure of effectiveness against which alternatives will be compared. The alternative with the optimum (maximum profit or minimum cost, for example) is the "best" alternative. In the crop planning example above, the objective function will be established by first defining the profits for each crop j, on a per-hectare basis, as c_j. The objective function can then be expressed as

$$\text{Maximize } z = c_1 x_1 + c_2 x_2 + c_3 x_3 \tag{1}$$

or

$$\text{Maximize } z = \sum_{j=1}^{3} c_j x_j$$

The value of the objective function is expressed in dollars because the units on c_j are dollars per hectare for crop j and the units on the decision variables (x_j) are hectares of crop j. Thus the combination of x's that yields the largest z is the optimum cropping plan.

The objective function could be increased without limit if there were no constraints on the problem. In this case, the farmer has only 200 ha of land and 300 h of labor time to devote to the crops in question. The constraints must be expressed in algebraic equations. In the case of the constraints, the amount of each resource needed to produce one unit of output must be known. The term a_{ij} is used to describe the amount of resource i needed to produce one unit of product j. The term b_i is used to represent the total amount of resource i available. Thus, the constraints for a linear programming problem, with $j = 1, 2, 3$ and $i = 1, 2$ (as in the above example), can be expressed as

$$a_{11}x_1 + a_{12}x_2 + a_{13}x_3 \leq b_1$$
$$a_{21}x_1 + a_{22}x_2 + a_{23}x_3 \leq b_2 \qquad (2)$$

The less-than-or-equal-to signs in these equations stem from the fact that the land and labor limits are not to be exceeded. In other words, the farmer may use less than the maximum of one resource if the optimum solution deems it best to do so.

If, in the example, it is assumed that each hectare of crop 1 requires 5 labor hours; each hectare of crop 2 requires 3 labor hours, and each hectare of crop 3 requires 4 labor hours, then the total linear programming formulation of the crop planning example can be stated as follows:

$$\text{Maximize } z = 50x_1 + 35x_2 + 40x_3$$

Subject to:

$$x_1 + x_2 + x_3 \leq 200 \qquad \text{(land constraint)}$$
$$5x_1 + 3x_2 + 4x_3 \leq 300 \qquad \text{(labor constraint)}$$

and

$$x_1 \geq 0, \qquad x_2 \geq 0, \qquad x_3 \geq 0$$

The latter constraint ($x_i \geq 0$, $i = 1, 2, 3$) is associated with the simplex method for solving linear programming problems, but in this case it also has a physical interpretation in that there will be land either planted or not planted.

Identifying and formulating linear programming problems is as much as an art as it is a science. The more one formulates linear programming problems, the better one's ability to formulate will become. The following examples serve to introduce students to the range of problems in agriculture that can be solved with linear programming and further their understanding of the terms and terminology used in formulating linear programming problems, especially the concepts of decision variables, objective function, and constraints.

II. LINEAR PROGRAMMING FORMULATION EXAMPLES

The crop mix problem described in Section I is often more commonly referred to as a "product mix" problem. In the situation described, the farmer had a number of crops being considered for production and a limited amount of resources. The question was, How much of each crop should be grown in order to maximize profits? This example is, of course, rather simple (e.g., small numbers of crops and resources considered) in order to illustrate the concept. The number of crops may always be limited due to equipment availability, but the number of different resources can often be quite large in a realistic problem.

A. The Diet Problem

The diet, or feed mix, problem deals with determining the amounts of alternative feeds to provide for livestock. The objective is to provide feeds that meet certain nutritional requirements at minimum cost. The decision variables are often the amounts of the various feeds to be included in the diet. The objective function coefficients define the cost per unit of each feed so that in minimizing the objective function one is minimizing the cost of feeding the livestock. The constraints express the need to meet minimum nutritional requirements. Each nutrient has its own constraint.

Consider the situation where a farmer, among his several enterprises, raises hogs for market. He has three types of feed available for his hog operation and is mainly concerned about nutrients. He has had each feed analyzed to determine the amounts of the three nutrients in a pound of each feed. He has also determined the minimum daily requirements of each nutrient for each hog and the cost per pound of each feed. He now wants to know how much of each feed should be fed to each hog if he is to meet the minimum nutritional requirements at a minimum cost. The data he has obtained are given in Table 1. Each entry under "feed" has units of nutritional value per pound of feed.

The decision variables for this problem can be defined as

x_j = pounds of feed j to be fed daily to each hog

While the cost of each pound of feed is known (from Table 1), the objective function can be written as

Minimize $z = 12x_1 + 6x_2 + 8x_3$

Table 1 Data for the Diet Problem

Nutrient	Daily nutritional requirement	Feed		
		Corn lb^{-1}	Tankage lb^{-1}	Alfalfa lb^{-1}
1	150	60	18	40
2	260	55	40	75
3	50	5	8	12
Cost (c)		12	6	8

Each nutrient will have a constraint that expresses the fact that its daily requirements must be met. Thus, the constraints are

$$60x_1 + 18x_2 + 40x_3 \geq 150 \quad \text{(nutrient 1 constraint)}$$

$$55x_1 + 40x_2 + 75x_3 \geq 260 \quad \text{(nutrient 2 constraint)}$$

$$5x_1 + 8x_2 + 12x_3 \geq 50 \quad \text{(nutrient 3 constraint)}$$

and, as before, x_1, x_2, and x_3 must be nonnegative.

This example illustrates how inequalities ($\geq$) and minimum objective functions can arise in linear programming problems.

B. The Transportation Problem

A soybean producer grows soybeans in four different fields located around the county. She stores her beans at two different sites. With rising energy costs, she wishes to determine the amount of beans from each field that should be stored at each storage area if transportation costs are to be minimized.

The first storage location has a 725 m^3 storage capacity, the second storage facility can hold 510 m^3, and the transportation costs are estimated to be $0.095/(m$^3 \cdot$km). The trucks are assumed to carry a full load on each trip.

The production of each field and the distances between the fields and the storage sites are given in Table 2.

The decision variables for this problem can be defined as

$$x_{ij} = \text{m}^3 \text{ soybeans shipped from field } i \text{ to storage location } j$$

The cost of shipping a cubic meter of soybeans from field i to field j is defined as

Table 2   Data for the Transportation Problem

Field no.	Field size (ha)	Expected yield		Hauling distance	
		m³/ha	Total (m³)	Storage 1	Storage 2
1	150	1.7	255	14	8
2	80	1.9	152	6	10
3	175	1.7	298	12	29
4	210	2.1	441	18	9

c_{ij} = cost per cubic meter per kilometer [$\$/(m^3 \cdot km)$] times the distance from field i to storage location j in km

Thus,

$c_{11} = \$0.095/(m^3 \cdot km) \times 14$ km $= \$1.33/m^3$

With these costs of shipping, the objective function becomes Minimize

$z = 1.33x_{11} + 0.76x_{12} + 0.57x_{12} + 0.95x_{22} + 1.14x_{31}$
$+ 2.76x_{32} + 1.71x_{41} + 0.86x_{42}$

Constraints on the problem relate to the need to store all the beans and the limits on storage capacities. Thus, for the need to store, the following constraints apply.

$x_{11} + x_{12} = 255$ (field 1 supply constraint)

$x_{21} + x_{22} = 152$ (field 2 supply constraint)

$x_{31} + x_{32} = 298$ (field 3 supply constraint)

$x_{41} + x_{42} = 441$ (field 4 supply constraint)

Limits on storage capacities can be expressed as

$x_{11} + x_{21} + x_{31} + x_{41} \leq 725$ (storage site 1 constraint)

$x_{12} + x_{22} + x_{32} + x_{42} \leq 510$ (storage site 2 constraint)

As before, the x's ≥ 0.

This formulation will allocate the beans from each of the four fields to the two storage sites in such a way as to minimize the total transportation cost. The problem ensures that all the beans are shipped (first constraint set) and that the amount shipped to each storage location does not exceed its storage capacity (second constraint set).

III. GENERAL LINEAR PROGRAMMING PROBLEM STATEMENT

A. Problem Statement

The basic problem in linear programming is one of either maximizing or minimizing a function (objective function) of several variables (decision variables), with the value of the variables being subject to a number of constraints. In linear programming, both the objective function and the constraints must be linear functions. The mathematical statement of the general linear programming problem (with n variables and m constraints) is as follows:

$$\text{Maximize (Minimize) } z = c_1 x_1 + c_2 x_2 + \cdots + c_m x_n \tag{3}$$

Subject to:

$$a_{i1} x_1 + a_{i2} x_2 + \cdots + a_{in} x_n \ (\leq, =, \geq) \ b_i, \qquad i = 1, \ldots, m \tag{4}$$

and

$$x_1, x_2, \ldots, x_n \geq 0 \tag{5}$$

where only one of the operators $(\leq, =, \geq)$ holds for each of the m constraints. The quantities c_j, a_{ij}, and b_i, $i = 1, \ldots, m$ and $j = 1, \ldots, n$, are all known constants while the x_j's are the unknown decision variables. This general form fits the formulations developed in the previous examples, and the student should verify this fact.

The c_j's are often referred to as "cost coefficients" or "objective function coefficients"; the a_{ij}'s may be referred to as "A matrix coefficients" or "structural coefficients"; and the b_i are often referred to as "right-hand-side coefficients" or "b coefficients."

The "standard" form of the linear programming problem will often be referred to in later discussions. This is the specific form of the linear programming problem statement in which all constraints are equalities and the right-hand side of each constraint is nonnegative. Thus, the linear programming problem in standard form is as follows:

$$\text{Maximize (Minimize) } z = c_1 x_1 + c_2 x_2 + \cdots + c_n x_n \tag{6}$$

Subject to:

$$a_{i1} x_1 + a_{i2} x_2 + \cdots + a_{in} x_n = b_i, \qquad i = 1, \ldots, m \tag{7}$$

and

$$x_1, x_2, \ldots, x_n \geq 0 \tag{8}$$

The method used to solve linear programming problems, the simplex method, requires, among other things, that the linear programming problem be in standard form. This requires that constraints expressed as inequalities be converted to equalities.

B. Slack and Surplus Variables

Inequality constraints are converted to equations through the use of "slack" or "surplus" variables. A slack variable is added to the left-hand side of a less-than-or-equal-to constraint to take up the "slack" implied by the $\leq$ sign. On the other hand, a surplus variable is subtracted from the left-hand side of a greater-than-or-equal-to constraint to take off the "surplus" implied by the $\geq$ sign.

To illustrate, consider the $\leq$ constraint,

$$a_{11}x_1 + a_{12}x_2 + a_{13}x_3 \leq x_1 \tag{9}$$

To make this inequality an equation, the variable x_4 will be added, and it will assume whatever value is necessary for the equation to hold. Thus, adding x_4 to Eq. (9) results in

$$a_{11}x_1 + a_{12}x_2 + a_{13}x_3 + x_4 = b_1 \tag{10}$$

For Eq. (10) to hold, x_4, like the other variables, must be nonnegative. From the standpoint of meaning relative to the initial problem, x_4 can be interpreted, for example, to be the resources (represented by the constraint) not used in the problem, i.e., the slack resources.

For the $\geq$ constraint, consider the following:

$$a_{11}x_1 + a_{12}x_2 + a_{13}x_3 \geq b_1 \tag{11}$$

In this case a surplus variable is subtracted from the left-hand side. It then assumes the value needed to make the inequality an equation.

$$a_{11}x_1 + a_{12}x_2 + a_{13}x_3 - x_4 = b_1 \tag{12}$$

Again, for Eq. (12) to hold, the surplus variable, x_4 must be nonnegative. To see the physical meaning, recall the diet example. In that situation the surplus variable represents the nutrients supplied over the minimum required.

Slack and surplus variables, while added to convert in inequalities to equations, do have meanings with respect to the original problem and are therefore a part of the problem. They carry valuable information about the problem that will be explored in more detail later.

When all the inequalities that exist in a set of constraints are converted to equalities, the set of constraints is in the form of a set of linear equations. Thus, linear programming can be thought of as a problem of optimizing a linear function in which the variables are constrained to be nonnegative and satisfy a set of simultaneous equations.

IV. GRAPHICAL VIEW OF LINEAR PROGRAMMING

In trying to understand how linear programming optimizes a linear function (objective function) in which the variables must satisfy a set of simultaneous linear equations, it is useful to graphically view what is happening. Since no combination of variable values that does not satisfy the constraints can be considered, it is of interest to graphically see this set of solutions and how it related to the objective function.

A. Unique Linear Programming Solution Example

Consider the problem

Maximize $z = 80x_1 + 90x_2$

Subject to:

$$x_1 + x_2 \leq 10$$
$$2x_1 + x_2 \leq 16$$
$$3x_1 + 5x_2 \leq 40$$
$$x_1, x_2 \geq 0$$

The first step in the graphical solution is to determine the set of numbered pairs (x_1, x_2) that are the feasible solutions. On the x_1, x_2 coordinate plane (Fig. 1) the first quadrant yields the set of points that satisfy the nonnegative restrictions $x_1 \geq 0$ and $x_2 \geq 0$. To find the set of points in the first quadrant satisfying the constraints, we must interpret geometrically inequalities such as

$$x_1 + x_2 \leq 10$$

If the equal sign holds, i.e.,

$$x_1 + x_2 = 10$$

we have the equation for a straight line, and any point on that straight

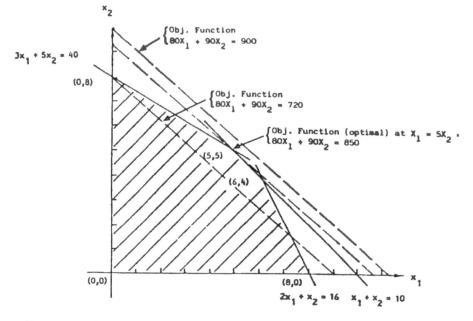

Figure 1 Graphical solution to a linear programming problem with a unique solution.

line satisfies the equation. Notice that any point lying in the half-plane below the line satisfies the constraint

$$x_1 + x_2 \leq 10$$

However, only those below the line and in the first quadrant satisfy both the constraint and the nonnegativity restrictions. Consider the other constraints in a similar manner, and we find a set of points that satisfy all the restrictions of the problem. This set of points is represented by the cross-hatched area of Fig. 1. All points in this shaded area are feasible solutions since they satisfy all the restrictions. This set of points is a convex set,* and it can be shown that the set of

---

*A set is convex if all points on a line connecting any two points in the set is also in

the set, e.g.,   is a convex set.   is not a convex set.

feasible solutions to a linear programming problem is always a convex set.

Now consider the objective function. To solve the linear program, we must find the point or points in the region of feasible solutions that give the largest value of the objective function. For any fixed value of z, say z_1,

$$z_1 = 900 = 80x_1 + 90x_2$$

is a straight line. Any point on this straight line will give a value of $z_1 = 900$. For each different value of z we obtain a different straight line, but the lines are all parallel. We wish to find the straight line with the largest value of z that has at least one point in the region of feasible solutions. Notice (Fig. 1) that the line

$$z_1 = 80x_1 + 90x_1 = 900$$

has no points in the region of feasible solutions and thus is not the optimal feasible solution to the problem. Let $z_2 = 720$, and note that the line

$$80x_1 + 90x_2 = 720$$

has points in the region of feasible solutions but that z can have a larger value and still be in the region of feasible solutions. By drawing lines parallel to z_1 and z_2 we find that the maximum value that z can attain and remain in the region of feasible solutions is found where the lines $x_1 + x_2 = 10$ and $3x_1 + 5x_2 = 40$ intersect, or at $x_1 = 5$, $x_2 = 5$. The value of z at the point is $80(5) + 90(5) = 850$. Note that the maximum value of z occurs at an extreme point or corner of the region of feasible solutions. It can be shown that the optimal solution of a linear programming problem always occurs at an extreme point of the convex set. In this example the optimal solution is unique, i.e., only one (x_1, x_2) pair lies in the region of feasible solutions and optimizes the objective function. The following example is one for which there are an infinite number of optimum feasible solutions.

B. Multiple Optima Example

Let's now look at the graphical solution of the linear programming problem

$$\text{Minimize } z = -4x_1 + 4x_2$$

Subject to:

$$x_1 - 2x_2 \geq -4$$
$$2x_1 - 3x_2 \leq 13$$
$$x_1 - x_2 \leq 4$$
$$x_1, x_2 \geq 0 \tag{13}$$

The region of feasible solutions (Fig. 2) is constructed in a manner similar to that in the previous example.

Notice that the objective function

$$z = -4x_1 + 4x_2$$

is parallel to the line

$$x_1 - x_2 = 4$$

between the lines

$$x_2 = 0 \quad \text{and} \quad 2x_1 + 3x_2 = 13$$

and that the minimum value of z occurs along this line. Although this is an extreme point of the convex set, the objective function has the same minimum value of -16 anywhere along the line

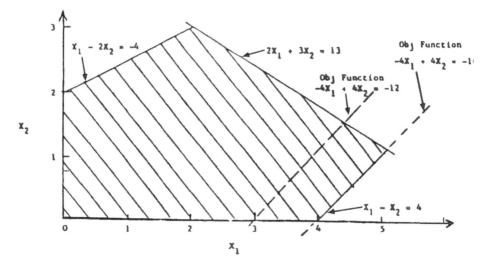

FIGURE 2   Graphical solution to a multiple optima problem.

$$x_1 - x_2 = 4$$

between the lines

$$x_2 = 0 \quad \text{and} \quad 2x_1 + 3x_2 = 13$$

There are an infinite number of (x_1, x_2) pairs along this line and consequently an infinite number of solutions to the problem.

Now let's turn our attention to the graphical representation of problems for which no solutions exist.

C. Unbounded Solution Example

Maximize $z = 2x_1 - x_2$

Subject to:

$$x_1 - x_2 \geq -1$$
$$x_1 - 2x_2 \leq 2$$
$$x_1, x_2 \geq 0$$

Proceeding as before, we sketch the set of points that satisfy the restrictions of the problem (Fig. 3). When we draw a typical objective function

$$z_1 = 2x_1 + x_2 = 4$$

we see that it lies in the region of feasible solutions. We also see that for any other value, z_2, of the objective function such that $z_2 > z_1$ the line

$$2x_1 + x_2 = z_2$$

also lies in the region of feasible solutions. This means that the objective function does not attain a maximum value on the set of points satisfying the constraints. This set is unbounded, and the problem has no solution. The situation is also referred to as a problem with an unbounded solution.

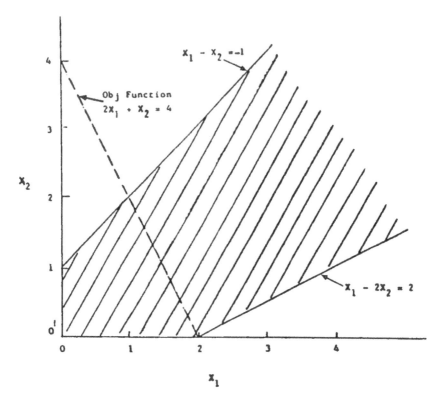

FIGURE 3   Graphical representation of an unbound solution problem.

D.   No Solution Example

Minimize $z = 2x_1 - 3x_2$

Subject to:

$$x_1 - 2x_2 \leq -4$$
$$x_1 + x_2 \leq 5$$
$$x_1 - x_2 \geq 2$$
$$x_1, x_2 \geq 0$$

When we draw the straight lines for the constraints and attempt to find the intersection of half-planes that satisfy the constraints, we find that no region of feasible solutions exists, i.e., there are no num-

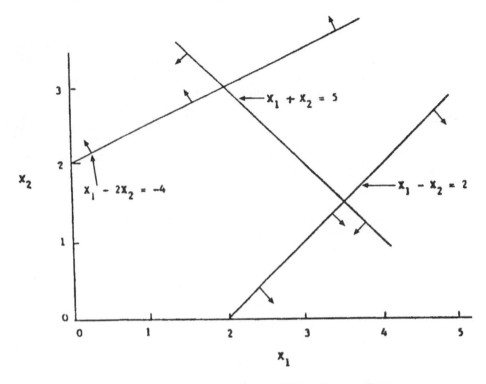

FIGURE 4 Graphical representation of a problem with no solution.

bered pairs (x_1, x_2) in the coordinate plane that satisfy all the restrictions. Therefore, the problem has no solution. (See Fig. 4.)

Summarizing, we have looked at examples of linear programming problems for which (1) there is a unique optimal solution, (2) there are an infinite number of optimal solutions, (3) the solution is unbounded, or (4) there is no solution.

V. LINEAR PROGRAMMING AND SIMULTANEOUS EQUATIONS

A. Linear Programming

The graphical view of linear programming portrays very clearly the formation of the region of feasible solutions using linear equations. It also demonstrates that the optimal solution of a linear programming problem is a solution to the set of simultaneous linear equations derived from the constraint inequalities and equalities. Thus, linear programming can be viewed as mainly concerned with the solution to simultaneous linear equations that arise from the restrictions on the variables (the constraints). The goal is to find the solution to the simultaneous equations for which all variables are nonnegative and that results in the maximum (or minimum) value of the objective function.

Through the addition of slack and surplus variables, a constraint set of inequalities can be converted to a set of simultaneous linear equations. In trying to solve this set of equations, however, there are normally more unknowns than equations. In this case, there are a large number of solutions.

One way to obtain a set of solutions to a system of m independent equations and n unknowns (where $m < n$) is to arbitrarily select $n - m$ variables, set them equal to zero, and solve the remaining m equations in m unknowns. By this method one can find $\binom{n}{m}$ solutions, where $\binom{n}{m}$ is the number of ways n things can be taken m at a time, i.e.,

$$\binom{n}{m} = \frac{n!}{m!(n - m)!}$$

Linear programming arrives at a solution in this manner. In addition, linear programming ensures that the next solution will be at least as good (i.e., $\geq$ if a maximization problem and $\leq$ in minimization) as the previous solution. Therefore in most cases the number of solutions to be considered will be less than $\binom{n}{m}$.

At this stage it would perhaps be instructive to see a specific example of a linear programming problem solved using this procedure.

B. Simultaneous Equation Example

Suppose a farm tool manufacturer has two primary resources, machine time and labor hours. In a certain period of time he has 200 machine-hours and 300 labor-hours to devote to three products, x_1, x_2, and x_3. Product x_1 takes 15 machine-hours and 10 labor-hours per unit. Product x_2 takes 10 machine-hours and 25 labor-hours per unit, and product x_3 takes 10 machine-hours and 20 labor-hours per unit. The manufacturer wishes to determine the mix of products that will maximize profits while not exceeding the machine-hours available. He wishes to provide full employment for his workforce and therefore requires that all available man-hours be used. This is a competing objective, for the maximum profit might be greater if the manager were not so concerned about his employees. The two constraints and nonnegativity restrictions in algebraic form are

$$15x_1 + 10x_2 + 10x_3 \leq 200$$

$$10x_1 + 25x_2 + 20x_3 = 300$$

$$x_1, x_2, x_3 \geq 0$$

The profits to the manufacturer will be \$5 per unit of x_1, \$10 per unit of x_2, and \$12 per unit of x_3. Thus, the optimizing to be performed is

Maximize $5x_1 + 10x_2 + 12x_3$

Since the first constraint is a $\leq$ inequality, we must add a slack variable and rewrite the set of constraints as

$$x_1 + 10x_2 + 10x_3 + x_4 = 200$$

$$10x_1 + 25x_2 + 20x_3 = 300$$

where x_4 does not need to be included in the objective function since it has a coefficient of zero.

Since $n = 4$ and $m = 2$, there are $\binom{4}{2} = 6$ solutions that can be obtained by setting variables equal to zero two at a time and solving for the remaining two variables. First setting x_3 and x_4 equal to zero, we get

$$15x_1 + 10x_2 = 200$$

$$10x_1 + 25x_2 = 300$$

or $x_1 = 7.2727$ and $x_2 = 9.0909$. In a similar manner we can obtain the five other solutions. Table 3 shows all the possible solutions and the value of the objective function for each.

The solutions (3 and 4) that are marked with an asterisk in Table 3 involve a negative variable, which violates the nonnegativity restriction. The objective function is not evaluated for such solutions. These solutions are said to be *infeasible* because they do not satisfy the conditions of the problem. The other solutions that do not satisfy the conditions of the problem are said to be feasible. Solution 6 is the feasible solution that maximizes the function and is said to be the *maximum feasible solution* or *optimal solution*.

It must be emphasized here that by a solution to the problem we are referring to a combination of the n variables and not to the maximum value of the function. This is because we are basically concerned with solutions to simultaneous equations. This is logical be-

Table 3 Six Solutions to Farm Tool Manufacturer Problem

Solution no.	x_1	x_2	x_3	x_4	Objective function $(5x_1 + 10x_2 + 12x_3)$
1	7.2727	9.0909	0	0	127.27
2	5	0	12.5	0	175
3*	30	0	0	−250	—
4*	0	−20	40	0	—
5	0	12	0	80	120
6	0	0	15	50	180

*See text.

cause the manufacturer is primarily trying to determine what products to make to maximize profits with the resources available.

VI. SIMPLEX METHOD

The simplex method or algorithm is an iterative procedure that moves from one corner of the region of feasible solutions to another, always improving the value of z, or at least not making it worse. Viewed another way, the simplex algorithm is an iterative procedure that moves from one basic feasible solution (of the system of simultaneous linear equations) to another, again always improving z. At each iteration one variable is removed from the solution and another inserted. At each iteration the solution is tested for optimality. If optimality is reached, the iterations end. If optimality has not been obtained, a new iteration is performed. This continues until optimality is reached or until an unbounded or infeasible solution is obtained.

The manner in which variables are chosen to leave solution ensures that the next solution will be feasible (all x's $\geq$ 0), and the manner in which variables are chosen to enter solution ensures that the next solution has the largest improvement in x possible. Optimality is reached when it is no longer possible to improve z.

The simplex method needs a *basic feasible solution* in order to begin. A basic feasible solution, recall, is a solution to the linear constraint equations in which the values of the variables in solution (the basic variables) are nonnegative. With this initial solution, the simplex proceeds as follows.

Step 1. Check the optimality of the current solution. This is accomplished by first determining the value of $z_j - c_j$ for the nonbasic variables. This is the relative effect of these nonbasic variables (those not in solution) on the objective function if they were to be brought into the solution. If none of the nonbasic variables has an improving effect on the value of z, the optimality has been reached. The value of c_j is simply the objective function coefficient for the jth variable. z_j is defined as the product of the objective function coefficients of the variables in solution at this iteration (c_{Bi}) and the value of the **A** matrix coefficients associated with variable j at this iteration (y_{ij}).

$$z_j = \sum c_{Bi} y_{ij}$$

where summation on i is over all variables in solution.

Step 2. If some $z_j - c_j < 0$, then the nonbasic variable having the smallest (largest absolute) $z_j - c_j$ is selected to enter solution. Denote the column in which the incoming variable is located as k.

Step 3. The variable to leave solution is selected by examining the constraints to see how large the nonbasic variable can be made before one of the current variables in solution goes to zero. The variable that goes to zero first is removed. If the variable that is to be removed is located in the rth row, the objective of the examination of the constraints is to find r such that

$$\frac{x_{Br}}{x_{rk}} = \min_i \left[\frac{x_{Bi}}{y_{ik}} > 0 \right]$$

where x_{Bi} is the value of the variables in solution at each iteration. The rationale for this "minimum ratio rule" is discussed later. The condition $y_{ik} > 0$ must be observed in simplex calculations, and this is where many students in their first simplex computations make a mistake—be careful!

Step 4. With the variables coming in and going out decided, the next step is to compute the new solution. This is accomplished by "pivot operations"—a sequence of elementary calculations that results in a system of simultaneous linear equations being reduced to an equation having a unit coefficient and all other variables in the equation having coefficients of zero. This solution of systems of linear equations is often referred to as the Gauss–Jordan elimination technique. The result of the pivoting operation is to have the identity matrix associated with the variables in the solution.

A. Tableau Format of Simplex Method

To further explain the mechanics of the simplex procedure, the tableau format is now described and applied to a simple maximization problem. The tableau format is basically a means of organizing the calculations and the results of the simplex procedure. Further, by expressing the pivoting operation in simple formulas, the entire simplex procedure becomes rather mechanical, and thus readily programmable on the computer.

Consider the following linear programming problem where $a_{1,n+1}$ through $a_{n,n+m}$ are slack variables.

$$\text{Maximize } z = c_1 x_1 + c_2 x_2 + \cdots + c_n x_n$$

Subject to:

$$a_{11}x_1 + a_{12}x_1 + \cdots + a_{1n}x_n + a_{1,n+1}x_{n+1} = b_1$$
$$a_{21}x_1 + a_{22}x_1 + \cdots + a_{2n}x_n + a_{2,n+2}x_{n+2} = b_2$$
$$\vdots \qquad \vdots \qquad\qquad \vdots \qquad\quad \vdots$$
$$a_{m1}x_1 + a_{m3m}x_1 + \cdots + a_{mn}x_n + a_{m,n+m}x_{n+m} = b_m$$
$$x_1, x_2, \ldots, x_n \geq 0$$

The *initial* tableau, using the above notation, is shown in Table 4. In this tableau, x_{Bi} represents the value of the variables in solution (initially $x_{Bi} = b_i$); $z_j = c_{Bi}y_{ij}$, where summation on i over variables in solution; and $z_j - c_j$ represents the relative measure of effectiveness coefficient (sometimes referred to as r_j).

The initial tableau, after a few iterations, can be generalized as shown in Table 5. In the general tableau, the c_i are the variables in solution at a given iteration and y_{ij} are the values of a_{ij} at an iteration after the initial tableau.

Table 4 Initial Tableau for the Simplex Method

c_j		c_1	c_2	$\cdots$	c_n	0	0	$\cdots$	0	x_i
c_{Bi}	Variable in soln	x_1	x_2	$\cdots$	x_n	x_{n+1}	x_{n+2}	$\cdots$	x_{n+m}	
c_{B1}	x_{n+1}	a_{11}	a_{12}	$\cdots$	a_{1n}	$a_{1,n+1}$	0	$\cdots$	0	b
c_{B2}	x_{n+2}	a_{21}	a_{22}	$\cdots$	a_{2n}	0	$a_{2,n+2}$	$\cdots$	0	b
$\vdots$										
c_{Bm}	x_{n+m}	a_{m1}	a_{m2}	$\cdots$	a_{mn}	0	0	$\cdots$	$a_{m,n+m}$	b
	z_j	z_1	z_2	$\cdots$	z_n	z_{n+1}	z_{n+2}	$\cdots$	z_{n+m}	z
	$z_j - c_j$	$z_1 - c_1$	$z_2 - c_2$	$\cdots$	$z_n - c_n$	$z_{n+1} - c_{n+1}$	$z_{n+2} - c_{n+2}$	$\cdots$	$z_{n+m} - c_{n+m}$	

Table 5 General Tableau for the Simplex Method

c_j		c_1	c_2	...	c_n	0	0	...	0	x_{Bi}
c_{Bi}	Variable in soln	x_1	x_2	...	x_n	x_{n+1}	x_{n+2}	...	x_{n+m}	
c_{B1}	c_1	y_{11}	y_{12}	...	y_{1n}	$y_{1,n+1}$	0	...	0	b_1
c_{B2}	c_2	y_{21}	y_{22}	...	y_{2n}	0	$y_{2,n+2}$	...	0	b_2
⋮										
c_{Bm}	c_m	y_{m1}	y_{m2}	...	y_{mn}	0	0	...	$y_{m,n+m}$	b_m
	z_j	z_1	z_2	...	z_n	z_{n+1}	z_{n+2}	...	z_{n+m}	z
	$z_j - c_j$	$z_1 - c_1$	$z_2 - c_2$	...	$z_n - c_n$	$z_{n+1} - c_{n+1}$	$z_{n+2} - c_{n+2}$	...	$z_{n+m} - c_{n+m}$	

The specific mechanics of performing the four previously discussed steps of the simplex procedure are as follows, where the terminology is that of the above tableaus.

Step 1. Check optimality of current solution. Check to see if all $z_j - c_j \geq 0$. If they are all positive or zero, the optimality has been reached; there is no variable that can be brought into solution and cause z to improve.

Step 2. If one or more of $z_j - c_j \leq 0$, then the solution can be improved by bringing in a new variable. Select the column having the smallest (largest absolute value) $z_j - c_j$ and call this column k. The variable in the kth column (i.e, the kth variable) will enter solution at the next iteration.

Step 3. The variable to leave solution must be determined. As noted, before, the variable to be removed from solution is determined by finding r such that

$$\frac{x_{Br}}{y_{rk}} = \min_j \left[\frac{x_{bi}}{y_{ik}}, \ y_{ik} \geq 0 \right]$$

The variable in the rth row is removed.

Step 4. The new tableau is computed by pivoting. Pivoting is nothing more than establishing the identify matrix with the variables in solution. As noted before, this simply means that in each column associated with a variable in solution, there must appear one unit coefficient and the rest zeros. This pivoting can be expressed through the equation

$$\hat{y}_{ij} = y_{ij} - y_{ik}\left[\frac{y_{rj}}{y_{rk}}\right], \qquad i \neq r; j = 1, \ldots, n + m; i = 1, \ldots, m$$

where $\hat{y}_{ij}$ and $\hat{y}_{rj}$ are the entries in the new tableau.

The entries is the new tableau for x_{Bi} and $z_j - c_j$ can be computed by applying the above pivoting equations to those locations in the tableau. In this case the $z_j - c_j$ would be treated as $y_{m+1,j}$ and the pivoting equations used. The new $z_j - c_j$ could also be computed from their definitions. In this case the z_j row in the tableau is needed; otherwise it is not needed. Likewise, the x_{Bi} are treated as $y_{i,n+m+1}$ and computed using the pivoting equations.

Once the new tableau is established, the procedure returns to step 1. Iterations of the procedure continue until either optimality is reached, an unbounded solution is found, or an infeasible solution is encountered. These other possible terminations of the simplex procedure are discussed later.

B. Example of Simplex Tableau Procedure

Consider the problem

Maximize $z = 2x_1 + 5x_2 + 4x_3$

Subject to:

$$2x_1 + x_2 + x_3 \leq 6$$
$$4x_1 + 2x_2 + 2x_3 \leq 5$$
$$x_1, x_2, x_3 \geq 0$$

First, the inequalities must be converted to equalities by adding slack variables. Thus, the constraints become

$$2x_1 + x_2 + x_3 + x_4 = 6$$
$$4x_1 + 2x_2 + 2x_3 + x_5 = 5$$

The initial tableau, using the above data, is shown in Table 6.

Here x_4 and x_5 are the variables in solution as they are readily evaluated when x_1, x_2, and x_3 are set to zero. So we have an initial basic feasible solution to begin the simplex procedure. The z_j and $z_j - c_j$ were computed using the data in the tableau from the initial statement of the problem.

Proceeding to step 1, a review of the $z_j - c_j$ row reveals that optimality has not been reached. All nonbasic variables (x_1, x_2, and x_3) have negative values, indicting that any one could be brought into solution in step 2 and result in an improvement in the objective func-

Table 6 Initial Tableau for Simplex Example

c_j		2	5	4	0	0	x_{Bi}
c_{Bi}	Variable in soln	x_1	x_2	x_3	x_4	x_5	
0	x_4	2	1	1	1	0	6
0	x_5	4	2	2	0	1	5
	z_j	0	0	0	0	0	0
	$z_j - c_j$	−2	−5	−4	0	0	0

tion. The value of $z_2 - c_2$ is the smallest, indicating that bringing x_2 into solution will result in the greatest improvement of z. Thus, x_2 is to enter solution at the next iteration (in the next tableau), and therefore $k = 2$.

In step 3, the variable to leave solution is determined by the minimum ratio rule. The row of the variable to leave solution (r) is determined by selecting the row with the smallest ratio of x_{Bi}/y_{ik} where $y_{ik} > 0$. For row 1,

$$\frac{x_{B1}}{y_{12}} = \frac{6}{1} = 6$$

and for row 2,

$$\frac{x_{B2}}{y_{22}} = \frac{5}{2} = 2.5$$

Thus, row 2 ($r = 2$) contains the variable to leave solution (x_5). The variable y_{22} becomes the pivot element as this is y_{rk} in the pivoting equations of step 4—it is involved in all pivoting operations.

To pivot (step 4), the equations noted earlier can be used noting that $k = 2$ and $r = 2$. Thus, each new $\hat{y}_{ij}$ can be computed directly from the equations. For example, $\hat{y}_{12}$ (with $i = 1$ and $j = 2$) is computed:

$$\hat{y}_{12} = y_{12} - y_{12} \left(\frac{y_{22}}{y_{22}}\right)$$

$$\hat{y}_{12} = 2 - 2 \left(\frac{4}{4}\right)$$

$$\hat{y}_{12} = 0$$

Table 7   Second and Final Tableau for Simplex Example

c_j		2	5	4	0	0	x_{Bi}
c_{Bi}	Variable in soln	x_1	x_2	x_3	x_4	x_5	
0	x_4	0	0	0	1	$-1/2$	$7/2$
5	x_2	2	1	1	0	$1/2$	$5/2$
	z_j	10	5	5	0	$5/2$	$25/2$
	$z_j - c_j$	8	0	1	0	$5/2$	

In a similar manner, using the appropriate pivoting equations, all new entries in the next tableau can be computed.

It is also instructive to note that the above equations will always result in the identity matrix associated with the variables in solution. Thus, the pivoting could be performed by simply performing the row manipulations that result in the identity matrix being associated with the variables row in solution, without memorizing the equations. The equations do the same thing. Thus, the tableau with x_4 and x_2 in solution is as shown in Table 7. The z_j row can be computed from its definition, and it then is used to compute $z_j - c_j$, or the $z_j - c_j$ can be computed by simply pivoting on the values from the previous tableau.

From the tableau of Table 7, it can be seen that all the $z_j - c_j$ entries are greater than or equal to zero; therefore, optimality has been reached at this tableau. The optimal solution to the problem is

Maximize $z = 25/2$

$$x_1 = 0; \qquad x_2 = \frac{5}{2}; \qquad x_3 = 0; \qquad x_4 = \frac{7}{2}; \qquad x_5 = 0$$

Of course, it is rare for a problem to be solved in one iteration, but the purposes here are to illustrate the simplex mechanics.

C. Example of Linear Programming Modeling and Solution via the Simplex Method

To further illustrate the modeling and solution of linear programming problems, consider the situation where a farmer has 100 ha of land, 400 labor hours that can be devoted to crop production on these 100 ha, and $25,000 of capital to fund the production on this land. The farmer has the equipment to produce three different crops and wants to know which crops, and how many hectares of each, to produce in order to maximize profits. The farmer has the crop data shown in Table 8.

The decision variables can be defined as follows.

x_i = number of hectares of crop i produced

The objective function represents the profit maximization goal if profits per hectare are multiplied by number of hectors (x_i) for each crop. Thus, the objective function is

Maximize $z = 60x_1 + 55x_2 + 68x_3$

subject to constraints on land, labor, and capital. Being careful to ensure that units are compatible across the inequalities, the constraints can be modeled in terms of the amounts of each crop produced as follows:

$$x_1 + x_2 + x_3 \leq 100 \qquad \text{(land)}$$
$$3x_1 + 4x_2 + 5x_3 \leq 400 \qquad \text{(labor)}$$
$$300x_1 + 260x_2 + 230x_3 \leq 25{,}000 \qquad \text{(capital)}$$

Also, the nonnegativity restrictions apply, so these three inequalities are a part of the model:

$$x_1 \geq 0, \qquad x_2 \geq 0, \qquad x_3 \geq 0$$

The land constraint could be an equality if the farmer wants to use all the land and maximize profits. However, the assumption is made that maximizing profits is the objective subject to a limit of 100 ha of land being available.

To solve using the simplex procedure, the constraint inequalities are corrected to equalities via the addition of slack variables. These slack variables represent unused land (x_4), unused labor (x_5), and unused capital (x_6).

Table 8 Data for Crop Mix Problem

	Labor hours required per hectare	Capital needed per hectare	Profit per hectare
Crop 1	3	300	60
Crop 2	4	260	55
Crop 3	5	230	68

$$x_1 + x_2 + x_3 + x_4 = 100$$

$$3x_1 + 4x_2 + 5x_3 + x_5 = 400$$

$$300x_1 + 260x_2 + 230x_3 + x_6 = 25{,}000$$

The initial simplex tableau can now be established using as the initial basic feasible solution the values of x_4, x_5, and x_6 when $x_1 = x_2 = x_3 = 0$. These variables are placed in the solution initially, and the basic solution is immediately available.

x_4 = 100 ha of land

x_5 = 400 h of labor

x_6 = 25,000 dollars of capital

Thus, the initial tableau is as given in Table 9. From this tableau, it can be seen that optimality has not been reached. Since x_3 has the smallest $z_j - c_j$ value, it is chosen to enter solution. Since x_5 has the minimum ratio, it is selected to leave solution.

The second tableau, after pivoting, is shown in Table 10. Optimality has not been reached as $z_1 - c_1$ and $z_2 - c_2$ are still negative. Since x_1 has the smallest $z_j - c_j$, it enters solution. In determining the variable to leave, x_4 cannot be considered as it has a negative y_{ij} under the x_i column. Since row 3 has the smallest ratio, x_6 leaves solution.

The third tableau, after pivoting, is shown in Table 11. At this tableau, optimality has been obtained. The optimal solution is

Maximize z = \$6222 profit

Table 9 Initial Tableau for Crop Mix Problem

c_j		60	55	68	0	0	0	x_{Bi}
c_{Bi}	Variable in soln	x_1	x_2	x_3	x_4	x_5	x_6	
0	x_4	1	1	1	1	0	0	100
0	x_5	3	4	5	0	1	0	400
0	x_6	300	260	230	0	0	1	25,000
	z_j	0	0	0	0	0	0	0
	$z_j - c_j$	−60	−55	−68	0	0	0	

Table 10   Second Tableau for Crop Mix Problem

c_j		60	55	68	0	0	0	x_{Bi}
c_{Bi}	Variable in soln	x_1	x_2	x_3	x_4	x_5	x_6	
0	x_4	0.40	0.20	0	1	−0.20	0	20
68	x_3	0.60	0.80	1	0	0.20	0	80
0	x_6	162	76	0	0	−46	1	6600
	z_j	40.8	54.4	68	0	19.66	0	5400
	$z_j - c_j$	−19.2	−0.6	0	0	13.6	0	

with

  40.74 ha of crop 1 grown
  55.56 ha of crop 3 grown
     0 ha of crop 2 grown

and with 3.70 ha of land not utilized. These hectares could not be used because all the labor and capital were exhausted (x_5 and x_6 have zero values).

D. Artificial Variables

Up to this point in presenting the simplex method, an initial basic feasible solution has been readily available. This has been the case because all the problems up to now have required the addition of

Table 11   Third and Final Tableau for Crop Mix Problem

c_j		60	55	68	0	0	0	x_{Bi}
c_{Bi}	Variable in soln	x_1	x_2	x_3	x_4	x_5	x_6	
0	x_4	0	0.01	0	1	−0.09	−0.002	3.70
68	x_3	0	0.52	1	0	0.37	−0.004	55.56
60	x_1	1	0.47	0	0	−0.28	0.006	40.74
	z_j	60	63.42	68	0	8.36	0.09	6222
	$z_j - c_j$	0	8.42	0	0	8.36	0.09	

slack variables to all constraints. What happens if there are greater-than-or-equal-to constraints or equations as constraints in the initial formulation? For example, suppose that in the previous problem, land was an equality in the initial formulation. This would have meant that x_4 would not be needed. Thus, the initial basic feasible solution would have been difficult to determine.

To easily find the first basic feasible solution needed to initiate the simplex procedure, *artificial variables* are used. These are variables that are added to the problem after all inequalities have been converted to equalities. They are therefore added to equations! They are added simply to get a basic feasible solution and to get the simplex under way with a minimum of effort. Since they are added to equations, however, they cannot be in the final solution at a positive level. They have no meaning to the problem (as slack and surplus variables do)—they are simply added to facilitate solution of the problem.

Since artificial variables are added for solution purposes only, they must be removed from the problem. One technique used to "force" artificial variables out of the solution to a linear programming problem is called the Big M method.

The Big M method "forces" artificial variables out of solution by giving them a large (Big M) positive or negative objective function coefficient depending upon whether the problem is a minimization or a maximization problem, respectively. In this manner, it is highly desirable, from the standpoint of optimizing the problem, to remove the artificial variable and not let it return to the solution.

To illustrate the procedure, consider the problem

Maximize $z = 3x_1 + 2x_2$

Subject to:

$$2x_1 + x_2 \leq 4$$
$$x_1 + x_2 = 1$$

Converting the constraints to equalities results in two equations but no readily apparent basic feasible solution.

$$2x_1 + x_2 + x_3 = 4$$
$$x_1 + x_2 = 1$$
$$x_1, x_2 \geq 0$$

In this simple problem, it would be possible to easily determine an initial basic feasible solution; however, in most practical problems it is not so simple. Consequently, an artificial variable will be added to the second constraint to illustrate how an initial basic feasible can be easily obtained. Thus, the constraints for the problem are now

$$2x_1 + x_2 + x_3 = 4$$
$$x_1 + x_2 + x_4 = 1$$

Artificial variable x_4, however, must not appear in the final solution at a value other than zero. If it does, the solution is not feasible because the second constraint is violated!

In setting up the first tableau for solving this problem by the simplex method, x_4 will be assigned a large negative objective function coefficient, $-M$. In this manner the simplex procedure will find that once x_4 is removed from solution, it will be highly undesirable to bring it back into solution. Thus, in the initial tableau, the objective function coefficients will be $(3, 2, 0, -M)$ for the respective variables. The first tableau is shown in Table 12.

x_3 and x_4 are in the initial basic feasible solution, and their values in the solution are the right-hand-side values from the two constraint equations, as before. The only difference is that x_4 is an artificial variable.

Since x_1 has the smallest value of $z_j - c_j$, it is to enter solution on the next iteration. The minimum ratio rule results in x_4 leaving solution. Pivoting results in the second tableau, as shown in Table 13. Since all $z_j - c_j$ are positive, optimality has been reached. With x_4 out of solution, the solution is feasible. Thus, x_4 permitted the simplex procedure to be easily initiated, and the large negative cost coefficient helped ensure that, once removed, the variable would not reenter solution. Again, if x_4 had been in the optimal solution at a zero value,

Table 12 Initial Tableau for Example of Big M Method

c_j		3	2	0	$-M$	x_{Bi}
c_{Bi}	Variable in soln	x_1	x_2	x_3	x_4	
0	x_3	2	1	1	0	4
$-M$	x_4	1	1	0	1	1
	z_j	$-M$	$-M$	0	0	$-M$
	$z_j - c_j$	$-M - 3$	$-M - 2$	0	0	

Table 13   Second and Final Tableau for Example of Big M Method

c_j		3	2	0	$-M$	x_{Bi}
c_{Bi}	Variable in soln	x_1	x_2	x_3	x_4	
0	x_3	0	-1	1	-2	2
3	x_1	1	1	0	1	1
	z_j	3	3	0	3	3
	$z_j - c_j$	0	1	0	$3 + M$	

the solution would be feasible, but if x_4 had a value in the optimal solution, the solution would be infeasible.

VII.   SIMPLEX ALGORITHM

There are a large number of computer routines available for solving linear programming problems. These routines vary greatly in capability. Some handle rather small problems, while others are designed to handle very large problems. Some simply give the optimal solution, while others provide considerable additional information for further analysis of the solution. Some routines are strictly set up to solve only minimization or maximization problems, while others can solve either. (By simply reversing the signs of the objective function coefficients, a minimization algorithm can be used to solve a maximization problem. Students are asked to verify this.)

In most routines, the user must inform the computer how many variables there are, how many constraints and of what type, and whether the problem is maximization or minimization. The objective function coefficients, the constraint matrix coefficients, and the right-hand sides are input to the computer using the specified formats. The coefficients may all have to be specified, or the user may be able to enter only nonzero coefficients (an advantage for large problems with many zero coefficients in the A matrix).

Output detail may be at the user's option. The solution only may be printed, or additional detail may be requested along with the solution. Many algorithms provide the option of modifying the problem and solving again.

In using a computer routine to solve a linear programming problem, the student should carefully read the routine's documentation. The specifics of how the routine operates, how data are input, and what is output are described. The understanding of linear programming obtained to this point will permit students to use most routines to solve a linear programming problem.

VIII. SOME MORE AGRICULTURAL APPLICATIONS

A. Machinery Management and Crop Production Problem

This example draws on Von Bargen (1980).

A 680 acre dryland cash-grain farm called Mechanized Acres is operated by one individual. This farm is located in eastern Nebraska and can produce corn, soybeans, grain sorghum, and winter wheat. A fixed complement of equipment is assumed for the crop production and grain handling system. Supplemental hired labor is available when needed such as at harvest time.

The objective for the crop production system analysis is to determine the acreage for each crop to maximize profit with respect to the costs that are proportional to the acreage. Fixed equipment costs of \$26,500 per year and a fixed labor expense of \$12,000 per year for the operator are subtracted from the profit function to determine the net profit.

The following decision variables are defined:

x_1 = acres of corn

x_2 = acres of soybeans

x_3 = acres of grain sorghum

x_4 = acres of wheat

x_5 = dummy variable associated with fixed cost

In addition to limitations on acres of cropland available, it is assumed that there are also limitations on time available in the spring for row crop planting and in the fall for row crop harvesting. These times are 90 h and 160 h, respectively, both at an 85% probability level. Time available for planting and harvesting wheat is assumed to be unlimited. There is also a minimum wheat acreage of 40 acres. This is the minimum required to justify special wheat equipment.

In order to develop the a_{ij} coefficients for the constraints involving row crop planting and harvesting time it is necessary to relate machine parameters to the operations (i.e., a six-row planter equipped for 30 in. rows, traveling at 4.2 mi/h, with a field efficiency of 0.6 is assumed). The a_{ij} coefficients must have units of hours per acre, or the inverse of field capacity. In the case of the six-row corn planter, the a_{ij} coefficient is

$$\frac{8.25}{\text{speed} \times \text{width} \times \text{field efficiency}} = \frac{8.25}{4.2 \text{ mi/h} \times 15 \text{ ft} \times 0.6}$$
$$= 0.22 \text{ h/acre}$$

All other a_{ij} coefficients in the planting and harvesting constraints are determined in a similar manner.

Based on the above information and the original statement of the problem, the following constraints can be written:

$x_1 + x_2 + x_3 + x_4 \le 680$	(land constraint)
$0.22x_1 + 0.19x_2 + 0.20x_3 \le 90$	(row crop planting time constraint)
$0.60x_1 + 0.57x_2 + 0.55x_3 \le 160$	(row crop harvesting time constraint)
$x_4 \le 40$	(wheat acreage constraint)
$x_5 = 38{,}500$	(fixed cost constraint)
$x_1, x_2, x_3, x_4, x_5 \ge 0$	(nonnegativity restrictions constraint)

The c_j coefficients for all decision variables except x_5, fixed cost, represent the net profit per acre. For each unit of x_5 the net profit will be reduced by one unit; consequently its cost coefficient is -1. The objective function for this problem is

Maximize $z = 84.50x_1 + 93.90x_2 + 93.00x_3 + 45.10x_4 - x_5$

Solving this problem we find the maximum net profit to be $6102.54. The optimal crop mix is 291 acres of grain sorghum and 389 acres of wheat.

B. Crop/Irrigation Planning

This example is from Hilliez and Lieberman (1974).

A certain farming organization operates three farms of comparable productivity. The output of each farm is limited both by the usable acreage and by the amount of water available for irrigation. The organization is considering three crops for planting that differ primarily in their expected profit per acre and in their consumption of water. Furthermore, the total acreage that can be devoted to each of the crops is limited by the amount of appropriate harvesting equipment available. The data for the upcoming season are given in Table 14.

In order to maintain a uniform workload among the farms, it is the policy to use some of the usable acreage as conservation acreage or wildlife habitat, and the percentage of usable acres saved or "set aside" should be the same on each farm. This percentage is to be determined by the other constraints and this one. However, any combination of the crops may be grown at any of the farms. The organization wishes to know how much of each crop should be planted at the respective farms in order to maximize expected profit.

To begin the modeling of this problem, it is quite clear that the decision variables, x_{ij} ($i = 1, 2, 3; j = 1, 2, 3$), should be the number of acres of the ith crop grown on the jth farm. Therefore, the objective is to maximize total profit in dollars, where

$$z = 400(x_{11} + x_{12} + x_{13}) + 300(x_{21} + x_{22} + x_{23})$$
$$+ 100(x_{31} + x_{32} + x_{33})$$

subject to $x_{ij} \geq 0$ (for $i = 1, 2, 3$ and $j = 1, 2, 3$) (no negative acres) and the restrictions modeled below.

The restrictions on usable acreage at each farm are

$$
\begin{array}{llll}
x_{11} & + x_{21} & + x_{31} & \leq 400 \\
x_{12} & + x_{22} & + x_{32} & \leq 600 \\
x_{13} & + x_{23} & + x_{33} & \leq 300
\end{array}
$$

The restrictions on water availability are

$$
\begin{array}{llll}
5x_{11} & + 4x_{21} & + 3x_{31} & \leq 1500 \\
5x_{12} & + 4x_{22} & + 3x_{32} & \leq 2000 \\
5x_{13} & + 4x_{23} & + 3x_{33} & \leq 900
\end{array}
$$

The restrictions on crop acreage are

$$
\begin{array}{ll}
x_{11} + x_{12} + x_{13} & \leq 700 \\
x_{21} + x_{22} + x_{23} & \leq 800 \\
x_{31} + x_{32} + x_{33} & \leq 300
\end{array}
$$

To model the policy of a uniform set-aside acreage, the total percentage of planted acreage per farm must be equal for each pair of the three farms. Therefore, the equations

Table 14 Data for Crop Irrigation Problem

Farm	Usable acreage	Water available (acre·ft)
1	400	1500
2	600	2000
3	300	900

Crop	Maximum acreage	Water consumption (acre·ft/acre)	Expected profit ($/acre)
1	700	5	400
2	800	4	300
3	300	3	100

$$\frac{x_{11} + x_{21} + x_{31}}{400} = \frac{x_{12} + x_{22} + x_{32}}{600}$$

$$\frac{x_{12} + x_{22} + x_{32}}{600} = \frac{x_{13} + x_{23} + x_{33}}{300}$$

$$\frac{x_{11} + x_{21} + x_{32}}{400} = \frac{x_{13} + x_{23} + x_{33}}{300}$$

must be satisfied. Since the first two equations imply the third, the third equation can be omitted from the model. Furthermore, these equations are not yet in a convenient form for a linear programming model, as all of the variables are not on the left-hand side. The final forms of the uniform workload restrictions are

$$3(x_{11} + x_{21} + x_{31}) - 2(x_{12} + x_{22} + x_{32}) = 0$$

$$(x_{12} + x_{22} + x_{32}) - 2(x_{12} + x_{23} + x_{33}) = 0$$

The optimal solution of this problem is as follows:

$x_{11} = 300$ acres; $x_{12} = 200$ acres; $x_{22} = 250$ acres;

$x_{23} = 225$ acres

Maximize $z = \$342,500$

C. Cost-Sharing Conservation Problem

This example is taken from Loftis and Ward (1980).

Many government programs for conservation in agriculture involve payments to farmers for a part of the cost of conservation structure or practices. Two necessary policy decisions in such programs have to do with how a given budget is to be divided between alternative conservation measures and what fraction of the total cost of each measure should be funded by the government.

If one is willing to make the following qualifying assumptions, this problem may be modeled as a set of linear constraints on resources and a linear objective function and be solved by linear programming.

1. The objective of the program is to maximize regional benefits within given levels of government and/or total expenditure.
2. Each dollar spent on a given conservation measure returns a known and constant dollar amount in regional benefits.
3. The program will be sufficiently attractive to farmers that all of the government monies included in the budget will be spent.

The objective function is therefore a linear function of the total amounts spent on all conservation measures included in the program. The constraint set is largely determined by political considerations and may include any or all of the following.

1. A budgetary limitation on the total size of the program or on the government share
2. Maximum and minimum values of the cost-share fractions for each conservation measure
3. Limitations on the amount spent on certain measures in relation to others

An example of the last type of constraint would be a requirement that twice as much be spent on nonstructural measures as on structural measures.

Consider a hypothetical situation in which a federally funded program is being developed to reduce erosion and sediment yield in the Great Plains regions through four conservation management practices. The practices under consideration are listed in Table 15 along with maximum and minimum cost-sharing fractions and expected benefits of each.

The total size of the program is to be limited to $400 million, and the federal share is not to exceed $100 million. The program is also limited by the following political constraints:

Table 15 Description of Alternative Practices in Cost-Sharing
Conservation Problem

| | Federal share | | Expected benefit |
Conservation practice	Min	Max	($ benefit/$ spent)
1. Planting and managing con- servation crops	0.2	0.5	2.00
2. Managing existing rangeland	0.3	0.7	1.75
3. Construction of sediment reservoirs	0.1	0.4	2.25
4. Construction of small on- farm measures such as terraces, ponds, and grassed waterways	0.2	0.8	2.50

(a) At least twice as much must be spent by the government
on practices 1 and 2, nonstructural measures, as on practices 3 and 4, structural measures.

(b) Federal expenditures for either one of the nonstructural
measures must not exceed three times the federal expenditure for the other nonstructural measure.

(c) Federal expenditures for either one of the structural measures must not exceed three times the federal expenditure
for the other structural measure.

These constraints ensure that nonstructural practices are favored
over structural practices and that there is a fairly equitable distribution of funds within each of the two categories. Limitations on spend-
ing for individual practices could be imposed as well but are not
included in the present example.

Eight decision variables are required for this problem. Let

x_1 = federal amount spent on conservation crops

x_2 = amount spent by farmers on conservation crops

x_3 = federal amount spent on range management

x_4 = amount spent by farmers on range management

x_5 = federal amount spent on sediment reservoirs

x_6 = amount spent by farmers on sediment reservoirs

x_7 = federal amount spent on small on-farm structures

x_8 = amount spent by farmers on small on-farm structures

Convenient units for the x's are 1 unit = \$10 million. The objective function coefficients are defined as follows:

c_j = the regional return in dollars per dollar spent on decision j;

$j = 1, 2, \ldots, 8$

The objective function is then

$$z = c_1x_1 + c_2x_2 + c_3x_3 + c_4x_4 + c_5x_5 + c_6x_6 + c_7x_7 + c_8x_8$$

The numerical values for the cost coefficients as found in Table 15 may be put into the above expression to yield

$$z = 2.0x_1 + 2.0x_2 + 1.75x_3 + 1.75x_4$$
$$+ 2.25x_5 + 2.25x_6 + 2.50x_7 + 2.50x_8$$

The linear programming algorithm is to maximize the above objective subject to the constraints described earlier, which may be formulated as follows.

The total program must not exceed \$400 million:

$$x_1 + x_2 + x_3 + x_4 + x_5 + x_6 + x_7 + x_8 \leq 40$$

The federal share is not to exceed \$100 million:

$$x_1 + x_3 + x_5 + x_7 \leq 10$$

Political constraint (a) above:

$$x_1 + x_3 \leq 2x_5 + 2x_7$$

or

$$x_1 - x_3 + 2x_5 + 2x_7 \leq 0$$

Political constraints as described in (b) above:

$$x_1 - 3x_3 \leq 0$$

and

$$-3x_1 + x_3 \leq 0$$

Political constraints as described in (c) above:

$$x_5 - 3x_7 \leq 0$$

and

$$-3x_5 + x_7 \leq 0$$

An additional set of constraints ensures that the cost-sharing fractions will remain within the allowable limits. The federal cost-share fraction for the alternative (Table 15) planting and managing conservation crops may be expressed as $x_1/(x_1 + x_2)$. Since the minimum value of this fraction is 0.2 (from Table 15), the following constraint is imposed.

$$\frac{x_1}{x_1 + x_2} \geq 0.2$$

which may be rewritten as

$$0.8x_1 - 0.2x_2 \geq 0$$

The maximum value of the same cost-share fraction is given as 0.5. The corresponding constraint is

$$0.5x_1 - 0.5x_2 \leq 0$$

Similar constraints that exist for the other three conservation measures are written below.

For managing existing rangelands,

$$0.7x_3 - 0.3x_4 \geq 0$$

$$0.3x_3 - 0.7x_4 \leq 0$$

For construction of sediment reservoirs,

$$0.9x_5 - 0.1x_6 \geq 0$$

$$0.6x_5 - 0.4x_6 \leq 0$$

For construction of small on-farm conservation measures,

$$0.8x_7 - 0.2x_8 \geq 0$$

$$0.2x_7 - 0.8x_8 \leq 0$$

After adding nonnegativity conditions,

$$x_1, x_2, x_3, x_4, x_5, x_6, x_7, x_8 \geq 0$$

the LP formulation is complete. The following is the optimal solution:

$$x_1 = 5.00; \quad x_2 = 5.00; \quad x_3 = 1.67; \quad x_4 = 0.72$$

$$x_5 = 2.19; \quad x_6 = 19.71; \quad x_7 = 1.14; \quad x_8 = 4.57$$

Maximum objective $z = 87.74$.

From these results the optimal distribution of funds and cost-share fractions were found to be as shown in Table 16. At the optimum, both of the budgetary constraints were tight, and a total benefit of $877.4 million would be expected from the program.

Table 16 Optimal Distribution of Funds and Cost-Share Fractions

Practice	Total expenditure, optimal ($/million)	Federal share, optimal
1. Conservation cropping	100.0	50%
2. Range management	23.9	70%
3. Sediment reservoirs	219.0	10%
4. On-farm structures	57.1	20%
Total	400.0	

From a practical viewpoint, one might ask, If the above problems were developed from an actual government conservation program, would the administrator allocate the money as noted in the solution? The answer to this question depends upon how well the LP formulation, with all its assumptions, describes the real-world problem. The administrator, without any further work on the LP solution, must try to evaluate (somewhat subjectively) the validity of the assumptions, the probability that the linear programming coefficients will remain the same over time, and other factors not included in the model (e.g., the likelihood that all structural measures will be used in appropriate situations and adequately maintained and the likelihood that nonstructural measures will be adequately maintained). Such an evaluation might cause the administrator to modify the linear programming results somewhat in deciding the appropriate funding levels between alternatives and the fraction of the cost of each measure to be funded by the government.

It is possible, via sensitivity analysis, to examine the effects of changes in the assumptions or benefit estimates on the solution. If the model results are greatly affected by small changes in benefit estimates or by removal or modifications of assumptions, then the administrator would need to have confidence in the validity of his assumptions in order to have confidence in the results.

D. Land Forming Design

This example is taken from Sowell et al. (1980).

A field with initial elevations as shown in Table 17 is to be landformed for improved water management. Based on soil characteristics and crops to be grown, it has been determined that in its final

Table 17 Data for Land Forming Design Problem

| | Original field elevations[a] | | |
| | Cross-row direction | | |
Row direction	1	2	3
1	10.01	10.12	9.97
2	10.10	10.03	10.03
3	10.06	9.75	9.76

[a]Elevations are determined at stations on 100 ft grids, i.e., spaced 100 ft apart in both the row and cross-row directions.

design the row slope of the field should be in the range of 0.1–0.5% and the cross-row slope in the range of 0–1% for optimum water management. The shrinkage characteristics of the soil are such that the volume of cuts in the field should exceed the volume of fills by a factor of 1.25–1; a tolerance of ±0.05 is permitted. It is desired to form the field to meet the above requirements while minimizing the amount of earth moved and consequently the cost of land forming.

There are several types of land forming designs depending upon whether constant or varying slopes are desired and where cross-row drainage is desired (Sowell et al., 1980). For this example we choose the design that assumes constant slopes in both the row and cross-row directions.

Formulation of the land forming design problem as a linear program is based on the assumption that the volume of cut or fill around a field station is a linear function of the depth of cut or fill at that station. The objective of a land forming design is to form the field to meet the specified range of slopes with a minimum of earth movement, which results in minimum cost. Correspondingly, the linear programming objective function is expressed in terms of minimizing the sum of cuts in the field. Before going further with the formulation of the linear program it is necessary to define the notation that will be used in the model.

x_{ij} = cut at station $[i, j]$, $i = 1, 2, 3; j = 1, 2, 3$

y_{ij} = fill at station $[i, j]$, $i = 1, 2, 3; j = 1, 2, 3$

u = slope in the row direction, ft/100 ft

v = slope in the cross-row direction, ft/100 ft

Using this notation, the objective function is

$$\text{Minimize } z = x_{11} + x_{12} + x_{13} + x_{21} + x_{22} + x_{23} + x_{31} + x_{32} + x_{33}$$

It was previously stated that fills and cuts are directly proportional; therefore, fills are not included in the objective function for they will be minimized when cuts are minimized.

There are five groups of constraints in the linear programming formulation of the land forming design problem. The first two are similar, one for the cross-row direction and one for the row direction, and represent the majority of the constraints. They are based on the arithmetic of the differences between the elevations of two adjacent field stations after the land has been formed. (Recall that the stations are spaced 100 ft apart in both the row and cross-row directions.) Consider the first two stations in the first cross row of the example (Table 17). The difference between elevations will be equal to the slope between the two stations, i.e.,

$$(10.01 - x_{11} + y_{11}) - (10.12 - x_{12} + y_{12}) = v$$

The first part of the equation in the parentheses represents the design elevation at station [1, 1], i.e., the original elevation minus the cut at the station plus the fill at the station. Obviously, either the cut or the fill will be equal to zero. The quantity in the second set of parentheses represents the design elevation of station [2, 1]. There is a similar constraint representing the difference in elevation of all adjacent stations in the cross-row direction and all in the row direction. Rearranging terms so that the decision variables are on the left-hand side of the equation and constants on the right-hand side, the first two groups of constraints are as shown in Fig. 5.

In the first two groups of constraints, the row slope u and cross-row slope v are decision variables. It was indicated earlier that these variables have upper and lower limits. The third group of constraints impose those limits.

Group 3 constraints are the limits on row and cross-row slopes:

$$v \geq 0.1; \qquad v \leq 0.5; \qquad u \leq 1.0$$

It is not necessary to have the constraint $u \geq 0$, since this is one of the nonnegativity restrictions.

The fourth group of constraints ensures that the ratio of cuts to fill is

$$1.25 \pm 0.05$$

Group 1 Constraints: Cross-row direction

$$-x_{11} + x_{12} \qquad\qquad + y_{11} - y_{12} = -0.11$$
$$-x_{12} + x_{13} \qquad\qquad + y_{12} - y_{13} = -0.15$$
$$-x_{21} + x_{22} \qquad\qquad + y_{21} - y_{22} = -0.07$$
$$-x_{22} + x_{23} \qquad\qquad + y_{22} - y_{23} = 0.0$$
$$-x_{31} + x_{32} \qquad\qquad + y_{31} - y_{32} = -0.31$$
$$-x_{32} + x_{33} \qquad\qquad + y_{32} - y_{33} = -0.01$$

Group 2 Constraints: Row direction

$$-x_{11} \qquad\qquad + y_{11} = -0.09$$
$$-x_{21} + x_{11} \qquad\qquad + y_{21} - y_{11} = -0.04$$
$$-x_{31} + x_{21} \qquad\qquad + y_{31} - y_{21} = -0.09$$
$$-x_{12} \qquad\qquad + y_{12} = -0.28$$
$$-x_{22} + x_{12} \qquad\qquad + y_{22} - y_{12} = -0.06$$
$$-x_{32} + x_{22} \qquad\qquad + y_{32} - y_{22} = -0.27$$

Group 3 Constraints: Limits on row and cross row sloped

$$-x_{13} + x_{23} \qquad\qquad - y_{13} \ge 0.1$$
$$+x_{13} \qquad\qquad + y_{13} \le 0.5$$
$$-x_{33} \qquad\qquad + y_{33} \le 1.0$$

Group 4 Constraints: Cut/fill ratio

$$x_{11} + x_{12} + x_{13} + x_{21} + x_{22} + x_{23} + x_{31} + x_{32} + x_{33} - 1.2y_{11} - 1.2y_{12} - 1.2y_{13} - 1.2y_{21} - 1.2y_{22} - 1.2y_{23} - 1.2y_{31} - 1.2y_{32} - 1.2y_{33} \ge 0$$
$$x_{11} + x_{12} + x_{13} + x_{21} + x_{22} + x_{23} + x_{31} + x_{32} + x_{33} - 1.3y_{11} - 1.3y_{12} - 1.3y_{13} - 1.3y_{21} - 1.3y_{22} - 1.3y_{23} - 1.3y_{31} - 1.3y_{32} - 1.3y_{33} \le 0$$

FIGURE 5   Constraints for the land forming design problem.

Table 18   Design Cuts and Fills for Land Forming Design
Example[a]

Row direction	Cross-row direction		
	1	2	3
1	+0.18	−0.04	0.00
2	−0.02	−0.06	−0.17
3	−0.08	+0.12	0.00

[a]Positive number indicates fill, negative indicates cut.

These constraints would be as follows:

$$\sum_{\substack{i=1,3 \\ j=1,3}} x_{ij} - 1.2 \sum_{\substack{i=1,3 \\ j=1,3}} y_{ij} \geq 0 \quad \text{and} \quad \sum_{\substack{i=1,3 \\ j=1,3}} x_{ij} - 1.3 \sum_{\substack{j=1,3 \\ j=1,3}} y_{ij} \geq 0$$

The fifth group of constraints are the nonnegativity restrictions.

$x_{11}, x_{12}, x_{13}, x_{21}, x_{22}, x_{23}, x_{31}, x_{32}, x_{33}, y_{11}, y_{12}, y_{13}, y_{21}, y_{22}, y_{23}, y_{31}, y_{32},$
$y_{33}, u, v \geq 0$

The solution to the land forming design problem, involving original field elevations as shown in Table 17, is given in Tables 18 and
19. The design slopes in the row and cross-row directions are 0.105
ft/100 ft and 0.109 ft/100 ft, respectively. The cuts or fills at the individual field stations are given in Table 18, and the final design
elevations are given in Table 19.

Table 19   Design Elevations for Land Forming Design Example

Row direction	Cross-row direction		
	1	2	3
1	10.19	10.08	9.97
2	10.08	9.97	9.86
3	9.98	9.87	9.76

REFERENCES

Hillier, F. S., and G. J. Lieberman. 1974. *Operations Research: Principles and Practice.* New York: Wiley.

Loftis, J. C., and R. C. Ward. 1980. Optimal cost sharing through linear programming. ASAE Paper No. 80539. Presented at the 1980 Annual Meeting of ASAE, San Antonio, TX, June 15–18.

Sowell, R. S., R. C. Ward, and L. H. Chen. 1980. Linear programming: Basic concepts, simultaneous equations and graphical solutions. ASAE Paper No. 80-5032. Presented at the 1980 Annual Meeting of ASAE, San Antonio, TX, June 15–18.

Von Bargen, K. L. 1980. Linear programming applied to a machinery management and crop production situation. ASAE Paper No. 80-5037. Presented at the 1980 Annual Meeting of ASAE, San Antonio, TX, June 15–18.

6

Expert Systems for Self-Adjusting Process Simulation

Steven J. Thomson

Agricultural Research Service, U.S. Department of Agriculture, Stoneville, Mississippi

I. INTRODUCTION

Much research in agricultural process management has focused on the development of simulation models. These models are used primarily to test the effect of changing inputs on one or more output state variables. Depending on the process involved, simulations vary in their absolute accuracy and rarely achieve it under most conditions. Input parameters are difficult to properly characterize, and complex processes may require significant simplification. For all these reasons simulations are used primarily in the research environment to test different management schemes in a relative sense.

Recent research has focused on a new method to permit simulations to be used as management tools. An approach documented herein uses readings from sensors to provide adjustments to input parameters of a simulation. The model was essentially parameterized "on the fly." The hope was that the model could make better estimates of temporal water status as readings from sensors provided appropriate feedback. Of course, sensors cannot accomplish these adjustments by themselves. A bit of intelligence derived from field experience and keen practical knowledge of the processes involved must be programmed into the system.

The following questions could be raised. Why might it be necessary to combine sensor- and model-based approaches? Why not dispense with the model and use sensors with appropriate algorithms as the primary instruments of decision making? The latter approach is appropriate in many cases. For example, high volume instruments (HVI) for cotton grading use sophisticated sensing systems that are continually being improved upon (Thomasson, 1993). Although the output could be used in a larger simulation of machinery sequences, nothing needs to be modeled to determine grade because sensing systems by themselves can do the job and do it well.

Soil water sensors, used for years to help the farm manager schedule irrigations, are inexact in their characterization of crop water stress. In their simplest mode, these sensors placed in the soil can help the farmer determine when to irrigate on the basis of a preset "trigger point," which usually indicates soil water potential or the amount of energy the plant exerts to extract water from the soil. Soil water sensors have been around for years, but farmers are still reluctant to use them for a variety of reasons. Some types of sensors respond well to temporal changes in water content but require servicing. Other types do not require servicing but respond more slowly to changes in actual soil water content. For proper irrigation scheduling, a sensor must "track" the soil water conditions in situations requiring frequent replenishment of water as in the drip irrigation of tomatoes (Smajstrla, 1985). Low frequency irrigation systems such as those of the center pivot type can, however, be scheduled with sensor readings that cannot respond immediately to soil drying (Thomson and Threadgill, 1987). This is because sensors can be read once a day (in the morning) for successful control.

Realizing limitations of soil water sensors, researchers have spent much time refining crop water use models for actual irrigation scheduling. These models need to be calibrated for soil type and crop characteristics in a specific environment. Sensed weather data are input to the models in many cases. Accuracy of critical input parameters becomes more critical as the required interval to read sensors becomes shorter, as with crops requiring frequent replenishment. Another problem is that many of these models are site-specific or need to be recalibrated in new environments. Periodic soil sampling or soil water sensing is wise to ensure that the model stays on track, but this can be labor-intensive.

In situations like the above where there may be imperfections in both sensor- and model-based management methods, a new method of combining the two has shown promise. Sensor responses may give clues as to which model input parameters require adjustment—a sort of dynamic calibration. Algorithms can be in place to regulate the amount of adjustment, interpret responses to the adjustment, and take appropriate action. Ideally, sensors would not be needed as the system "learns" and provides more accurate control of the process.

The purpose of this chapter is to summarize the development of the method and the results from a study that used sensors and a water use model together. Other potential applications for the method are discussed as well as potential problems in developing the method for a particular application.

II. CASE STUDY: KNOWLEDGE SYSTEM DEVELOPMENT

Irrigation management can be accomplished using properly interpreted readings from sensors placed in the soil, water balance models, or both. As has been mentioned, both methods have drawbacks. Realizing this, an objective of combining the two to provide better control was implemented.

A. Model Background

The model chosen for the study was PNUTGRO 1.02 (Boote et al., 1989a). PNUTGRO is a process-oriented peanut crop growth model that is an adaptation of a soybean model, SOYGRO (Wilkerson et al., 1983; Jones et al., 1988). The model predicts dry matter growth, leaf area index (LAI), crop development, and final yield depending on daily weather data and specific soils. Primarily developed as a research tool, PNUTGRO allows scientists to test the effect of different soil, weather, cultivar, and management factors on crop yield. The model can be run on a time step of 1 day covering the entire growing season. Water balance components of this model were implemented in the system described herein.

The knowledge system described here could be used with any crop simulation model that uses the IBSNAT (1986) format for input and output files. In addition to PNUTGRO, models that follow the IBSNAT format include CERES-MAIZE (Jones and Kiniry, 1986),

CERES-WHEAT (Ritchie, 1985), and SOYGRO (Wilkerson et al., 1983; Jones et al., 1988). PNUTGRO 1.02 was chosen because lysimeter evaluations were performed in this study on Florunner peanuts.

B. Field Experiment

Two weighing lysimeters (300 × 220 cm) were instrumented to determine the soil water balance. Both lysimeters were installed at the Irrigation Research and Education Park (IREP) by Butts (1988). For this study, each lysimeter was instrumented with four load cells, a net radiometer, and thermocouples. Peanuts (Florunner) were planted by hand on May 20, 1988 in a moist soil 4 cm (1.5 in.) deep and 4 cm (1.5 in.) apart in each row. The plants were thinned to 8 cm (3 in.) apart in the north lysimeter when a suitable stand was observed 15 days after planting (DAPL), June 3. Plants in the south lysimeter lagged somewhat behind the ones in the north lysimeter in the early stages and were thus thinned 21 DAPL, June 9.

The equivalent of 560 kg/ha of a 5-10-15 mix of fertilizer was incorporated into the soil for each lysimeter. A decision was made not to incorporate a preplant herbicide because the lysimeters were small enough to be weeded manually. Watermark* soil water sensors (Irrometer, 1989) were placed at several depths (Fig. 1) on June 21 in one location of each lysimeter that showed a good stand population. These sensors were read manually at least once a day using the Watermark 30KTC meter. To obtain soil water tension from meter readings, a curve was derived for the Watermark 30KTC meter (Thomson, 1990). This curve was related to the sensors' ac resistance, which is a function of soil water tension and temperature (Thomson and Armstrong, 1987). To provide temperature input for compensation of the soil water sensors, thermocouples were placed 8, 20, 40, 60, and 90 cm below the surface, the same depths as the sensors. Since some sensors were placed under the growing canopy and some under exposed soil early in the season, two banks of thermocouples were placed in each lysimeter to represent both covered and exposed conditions. Thermocouples and load cells were read every half hour.

A manually activated sump pump and dual tank system were used to drain the lysimeters after large rain events. The quantity of water removed was measured and subtracted from the total water removal indicated by the load cell measurements. In this way, water

---

*Use of trade names in this publication does not imply endorsement by the authors of the products named or criticism of ones not mentioned.

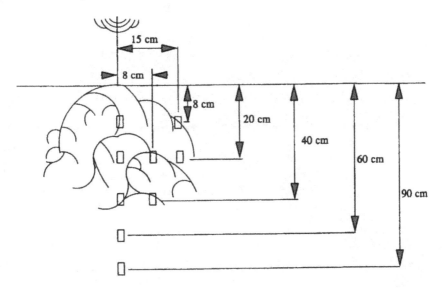

FIGURE 1   Soil water sensor placements in the root zone.

use by the crop during the period of sump pump operation could be estimated. Details of calibration procedures for the load cells and a description of the data acquisition system are given by Thomson et al. (1989) and Thomson (1990).

C. Program Flow Summary

The knowledge-based irrigation management system consists of five parts. These parts and the programming language or shell used to develop each part are

1. An expert system for sensor evaluation (VP-Expert)
2. Data input and calculation routines (Pascal)
3. Parameter adjustment routines applied to PNUTGRO input files (Pascal)
4. PNUTGRO code modifications (Fortran)
5. Irrigation scheduling routines (Pascal)

Some parts of the system are dependent on others and were thus developed concurrently. Each component is detailed below.

D. Expert System to Interpret Sensor Readings

An expert system to interpret soil water sensor readings was developed from previous knowledge about soil water sensor responses to wetting and drying as described in detail by Thomson (1990). The principal objective was to flag possible errors in sensor readings so they would not be used to adjust model input parameters. Decisions were based on

1. Whether or not enough drying had occurred to compare sensor responses and thus allow evaluation
2. Readings that may be out of the sensor's calibrated range of operation
3. Type of sensor (ranges of operation may be different)
4. A sensor's physical proximity to other sensors
5. A sensor's response to drying compared to that of other sensors at the same depth

Figure 2 briefly illustrates the expert system's use in interpreting sensor readings. These data were taken from Watermark granular matrix sensor readings performed in a lysimeter study. Figure 2a shows responses at the 8 cm depth both in the row and 15 cm away from the row. (Readings away from the row were used to gain knowledge about lateral root distribution.) Sensor readings at the same depth in the center of the row of each lysimeter show a slight temporal divergence. The expert system, however, allows a certain amount of variability, and both sensor readings in the row would be accepted in this case. Thus, readings in the row were averaged and could be used for parameter adjustment.

The bottom graph, Fig. 2b, shows a drying response at a much deeper location. Most of the readings shown are below the lower limit of tension (10 kPa) at which the sensors can be read accurately (Thomson and Armstrong, 1987; Pogue, 1991). If the sensors were tensiometers, readings would be accepted because they do not show an out-of-range condition at the low end of their operating range. Out-of-range conditions at the high end of operation are not presently treated in the knowledge base, although rules could be developed. (Tensiometers can "break tension" at high tensions.) Thus, it is assumed that the soil is not permitted to dry beyond the workable range of the sensor before irrigation. Although an out-of-range reading cannot be used to adjust soil properties, the program still uses an out-of-range reading that is increasing to infer water use in the deep-rooted horizons. Along with readings from other sensors, relative sensor responses can infer a root distribution and be used to adjust root weighting factors.

Daily sensor data were input to a Pascal routine (FIELDRAW) (Fig. 3) and imported by a spreadsheet macro into the spreadsheet

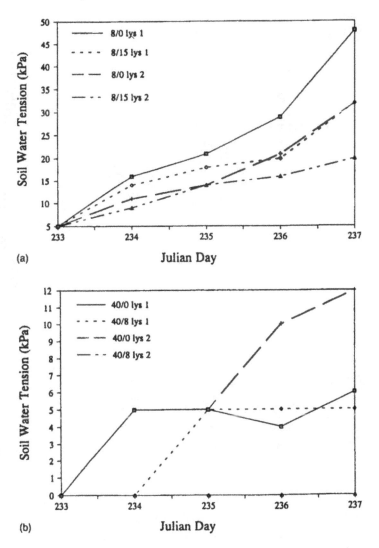

FIGURE 2   Sensor drying responses in August, at (a) 8 cm depth and (b) 40 cm depth.

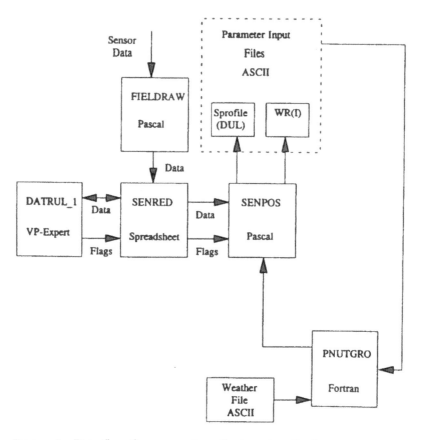

FIGURE 3   Data flow for parameter adjustment method.

SENRED.WKS. DATRUL_1, written in the rule development language of VP-Expert, evaluated sensor data taken from the spreadsheet and placed flags directly into the spreadsheet file, SENRED.WKS, next to readings that were questionable. Readings that were out of the sensor's range of operation received a 1 in the flag location. Sensors that were judged to be giving false readings based on criteria 4 and 5 above were given a 2. Different flags were placed for these two situations because SENPOS, the parameter modification program, still used out-of-range readings that were increasing to modify a root weighting function WR(I), which is discussed later. Absolute values of sensor readings that were out of the sensor's range of operation could not be used for other adjustments. Spreadsheet macros in SENRED were written to import ASCII sensor data from FIELDRAW, move data within the spreadsheet so DATRUL_1 could analyze a new set of sensor readings, and print data out to an ASCII file for use by SENPOS in adjusting PNUTGRO parameters.

E. Data Input and Calculation Routines

To construct the knowledge base for PNUTGRO parameter modification, routines were developed to modify the PNUTGRO weather file, accept soil water tension data from the user, extract soil water contents and root length densities from PNUTGRO output files, convert soil water contents to tensions, and display the results.

The user is asked to enter sensor data or run the system with default sensor files. If the default option is chosen, the sensor data acquisition routines are bypassed. If the user has sensor readings, the program asks how many sensor stations are in the field and requests values of the previous day's total solar radiation, maximum and minimum air temperatures, rainfall, and photosynthetically active radiation (PAR). These are necessary inputs to the PNUTGRO crop model. If the user does not have a reading for PAR, the program calculates an approximate value from solar radiation. (PAR is needed in photosynthesis routines and influences specific leaf area. PAR is not used directly for water balance calculations but does influence root growth and root water uptake.)

If the program is being run for the first time, the user is asked for the number of sensors being read at each station and for sensor depths and lateral spacings in sequence. The program presents the user with a graphical representation of the input sensor layout for approval. This allows correction of mistakes the user might have made in entering the layout. The user is then asked for the Julian day, time of day the sensors were read, and tension values for each sensor in sequence. The program stores each station's readings in a separate data file. Each data file keeps the latest 2 days of sensor readings.

When all sensor data have been entered, the program sends weather data to the input weather file for PNUTGRO at Julian day 1. The program uses weather data for a typical high stress day for several days in the future to forecast the earliest date irrigation should be required (Ross, 1984). Irrigation trigger level is based on a user-entered soil water tension value that is converted from water content using a programmed soil water characteristic curve.

The number of future days that use weather data for irrigation forecasting is variable. After the last forecast day, data for a typical season are entered into the data file. If data without rain were placed in the PNUTGRO weather file for the entire season, the crop would show significant stress during each run of PNUTGRO. Although the crop stressing later in the season would not affect the short-term irrigation forecast, other useful results (such as yields) could not be obtained. The program is thus designed to allow other PNUTGRO variables such as yield to be examined as desired using a prior weather file and daily updating.

To make later comparisons between soil water contents based on sensor data and on the crop model, the program weights soil water content and root length per unit area in the defined PNUTGRO soil horizons to zones around the field sensors. The program must do this because PNUTGRO soil horizons will not be at the same depths as zones that sensors represent. Even if the user consistently placed sensors at the midpoint of defined PNUTGRO horizons, those horizons would not be wide enough to accommodate the field sensor's zone of influence. Each field sensor has an assumed zone of influence that is automatically expanded or contracted depending on its proximity to other sensors in the zone. The initial zone of influence was set to 14 cm based on experience and can be changed by the user.

A procedure in the program also performs error checking if the user enters questionable weather data. If total radiation or max/min temperatures are outside a reasonable range, the user is alerted to take corrective action. This may involve checking the sensor or instrumentation. Under these conditions, the weather file is not modified until the problem is fixed. If the value for PAR is questionable, the user is alerted, but the weather file is modified with a value calculated from the total radiation. If the user enters zero rainfall for that day but observed an appreciable amount of rain, he or she is told to check the data collection system or instrumentation. In this case, the weather file is still modified if rainfall was the only questionable input value.

F. Parameter Adjustment Routines

The parameter adjustment algorithms were developed for soil characteristic and root distribution functions based on my knowledge and the experience of two other specialists. PNUTGRO was modified to accept ET data from the field test so differences in sensor- and model-based results could be attributed to different soil-water and rooting characteristics only. Complete program listings and further details on all PNUTGRO Fortran code modifications conducted for this study can be found in Thomson (1990).

1. Modification of Soil Parameters

Soil parameters were modified by changing parameters in SPROFILE.PN2, the soil characteristics input file to PNUTGRO. Thus, no modifications were needed to the PNUTGRO code itself. The Pascal program SENPOS modifies soil parameters as interpreted data from soil water sensors dictate. The soil input file contains soil characteristics data for 29 different soils. When it is run, PNUTGRO automatically picks the proper data set for the soil chosen.

Two specialists initially identified parameters in the soil characteristics file that may be different from actual values for the Millhopper soil used in the study (Smajstrla, 1989) and those that would influence zone water contents significantly (Hoogenboom, 1989). Based on these discussions, my familiarity with the processes involved, and several model runs to determine output sensitivity to changing parameters, evidence was strong that the DUL (drained upper limit) needed adjustment. The DUL is the water content below which drainage no longer occurs out of a soil layer. A DUL of 0.110 cm^3/cm^3 was felt to better represent sands in the area (including the Millhopper soil) than the value of 0.086 cm^3/cm^3 present in the data file for a Millhopper soil. (This value is close to 0.107 cm^3/cm^3, the value for a generic medium sand in the soil data file.)

Although not significant in the high (wet) ranges of modeled water content, the parameter describing the lower limit of water content (LL) can influence modeled water contents in drier ranges that occur most often in the shallow soil zones. For this reason, LL and DUL were adjusted together to approach the values for a medium sand. The discussions that follow focus on DUL adjustment although the adjustment method is the same for LL.

Rules were designed and implemented in the procedure SprofileModif to adjust parameters (see Table 1). The body of the procedure is executed only if the soil is drying. Decisions to adjust parameters are based on a comparison of model- and sensor-based representations of the composite tension in the water regulation zone. Composite tension from the model is determined by averaging the water contents (cm^3/cm^3) in each layer within the calculated regulation zone and converting the result to tension (kPa) using a published soil water characteristic curve for a Millhopper sand (Carlisle et al., 1985). Sensor-based composite tension is determined the same way, although no conversion was necessary because tension was read directly. Only morning readings of soil water tension were used to allow for soil water redistribution in the evening (Thomson and Threadgill, 1987).

Table 1   Knowledge System Rules in SENPOS

Procedure: SprofileModif
Rule
   IF (composite sensor suction) > (composite modeled suction + 10)
   THEN subtract value from DUL in soil characteristics input file.
Rule
   IF (composite modeled suction) < (composite sensor suction + 10)
   THEN add value to DUL in soil characteristics input file.
Remarks: In both cases, DUL is modified by a factor proportional to the
difference in modeled and sensor suction. User is informed of the modification.

Procedure: RootDepth
Rule
   IF (one day increase in sensor reading >5)
   THEN assign RootDepthSensor to sensor depth.
Rule
   IF (one day increase in modeled suction >2)
   THEN assign RootDepthModel to depth of sensor in modeled zone.
Remarks: Consequences are used in procedure WRModif to modify root
weighting factors. User is informed of rooting depths indicated by both the
model and the sensor.

Procedure: WRModif
Rule
   IF (RootDepthSensor) > (RootDepthModel)
   THEN subtract values from root weighting factors below sensor depth.
Rule
   IF (RootDepthSensor) < (RootDepthModel)
   THEN add value to root weighting factors below sensor depth in modeled
      zone.
Remarks: Root weighting factors are modified by a value proportional to the
relative magnitude of root weighting factor.

If the model-based representation of composite tension was
greater (or less) by 10 kPa than the composite sensor tension on any
portion of the drying cycle, a small water amount proportional to the
difference between the two tensions was added to (or subtracted
from) the DUL value in SPROFILE.PN2. This is represented by the
equation

$$C = 0.021(S_m - S_s)/AF \tag{1}$$

where

C = amount of water added to or subtracted from DUL (cm^3/cm^3)

S_m = model-derived composite tension value in root zone (kPa)

S_s = composite sensor tension in root zone (kPa)

AF = adjustment factor (or "gain" term) = 100

The constant (0.021) in the equation is the difference between the desired DUL (0.107) and the DUL in the soils file (0.086).

Using a small tension difference (10 kPa) as the criterion for DUL modification allowed parameter adjustments to occur in the early days of drying so the chance to modify the DUL would not be missed because of a possible rainfall event. It was noticed during the field study that extended periods of no rain were rare. If a larger tension difference were used as criterion for changing the DUL, the entire adjustment of 0.021 could conceivably be made all at once. However, a large tension difference might take several days to occur since tension increases nonlinearly (more slowly under wet conditions) due to the soil characteristic curve. Thus, if a rainfall event were to occur during that period, no adjustment could be made. Small increments were added to or subtracted from the DUL because a small observed tension difference (10 kPa) provided relatively weak evidence that the DUL needed adjustment.

2. Root Weighting Factors

In PNUTGRO, root water uptake is a strong linear function of root length density and contains a very small logarithmic component. Root length density, in turn, is weighted by a root length density factor that is a linear function of the root weighting factor WR(I). WR(I), therefore, can be used to modulate root water uptake. In determining WR(I), PNUTGRO accounts for the genetic capability of different crops or cultivars to proliferate in the root zone. The WR(I) function was derived for a sandy soil from experimental data of Robertson et al. (1980) and is specified for many fine sands in the soil characteristics input file SPROFILE.PN2. As the WR(I) function significantly influences the degree of modeled water uptake in the soil zones, reliable estimates of the function are crucial.

The root function derived as the starting point (before modification) was a composite of the Robertson et al. (1980) curve and an extrapolation (Fig. 4). Examination of soil water sensor data at the lysimeter indicated that water uptake observed using sensors oc-

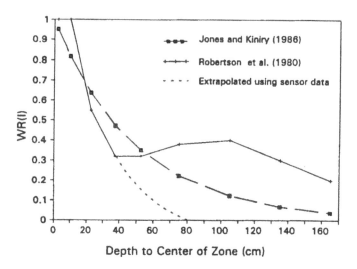

FIGURE 4   Root weighting function WR(I) as a function of depth.

curred at much shallower depths than the original curve indicated. The extrapolated curve was an approximation but was felt to be more consistent with observed sensor responses at the lysimeter installation. Another study by Boote et al. (1989b) also indicated a shallower rooting system for peanuts under similar conditions.

Two sets of rules were needed to modify the root weighting factors. A rule was needed to infer where water use was occurring based on sensor readings. A similar rule inferred where water use was occurring from the model. A resulting depth of water use was determined at the output of each rule. Based on a comparison of where water use was occurring, another set of rules modified the root weighting factor. Table 1 lists the rules described below in procedures RootDepth and WRModif.

The procedure RootDepth looped through all individual sensors in the field starting with the shallowest to determine if water uptake had been detected. If readings over 1 day increased more than 5 kPa, the variable RootDepthSensor was set to the corresponding sensor depth. Similar criteria were used for model outputs. If model-derived tension values increased more than 2 kPa over a 1 day period, the corresponding depth was assigned to the variable RootDepthModel. A greater tolerance was used for the sensor because readings in the field can fluctuate slightly. Human error in reading the sensor and temperature effects exposing small inaccuracies in sensor calibration equations derived for this study (Thomson and Armstrong, 1987) are potential sources of error.

If the sensor-determined rooting depth was unequal to the model-determined rooting depth, the procedure WRModif was activated. The procedure modified root weighting factors so that future model-based representations of water use distributions in sensor zones could converge on sensor-based representations. The equation to modify the root weighting factors was

$$Wr_{new} = Wr \pm F(Wr/Wr_{tot}) \tag{2}$$

where

Wr_{new} = new root weighting factor
Wr = previous root weighting factor
F = adjustment factor (or "gain" term) = 0.075
Wr_{tot} = sum of root weighting factors in root zone

As with modification of the DUL, the gain term was chosen to provide a conservative adjustment. This could be done because the root weighting function was reasonably close to the "sensor-inferred" function initially.

G. Irrigation-Scheduling Routines

Routines in the program ask the user about irrigation management objectives including the composite tension to trigger the irrigation system. The user is also asked what percentage of root mass to replenish with irrigation water and what type of irrigation system is being used. Solid set sprinkler or center pivot systems are the two choices available. Management differs greatly between the two systems, as center pivots require an early start-up so the crop in the last part of the field irrigated is not water-stressed.

An important calculation determines the water regulation zone. The user-entered value of percentage replenished root mass and model outputs of root length in each zone determine the depth of this zone. All composite water potentials calculated from the PNUTGRO crop model used to trigger irrigation are calculated within this regulation zone. Irrigation is also applied to the bottom of this zone. The user receives feedback from the program regarding suitability of sensor placements (if enough were placed in the zone) and is told to what depth the zone will be replenished corresponding to the entered value of percentage root mass.

A program procedure determines a projected irrigation date and calculates the amount of water to apply to the field. The projected irrigation date is based on model outputs of soil water content in each soil zone several days ahead and depends on the type of irrigation system. Soil water contents are converted to tensions using a published soil water characteristic curve [in this case, for a Millhopper sand (Carlisle et al., 1985)]. A composite tension in the regulation zone is calculated and compared to the user-entered value. When the user-entered trigger level is reached or exceeded, irrigation is recommended. The system assumes no rainfall as a worst-case condition. If rainfall occurs, the user-entered rainfall updates the prediction.

The amount of water to apply is based on the difference between the water content corresponding to the projected tension on that date and the soil water content at field capacity in the zone of water regulation. The soil water content at field capacity is assigned the value of drained upper limit (DUL) in the first five soil horizons listed in the soil characteristics file for the soil already hard-coded into the soil characteristics file. This soil characteristics file, SPROFILE.PN2, is an input file to PNUTGRO. The DUL can change as the knowledge base is run, as it is one of the parameters altered by sensor-based feedback. Procedures were written to provide water amount corrections for the next irrigation cycle. The deepest sensor to respond within 1 day of irrigation indicates the depth of water penetration. If the depth of penetration is different than the calculated zone of regulation, the procedure AdjustIrrig adds or subtracts a small amount (2.5 mm) of water to be applied to the next irrigation cycle. Since this adjusted amount must be used for later activations of the program, it is stored to and accessed from a permanent file.

A procedure, DayToIrrigate, provides different management for cases where the system applies water uniformly with time (as in a solid set system) or nonuniformly (as in a center pivot system). For a solid set system, it is assumed that the entire field is irrigated at once. There is, therefore, no lead time required to apply water. Center pivot systems provide added complexity if a model-based prediction is used to schedule irrigation. In DayToIrrigate, the user is asked to enter the amount of water the system applies and the time to traverse the entire field at the 100% travel. This information is used to determine the lead time needed to start the system so the crop is not stressed at the end of the cycle.

To apply the center pivot scheduling algorithm properly, some assumptions were made. First, the model-based predictions should represent conditions in the first part of the field irrigated. Likewise, all soil water sensors should be confined to the first part of the field.

Drying occurs after a saturating rain where field soil moisture conditions are assumed to be uniform. Nonuniform soil moisture conditions occur after the first irrigation cycle, so sensor readings cannot be averaged if the sensors were widely spaced.

For the present strategy, a primary requirement is that the system arrive at the last part of the field on time so the crop does not stress. If calculations show that the center pivot cannot apply the required amount of water in time, a lesser amount is applied proportional to the system capacity and the time required to reach the end of the field. The user is told how much water should be applied, how much will actually be applied, and the percentage timer setting required to run the system.

III. RESULTS AND DISCUSSION

A. Tests of the Parameter Adjustment Routines

To test the parameter adjustment methods of SENPOS, daily PNUTGRO runs using the IREP weather data were made. The purpose of the test was to determine if model-based representations of soil water potential converged on sensor readings as SENPOS adjusted the three selected model input parameters simultaneously. The test also verified proper operation of each parameter adjustment routine.

1. Soil Characteristic and Root Distribution Modifications

Drying periods July 26–30 and August 19–23 were selected to illustrate convergence of soil water potentials. To facilitate the test, shortcuts were taken to enter data quickly using a screen editor instead of running the system in the "user mode" (illustrated in Fig. 3).

Data from a field test weather file were imported by PNUTGRO. Soil water sensors began to show drying on July 27, Julian day 207 after a rain event. This was consistent with soil water drying trends noted in the PNUTGRO output file. PNUTGRO's own calculations of soil water redistribution after a rain event set the initial conditions. Root weighting factors were set to values derived from examination of soil water sensor data for the entire season. This is represented by the extrapolated curve of Fig. 4 and served as the initial "best guess" rooting function.

Soil water sensor data from the north lysimeter for July 27 and 28 were manually entered into a data file input to SENPOS. A weak drying response was noticed from two shallow sensors placed in the south lysimeter. This was probably due to the sensors not being close enough to the active roots. One sensor also indicated a rapid wetting response compared to the others, indicating that applied water was probably reaching the sensor through fissures created at installation. Thus, sensors in the south lysimeter had to be relocated due to improper installation and were not used for the July runs.

SENPOS was then run on July 28 (day 210). Both model- and sensor-derived values of soil water content and soil water tension in each sensor zone were recorded. After the July 28 run, PNUTGRO was run again with any parameters that might have been changed by SENPOS. During each PNUTGRO consultation, soil characteristics that are used in the simulation are displayed to the user. The user can see if any parameters have been changed by SENPOS. For this study, the parameters DUL and WR(I) in each soil zone were recorded before each new run of PNUTGRO. SENPOS and PNUTGRO were run alternately in this manner until July 31.

Other runs were implemented between August 19 and 23 (days 232–236). The first run of SENPOS was on August 21 (day 234) using sensor data for August 20 and 21. Parameters that were adjusted at the end of the July runs served as a starting point for the August 21 run. The August runs of SENPOS and PNUTGRO were carried out the same as the July runs except that sensor readings were averaged for the two lysimeters.

Figure 5 illustrates that the DUL for the top five soil zones (down to 60 cm) increased with each run of SENPOS until day 235. Composite tension values determined by the model were greater than composite sensor readings, necessitating a daily change in DUL values. Figure 6 indicates that the model-based tension began to converge on the sensor-based tension in July and continued convergence in August. The results of Fig. 6 represent the combined effects of changing root weighting factors [WR(I)] as well as the DUL and LL values in a 24 cm zone of water regulation. For this study, the water regulation zone corresponded to about 70% of the root mass (or root length assuming a linear relationship) as recommended by Richards and Marsh (1961) and Taylor and Ashcroft (1972). A procedure in SENPOS determined 28 cm to be the water regulation zone corresponding to 70% of root mass. A 24 cm zone was calculated in a SENPOS procedure to be the closest discrete point to 28 cm at the edge of the 20 cm sensor's zone of influence. (A SENPOS procedure calculated a 12 cm zone of influence around each sensor placed 8 or 20 cm deep.)

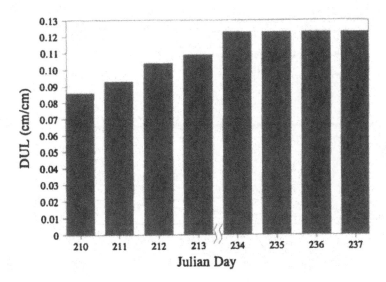

FIGURE 5   Drained upper limit (DUL) value of the top five soil zones as modified by parameter adjustment routine.

Figure 7 illustrates modifications to the root weighting factors, WR(I), in July (days 210–214) and August (days 235–238). The July plot (Fig. 7a) shows that weighting factors were reduced on day 212 in the 15–30 cm zone and were progressively reduced below 30 cm during the entire period. Progressive reductions below 30 cm did not influence water use calculations in the 24 cm zone of regulation because reductions in the lower zones were not added proportionately to the upper zones. The August plot (Fig. 7b) shows that WR(I) increased for day 235 in zones below 30 cm. The factors were reduced at shallower depths in August than any reductions that occurred in July. WR(I) reductions in this zone would influence model-based water use calculations in the 24 cm regulation zone.

Figure 8 illustrates how sensitive the output tensions (converted from PNUTGRO water content using the characteristic curve) are to changes in the root functions. The soil DUL and LL were set to 0.107 cm³/cm³ and 0.035 cm³/cm³, respectively (the desired convergence values), to create these graphs. The modified root function yields tension values that are closer to sensor representations in both cases before the adjustment algorithms further modify the function (as shown in Fig. 7). Of course, the user may not have the luxury of approximating the root function with field data. If the user were to

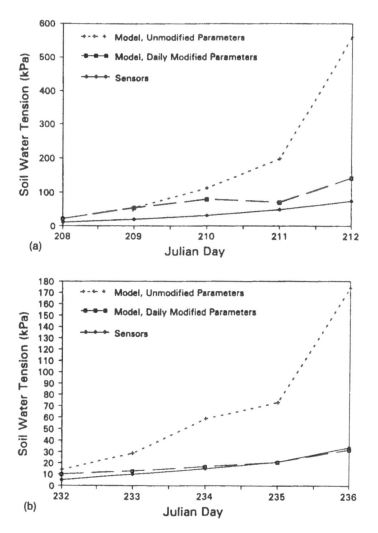

FIGURE 6 Composite soil water tension results from sensors; model using unmodified DUL, LL, and WR(I); and model using daily modified DUL, LL, and WR(I) for (a) July 26–30 and (b) August 19–23 in 24 cm zone of regulation.

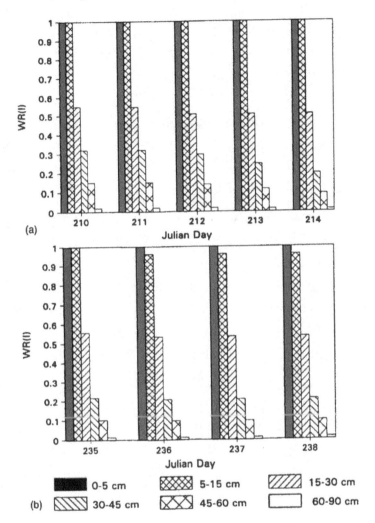

FIGURE 7   Root weighting factors in several soil zones modified by the parameter adjustment routine for (a) July and (b) August.

use the empirical function described by Jones and Kiniry (1986) as a first approximation, convergence represented in Fig. 6 could still be established in a short time. However, if the user were to run the model with the rooting function derived from Robertson et al. (1980), appropriate rules would be needed to alter the overall function rap-

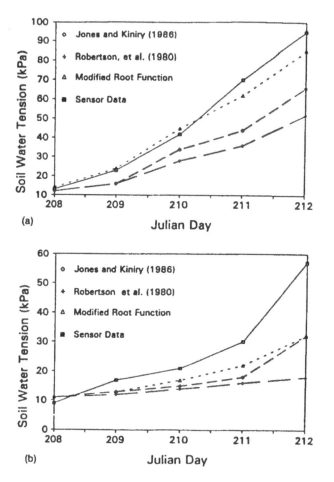

FIGURE 8 Comparison of modeled soil water tension using three root functions to sensor tension at (a) 8 cm depth and (b) 20 cm depth.

idly. Rapid adjustments such as this could cause oscillation as the root function attempts to "close in" on the correct value. In the absence of other rooting data for peanuts, it is suggested that a user running this knowledge system (or PNUTGRO alone) use the modified root function described herein. For other models following the IBSNAT (1986) format, the empirical function described by Jones and Kiniry (1986) should be used as a first approximation in the absence of reliable crop- or soil-specific data.

B. Field Test of Irrigation Scheduler

The irrigation scheduling algorithms were tested using data at the same lysimeter site that was used to evaluate the parameter adjustment algorithms. The lysimeters were irrigated using sensor readings as these algorithms had yet to be finalized. To test the irrigation-scheduling portion of the decision support system, the system was run using weather data from the lysimeter test and finalized algorithms. The following illustrates decisions made by the system regarding irrigation.

A test of the irrigation-scheduling portion of the decision support system was performed over one season on irrigated peanuts grown in two lysimeters. Rainfall, irrigation, soil water sensor potentials, and actual ET were measured. Figures 9–11 show soil water tension data from sensors for the growing season at the 8, 20, and 40 cm depths, respectively. The first major drying response that could indicate pending irrigation began on Julian day 210 as seen in the figures for the north lysimeter. A composite tension value for these three depths was calculated by the program. The composite sensor value was not great enough to warrant irrigation, and the lysimeters were not irrigated, as illustrated in Fig. 12b.

Although irrigation did not take place (Fig. 12), a model run on day 210 indicated that irrigation was needed because its calculated composite tension value was greater than a user-entered trigger value of 60 kPa. The model indicated that 1.31 cm of water was needed to replenish a water regulation zone of 20 cm. This 20 cm regulation zone was also calculated in the program. At this point, model-based representations of soil water tension were beginning to converge on sensor readings as a key soil parameter (drained upper limit or DUL) and root weighting factors were being adjusted, but convergence was not established until the next significant drying cycle (occurring in August). During July, irrigation was called for on days 210 and 212. Thus, the conservative parameter adjustments implemented could not permit the model to be used reliably by itself this early in the season. Based on convergence occurring in August, however, the model could be used reliably once it had been "trained" by the sensors during the "second round" of adjustments.

Drying cycles of a reasonable duration (necessary to activate the correction algorithms) were few and far between as indicated by a lengthy period (20 days) between the July drying cycle and the August drying cycle. As has been mentioned, convergence was related to the degree of adjustment allowed (or a "gain" term). Conservative

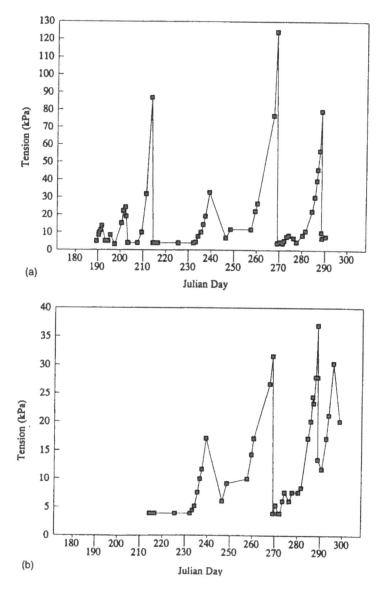

FIGURE 9   Soil water tension data at 8 cm depth for Watermark sensors in
(a) north lysimeter and (b) south lysimeter.

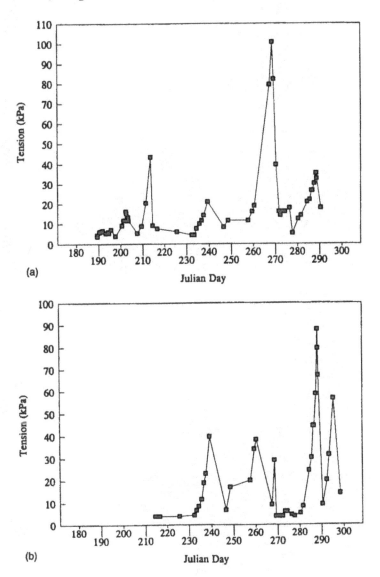

(a)

(b)

FIGURE 10   Soil water tension data at 20 cm depth for Watermark sensors
in (a) north lysimeter and (b) south lysimeter.

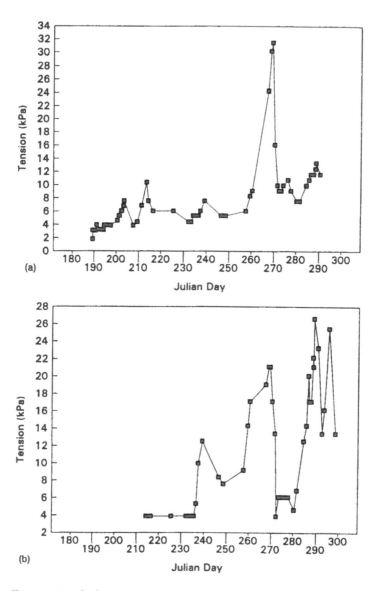

FIGURE 11   Soil water tension data at 40 cm depth for Watermark sensors
in (a) north lysimeter and (b) south lysimeter.

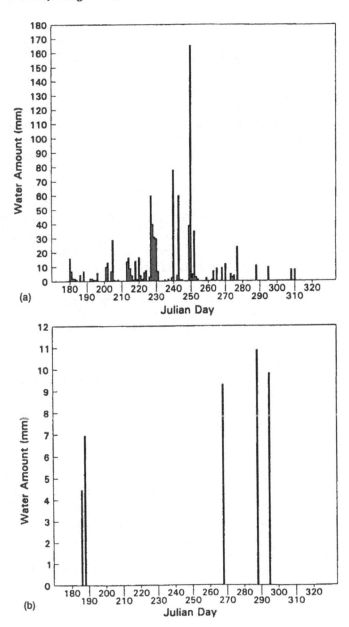

FIGURE 12 Water applications (a) from rainfall and (b) from supplemental irrigation.

adjustments have to be made under conditions like these when evidence is not very strong that a particular parameter needs adjustment. Running the system in a drier region could yield many more drying cycles, allowing full corrections to occur early. This would allow even more conservative adjustments than were already effected. A system that would work better under drier conditions is desirable because dry regions are, of course, more critical with regard to irrigation. For this study, it is fortunate that adjustments were made because model-based predictions left uncorrected would have signaled many unnecessary irrigations.

Environmental conditions and characteristics of the soil at the lysimeter site provided a stringent test of the method as evidenced by the next irrigation event at day 268. On this day, 0.931 cm of water was applied to the lysimeters based on sensor readings and a composite trigger level of 60 kPa (Fig. 12b). There was some question as to whether irrigation should be delayed because of forecast rainfall. For this reason, it was decided that the amount of water applied should be a little less than the amount needed to replenish a full 40 cm water regulation zone. (The sensors indicated significant water uptake at the 40 cm depth as seen in Fig. 11.) Therefore, the amount of water applied was calculated to be the amount needed to refill a 20 cm zone to field capacity (or the drained upper limit). This amount could allow for additional recharge by rainfall. Irrigation was decided on because sensor tension levels were climbing very rapidly due to the low water retention characteristics of the Millhopper sand. Indeed, enough rain to replenish the zone fell within a couple of days after irrigation commenced.

A run of the model also indicated that irrigation was warranted on day 268. This late in the season, parameters had long since settled on their "static" values, and model-based soil water tension was apparently tracking the sensors quite well. However, the amount of water to apply was calculated to be 2.5 cm, an amount required to replenish a zone of 40 cm. The model does not presently account for prevailing weather conditions and thus does not allow a deficit irrigation strategy. The user would need to override the calculated water amount if less water were desired.

C. Evapotranspiration Considerations

Although present parameter adjustment routines adjust only soil water parameters and the rooting function, evapotranspiration (ET) should be considered because inaccuracies in modeling crop water use can also influence results. Many trends can be observed by comparing model-based ET with measured ET.

Figure 13a compares simulated and measured ET rates over the season. By examining trends over a shorter time scale, more detailed observations can be made. Figure 13b illustrates an ET comparison over the period August 12–29 (days 225–242). Data for days 231–237 represent a drying cycle immediately after a high intensity rain, and ET measured during this period should be close to potential ET rates. During this period, measured ET was greater than simulated ET by an average of 28%. This might be partially explained by sensible heat advection occurring over exposed lysimeters. The quantity LE/Rn (latent heat of vaporization of water times ET rate divided by net radiation) can be used to ascertain sensible heat advection (Verma and Rosenberg, 1977). If LE/Rn > 1.0, sensible heat consumption is occurring at the soil surface. Over a sunny period of days 230–237, LE/Rn averaged 1.47. It showed a low of 1.09 on day 230 and a peak of 1.85 on day 236. Measured ET was significantly higher than simulated ET during periods of sensible heat consumption.

All days where simulated ET was greater than measured ET were characterized by low radiation loads (with or without rain). On days 226 and 238, for example, weather was cloudy with no precipitation. PNUTGRO used a form of the Priestley and Taylor (1972) ET model, which included an advection modification based on air temperature (Ritchie, 1985; Jones and Kiniry, 1986). Ritchie (1985) used a multiplier of 1.1 on the ET equation to account for the effects of unsaturated air and increased the multiplier to allow for advection when the maximum temperature was greater than 24°C. In PNUT-GRO, however, the multiplier was increased when the maximum temperature exceeded 34°C. Air temperatures on days 226 and 238 were well below this threshold, so potential ET on these days was simulated by the unmodified Priestley–Taylor equation. A study by Steiner et al. (1991) indicated that the unmodified Priestley–Taylor model under predicted ET at low ET rates and overpredicted ET during periods of high evaporative demand. The authors suggested that a more responsive advection function could improve predictions at both ends of the ET spectrum.

If, in the future, rules are developed to modify the ET model, methods should be applied separately to evaporation and transpiration components. This would be very important in the early stages of crop development because the evaporation component is a large part of the total ET. Although sensor-based feedback would be difficult to implement in the earliest parts of the season, information

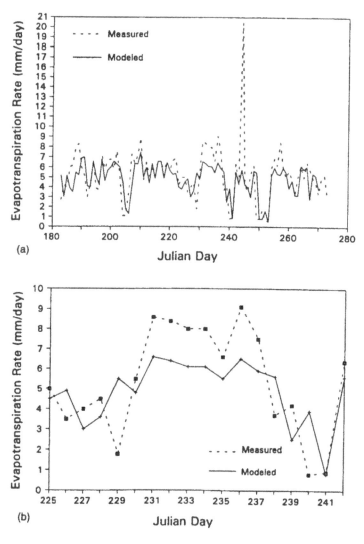

FIGURE 13   Comparison of measured and modeled ET rates (a) over peanut growing season and (b) from August 12 to August 29.

could soon be obtained as roots begin to proliferate. Soil water sensors cannot provide reliable information on soil evaporation in the shallowest soil horizons, but they can provide information on the combined evaporation and transpiration in horizons where roots proliferate.

IV. SUMMARY AND CONCLUSIONS FOR CASE STUDY

A sensor-based knowledge system was developed to adjust soil water and rooting parameters of a crop model. The objective of this study was to provide a better representation of temporal soil water conditions for better water and chemical management and irrigation scheduling. Less reliance on sensors could be achieved as the model makes better estimates. An expert system to evaluate the suitability of sensor readings used to calibrate PNUTGRO was coded in a commercial expert system shell, VP-Expert. Pascal routines adjusted drained upper limit (DUL), lower limit of water content (LL), and root weighting factors [WR(I)] that were input values to the PNUTGRO crop model. Sets of field data from July and August 1988 verified that model-based representations of temporal soil water status converged on sensor-based representations as parameters reached their new static values. It is envisioned that the approaches and procedures outlined in this chapter can be applied widely to systems that might benefit from sensor feedback to calibrate model input parameters that are difficult to characterize. Other potential applications are discussed in the next section.

Expert system rules that determined whether a sensor reading was valid compared a sensor's reading with readings from other sensors placed at the same depth. This did not consider the chance of the entire system drifting. It is conceivable that the entire system could be off if all sensors drift significantly off calibration, for example. Although significant drift was not noticed over a year of field installation, periodic calibration checks are recommended.

An unexpected sensor response to wetting was noticed during the lysimeter study resulting from a rain event. Figure 14 illustrates that one sensor (20/0, lysimeter 2) showed complete rewetting, although several others at the same depth showed no such response. In this case, there was evidence that the sensor might have been placed improperly, creating a fissure that established a direct line for water to saturate the sensor. Although cases like this may be infrequent, decision rules should be developed to handle them.

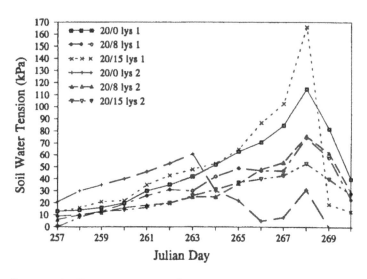

FIGURE 14   Sensor responses due to a 0.4 cm rain event (at day 263).

Additional rules could be developed to elaborate on the characteristics of different sensor types. For example, my experience with Watermark sensors indicates that the sensors themselves usually do not "fail" in the wet range of operation unless wires break or fray. If one of these sensors does not show water use from drying and other sensors at the same depth do indicate water use, for example, it is much more likely that the sensor needs to be relocated to the zone of active roots. These sensors show a very gradual degradation in the wet range of operation characterized by readings that do not "return to zero" when saturated (Pogue, 1991). Additional decision rules could be written to accommodate these cases.

It was also mentioned that tensiometers can "break tension" in the dry ranges of operation (approximately 80 kPa). As coded, there are no rules that presently accommodate this situation. Breaking tension would be an infrequent problem when scheduling irrigation in a field with sandy soil. Usually, a crop is never allowed to approach this stress level because a sandy soil holds very little water at this tension level. It is thus common practice to irrigate much sooner than 80 kPa. By contrast, data from a field study under corn in Shenandoah County, Virginia (Fig. 15) showed deep rooting and tension levels frequently exceeding 80 kPa in some horizons before irrigation was called for. Figure 15a indicates when tensiometers broke tension and

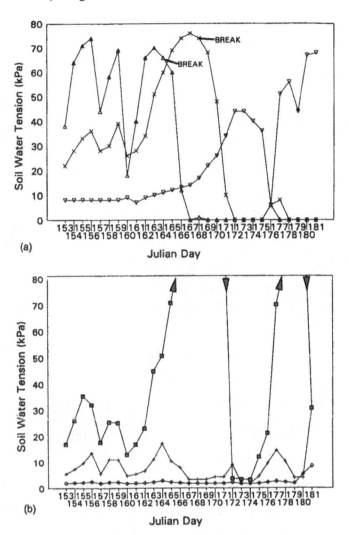

FIGURE 15 Soil water sensor responses under corn in Shenandoah County, VA (June 1993). (a) Tensiometer responses; (b) Watermark sensor responses. (△, ■ = 15 cm depth, ✕, + = 30 cm depth, and ▽, ◇ = 60 cm depth.)

required recharging. The soil in Shenandoah County was a Wolfgap loam. Although the 15 cm reading would be disregarded later in the season, uptake was pronounced at the 60 cm depth quite early in the season. This contrasts with past observations under corn in a Lakeland sand (Tift County, Georgia), for example. Under these conditions, I have not seen significant water uptake (as registered by sensors) at the 60 cm depth (Thomson, 1983). This was not seen to be a hard pan problem. Corn in the sandy soil required frequent irrigation, and roots were thus most prominent in the shallow 20–45 cm depths.

Figure 15b is a plot of Watermark sensor readings in the same row. These readings were taken with the Watermark 30KTC meter and converted to resistance using an equation found in Thomson (1990). Readings were further compensated for temperature and converted to tensions (up to 100 kPa only) using the relationship of Thomson and Armstrong (1987). It is interesting to note temporal differences between tensiometer and Watermark readings as shown in the figures. Tensiometers seemed more responsive than Watermark sensors to wetting and drying. A major drawback, however, was that they required periodic service as indicated on the graphs. For this reason, the farm manager taking the data was biased in favor of the Watermark sensor (Wilkins, 1993). Watermark sensors at the 30 and 60 cm depths seemed rather unresponsive this early in the season (June), which, initially, might indicate improper placement. Later in the season, however, sensors at these locations began to register good drying. These comparisons show clearly that further study is needed to better quantify sensor response characteristics. A well-tuned expert system could then do a better job interpreting these responses.

Based on observations noted, expert system rules for sensor selection might also be desirable. It is clear from the data of Fig. 15, for example, that one type of sensor might be better suited than another for reasons already mentioned. Another sensor that has not been mentioned is the gypsum block (Delmhorst, 1993). This sensor works on the same resistance principle as Watermark sensors and gives readings up to 1500 kPa tension. Although conventional crop management would not allow this stress level to be approached, resolution is said to be good from about 30 to 150 kPa, an acceptable management range for crops grown in medium and fine textured soils. These sensors need to be replaced yearly because they dissolve, but they are the least expensive of the three sensor types mentioned. The Watermark sensors can be reused, but removing resistance-type sensors from dense, compacted soils can be an arduous task. For this reason, a farm manager may opt to leave them in the soil and install new ones the next year.

PNUTGRO was modified to accept daily ET as input from the field study so differences in sensor- and model-based results could be attributed to different soil water and rooting characteristics only. For actual application, ET cannot be an input, and analysis presented in Thomson (1990) and herein indicates that the model tracked measured ET very well. For this study, modeled ET could have been used with little influence on the results.

Care must be taken when adjusting parameters that influence the same variable. The developer should be keenly familiar with the processes involved and how variables interact. Many model runs and sensitivity studies can increase the developer's knowledge and thus allow intelligent examination of variable interactions. For this study, conservative adjustments to parameters were made to achieve the desired goal. By analogy with control system theory, the output reflected somewhat of an "overdamped" response. This type of response was desired because, as indicated for the DUL, evidence was not strong enough to warrant large adjustments. In general, a conservative approach is probably the best course to follow until the knowledge base grows and more evidence (extracted and interpreted from field data) is presented. I believe that the system was quite stable with the adjustment factors used.

A key factor in acceptance of this method is how universally applicable it may be. Soil and rooting characteristics, for example, vary widely with the soil and crop. Running this system on fields with a few more soil–crop combinations would further strengthen the knowledge base. Other soil characteristics that were not thought to be significant here might require consideration. A major advantage is that once a robust knowledge base is constructed it can be self-calibrating. That is, many exhaustive field trials should not be needed because the rules should "know" what trends to look for. Machine learning could be used to ensure that the proper parameter was adjusted. If, for example, an inappropriate adjustment was made, later trends in data might give a clue that the adjustment was wrong. If an adjustment to the correct parameter were then made and subsequent water status data showed more consistent tracking, the system could recall this and apply the correct adjustment later if similar environmental conditions were encountered. Self-learning systems that use neural networks might be used for the parameter adjustment portion of this work and have shown promise in estimating model outputs with limited input data (Altendorf et al., 1992).

Except for early in the season, the system worked well as an irrigation scheduler under the conditions tested. Early in the season, parameter adjustments had just begun and the model indicated drier conditions than the sensors indicated. For this reason, the model called for irrigation when none was warranted. To effect parameter adjustments, this system requires drying cycles of sufficient duration that comparisons can be made between sensor tensions and model-based tensions. Based on the magnitude of the difference, selected parameters are adjusted with the objective of correcting the simulated soil water status. As has been stated, there were few drying cycles of sufficient magnitude at the lysimeter plots. By contrast, sensor readings during the summer of 1993 at Shenandoah County, Virginia indicated many drying cycles. Drying cycles of sufficient duration occurring early would allow parameters to adjust and stabilize early to their final static values. Thus, this system should work better under a drier climate.

V. OTHER POTENTIAL APPLICATIONS

As has been indicated in Thomson et al. (1993), this methodology could find its use in many applications. One application could be in grain or peanut drying. The following is an untried concept that could apply the methods of the irrigation problem to the grain drying problem.

In grain drying, especially in slower, more efficient in-bin drying, there is a problem in the measurement of moisture content of the grain at different levels in the grain mass, which may be as deep as 10 or 12 ft (3 or 4 m). Grain drying models have been developed that can accurately estimate the grain moisture content throughout the grain mass, given an initial grain moisture reading, airflow rate, temperature, and dew point of the entering airflow. The initial grain moisture can be measured accurately as the grain is placed in the bin, and the input air conditions can be measured accurately over time as the drying progresses. Airflow is difficult to measure accurately, however, and in addition it is felt in the industry that airflow may change as the grain shrinks during drying. So the question is: Can an initial estimate of airflow rate be adjusted during drying by an automatic expert system, as in the irrigation problem, so that grain moisture can be accurately estimated by a grain drying simulator? The requirements for solving this problem in a way analogous to the irrigation problem would require a method of sensing the grain moisture content throughout the bin, and this is manually difficult, as it is necessary to probe 1–4 m into wet grain. However, a similar approach could be taken using temperature along the depth of the grain as the measured parameter, as a thin probe containing many thermocouples for temperature measurement would not be difficult to install.

Then the drying simulation, using the initial estimate of airflow rate, could be run, and the temperatures at several depths would be compared with the measured values. Careful analysis of this problem could result in rules adjusting the assumed airflow rate to bring the simulated temperatures closer to the measured temperatures as the drying continues. Since the conditions of the ambient air going into the fan and bin are dynamically changing, the heat and mass transfer modeling in the drying simulation must be validated before confidence can be placed in this adjustment technique. However, several grain drying models exist, and this technique could allow the use of simulation to predict the actual drying situation in a bin. This would allow adjustments, such as turning the airflow off during periods when no drying is occurring, to be made by an expert system to reduce cost and improve the quality of the product.

The idea of using an expert system to implement the concept of adaptive control or machine self-learning can be realized by methods given in this chapter. Simulations of processes such as crop growth, evapotranspiration, and grain drying based on physical and biological principles can be extremely useful, especially if they can be tuned or adjusted to fit the particular environment in which they are being used. These environments, which change with location, include the water-holding properties of the soil in the irrigation problem and the properties of the grain mass that affect airflow in the grain drying problem. It is hoped that these principles can be applied by other researchers to improve the usefulness of agricultural system simulations.

REFERENCES

Altendorf, C. T., M. L. Stone, R. L. Elliot, and M. A. Kizer. 1992. Determining soil moisture using soil thermal properties. Paper No. 92-3020. St. Joseph, MI: American Society of Agricultural Engineers.

Boote, K. J., J. W. Jones, G. Hoogenboom, G. G. Wilkerson, and S. S. Jagtap. 1989a. PNUTGRO V1.02. IBSNAT Version. *User's Guide*. Gainesville, FL: University of Florida.

Boote, K. J., J. M. Bennett, J. W. Jones, and H. E. Jowers. 1989b. On-farm testing of peanut and soybean models in north Florida. Paper No. 89-4040. St. Joseph, MI: American/Canadian Societies of Agricultural Engineers.

Butts, C. L. 1988. Modeling the evaporation and temperature distribution of a soil profile. Ph.D. Dissertation, Agricultural Engineering Department, University of Florida, Gainesville.

Carlisle, V. W., M. E. Collins, F. Sodek III, and L. C. Hammond. 1985. Characterization Data for Selected Florida Soils. Soil Science Report No. 85-1. Gainesville, FL: University of Florida, Institute of Food and Agricultural Sciences, Soil Science Department Laboratory, p. 169.

Delmhorst. 1993. Application literature for gypsum soil blocks. Towaco, NJ: Delmhorst Instrument Company.

Hoogenboom, G. 1989. Personal communication. Gainesville, FL: Department of Agricultural Engineering, University of Florida.

IBSNAT. 1986. *Decision Support System for Agrotechnology Transfer (DSSAT).* Documentation for IBSNAT crop model input and output files. Version 1.0. Technical Report 5. Honolulu, HI: University of Hawaii, College of Tropical Agriculture and Human Resources.

Irrometer. 1989. Application literature for Watermark Model 200 soil moisture sensor. Riverside, CA: Irrometer Co.

Jones, C. A., and J. R. Kiniry (Eds.). 1986. *CERES-MAIZE: A Simulation Model of Maize Growth and Development.* College Station, TX: Texas A&M University Press.

Jones, J. W., K. J. Boote, S. S. Jagtap, G. Hoogenboom, and G. G. Wilkerson. 1988. SOYGRO V5.41. IBSNAT Version. *User's Guide.* Gainesville, FL: University of Florida.

Pogue, W. 1991. Personal communication. Riverside, CA: Irrometer Company.

Priestley, C. H. B., and R. J. Taylor. 1972. On the assessment of surface heat and evaporation using large-scale parameters. *Monthly Weather Rev. 100*(2): 81–92.

Richards, S. J., and A. W. Marsh. 1961. Irrigation based on soil suction measurements. *Soil Sci. Soc. Am. Proc. 25*(1):65–69.

Ritchie, J. T. 1985. A user oriented model of the soil water balance in wheat. In: *Wheat Growth and Modeling.* E. Fry and T. K. Atkin (Eds.), NATO-ASI Series. New York: Plenum, pp. 293–305.

Robertson, W. K., L. C. Hammond, J. T. Johnson, and K. J. Boote. 1980. Effects of plant-water stress on root distribution of corn, soybeans, and peanuts in a sandy soil. *Agron. J.* 72:548–550.

Ross, B. B. 1984. Irrigation scheduling with a farm computer. Microcomputer conference and trade show, Mar. 19–21. Blacksburg, VA: Virginia Polytechnic Institute and State University, pp. 23–32.

Smajstrla, A. G. 1985. Design and management of drip irrigation systems for tomatoes. Agric. Eng. Extension Mimeo Report 85-13. Gainesville, FL: Florida Cooperative Extension Service, IFAS, University of Florida.

Smajstrla, A. G. 1989. Personal communication. Gainesville, FL: Department of Agricultural Engineering, University of Florida.

Steiner, J. L., T. A. Howell, and A. D. Schneider. 1991. Lysimetric evaluation of daily potential evapotranspiration models for grain sorghum. *Agron. J.* 83(1):240–247.

Taylor, S. A., and G. L. Ashcroft. 1972. *Physical Edaphology: The Physics of Irrigated and Non-Irrigated Soils.* San Francisco, CA: W. H. Freeman.

Thomasson, J. A. 1993. Foreign matter effects on cotton color measurement: determination and correction. *Trans. ASAE* 36(3):663–669.

Thomson, S. J. 1983. Unpublished tensiometer and Watermark response data; water management under center pivot irrigation. Tifton, GA: Coastal Plain Experiment Station.

Thomson, S. J. 1990. Knowledge system for determining soil water status using sensor feedback. Ph.D. dissertation, University of Florida, Agricultural Engineering Dept., Gainesville, FL. Ann Arbor, MI: University Microfilms, Intl.

Thomson, S. J., and C. F. Armstrong. 1987. Calibration of the Watermark Model 200 soil moisture. *Appl. Eng. Agric.* 3(2):186–189.

Thomson, S. J., and E. D. Threadgill. 1987. Microcomputer control for soil moisture based scheduling of center pivot irrigation systems. *Comput. Electron. Agric.* 1(4):321–338.

Thomson, S. J., J. W. Mishoe, R. M. Peart, and F. S. Zazueta. 1989. IEEE-488 Interface and Instruments for High Resolution Measurement of Data. Paper No. 89-3560. St. Joseph, MI: American Society of Agricultural Engineers.

Thomson, S. J., R. M. Peart, and J. W. Mishoe. 1993. Parameter adjustment to a crop model using a sensor-based decision support system. *Trans. ASAE* 36(1):205–213.

Verma, S. B., and N. J. Rosenberg. 1977. The Brown–Rosenberg resistance model of crop evapotranspiration—modified tests in an irrigated sorghum field. *Agron. J.* 69:332–335.

Wilkerson, G. G., J. W. Jones, K. J. Boote, K. T. Ingram, and J. W. Mishoe. 1983. Modeling soybean growth for crop management. *Trans. ASAE* 26:63–73.

Wilkins, C. 1993. Personal communication. Wilkins' farm, Shenandoah County, VA.

7

Evaporation Models

John L. Monteith

Institute of Terrestrial Ecology, Penicuik, Midlothian, Scotland

I. HISTORICAL BACKGROUND

Although the evaporation of water from land surfaces and from vegetation is a process of profound importance for agriculture, ecology, and geophysics, mechanisms governing the process remained obscure until relatively late in the history of science. In 1867, G. J. Symons described evaporation as "the most desperate art of the desperate science of meteorology," and it was an art that made little progress until well into the 20th century.

Reviewing the partitioning of available energy at the earth's surface, Brunt (1939) wrote: "A fraction, *whose magnitude is not readily estimated* [italics mine], is used in evaporating water from the surface of grass, leaves, etc. and in high latitudes a portion is used in melting ice and snow." A few years later, however, the first measurements of fluxes of water vapor over uniform vegetation by Thornthwaite and Holzman (1939) signaled the start of major advances in both experimental and theoretical aspects of evaporation science. Progress was then interrupted by World War II, and it was not until 1948 that the publication of three seminal papers initiated a field of research that is still expanding 50 years later.

At Rothamsted, England, Penman (1948) used a set of small ly-simeters, installed many years previously, to measure evaporation from open water, bare soil, and grass. By combining a thermodynamic equation for surface heat balance and an aerodynamic equation for vapor transfer, he derived a general formula for the rate of evaporation from open water as a function of climatic elements (temperature, vapor pressure, wind, and radiation) measured at screen height. The key to this analysis was the elimination of surface temperature (a rarely measured and often unmeasurable quantity) which was achieved by assuming that saturation vapor pressure was a linear function of temperature over small ranges. If Penman's equation had been published a few decades later it would almost certainly have been known today as Penman's "model"!

Working independently in the USSR, Budyko (1948) used the same primary equations as Penman but estimated surface temperature by a process of iteration, thereby deriving estimates of evaporation more accurately than Penman but more laboriously. (This 1948 monograph was an English translation of the original Russian text that predated Penman's classic paper.)

Thornthwaite (1948) proposed a completely different solution to the problem of estimating evaporation from watersheds in the United States, given minimal climate records. Like Penman, he argued that evaporation was driven primarily by energy available from radiation, but, for the practical reason that very few measurements of radiation were available, he used air temperature as a surrogate variable. This argument is still valid in some circumstances, and a recent prediction of global evaporation by Mintz and Walker (1993) used Thornthwaite's method in preference to Penman's.

This review of evaporation model opens with a brief description of Thornthwaite's formula but deals mainly with developments from the theoretical base established by Budyko and Penman. One major concept emerging from their work was the distinction between "potential" rates of evaporation from vegetation or soil obtained when the supply of water was abundant and "actual" rates prevailing when the supply was restricted.

A. Thornthwaite's Empirical Formula

After examining a large number of hydrological records for climatically contrasting sites in the United States, Thornthwaite concluded that the potential evaporation in any month could be correlated with the mean monthly temperature $T(°C)$ by using a scaling factor I defined as

$$I = \sum (T/5)^{1.514} \tag{1}$$

where T is mean monthly temperature and the summation is over 12 months. The scaling factor was then used to define an exponent a such that

$$E = 16(10T/I)^a \tag{2}$$

where

$$a = (0.675I^3 - 77.1I^2 + 17{,}920I + 492{,}390) \times 10^6 \tag{3}$$

This set of equations contains eight numerical constants, all determined empirically and some quoted with unwarranted precision. As needed, the constants were modified to allow for the dependence of daylength on latitude and for monthly temperatures outside the range 0–26.5°C.

In a well-balanced critique of Thornthwaite's formula, Pelton et al. (1960) concluded that it was most reliable for periods of a month or longer because, over a year and at many sites, monthly mean temperature and potential transpiration were strongly correlated. However,

> Because of the severe limitations of mean temperature methods, they will yield correct *short-period estimates* of evapotranspiration only under fortuitous circumstances and cannot be relied on for general use . . . To expend effort improving mean temperature methods for short-period potential evapotranspiration estimates and "adjusting" the estimates to a given locale and crop when mean temperature has little or no physical foundation for this use appears to be "kicking a dead horse."

Despite this unequivocal warning, agronomists and engineers, especially in the United States, continued to refine and use empirical models for estimating potential evaporation from mean temperature (Blaney and Criddle, 1962) or diurnal temperature range (Hargreaves and Samani, 1982). When locally calibrated, they perform well considering how little input they need as shown in a recent comparison by Allen et al. (1994). However, because they are products of current climate they are prone to error when used to predict changes of evaporation rate as a consequence of global warming, particularly in dry climates as shown by McKenny and Rosenberg (1993). Empirical models based on solar radiation (e.g., Makkink, 1957; Jensen and Haise, 1963) also perform well in environments where they have been calibrated.

B. Penman's Mechanistic Equation

Starting from the basic physics of heat and vapor transfer, Penman (1948) developed a mechanistic model for evaporation from free water surfaces and extended it empirically to bare soil and grass. This model was a practical application of the First Law of Thermodynamics which requires that the net rate at which radiant energy is absorbed at any point on the earth's surface, often written R_n (W/m^2), must be balanced by the rate at which energy is (1) transformed to latent heat of evaporation, λE (W/m^2), where λ (J/g) is the latent heat of evaporation for water and E $[g/(m^2 \cdot s)]$ is a mass flux of water vapor, or (2) dispersed to the atmosphere by convection, or (3) stored in a substrate (soil, rock, water, etc.). (Energy stored by photosynthesis, essential for growth, is energetically negligible.)

Because instruments for measuring R_n were not available in the 1940s (and are still very rare at climatological stations 50 years later), Penman calculated this quantity from its components. Incoming short-wave (solar) radiation was estimated from hours of sunshine using a relation derived by Angstrom [cited by Brunt (1939)] and *net* long-wave radiation was estimated from air temperature and vapor pressure using an empirical formula derived by Brunt.

Further empiricism was needed to obtain (1) a vapor transfer coefficient (the flux of water vapor per unit difference in vapor pressure between a water surface and the air passing over it at some arbitrary height) and (2) a corresponding heat transfer coefficient. For both heat and vapor, the transfer coefficient used by Penman was obtained from measurements of evaporation from small (76 cm diameter) evaporation tanks and had the general form $f(u) = a(1 + bu)$, where a and b were empirical constants and u was wind speed measured at a standard height of 2 m. This function was used to estimate fluxes of both sensible heat and latent heat from a surface, given air temperature and vapor pressure (as measured at the height of a climatological screen) and the effective temperature and vapor pressure of a wet surface.

By eliminating surface temperature and vapor pressure from this set of equations, Penman was able to express the latent heat of evaporation from a wet surface as

$$\lambda E = (\Delta H + \gamma E_a)/(\Delta + \gamma) \tag{4}$$

where Δ is the rate of change of saturation vapor pressure with temperature (which increases rapidly with temperature from 45 to 145 Pa/K between 0 and 20°C) and γ is the psychrometer constant in the same units of pressure per unit temperature difference. (The fact that this "constant" is a weak function of both temperature and pressure is usually ignored. At a standard pressure of 101.3 kPa, it increases from 65 to 67 Pa/K between 0 and 20°C.

In Penman's 1948 paper, H was the amount of energy per unit of ground surface as received from radiation, less a fraction stored in the soil. (A few workers, including the writer, still observe this tradition, but most use the symbol H for sensible heat flux.) Penman assumed that over periods of a week or longer, soil heat storage could be neglected compared with the net amount of heat received from radiation.

Using contemporary units,* the composite variable E_a can be defined as

$$E_a = \rho\lambda D f(u) \tag{5}$$

where ρ is air density (g/m³), λ (J/g) is the latent heat of vaporization of water, D (kg/kg) is the mean specific humidity deficit of air at screen height, and $f(u)$ (m/s) is a linear function of wind speed at a standard height, usually 2 m.

Dividing the numerator and denominator of Eq. (4) by γ and writing $\varepsilon = \Delta/\gamma$ gives

$$\lambda E = \frac{\varepsilon H + \rho\lambda f(u)D}{\varepsilon + 1} \tag{6}$$

Penman's model combined robust thermodynamic and aerodynamic principles and provided a sound basis for theoretical developments in evaporation science. It could not be used in practice, however (to estimate irrigation need, for example), without recourse to a substantial amount of empiricism. Whereas Thornthwaite's model contained eight empirically determined constants, Penman's, in its original form contained six—four for estimating net radiation from

---

*The water vapor content of air is conveniently expressed as a dimensionless specific humidity q (kg of water vapor per kg of moist air), related to vapor pressure by $q = xe/[(p - e) + xe] \sim xe/p$, where $x = 0.622$ is the ratio of molecular weights for water vapor and air, e is vapor pressure, and p is air pressure in the same units. The approximation is valid because e is invariably two orders of magnitude less than p. The parameters Δ and γ then assume units of K^{-1} and $\gamma = c_p/\lambda$ (K^{-1}), where c_p[J/(kg·K)] is the specific heat of air at constant pressure.

sunshine hours and vapor pressure and two for the wind function. A seventh constant was needed to estimate the "actual" rate of evaporation from soil or vegetation as a fraction of the "potential" rate for open water in the same environment. Subsequent theoretical and instrumental developments, now reviewed, made much of this empiricism unnecessary.

II. EXTENSIONS OF PENMAN'S EQUATION

A. The Penman–Monteith Equation

A major generalization of Penman's original equation was made possible by introducing "resistances"—simple electrical analogs for the potential difference needed to drive unit flux in systems that involve the transport of momentum, heat, and water vapor (Monteith and Unsworth, 1990). (Such resistances have dimensions of time per unit length and are usually quoted in units of seconds per meter or per centimeter.)

What subsequently became known as the Penman–Monteith equation (PME hereafter), incorporating both an aerodynamic resistance r_a and a surface or stomatal resistance r_s, was first published in a somewhat obscure contract report from a group led by F. A. Brooks at Davis, California (Monteith, 1963), and a more extended treatment followed (Monteith, 1965). An almost identical contemporary model was developed wholly independently by Rijtema (1965).

In the PME, the empirical wind function $f(u)$ used in Penman's equation was replaced by the reciprocal of an aerodynamic resistance so that the second term in the numerator became

$$E_a^* = \rho\lambda D/r_a \tag{7}$$

and the psychrometer constant was modified to

$$\gamma^* = \gamma(r_s + r_a)/r_a \tag{8}$$

so that the PME could be written in Penman's format as

$$\lambda E = (\Delta H + \gamma E_a^*)/(\Delta + \gamma^*) \tag{9}$$

The appearance of a wind-dependent function in the denominator as well as in the numerator of Eq. (9) implies that the rate of evaporation calculated from the PME is always less dependent on wind speed than the rate from the corresponding Penman equation when other elements of climate are unchanged. In general, estimated rates are usually insensitive to the magnitude of r_a, and the error generated by neglecting the influence of buoyancy correction (see later) is often small. In contrast, evaporation rate is usually a strong function of surface resistance (Fig. 1).

By differentiating Eq. (9) with respect to $1/r_a$, it can be shown that $dE/d(1/r_a) = 0$ when the surface resistance is

$$r_s = \rho c_p (\Delta^{-1} + \gamma^{-1}) D / H \qquad (10)$$

and substitution of this value of surface resistance into Eq. (9) gives

$$\lambda E / H = \Delta / (\Delta + \gamma) \qquad (11)$$

It follows that when the surface resistance is close to the value set by Eq. (10), the evaporation will be insensitive to wind speed and close to $\Delta H / (\Delta + \gamma)$ (Fig. 1; see also Section IV).

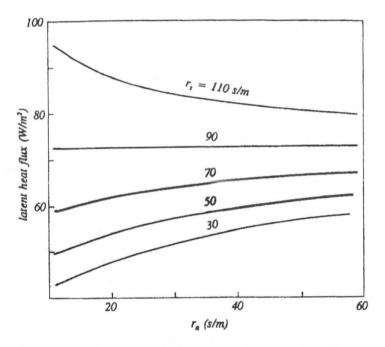

FIGURE 1 Dependence of latent heat of evaporation λE on aerodynamic resistance r_a and surface resistance r_s when total available heat $H = 105$ W/m^2, saturation vapor pressure deficit $D = 0.2$ kPa, and $\Delta = 0.15$ kPa/K. In most circumstances, evaporation rate decreases as r_s increases (top curve). However, there is critical value of r_s (90 s/m in this case) at which $\lambda E / H = \Delta / (\Delta + \gamma)$ [see Eq. (15)] so that E is independent of r_a. When r_s is smaller than this value, the evaporation rate increases as r_a increases.

When the concept of a "surface resistance" was first introduced, it was severely criticized by a group of Australian meteorologists who were convinced that the complex processes of heat and vapor exchange in a canopy could not and should not be modeled in such a simplistic way [see Appendix to Monteith (1963)]. Equation (9) has stood the test of time, however, and has been used on many scales to estimate evaporation from single leaves, from crop stands, and from extensive areas of uniform vegetation in models of the general circulation of the atmosphere.

B. Sources of Error

Because the net heat term ΔH is usually several times larger than ΔE_a, it is the main source of error in the numerator of the PME. This error has three components: the *net* flux of radiation received by vegetation over the period of application; the *net* amount of heat stored in the vegetation; and the *net* amount of heat stored in the soil. For fractions of a day or for forests with a large heat capacity, the magnitude of the storage term must be evaluated and retained if it is comparable with the net radiative flux over the same period [see Thom (1975), p. 93 et seq.].

Net radiation can either be measured directly or estimated from its components. Unfortunately, net radiometers are notoriously prone to several types of error, mainly associated with changes in the transmission properties of polythene domes during exposure (Field et al., 1992; Duchon and Wilk, 1994). Climatological measurements of net radiation are therefore both rare and unreliable. Records of solar radiation are more trustworthy and more readily available but must be complemented by estimates of long-wave incoming radiation (usually obtained from empirical formulas) and from outgoing radiation assumed equal to blackbody radiation at air temperature because the radiative surface temperature is rarely available. However, the difference between air and surface temperatures can be accounted for by defining an effective radiative resistance as $r_R = \rho c_p/4\sigma T_a^3$, where σ is Stefan's constant and T_a is the air temperature in Kelvin (Monteith and Unsworth, 1990). The value of this resistance decreases from 245 s/m at 10°C to 186 s/m at 30°C. If the effective resistance for heat transfer by convection and radiation, treated as parallel processes, is taken as $(r_R^{-1} + r_a^{-1})^{-1}$, it is legitimate to estimate net radiation as the flux density R_{ni} (W/m²) for a surface at air temperature T_a (°C) (sometimes referred to as the "isothermal" net radiation). For clear skies an appropriate algorithm for the long-wave component of net radiation R_{nl} (Monteith and Unsworth, 1990) is

$$R_{nl} = 107 - 0.3T_a \tag{12}$$

C. Applications of the Penman–Monteith Equation

Within comprehensive models of crop growth and yield, the PME can readily be exploited as a submodel for transpiration either at the scale of a single leaf or at canopy level provided the relevant physical and physiological variables are specified a priori or are available as output from other parts of the model. Following an exhaustive comparison of lysimeter measurements of evaporation and predictions from a range of formulas, Allen et al. (1994) concluded that the PME was more reliable than other models over a wide range of climates. The PME has therefore been adopted by FAO as a substitute for the Penman equation modified by Doorenbos and Pruitt (1977) in an irrigation manual usually referred to as FAO-24. Several other evaporation models, including the original Penman equation, fitted Allen's measurements data almost as well as the PME and can therefore be confidently used for practical applications when the surface resistance is known.

Doorenbos and Pruitt altered Penman's original formula by introducing an alternative wind function that predicts a smaller value of the aerodynamic resistance at a given wind speed. However, Thom and Oliver (1977) demonstrated that an error in the original wind function (which was appropriate for a smooth water surface rather than for rough vegetation) fortuitously compensated for the assumption that the surface resistance was effectively zero. Eventually, Allen et al. (1994), using measurements from high-class lysimeters, demonstrated conclusively that when this compensation was removed, the "improved" formula overestimated evaporation by 10–30% (the precise figure depending on the relative magnitude of radiation and humidity deficit terms).

Since 1981, the British Meteorological Office has used the PME operationally over the whole of Britain to provide weekly and monthly estimates of evaporation and soil water deficit (Thompson et al., 1981). The correct type of wind function was used along with empirical relations between surface resistance, leaf area (Grant, 1975), and soil water potential (Russell, 1980).

III. RESISTANCE COMPONENTS

It has been shown that both aerodynamic and surface resistances play an important role in establishing rates of evaporation from land surfaces. Their properties are now examined more closely for single leaves, for uniform stands of vegetation, and for bare soil.

A. Single Leaves

For the simplest case of a single transpiring leaf, the total resistance to the diffusion of water vapor from substomatal cavities to the external airstream is analogous to an electric current passing through two resistances in series, usually referred to as "stomatal" and "aerodynamic" resistances.

The magnitude of the stomatal resistance can be estimated in principle from the number of stomata per unit leaf area and from the mean diameter and length of pores. These are the main variables of diffusion models for stomata (e.g., Penman and Milthorpe, 1967), but because they are rarely known and are difficult to determine with precision, stomatal resistances are usually calculated from measured rates of transpiration and estimated gradients of vapor concentration. When stomata are closed, leaves of most species continue to lose water very slowly by diffusion through waxy cuticles, but their resistance is so large compared with stomatal resistance that it is usually assumed to be infinite.

There is an aerodynamic resistance r_a to diffusion in the boundary layer surrounding the leaf within which the transfer of heat, water vapor, etc. proceeds at a rate governed by molecular diffusion. Provided the wind speed is great enough and the temperature difference between leaf and air is small enough to ensure that transfer processes are not affected by gradients of air density, the boundary layer resistance depends on air velocity and on the size, shape, and attitude of the leaf with respect to the airstream. In very light wind, however, rates of transfer are determined mainly by gradients of temperature and therefore of density, so that the aerodynamic resistance depends more on the mean leaf–air temperature difference than on wind speed. Empirical formulas for estimating leaf boundary later resistance as a function of wind speed and characteristic dimension can be obtained from the engineering literature as presented by Monteith and Unsworth (1990) and by Jones (1992), for example.

To obtain the correct total (stomatal plus aerodynamic) resistance to vapor diffusion, r_{st}, it is necessary to distinguish between "amphistomatous" leaves with stomata on both abaxial (ab) and adaxial (ad) surfaces and "hypostomatous" leaves with stomata on adaxial surfaces only. For the amphistomatous class, the total resistance is found by combining the resistance for the two surfaces in parallel to give

$$r_{st} = [1/(r_s + r_a)_{ab} + 1/(r_s + r_a)_{ad}]^{-1} . \tag{13}$$

For the hypostomatous class, the resistance is simply $r_s + r_a$.

Models for the rate of evaporation from a single leaf (or of an individual plant treated as an assembly of leaves) have been used by glasshouse engineers seeking an optimum thermal environment for growth or production by taking account of the balance between the market value of a crop and production costs, including heating or cooling. Such models are usually based on engineering literature for heat transfer from small plates of regular shape (e.g., circular or rectangular). For example, Seginer (1984) suggested that the resistance to heat transfer by free convection from rose leaflets (both surfaces) in virtually still air could be expressed (in s/m) as $660/(\delta T/d)^{0.25}$, a quantity obtained from an appropriate Grashof number (as given in the ASHRAE handbook, for example). The quantity δT is the mean temperature difference (K) between the surface of the leaf and the air surrounding it (as measured with an infrared thermometer, for example) and d is a characteristic leaf dimension.

The corresponding resistance to water vapor transfer r_{av} depends, in principle, on the ratio of molecular diffusion coefficients for heat and the gas in question, but the difference between resistances for heat and water vapor is usually small enough to neglect.

The same procedures are valid for single leaves outdoors with the obvious complication that wind speed and direction are usually very variable. In this case, the aerodynamic resistance may be estimated from a Nusselt number depending on the characteristic dimension of the leaf in the average direction of the wind and the component of average wind speed parallel to the surface (Monteith and Unsworth, 1990).

B. Land Surfaces

In the field, aerodynamic resistances for homogeneous surfaces such as bare soil or a crop canopy are large-scale analogs of the boundary layer resistances for individual leaves and other organs. In principle, resistances for heat, water vapor, and other scalars can be estimated as the gradient needed to maintain unit flux, but in practice they are often derived from measurements of wind speed and from a knowledge of the aerodynamic properties of the surface, using the following procedure.

When the atmosphere above an extensive uniform stand of vegetation is in a state of neutral stability (i.e., when the temperature gradient is close to the dry adiabatic lapse rate or $-1°C/100$ m), the increase of wind speed u with height z above the ground is given by

$$u(z) = (u_*/k)\ln[(z - d)/z_0)] \tag{14a}$$

The so-called friction velocity u_*(m/s) is defined by

$$u_* = (\tau/\rho)^{0.5} \tag{14b}$$

where τ is the momentum flux from atmosphere to surface (often referred to as a "shearing stress") and ρ is air density. Von Karman's constant k, as determined experimentally, was assumed for many years to be 0.41, but recent measurements suggest that 0.397 ± 0.010 may be a more appropriate value [see, e.g., Frenzen and Vogel (1995)].

The ability of a surface such as the canopy of a crop to absorb momentum (at a given wind speed) is specified by a "roughness length" z_0 measured not from the ground surface but from the *effective* height of the canopy as defined by a zero plane displacement d. For many years z_0 and d were estimated empirically as fractions of crop height h (e.g., $z_0 = 0.1h$, $d = 0.7h$). A model for momentum transfer within canopies developed by Shaw and Pereira (1982) provides a more mechanistic basis for estimating z_0 and d, and an application of this model by Choudhury and Monteith (1988) was successfully tested by van den Hurk et al. (1995). The main structural parameter of the model is the product X of leaf area index L and the mean drag coefficient of individual leaves c_d. Zero plane displacement and roughness length are then given by

$$d = 1.1h(1 + X^{0.25}) \tag{15}$$

and

$$z_0 = z_{0'} + 0.3hX^{0.5} \quad \text{when} \quad 0 \leq X \leq 0.2 \tag{16a}$$

or

$$z_0 = 0.3h(1 - d/h) \quad \text{when} \quad 0.2 \leq X \leq 1.5 \tag{16b}$$

and z_0' is the roughness length for the soil surface.

The resistance to momentum transfer r_{aM} between the surface and a height z can be obtained from Eqs. (16a) and (16b) because it is the momentum difference across the resistance or $\rho[(u(z) - u(z - d)] = \rho u(z)$ divided by the momentum flux ρu_*^2. This implies that

$$r_{aM} = \{\ln[(z - d)/z_0]\}^2/k^2 u(z) \tag{17a}$$

so that r_{aM} is inversely proportional to $u(z)$ provided other parameters are independent of wind speed. In practice, because most vegetation has leaves, petioles, or stems that bend in the wind, d tends to decrease with increasing wind speed [as shown by Vogt and Jaeger

(1990), for example], whereas z_0 may either decrease or increase depending on changes in foliage density. Businger (1956) suggested that Tanner and Pelton (1960) confirmed that the reciprocal of r_{aM} is equivalent to the wind function $f(u)$ in Eq. (5).

When foliage is warmer or cooler than the air above it, mechanically generated turbulence is respectively enhanced or inhibited by the action of buoyancy. For these so-called non-neutral conditions, Eq. (17a) can, in principle, be modified by introducing a stability parameter such as the Richardson number or the Monin–Obukhov length (Thom, 1975; Monteith and Unsworth, 1990), but the former requires measurements of temperature and wind speed gradients that are not generally available and the latter requires a sensible heat flux that cannot be obtained without iteration.

An aerodynamic resistance for the transfer of scalars such as heat or water vapor can be obtained by adding to the momentum resistance r_{aM} an additional component r_b to account for the absence of a process equivalent to the bluffbody forces that enhance momentum transfer (Monteith and Unsworth, 1990). For arable crops and with u_* in the range 0.1–0.5 m/s, Thom (1975) suggested the empirical relation $r_b = 6.2u_*^{-0.67}$. Alternatively, heat and vapor resistances can be obtained directly from Eq. (17a) by replacing z_0 with scalar-specific roughness lengths such as z_{0H} for heat or z_{0V} for vapor in place of the momentum roughness usually written z_0. Equation (17a) can then be rewritten for heat or water vapor as

$$r_{aH,V} = \{\ln[(z - d)/z_{0H,V}]\}^2/k^2u(z)$$
$$= \{\ln[(z - d)/z_0] + \ln(z_0/z_{0H,V})\}^2/k^2u(z) \qquad (17b)$$

Garratt (1992) suggested that $\ln(z_0/z_{0H}) \approx \ln(z_0/z_{0V}) \approx 2$ was appropriate for diverse types of vegetation, but from the analysis of many measurements over short grass, Duynkerke (1992) reported values of $\ln(z_0/z_{0H})$ ranging from about -2 to 13 and increasing both with leaf area index and with u_*. Precise modeling of evaporation rates therefore requires a careful assessment of differences in the flux–gradient relation for heat, for vapor, and for momentum.

Because it is possible to identify an aerodynamic resistance for extensive uniform vegetation [using Eq. (17a)], it is also possible to identify a "surface" resistance imposed by stomata, assuming that a complete canopy is endowed with the properties of a single "big leaf." As for individual leaves, the "surface" resistance of a canopy is usually estimated indirectly from fluxes and gradients using a procedure discussed in the next section.

C. Inversion of the Penman–Monteith Equation

Whereas the original Penman model has been used mainly to estimate the irrigation need of crops or seasonal changes in the water
balance of catchments, a common use of the PME is to estimate the
surface resistance of vegetation when its rate of transpiration has been
measured independently. The magnitude of this resistance can then
be used as an index of ground cover, water stress, nutrient deficiency,
etc. The procedure is to invert Eq. (9) to obtain

$$r_s = \frac{r_a}{\gamma \lambda E} \left[\Delta(H - \lambda E) + \gamma(E_a^* - \lambda E) \right] \tag{18}$$

where variables on the right-hand side of the equation are measured
or estimated. In some of the first published measurements of surface
resistance (Monteith, 1963), r_s for irrigated grass at Davis, California,
achieved minimum values in the range 0.5–1 s/cm in the middle of
the day. Morning values were somewhat larger, and late afternoon
values were substantially larger. Such behavior suggested that the
value of r_s as evaluated from Eq. (18) encapsulated the dependence
of stomatal aperture on weather and water supply. Many subsequent
measurements supported this conclusion and confirmed that the
range of resistances observed at Davis was characteristic of many
types of vegetation [as cited by Kelliher et al. (1995), for example].

D. Submodels of Surface Resistance

As a basis for estimating surface resistance in prescribed weather or
interpreting measurements in terms of weather variables, Jarvis
(1976) introduced the concept of a maximum conductance g_m (the
reciprocal of a minimum resistance). This conductance is achieved in
principle (but rarely, if ever, in practice) at simultaneously optimal
values of solar radiation S incident on a canopy; air temperature T_a
and vapor pressure deficit D both measured at screen height; and soil
water content averaged over the root zone (θ). The actual conductance
of a leaf at any time is given by

$$g = g_m f_1(S) f_2(T_a) f_3(D) f_4(\theta) \tag{19}$$

where each of the component functions f_1, f_2, etc. has a maximum
value of 1. The general form of these functions is known from laboratory measurements on single leaves (e.g., f_1 is commonly a rectangular hyperbola or negative exponential function), but the parameters
defining the functions have to be determined experimentally by using
Eq. (19) over as wide a range of environmental conditions as possible.

Equation (19) has been used to relate the effective stomatal/ surface resistance of canopies to variables measured above the canopy notwithstanding the fact that stomata respond to the temperature of leaf tissue rather than to air temperature, to the vapor pressure deficit at the leaf surface rather than at some undefined distance from foliage, and to the whole range of solar irradiance experienced *within* a canopy rather than the irradiance measured *above it*. These anomalies can be partly but not completely circumvented by treating foliage as a system with n discrete horizontal layers, each with a leaf area index of $l(n)$ and a conductance $g(n)$ depending on local illumination and temperature, age, etc. For the simplest case of layers in parallel, the bulk conductance is $\Sigma g(n)l(n)$ provided the aerodynamic conductance between layers is much greater than $g(n)$ (McNaughton and Jarvis, 1983). This is rarely a realistic model, however, so conductance is usually estimated by inverting the PME applied to the whole canopy as already discussed.

A second type of stomatal model developed by Ball et al. (1987) was based on the thesis that stomatal aperture is tightly coupled to the net rate of photosynthesis as determined by biochemical processes that are functions of irradiance, temperature, etc. In this treatment, stomatal conductance of single leaves is expressed as

$$g = g_0 + kAh_s/c_s \qquad (20)$$

where h_s and c_s are respectively the relative humidity and carbon dioxide concentration at the leaf surface, A is the net rate of photosynthesis, g_0 is the cuticular conductance of the leaf as measured when stomata are shut so that A is virtually zero, and k is a constant with appropriate dimensions.

Though based on careful laboratory work and much used, this model is mechanistically unsound because there is no evidence that stomata respond to relative humidity as distinct from absolute humidity, a quantity that is often an acceptable surrogate for transpiration rate. This anomaly was resolved by Leuning (1995), who showed that laboratory measurements of photosynthesis and stomatal conductance were consistent with an equation proposed by Lohammar et al. (1980), namely,

$$g = g_0 + \frac{kA}{c_s(1 + D_s/D_{s0})} \qquad (21)$$

where D_s is the specific humidity deficit at the leaf surface and D_{s0} is a scaling factor. As the transpiration rate E is given by gD_s (assuming negligible temperature gradient across the cuticle), the stomatal component of resistance is given by

$$g - g_0 = \frac{kA}{c_s(1 + E/gD_{s0})} \tag{22}$$

For the common condition that g is much larger than g_0, rearrangement of Eq. (22) gives

$$g = g_m(1 - E/E_m) \tag{23}$$

where g_m ($\approx kA/c_s$) is the apparent maximum conductance when $E = 0$ and E_m ($\approx g_m D_{s0}$) is an apparent maximum transpiration rate achieved when stomata are closed. Reanalysis of many laboratory measurements (Monteith, 1995) confirmed the general validity of Eq. (23) within the experimental range of saturation deficit and Bunce (1997) has suggested that peristomatal transpiration is implicated.

Notable applications of Eq. (19) were analyses of evaporation over pine forest at Thetford in southeast England (Stewart, 1988) and in the Les Landes region of southwest France (Gash et al. 1989). Although multiple regression analysis applied to both sets of measurements yielded similar sets of functions as shown in Fig. 2, this agreement does not guarantee that all three functions are appropriate. The radiation function has the general hyperbolic form observed in many laboratory experiments. The temperature response resembles (but

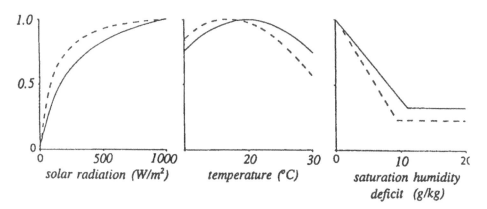

FIGURE 2 Canopy conductance (as fraction of maximum value) for Scots Pine forest in Thetford, southeast England (----) and Maritime Pine forest in the Les Landes district of southwest France (—). Plots represent statistical fits to functions assumed for dependence of conductance on solar radiation, temperature, and specific humidity deficit. (From Gash et al., 1989).

cannot be compared directly with) laboratory measurements, which are almost invariably referred to the temperature of foliage rather than that of the ambient air. The humidity response implies that when saturation deficit increases from zero, the evaporation rate increases as stomatal conductance decreases, and this behavior has been observed in the laboratory. Above a saturation deficit of 10 g/kg, however, the surface resistance remains fixed as the evaporation rate continues to increase in response to an increasing deficit. As this is inconsistent with all the laboratory evidence on which Eq. (23) is based, it is likely that the humidity response incorporates a fallacious artifact.

Other examples of deriving stomatal functions from field measurements can be found in the work of Dolman (1993), who used the Jarvis function to analyze measurements over Amazonian forest, and of Dougherty et al. (1994), who fitted measurements of stomatal conductance on C4 grasses to the original Ball–Berry model and five other superior models derived from it. In the best model, conductance was shown to be proportional to a power of the function A/Dc_s, which has several features in common with Eq. (21).

The substantial amount of scatter observed when measured values of surface conductance are correlated with values calculated from a set of regression functions for individual parameters has several potential sources: errors of observation, weakness of the assumption that normalizing functions can be treated as independent multipliers [Eq. (19)], neglect of heat and vapor storage in canopies, and the adoption of a physiologically unsound function for the putative response to saturation deficit.

If, as demonstrated in the studies cited, the response of the apparent surface resistance or conductance of canopies to environmental factors is similar to the response of single leaves, some measure of correlation would be expected between maximum values of conductance observed on single leaves in the laboratory (g_{max}) and maximum values of canopy conductance observed in the field (G_{max}). Kelliher et al. (1995) demonstrated that such congruence does exist, assisted by the fact that the rate of evaporation used to estimate G_{max} often includes a component of soil evaporation. This decreases as leaf area index increases in such a way that G_{max} is less dependent on leaf area than might be expected. To summarize conclusions from the Kelliher et al. study, mean values of g_{max} and G_{max} ranged from 6 to 20 mm/s for natural vegetation and from 12 to 23 mm/s for agricultural crops. The ratio G_{max}/g_{max} was about 3 ± 1 (see Fig. 3).

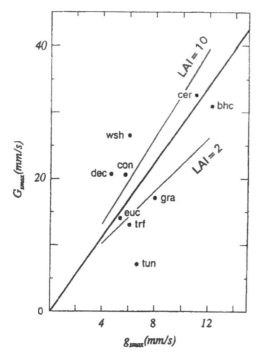

FIGURE 3 Maximum canopy conductance $G_{s,max}$ for a range of vegetation types plotted against maximum leaf conductance $g_{s,max}$ with (bold) line of best fit through origin. Thin lines are model estimates of the relation for leaf area indices of 2 and 10. Key: tun, moss and tundra/lichen; gra, grassland; cer, wheat; bhc, arable crops; wsh, heather and macchia shrub; euc, eucalyptus forest; dec, deciduous forest; con, coniferous forest; and trf, -tropical forest. (From Kelliher et al., 1995.)

E. Submodels of Capacitance

The model used above to explore the exchange of heat and water between vegetation and the atmosphere was essentially an electrical analog containing a network of resistances but neglecting capacitances—elements of the system capable of storing heat or water for periods usually longer than an hour but less than a day. Their properties are now considered briefly.

1. Heat Storage in Soil and Vegetation

Diurnal changes in the storage of heat in the soil are needed for precise hourly estimates of evaporation using Penman-type equations.

They can be accounted for in a somewhat crude way by making the heat flux at the soil/air interface a sinusoidal function of time as in the Choudhury–Monteith (1988) model. However, van den Hurk and McNaughton (1995) showed that this leads to substantial errors that are avoided when the "force-restore" method of Deardorff (1978) is used to estimate soil heat flux.

Storage of heat in vegetation is usually negligible for arable crops but can be a significant component of the heat balance in forests. Following Thom (1975), the maximum rate of change of mean biomass temperature may be assumed equal to the change of air temperature $\partial T_a / \partial t$ (K/h), and the specific heat of vegetation may be taken as 70% of the value for water, i.e., 2.9 kJ/(kg·K). Then for a biomass of m kg/m^2, the maximum change of heat storage is equivalent to a flux of about 0.8 $m\, \partial T_a / \partial t$. This term will always be negligible for short vegetation with $m \approx 10$ g/m^2. But for a representative change in temperature of 3 K/h, the corresponding flux for a forest with $m \approx 10$ kg/m^2 is 24 W/m^2, a significant quantity with the same order of magnitude as fluxes of sensible heat.

2. Water Storage in Soil

Passioura (1983) suggested that when the roots of a uniform crop start to extract water at a specified depth, the volumetric water content θ can often be expressed as a negative exponential function of time, e.g.,

$$\theta = \theta_m \exp(-t/\tau) \tag{24}$$

where θ_m is the water "available" at $t = 0$ and τ is the time constant of the system analogous to the time constant CR of an electric circuit containing a capacitance C and a resistance R.

A time constant for roots at a specific depth can be estimated by measuring the local water content of soil using a neutron probe, for example, provided the profile is not rewetted by rain. When θ_m was estimated from the difference between water held at field capacity and permanent wilting point (as determined from laboratory samples), values of τ for a stand of sorghum growing on a vertisol at Hyderabad (India) ranged from 35 to 60 days compared with 10 days for millet on an alfisol (Monteith, 1986). Subsequently, when Robertson et al. (1993) and Singh et al. (in press) used the difference between maximum and minimum values of θ as observed in the field instead of arbitrary conventional limits for "available" water, they obtained values of τ in the range 15–20 days.

Models of extraction of water by plant roots are still at a primitive stage of development compared with models for the evaporation of extracted water. When the rate of evaporation is determined by weather, this disparity is usually tolerable, but when growth is limited by the availability of water to roots, most models resort to somewhat crude empiricism and predictions are therefore unreliable.

3. Water Storage Within Plants

The concept of capacity is also relevant to the flow of water through plants because cells in stems and leaves contribute to the loss of water by evaporation during the day and are recharged at night. This process must be taken into account in any detailed model of the time course of water loss by plant stands, but few relevant measurements or models are available. Hunt et al. (1991) provided detailed estimates of hydraulic resistances and capacities for roots, stems, and leaves of different types of vegetation and computed corresponding time constants from the product of resistance and capacitance. These range from a minimum of about 20 s for grass with a capacitance of 1 m^3/MPa to about 5 h for trees with a capacitance of 1000 m^3/MPa.

4. Water Storage Within Canopies

When rain starts to fall on vegetation, leaves and other surfaces intercept a fraction that depends on factors such as area, angular distribution, and roughness. Eventually the rate of interception is balanced by the rate of loss by dripping from wet surfaces, and the amount of water retained after saturation is known as the canopy storage capacity (S), a quantity usually lying in the range 0.5–2.0 mm. Rutter et al. (1975), Calder (1977, 1986), and others explored the relation between interception and evaporation rates by writing the net rate of change of intercepted water (dI/dt) as a balance between (1) the rate at which water is intercepted, assumed to be a fraction x of the precipitation rate P, and (2) the rate at which it is subsequently lost either by evaporation (E) or by dripping [assumed to be a function $f(I)$ of I]. It follows that

$$\frac{dI}{dt} = xP - E - f(I) \tag{25}$$

Rutter et al. (1975) assumed that $f(I)$ was a positive exponential function of I and that E was equivalent to the potential rate E_p obtained from the Penman–Monteith equation multiplied by I/S when I was less than S. Calder used E_p in all conditions and showed that measurements in a spruce forest appeared to fit the simple linear model $f(I) = bI$. The value of b depended on the choice of parameters in the Penman–Monteith equation used to estimate E_p and also showed a weak dependence on the value of I.

IV. THE PRIESTLEY–TAYLOR EQUATION

The Penman equation implies that when energy is supplied at a rate H to a wet surface exposed to saturated air ($D = 0$), then the latent heat of evaporation is

$$\lambda E = \Delta H/(\Delta + \gamma) \tag{26}$$

and is therefore independent of wind speed [see Eq. (11)]. More generally, provided the saturation deficit of air is constant with height so that air in contact with foliage has the same value (D) as air at screen height, the latent heat of evaporation is

$$\lambda E = \rho c D/\gamma r_s \tag{27}$$

When this equation is used to eliminate D from Eqs. (8) and (10), rearrangement of terms again gives Eq. (26).

In practice, saturation vapor pressure deficit is not constant with height. In the absence of cloud and for reasons outlined below, it increases with height within a convective boundary layer (CBL) that is often several hundred meters deep at dawn, expanding to several thousand meters in depth as the day progresses.

The CBL is capped by an inversion through which relatively warm dry air is mixed into the boundary layer from above. At the ground, however, air passing over vegetation receives water vapor by transpiration and may be heated or cooled depending on the difference between surface and air temperature. Commonly, the net flux of water vapor into the CBL is small, so vapor pressure changes little during the day. In contrast, temperature invariably increases in response to radiative heating and to an input of sensible heat from above the CBL that is modulated but never reversed by sensible heat exchange at the ground. An important consequence of these intimately linked processes is that saturation vapor pressure deficit increases with height in such a way that rates of transpiration exceed predictions from Eq. (26) by 20–30%. To deal with this excess empirically, Priestley and Taylor (1972) suggested that the equation should be modified by introducing a multiplier (α) such that

$$\lambda E = \alpha \Delta H/(\Delta + \gamma) \tag{28}$$

Using this Priestley–Taylor equation (PTE) as a definition of α, the PME, Eq. (9), can be rewritten in terms of α by multiplying both sides of the equation $(\Delta + \gamma)/\Delta H$ to give

$$\alpha = \frac{1 + N}{1 + R} \tag{29}$$

where $N = \rho c_p D / \Delta H r_a$ is a nondimensional climate number depending on radiation, temperature, vapor pressure, and wind speed; and

$$R = \frac{\gamma}{\Delta + \gamma}\left(1 + \frac{r_s}{r_a}\right) \tag{30}$$

γ is a nondimensional resistance ratio depending primarily on the relative size of surface and aerodynamic resistances but with a weak temperature dependence through the value of Δ.

Many early attempts to test the PTE appeared to be successful insofar as values of α close to 1.26 were obtained both for water surfaces and for diverse types of well-watered vegetation. Eventually, however, it was demonstrated that α could achieve values as large as 11 for spruce forest with wet foliage (Shuttleworth and Calder, 1979) and values much smaller than 1.26 over vegetation with a restricted water supply as discussed later (McNaughton and Spriggs, 1989). Considerable caution is therefore needed when Eq. (29) is used with some arbitrary value of N (usually between 0.2 and 0.3) to estimate a "potential" rate of evaporation. In the CERES family of crop models, this potential rate is multiplied by a ratio of actual to potential evaporation obtained from the PME. However, the PTE and PME yield the same estimate of potential evaporation only when the surface is completely decoupled from the atmosphere (see below).

Jury and Tanner (1975) drew attention to anomalously large values of α obtained in the presence of advection, e.g., when air from dry land passed over an irrigated field. They therefore suggested that the Priestley–Taylor coefficient should be expanded to include an empirical saturation deficit term. This procedure has not been widely adopted, however, partly because it departs from the simplicity of Eq. (26) without improving on the physics behind Eq. (9), and partly because major errors arise when empirical constants derived in one climate are transferred to another where the rainfall and humidity regime are different.

A. The Concept of Coupling

McNaughton and Jarvis (1983) explored the advantages of combining the PTE and PME by writing

$$\lambda E = \Omega \, \frac{\Delta H}{\Delta + \gamma} + (1 - \Omega) \, \frac{\rho c_p D}{\gamma r_s} \qquad (31)$$

where

$$\Omega = \left[1 + \frac{\gamma}{\Delta + \gamma} \left(\frac{r_s}{r_a} \right) \right]^{-1} \qquad (32)$$

In a system where r_a is very large compared with r_s so that $\Omega \approx 1$, the latent heat flux is given by $\Delta H / (\Delta + \gamma)$, which is the flux obtained from the PTE with $\alpha = 1$. In this case, the vegetation is regarded as effectively "decoupled" from the atmosphere above it, implying that the saturation deficit at the surface is controlled by physical processes at the surface. Conversely, when r_a is very small compared with r_s, $\Omega \approx 0$, the latent heat flux is given by the second term in Eq. (31), and the saturation deficit is imposed on the system by the state of the air mass passing over it.

This type of model has helped to explore and interpret the contrasting behavior of evaporation from vegetation with relatively smooth surfaces such as short grass (decoupled from the air passing over it) and from very rough surfaces such as forests (tightly coupled). In particular, it has thrown light on the significance of evaporation from water intercepted by foliage. The relatively small amount of water intercepted by short vegetation evaporates very slowly because the total rate of evaporation plus transpiration is controlled by available energy. Forests intercept much larger amounts of water, which evaporate rapidly because of rapid mixing, and this process is now considered in more detail.

B. Boundary Layer Models and Feedback

Most models of evaporation incorporate climatic variables such as temperature and vapor pressure that are assumed to be independent of the state of the underlying surface. In practice, however, these variables depend on the way air mass properties determined by weather on a synoptic scale interact with surface properties on a local scale. Feedback generated by these interactions can be explored within models for the CBL.

Complete CBL models include dynamical elements, but McNaughton and Spriggs (1989) showed that a simpler thermodynamic model of CBL growth was consistent with one of the best sets of measurements available from a site in the Netherlands (Driedonks, 1981). They then used a subset of the same measurements to predict how the daily loss of water by evaporation should change in response to changes in surface resistance, assumed constant throughout the day. To present this relation, they plotted the Priestley-Taylor coefficient α as a function of the logarithm of surface resistance r_s to obtain a set of sigmoid curves. However, it is more informative to plot the reciprocal of α against r_s because this yields a set of straight lines (Fig. 4) defined by an intercept α_0 and a slope $m = d(1/\alpha)/dr_s$ so that

$$\alpha^{-1} = \alpha_0^{-1} + mr_s \tag{33}$$

For evaporation from an extensive region, the climate number N (Eq. 29) is expected to increase as surface resistance increases (during a spell of dry weather, for example) because of an increase in

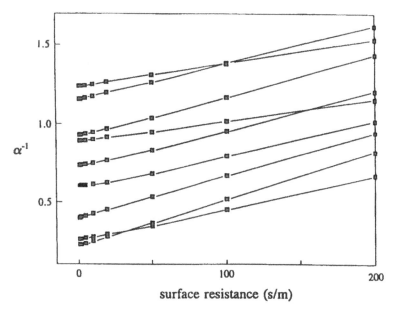

FIGURE 4   Reciprocal of the Priestley–Taylor coefficient, α, estimated as a function of surface resistance by McNaughton and Spriggs (1989), who coupled a boundary layer model to meteorological measurements for 9 days over grassland at Cabauw, Netherlands (Driedonks, 1981). Nine lines join sets of points at seven arbitrary values of surface resistance assumed constant throughout the day so that each set represents daily mean estimates $(\alpha)^{-1}$ as a function of resistance for daylight hours on 1 day. Sets of points for each day are displaced vertically to avoid confusion. Values of α for zero resistance ranged from 1.30 to 1.57 with a mean of 1.40 ± 0.1.

vapor pressure deficit. The resistance number R Eq. (30) is a function of aerodynamic as well as surface resistance.

By eliminating α from Eqs. (29) and (33), it can be shown that when the surface resistance is r_s, the *potential* rate of evaporation (expressed as a Priestley–Taylor coefficient) is given by

$$PE = 1 + N = \frac{\alpha_0[1 + f(1 + r_s/r_a)]}{1 + \alpha_0 m r_s} \tag{34}$$

where $f = \gamma/(\Delta + \gamma)$.

From Eq. (33), the *actual* rate of evaporation is

$$AE = \frac{\alpha_0}{1 + \alpha_0 m r_s} \tag{35}$$

Alternatively, by eliminating r_s from Eqs. (29) and (33), the Priestley–Taylor coefficient for actual evaporation can be expressed as

$$\alpha = \frac{1 + N - X}{1 + f - \alpha_0 X} \tag{36}$$

where $X = f/m r_a$ can be regarded as a climatological variable depending on mean air temperature and wind speed and on the structure of the PBL.

V. MODELS FOR EVAPORATION FROM SOIL

The rate at which water evaporates from wet soil depends partly on the state of the atmosphere as specified by radiation, wind speed, humidity, etc. and partly on the state of the soil as established by the distribution of water held in pores and by vertical gradients of temperature and vapor pressure. Fully comprehensive models of evaporation from soil, as described by ten Berge (1990), for example, account rigorously for the way mass and heat transfer are coupled within soil and at the soil/air interface.

Van Bavel and Hillel (1976) described a representative model of soil evaporation in which a loam soil with a depth of 1.13 m was treated as 14 layers of increasing depth starting with a 1 cm layer at the surface. Soil water potential and hydraulic conductivity were specified functions of water content, and fluxes of water and heat were assumed to be independent rather than coupled as in the more rigorous model of Philip and de Vries (1957). The short-wave reflectivity and long-wave emissivity of the soil surface were functions of

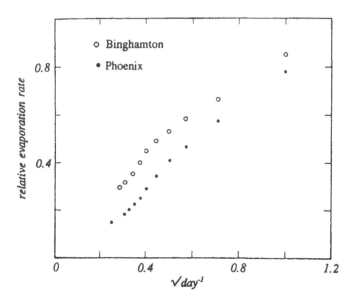

FIGURE 5   Rate of evaporation from bare soil estimated from the model of
Van Bavel and Hillel and plotted (as fractions of maximum rate) against the
inverse square root of time from the onset of drying. Sites chosen for climatic
contrast were Binghamton, New York, and Phoenix, Arizona. (From Van
Bavel and Hillel, 1976.)

the water content of the top layer, and the aerodynamic resistance
between the surface and a reference height of 2 m was found by using
conventional equations incorporating wind speed, roughness, and a
stability correction. Climate records for seven contrasting stations
within the United States were used to explore the behavior of the
model when run with an hourly time step in a CSMP program. During the so-called "first stage" of drying when the soil surface was
continuously wet, daily values of potential evaporation estimated by
the model were very close to values obtained from an equation of the
Penman type.

   Lee and Pielke (1992) and Mihailovic et al. (1995) reviewed a
range of empirical models in which the evaporation rate was estimated from the aerodynamic resistance r_a between a uniform surface
and the air at a reference height z; and from the difference between
the specific humidity of air at z ($q(z)$) and at the soil surface, assumed
to be the saturation value ($q_s(0)$) at mean surface temperature. In these
models, the reduction of evaporation rate during the second stage of

drying is taken into account by introducing an empirical factor depending on water content at the surface. Some workers suggest that the surface humidity should be reduced by a fraction x so that

$$E = (xq_s(0) - q(z))/r_a \qquad (37)$$

Others place x outside the brackets implying that the surface water content determines the evaporation rate directly rather than indirectly through the surface humidity. In this second category, the model developed by Deardorff (1978) performed well in a comprehensive field test of 9 models by Mihailovic et al. (1995).

Less sophisticated methods can be used to estimate daily rather than hourly rates of evaporation from soil. For example, a simple linear relation between evaporation rate and the square root of time (in days) is characteristic of a system in which an upper layer of soil is treated as completely dry and a lower layer as saturated (Monteith, 1981). Predictions from such a model are very similar to those obtained from the complete Van Bavel–Hillel model (Fig. 5). In particular, the **relative** rate of actual evaporation from a hot dry station (Phoenix, Arizona) was less than the rate at a humid station (Binghamton, New York) whereas the potential rate of evaporation was greater. This is because the soil resistance increases more rapidly at the dry site.

VI. MULTIPLE-SOURCE MODELS OF EVAPORATION

Previous sections reviewed models of evaporation from vegetation completely covering the ground or else from bare soil. In most agricultural or horticultural systems, however, as well as in natural ecosystems, transpiration and soil evaporation occur simultaneously and are significant components of total evaporation. This type of compound system is much more difficult to model, partly because extraction of water by root systems competes with the upward diffusion of water through soil pores and partly because the supply of heat to the soil surface from above depends on the attenuation by foliage of radiation and on turbulent mixing within the canopy. Several models tackling this problem in different ways are now reviewed briefly.

In the Penman–Monteith equation (PME), vegetation is treated as a simple plane surface with uniform properties, the discrete contributions of evaporation from soil and transpiration from leaves are ignored, and all fluxes are treated as one-dimensional. The first major attempt to escape from these constraints was made by Shuttleworth and Wallace (1985), who constructed a two-dimensional network of resistances (Fig. 6) to account for the separate contributions of soil and foliage to vertical fluxes of sensible and latent heat. Inevitably, equations derived for the two components of evaporation are more cumbersome than Eq. (9) and contain additional variables (e.g., aerodynamic resistances to diffusion for leaf-to-canopy transfer and for soil to canopy level transfer). With this model, it was possible to distinguish between the fraction of water transpired and the fraction lost directly from the soil.

There are essentially two layers in the Shuttleworth–Wallace model. One extends downward from a reference height above an extensive uniform canopy to an apparent source/sink of heat and vapor within it; and the second extends from the canopy to the soil surface

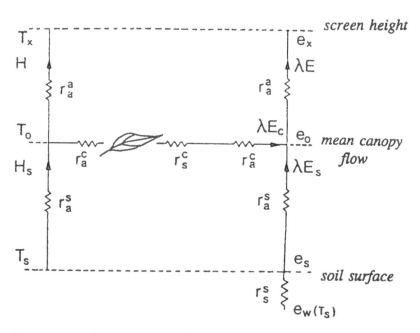

FIGURE 6   Resistance network for simultaneous transpiration from foliage and evaporation from bare soil in model of Shuttleworth and Wallace (1985). (Subscripts a and s, refer to aerodynamic and soil components. Subscripts x and 0 refer to screen and mean foliage height; superscripts a, s, and c refer to transfer processes between canopy and screen height, between soil surface and canopy, and between canopy and foliage, respectively. λE and H are latent and sensible heat fluxes, and e is vapor pressure.)

below which a resistance to vapor diffusion and a heat flux were prescribed. The canopy resistance was also prescribed.

Choudhury and Monteith (1988) developed a more complete four-layer model incorporating an upper layer of completely dry soil and a lower wet layer as described earlier. Leaf resistance was proportional to irradiance at leaf level. Predictions from the model agreed well with measurements of surface temperature and evaporation rate for a stand of wheat grown at Phoenix, Arizona.

A second type of four-layer model containing two types of vegetation, vertically discrete, was developed by Dolman (1993) for use in general circulation models and was calibrated for tropical rainforest, savannah, and an agricultural crop. Leaf resistance was estimated as a function of weather and soil water, using the formulation of Jarvis already described, and soil resistance was a function of soil water content.

Van den Hurk et al. (1995), compared the performance of three vegetation models with measurements made in a semiarid sparsely vegetated vineyard in Spain during the EFEDA campaign. The models were selected from the work of Choudhury and Monteith (1988) already described, Deardorff (1978), and Viterbo and Beljaars (1995). This exercise revealed that all three models had specific strengths and weaknesses, suggesting that hybrid vigor and rigor might be created by combining their best components. This was a commendable exercise that would be worth repeating over a range of environments. Over the past decade or two, progress has often been inhibited by the failure of modelers and measurers to communicate with each other.

The Shuttleworth–Wallace and Choudhury–Monteith models represent sources of evaporation as plane surfaces parallel to the ground with characteristics specified by a set of simple parameters such as a surface resistance or a roughness length. In the real world, foliage is vertically distributed in a somewhat chaotic way, but a broad impression of how sources of water vapor are distributed vertically can be obtained from vertical profiles of water vapor. Disconcertingly, however, there is evidence that fluxes within canopies do not always move in the direction expected from corresponding gradients, a phenomenon peculiar to systems in which the scale of turbulence is comparable with singularities in the corresponding concentration profile.

To explore this problem in a wider context, Raupach (1989) developed a new type of model for the transport of scalars within crop canopies that, in principle, could be used to predict the vertical gradient of source strength for water vapor (or any other scalar quantity) from the corresponding profile. The essence of his theory is a distinction between the systematic movement of small but discrete elements of water vapor close to the point of evaporation ("near-field" behavior) and the subsequent random movement of elements under the influence of turbulence ("far-field" behavior).

Van den Hurk and McNaughton (1995) have shown that near-field behavior can be simulated by introducing an additional resistance in a lumped canopy model of the type described by Shuttleworth and Wallace. The resistance is defined as the increase in concentration per unit of flux attributable to displacement within the near field. In Fig. 7, such a resistance would therefore be added in series with the resistance. Van den Hurk and McNaughton tabulated plausible values of the additional resistance that lie mainly between $0.3/u^*$ and $0.4/u^*$, where u^* is the traditional friction velocity. They then showed that predicted rates of evaporation from the canopy and from wet soil beneath are not very sensitive to the introduction of the near-field resistance because it is an order of magnitude less than the resistances already included in the Shuttleworth–Wallace model.

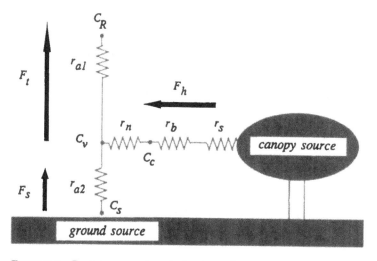

FIGURE 7   Resistance network for transfer processes in vegetation incorporating a "near-field" resistance r_n in series with a bulk boundary layer resistance r_b (equivalent to r_a' in Fig. 6) and a physiological resistance r_s (After van den Hurk and McNaughton, 1995.) (C is the concentration of an entity at screen height, vegetation height, or the soil surface, subscripts R, v, and s, respectively. F is a flux from the soil, from vegetation, or from the total soil–vegetation system, subscripts s, h, and t, respectively.)

Raupach's analysis represents major progress in attempts to relate vertical profiles of water vapor concentration within a canopy to local losses by transpiration that play a major role in the water economy of leaves; and gains by condensation that promote the spread of fungal diseases.

Multiple-source models of evaporation have also been used to predict and to optimize the rate of drying of hay (e.g., Tuzet et al., 1993).

VII. EVAPORATION AND ADVECTION

In earlier sections of this review, it was assumed implicitly that measurements of climate obtained at some standard height and subsequently used to estimate evaporation rates were representative of the surrounding area. This assumption is invalid when the physical properties of the surface are discontinuous as a result of major changes in vegetation height, for example, or in the availability of water following irrigation. Several models have been developed to explore changes in evaporation rate and temperature when air in equilibrium with a relatively hot dry surface such as bare soil encounters the cool wet surface of irrigated vegetation. In early work (e.g., Philip, 1959, 1987; McNaughton, 1976), surface resistance was assumed to be independent of the imposed climate, but Kroon and de Bruin (1993) described a numerical model in which resistance was a function of vapor pressure deficit and temperature and in which changes of surface roughness were accounted for.

Itier and Perrier (1976) and Brunet et al. (1994) derived a function for the change in concentration of a scalar downwind of a step change in the scalar flux. They applied this to the case of an irrigated field downwind of dry land with step changes in temperature and humidity both specified and with the net flux of heat assumed constant throughout the system. They also argued that the evaporation rate downwind of the discontinuity could be assumed constant because many laboratory measurements appear to show that stomatal resistance increases with saturation vapor pressure deficit. However, it is more likely that stomata respond to the flux of water vapor through them than to a vapor pressure deficit [Eq. (23)].

VIII. MODELS FOR REMOTE SENSING OF EVAPORATION RATES

Crop canopies emit radiative signals that can be used to derive information about their rates of transpiration and water status. Attention has been concentrated mainly on (1) radiation emitted in the long-wave spectrum from which a radiative surface temperature can be obtained as a step in estimating the availability of water and (2) relative irradiance in the red and near-infrared spectrum from which vegetation indices may be derived as a step toward estimating ground cover.

In terms of aerodynamic processes, the effective surface temperature of a canopy, T_e is defined by the equation that relates sensible heat flux C to air temperature at a reference height, T_a, and aerodynamic resistance r_a:

$$C = \rho c_p (T_e - T_a)/r_a \tag{38}$$

The mean rate of evaporation can then be estimated (in mm/day) as

$$E = [H - \rho c_p (T_e - T_a)/r_a]/\lambda + \omega \tag{39}$$

where H is total heat flux and ω is an error term that includes the soil heat flux. According to Choudhury and de Bruin (1995), typical values of these parameters estimated by remote sensing over periods on the order of a day are $\omega = -0.1$ mm/day (range -0.1 to $+0.1$ mm/day) and $\rho c_p/r_a = 0.37$ mm/(day·K) [range 0.25–0.64 mm/(day·K)]. This technique is valid only when the surface is unobscured by cloud, and the authors warn that "correction of infrared temperature observations for the surface emissivity and atmospheric effects remains a challenging problem." Another unresolved problem is the existence of systematic differences between radiative surface temperature and aerodynamic surface temperature as observed by Huband and Monteith (1986), for example.

By eliminating E from Eqs. (9) and (39), the aerodynamic surface temperature of uniform vegetation can be expressed as

$$T_e = T_a + \frac{\gamma^* r_a R_n/\rho c_p - D}{\Delta + \gamma^*} \tag{40}$$

Equation (40) is the theoretical basis of an empirical relation between radiative surface temperature T_r and saturation deficit referred to by Idso (1982) as a "non-water-stressed baseline." When T_r was measured with a radiometer held over well-watered vegetation in bright sunlight, and D was measured at some reference height above the canopy, Idso demonstrated that radiative surface temperature decreased linearly with increasing D as predicted by Eq. (40) when R_n and γ^* are effectively constant. The baseline equation can therefore be written as

$$T_r = T_{r0} - aD \tag{41}$$

where T_{r0} is a maximum surface temperature estimated by extrapolation for the (unattainable) condition $D = 0$. Once the parameters of this relation have been established, it can be used to detect anomalously large values of surface temperature that are an indication of water stress.

Idso (1987) later suggested that because this simple linear model was valid within narrow limits for a wide range of crop types, the canopy resistance must be effectively constant during much of the day, in conflict with extensive evidence from porometers. However, it can be shown that the closure of stomata in response to increasing demand for water [Eq. (23)] is compatible with a linear relation between T_e and D.

This type of analysis is valid only when ground cover is complete so that radiative signals from foliage are not contaminated by emission from the soil. More generally, the spatial distribution of evaporation rate and of surface resistance can be estimated from satellite images that provide estimates of ground cover, surface temperature, and reflectivity, supplementing surface-based measurements of air temperature and aerodynamic resistance (e.g., Bastiaasen et al., 1994).

IX. CONCLUSION

Crop plants respond to their environment in ways that often seem too complex for rigorous quantitative descriptions as a basis for model building and prediction. Fortunately, although transpiration is always under physiological control, it is essentially a physical process and is therefore subject to aerodynamic and thermodynamic laws that provide a robust basis for models of the type described in this chapter. The use of such models to predict crop water use has therefore made substantial progress over the past 50 years and must continue to expand and diversify as the world's water becomes an increasingly scarce and precious resource.

REFERENCES

Allen, R. G., M. Smith, A. Perrier, and L. S. Pereira. 1994. An update for the definition of reference evapotranspiration. International Commission for Irrigation and Drainage Bulletin 43, No. 2.

Ball, J. T., I. E. Woodrow, and J. A. Berry. 1987. A model predicting stomatal conductance. In: *Progress in Photosynthesis Research*. Vol. IV. (J. Biggins (Ed.). Dordrecht: Martinus Nijhof, pp. 221–234.

Bastiaasen, W. G. M., D. H. Hoekman, and R. A. Roebeling. 1994. A methodology for the assessment of surface resistance and soil water storage variability at mesoscale based on remote sensing measurements. IAHS Special Publication No. 2. Wageningen: Wageningen Agricultural University.

Blaney, H. F., and W. D. Criddle. 1962. Determining Consumptive Use and Irrigation Water Requirements. Tech. Bull. 1275. Washington, DC: Agricultural Research Service, USDA.

Brunet, Y., B. Itier, K. J. McAnaney, and J. P. Lagouarde. 1994. Downwind evolution of scalar fluxes and surface resistance under conditions of local advection. I. A reappraisal of boundary conditions. *Agric. Forest Meteorol.* 71:211–225.

Brunt, D. 1939. *Physical and Dynamical Meteorology.* London: Cambridge Univ. Press.

Budyko, M. I. 1948. *Evaporation Under Natural Conditions.* Leningrad: GIMIZ. English Transl.: Jerusalem: Israel Program Sci. Transl., 1963.

Bunce, J. A. 1997. Does transpiration control stomatal response to vapor pressure deficit? *Plant, Cell and Env.* 20:131–135.

Businger, J. 1956. Some remarks on Penman's equation for evapotranspiration. *Neth. J. Agric. Sci.* 4:77–80.

Calder, I. R. 1977. A model of transpiration and interception loss from a spruce forest in Plynlimon, Central Wales. *J. Hydrol.* 33:247–265.

Calder, I. R. 1986. The influence of land use on water yield in upland areas of the U.K. *J. Hydrol.* 88:201–211.

Choudhury, B. J., and H. A. R. de Bruin. 1995. First order approach for estimating unstressed transpiration from meteorological satellite data. *Adv. Space Res.* 16:10167–10176.

Choudhury, B. J., and J. L. Monteith. 1988. A four-layer model for the heat budget of homogeneous land surfaces. *Quart. J. Roy. Meteorol. Soc.* 114: 373–398.

Deardorff, J. W. 1978. Efficient prediction of ground surface temperature and moisture with inclusion of a layer of vegetation. *J. Geophys. Res.* 83:1889–1903.

Dolman, A. J. 1993. A multiple-source surface energy balance model for use in general circulation models. *Agric. and Forest Meteorol.* 65:21–45.

Doorenbos, J., and W. O. Pruitt. 1977. Crop water requirements. Irrigation and Drainage paper, No. 24 (revised). FAO, Rome.

Dougherty, R. L., J. A. Bradford, P. I. Coyne, and P. L. Sims. 1994. Applying an empirical model of stomatal conductance to three C-4 grasses. *Agric. Forest Meteorol.* 67:269–290.

Driedonks, A. G. M. 1981. Dynamics of the Well-Mixed Boundary Layer. Sci. Rep. WR 81-2, KNMI, de Bilt, Netherlands.

Duchon, C. E., and E. Wilk. 1994. Field comparisons of direct and component measurements of net radiation under clear skies. *J. Appl. Meteorol.* 33: 245–251.

Duynkerke, P. G. 1992. The roughness length for heat and other vegetation parameters for a surface of short grass. *J. Appl. Meteorol.* 31:579–586.

Field, R. T., L. J. Fritschen, E. T. Kanemasu, E. A. Smith, J. B. Stewart, S. B. Verma, and W. P. Kustas. 1992. Calibration, comparison and correction of net radiation instruments used during FIFE. *J. Geophys. Res.* 97(D17): 18681–18695.

Frenzen, P. and C. A. Vogel. 1995. A further note on the magnitude and apparent range of vanation of the von Karman constant. *Boundary-Layer Meteorology* 75:315–317.

Garrat, J. R. 1992. *The Atmospheric Boundary Layer.* London: Cambridge Univ. Press.

Gash, J. H. C., W. J. Shuttleworth, C. R. Lloyd, J. C. Andre, and J. P. Goutorbe. 1989. Micrometeorological measurements in Les Landes forest during Hapex-Mobilhy. *Agric. Forest. Meteorol.* 46:131–147.

Grant, D. 1975. Comparison of evaporation from barley with Penman estimates. *Agric. Forest Meteorol.* 15:49–60.

Hargreaves, G. H., and Z. A. Samani. 1982. Estimating potential evapotranspiration. Tech. Note. *J. Irrig. Drainage, ASCE* 108:225–230.

Huband, N. D. S., and J. L. Monteith. 1986. Radiative surface temperature and energy balance of a wheat canopy. I. Comparison of radiative and aerodynamic canopy temperature. *Boundary-Layer Meteorol.* 36:1–17.

Hunt, E. R., S. W. Running, and C. A. Federer. 1991. Extrapolating plant water flow resistances and capacitances to regional scales. *Agric. Forest Meteorol.* 54:169–195.

Idso, S. B. 1982. Non-water-stressed baselines: A key to measuring and interpreting plant water stress. *Agric. Meteorol.* 27:59–70.

Idso, S. B. 1987. An apparent discrepancy between porometry and infrared thermometry relative to the dependence of stomatal conductance on air vapour pressure deficit. *Agric. Forest Meteorol* 40:106–107.

Itier, B., and A. Perrier. 1976. Presentation d'une etude analytique de l'advection. I. Advection liee aux variations horizontales de concentration et de temperature. *Ann. Agron.* 27:111–140.

Jarvis, P. G. 1976. The interpretation of the variations in leaf water potential and stomatal conductance found in canopies in the field. *Phil. Trans. Roy. Soc. Lond.* B 273:593–602.

Jensen, M. E., and H. R. Haise. 1963. Estimating evapotranspiration from solar radiation. *J. Irrig. Drainage, ASCE* 89:15–41.

Jones, H. G. 1992. Plants and Microclimate (2nd Ed.). Cambridge University, Cambridge.

Jury, W. A., and C. B. Tanner. 1975. Advection modification of the Priestley and Taylor evapotranspiration formula. *Agron. J.* 67:840–842.

Kelliher, F. M., R. Leuning, M. R. Raupach, and E. D. Schulze. 1995. Maximum conductances for evaporation from global vegetation types. *Agric. Forest Meteorol.* 73:1–16.

Kroon, L. J. M., and H. A. R. de Bruin. 1993. Atmosphere–vegetation interaction in local advection conditions. *Agric. Forest Meteorol.* 64:1–28.

Lee, T. J., and R. A. Pielke. 1992. Estimating the soil surface specific humidity. *J. Appl. Meteorol.* 31:480–484.

Leuning, R. 1995. A critical appraisal of a combined stomatal-photosynthesis model for C₃ plants. *Plant Cell Environ.* 18:339–355.

Lohammar, T., S. Larsen, S. Linder, and S. O. Falk. 1980. FAST simulation model of gaseous exchange in Scots pine. *Ecol. Bull.* 32:505–523.

McKenny, M. S., and N. J. Rosenberg. 1993. Sensitivity of some potential evapotranspiration estimation methods to climate change. *Agric. Forest Meteorol.* 64:81–110.

McNaughton, K. G. 1976. Evaporation and advection. II. Evaporation downwind of a boundary separating regions having different surface resistances and available energies. *Quart. J. Roy. Meteorol. Soc.* 102:193–202.

McNaughton, K. G., and P. G. Jarvis. 1983. Predicting effects of vegetation changes on transpiration and evaporation. In: *Water Deficits and Plant Growth.* Vol. 7. T. Kozlowski (Ed.). New York: Academic Press.

McNaughton, K. G., and T. W. Spriggs. 1989. A mixed-layer model for regional evaporation. *Boundary-Layer Meteorol.* 34:243–262.

Makkink, G. F. 1957. Testing the Penman formula by means of lysimeters. *J. Inst. Water. Eng.* 11:277–288.

Mihailovic, D. T., B. Rajkovic, L. Dekic, R. A. Pielke, T. J. Lee, and Y. Zhopia 1995. The validation of various schemes for parameterising evaporation from bare soil. *Boundary-Layer Meteorol.* 76:259–289.

Mintz, Y., and G. K. Walker. 1993. Global fields of soil moisture and land surface evapotranspiration derived from observed precipitation and surface air temperature. *J. Appl. Meteorol.* 32:1305–1333.

Monteith, J. L. 1963. Calculating evaporation from diffusive resistances. In: Investigation of Energy and Mass Transfers Near the Ground. Final Report under Contract DA-36-039-SC-80334. U.S. Army Electronic Proving Ground's Technical Program.

Monteith, J. L. 1965. Evaporation and the environment. *Symp. Soc. Exper. Biol.* 19:205–234.

Monteith, J. L. 1981. Evaporation and surface temperature. *Quart J. Roy. Meteorol. Soc.* 107:1–27.

Monteith, J. L. 1986. Manipulation of water by crops. *Phil. Trans. Roy. Soc. Lond. A* 316:245–259.

Monteith, J. L. 1995. A reinterpretation of stomatal responses to humidity. *Plant, Cell Environ.* 18:357–364.

Monteith, J. L., and M. H. Unsworth. 1990. *Principles of Environmental Physics.* London: Edward Arnold.

Passioura, J. B. 1983. Roots and drought resistance. *Agric. Water Manage.* 7: 265–280.

Pelton, W. L., K. M. King, and C. B. Tanner. 1960. An evaluation of the Thornthwaite and mean temperature methods for determining potential evapotranspiration. *Agron. J.* 52:387–395.

Penman, H. L. 1948. Natural evaporation from open water, bare soil, and grass. *Proc. Roy. Soc. Lond. A193*:120–145.

Penman, H. L., and F. L. Milthorpe. 1967. The diffusive conductance of wheat leaves. *J. Exp. Bot. 18*:442–457.

Philip, J. R. 1959. The theory of local advection I. *J. Meteorol. 16*:535–547.

Philip, J. R. 1987. Advection, evaporation and surface resistance. *Irrig. Sci. 8*: 101–114.

Philip, J. R., and D. A. de Vries. 1957. Moisture movement in porous materials under temperature gradients. *Trans. Am. Geophys. Union 38*:222–232.

Priestley, C. H. B., and R. J. Taylor. 1972. On the assessment of surface heat flux and evaporation using large-scale parameters. *Mon. Weather Rev. 100*: 81–92.

Raupach, M. R. 1989. A practical Lagrangian method for relating scalar concentrations to source distributions in vegetation canopies. *Quart. J. Roy. Meteorol. Soc. 115*:609–632.

Rijtema, P. E. 1965. An Analysis of Actual Evapotranspiration. Report 659. Wageningen: Centre Agric. Publ. Agricultural Research.

Robertson, M. J., S. Fukai, M. M. Ludlow, and G. L. Hammer. 1993. Water extraction by grain sorghum in a humid environment. I. Analysis of water extraction pattern. *Field Crops Res. 33*:81–97.

Russell, G. 1980. Crop evaporation, surface resistance, and soil water status. *Agric. Meteorol. 21*:213–226.

Rutter, A. J., A. J. Morton, and P. C. Robins. 1975. A predictive model of rainfall interception in forests. 2. Generalization of the model. *J. Appl. Ecol.*, 12:367–380.

Seginer, I. 1984. Night transpiration of greenhouse roses. *Agric. Forest Meteorol. 30*:257–268.

Shaw, R. H., and A. R. Pereira. 1982. Aerodynamic roughness of a plant canopy: a numerical experiment. *Agric. Meteorol. 26*:51–65.

Shuttleworth, W. J., and I. R. Calder. 1979. Has the Priestley–Taylor equation any relevance to forest evaporation? *J. Appl. Meteorol. 18*:639–646.

Shuttleworth, W. J., and J. S. Wallace. 1985. Evaporation from sparse crops —An energy combination theory. *Quart. J. Roy. Meteorol. Soc. 111*:839–855.

Singh, P., J. L. Monteith, K. K. Lee, T. J. Rego, and S. P. Wani. 1996. Response to fertilizer nitrogen and water of post rainy-season sorghum on a vertisol. *J. Agri. Sci. (in press)*.

Stewart, J. B. 1988. Modelling surface conductance of a pine forest. *Agric. Forest Meteorol. 43*:19–37.

Tanner, C. B., and W. L. Pelton. 1960. Potential evapotranspiration estimates by the approximate energy balance method of Penman. *J. Geophys. Res.*, 65:3391–3396.

ten Berge, H. F. M. 1990. Heat and water transfer in bare topsoil and the lower atmosphere. Wageningen: PUDOC.

Thom, A. S. 1975. Momentum, mass, and heat exchange in plant communities. In: *Vegetation and the Atmosphere*. Vol. 1. J. L. Monteith (Ed.). London: Academic Press.

Thom, A. S., and H. R. Oliver. 1977. On Penman's equation for estimating regional evaporation. *Quart. J. Roy. Meteorol. Soc. 103*:345–357.

Thompson, N., I. A. Barrie, and A. Ayles. 1981. The Meteorological Office rainfall and evaporation calculation system: MORECS (July 1981). Meteorological Office Bracknell, UK, Hydrological Memorandum No. 45.

Thornthwaite, C. W. 1948. An approach towards a rational classification of climate. *Geogr. Rev.* 38:55–94.

Tuzet, A., A. Perrier, and A. K. Oulid Aissa. 1993. A prediction model for field drying of hay using a heat balance method. *Agric. Forest Meteorol.* 65: 63–89.

Van Bavel, C. H. M., and D. I. Hillel. 1976. Calculating potential and actual evaporation from a bare soil surface by simulation of concurrent flow of water and heat. *Agric. Meteorol.* 17:453–476.

Van den Hurk, B. J. J. M., and K. G. McNaughton. 1995. Implementation of near-field dispersal in a simple two-layer resistance model. *J. Hydrol.* 166: 293–311.

Van den Hurk, B. J. J. M., A. Verhoef, A. R. van den Berg, and H. A. R. de Bruin. 1995. An intercomparison of three vegetation/soil models for a sparse vineyard canopy. *Quart J. Roy. Meteorol. Soc.* 121:1867–1889.

Viterbo, P. and A. C. M. Beljaars. 1995. An Improved Land Surface Parameterization Scheme in the ECMWF Model and Its Validation. Tech. Report TR-75. Reading, UK: ECMWF.

Vogt, R., and L. Jaeger. 1990. Evaporation from a pine forest. *Agric. and Forest Meteorol.* 50:39–54.

8

Simulation in Crop Management: GOSSYM/COMAX

Harry F. Hodges, Frank D. Whisler, Susan M. Bridges, and K. Raja Reddy

Mississippi State University, Mississippi State, Mississippi

James M. McKinion

Agricultural Research Service, U.S. Department of Agriculture, Mississippi State, Mississippi

I. INTRODUCTION

Agriculturalists and plant scientists have been interested in predicting plant responses to their environment for centuries. Even as late as the middle of the twentieth century it became apparent that some relatively simple relationships between plant development and temperature could be established that aided modern agriculture. Accumulated heat units, expressed as growing degree-days, can be used to predict the maturation of crops and therefore can be useful in scheduling their harvest. By the early 1970s, it became apparent that high speed computing capabilities were imminent. Crop scientists, entomologists, and other agriculturalists began to seriously consider the possibility of modeling growth and developmental processes in both crops and insects. The variability among soils imposed addi-

235

tional difficulties, as these differences caused temperature variability as well as different water and nutrient conditions. These, in turn, caused plant responses to be different even with similar above-ground weather conditions.

The objective of this chapter is to describe a process-level cotton crop simulation model, how such a model led to an expert system to expedite its use, and how these technologies can be used in crop management and to augment future crop production technologies.

Crop models are needed that simulate plant response to weather as well as soil processes and their effects on plants. Weather conditions interact with soil water availability and affect water status in plants. The water status conditions influence how several plant processes respond to other environmental factors. Unless these processes and their intraplant interactions have the appropriate relationships defined and modeled correctly, the simulations will not be accurate enough to be used in real-world situations. It was apparent that whole crop production systems could and should be modeled. Unfortunately, there is an inclination among many scientists to keep or develop models based on simple relationships that are known, but those relationships fail to predict growth and development when environmental stresses become important in the production process. Several crop simulation models were developed in the late 1970s and 1980s (Baker, 1980; Whisler et al., 1986). These models have been discussed in several places, but most have been used primarily for academic purposes or found useful for only some limited aspect of production management. GOSSYM, a cotton crop model, was developed during that period (Baker et al., 1983).

A.   The Simulation Model—GOSSYM

GOSSYM was intended to simulate crop responses to physical conditions and provide producers with information that would aid in making management decisions. To do that, a comprehensive understanding of the cotton plant itself is needed, including mathematical relationships among weather and soil and the responses of various plant processes to those physical conditions. The information was organized into relatively small closely related packages sometimes called subroutines or modules. GOSSYM's program flow and model structure are presented in Fig. 1.

GOSSYM is the main program from which all of the subroutines vertically below it in the diagram are called. CLYMAT reads the daily weather information and calls DATES, which keeps track of both Jul-

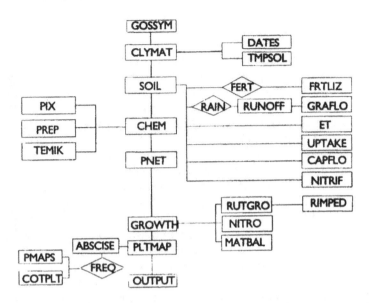

FIGURE 1 Flow diagram of the subroutines and structure of GOSSYM, a cotton crop simulator. See text for details.

ian day number and the calendar date being simulated. CLYMAT then calls TMPSOL, which calculates the soil temperatures by soil layer. SOIL is a mini main program, which calls the soil subprograms (Boone et al., 1995). The soil routines provide the plant model with estimates of soil water potential in the rooted portion of the soil profile, an estimate of the nitrogen entrained in the transpiration stream available for growth, and an estimate of metabolic sink strength in the root system.

The below-ground processes are treated in a two-dimensional grid. The material balances of water and nitrate, ammonia, and organic matter are maintained and updated several times each day. FERTLIZ distributes ammonium, nitrate, and urea fertilizers in the soil profiles. GRAFLO simulates the movement of both rain and irrigation water into the soil profile by gravitational flow. ET estimates the rate of evaporation from the soil surface and transpiration from the plant. UPTAKE calculates the amount of the nitrogen and water taken from the soil region where roots are present. CAPFLO estimates the rewetting of dry soil from wetter soil by capillary flow. NITRIF calculates the conversion of ammonium to nitrates by bacterial action in the soil medium. CHEM is also a mini main program, one that calls two subprograms to calculate the effect of chemicals on plant physiological processes. These programs are PIX (K. R. Reddy et al., 1995a) and PREP (V. R. Reddy, 1995). PIX deals with the effects of the plant growth regulator mepiquat chloride, and PREP deals with the effect of a boll opener, ethephon, and defoliant chemicals.

In PNET, leaf water potential, canopy light interception, photosynthesis, and respiration are calculated. Then, in GROWTH, potential dry matter accretion of each organ is calculated from temperature. These potential organ growth rates are adjusted for turgor and nitrogen availability. Photosynthates and any reserve carbohydrates are partitioned to the various organs in proportion to growth requirements. The partition control factor is the carbohydrate supply/demand ratio. RUTGRO calculates the potential and actual growth rates of roots. RIMPED calculates the effect of increasing soil bulk density and penetration resistance on the capability of roots to elongate. NITRO calculates the partitioning of nitrogen in the plant.

MATBAL keeps track of the nitrogen and carbon material balance in all parts of the plant and soil complex. In PLTMAP, fruit loss and developmental delays are calculated using both carbohydrate and nitrogen supply/demand ratios. These developmental delays are used to allow the simulator to slow the plastochron intervals that are calculated as functions of temperature depending on the intensity of the stress. ABSCISE estimates the abscission rate of fruit, squares (cotton flower buds), and leaves caused by nutrient or water deficits. PMAPS, COTPLT, and OUTPUT print various user-selected reports from the model. The program cycles through these subroutines one day at a time from emergence to the end of the season. A more complete description of the subroutines and how GOSSYM works can be found in Baker et al. (1983) and Baker and Landivar (1991).

GOSSYM simulates both crop and soil responses to weather and cultural practices. It was first field tested on private farms in 1984. It quickly became apparent that, although the in-season information GOSSYM could provide was interesting and useful, interpretation of the tabular data results and the time required to provide the model with weather data and cultural practice information was too time-consuming for the busy producer. Therefore, an expert system COMAX (CrOp MAnagement eXpert) was developed and coupled with GOSSYM to provide assistance for both of those activities and to automate some management decision aids.

This combination of a simulation model and an expert system that automated running the model reduced the operator time required to obtain a model-aided management decision. Today, if long-term weather records are available, the producer may run the model with many weather and management scenarios to increase the probability of selecting the management practice that most nearly fits his or her objectives.

Such a model, based on important physiological processes and their responses to physical conditions, should be able to correctly simulate unique combinations of conditions that were not explicitly represented in the model development process. Such model capability provides the characteristics that are needed in combination with precision agriculture. Precision agriculture will cause producers to question why certain areas yield more than other areas. Process-level mechanistic models will help provide that information.

II. PLANT PROCESSES

One needs data that provide information at an appropriate level of detail to model the way the crop responds to environmental conditions. We generally need biological rate functions on the whole plant or canopy level rather than on a more detailed physiological or biochemical process level. An understanding of the physiology and biochemistry is desirable; however, modeling at that level requires too much simulation time and often requires assumptions that many scientists are not willing to make.

To obtain suitable data for developing GOSSYM, naturally lighted plant growth chambers known as Soil-Plant-Atmosphere-Research (SPAR) units were designed and constructed that had temperature, CO_2, water, and nutrients controlled electronically. Although their design has evolved over years of use, their basic structure is essentially as described by Phene et al. (1978), Parsons et al. (1980), and McKinion and Hodges (1985). This facility allowed us to determine the relationships of the varied factor with crop response. These experiments were conducted in environments in which major effects were made to allow only the variable being tested to be limiting. Plants were typically grown in well-watered sand and adequately fertilized with all known mineral nutrients (K. R. Reddy et al., 1992a, 1992b). Plants were also grown in higher CO_2 to enhance photosynthesis to avoid carbon deficits, and the plants were not subjected to disease or insect infestations. Thus, the plants were grown at their potential rates and limited only by the variable being tested. Potential rates may be estimated from relationships developed in this manner and then corrected by stress factors known to occur in the natural environment. Thus, potential rates may be estimated from the temperatures that occurred and actual rates simulated by reducing the potential rates with "stress factors."

To comprehend a plant's responses to its environment, one needs to understand the morphological characteristics of the crop and its responses to temperature, water, and nutrient supply. It is necessary to describe the plant as a whole and each facet of growth at the organ level. The growth and development of each plant organ are also influenced by competition from other organs as well as by environmental conditions. Mauney (1986) described the anatomy and morphology of cultivated cotton. The primary axis of the cotton plant results from elongation and development of the embryo. In the dormant seed, the primary axis consists of a radicle, a hypocotyl, and a poorly developed epicotyl. The epicotyl contains one true leaf initial and a dome of meristematic cells. The growing cotton plant actively proliferates new cells on many fronts. Thus, all the differentiation of the vegetative framework above the cotyledons takes place after germination.

Cotton is indeterminate in growth habit, in that the mainstem apex continuously initiates leaves and axillary buds. The axillary buds on the lower nodes develop into vegetative or monopodial branches if conditions are favorable. The axillary buds in the upper nodes, normally above node 5, develop into fruiting or sympodial branches. Vegetative branches behave much like the mainstem in that they produce both vegetative and fruiting branches. Fruiting branches, on the other hand, initiate one true leaf and then terminate as a flower. Branch elongation is accomplished by growth of axillary buds producing a sympodial zigzag structure (Mauney, 1984; Mutsaers, 1983).

A. Phenology

Phenology is defined as the length of time required for a particular developmental event to occur or the period between events. For example, the length of time from emergence to first flower bud (commonly called *square* in cotton) formation is a phenological event. Also, the time between unfolding of leaves on the mainstem or fruiting branches are phenologically meaningful periods, as is the duration of a leaf or internode expansion.

1. Days to First Square

Floral initiation, defined as the time between plant emergence and the appearance of a square 3 mm in length, is very strongly influenced by temperature. In GOSSYM, daily progress to floral bud emergence (the reciprocal of time is defined as daily developmental rate) is calculated by summing the values for average daily temperature (Fig. 2). The model predicts that a square will be produced when the summed values equal 1 or greater. About 40 days are required at 20°C and only 23 days at 30°C. Temperatures above 28°C do not speed the floral initiation process, nor does additional photosynthates (K. R. Reddy et al., 1993a). Progress toward square formation actually slows at temperatures above 28°C, suggesting that injury occurs at higher temperatures. No progress toward floral development occurs below 15°C.

When Baker et al. (1983) developed GOSSYM, the days to first square value was based on data of Moraghan et al. (1968). That model predicted 20% more time to produce first square than was observed in field conditions. They calibrated the model by introducing a multiplier into the temperature response function for this phenological event. This adjustment improved the predictions at near-optimum temperature, but predicted rates were too fast at below- and above-optimum temperatures. We have since learned that modern cultivars respond to temperature differently than those grown several years

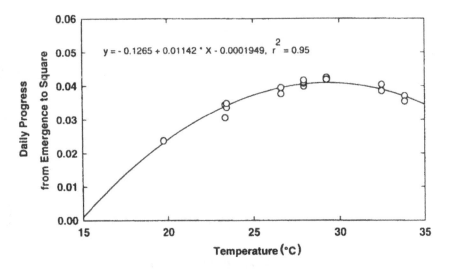

FIGURE 2 Daily progress of upland cotton from emergence to first flower bud (square) as a function of average daily temperature. Appearance of a square is predicted when the sum of the daily progress values equals 1. (From K. R. Reddy et al., 1993a.)

ago. Modern cultivars initiate squares much earlier than older cultivars and are able to produce squares and fruiting structures at higher
temperatures (K. R. Reddy et al., 1993a).

Others have used growing degree-days as a way of summarizing temperature data to predict time of first square (Jackson, 1991).
This procedure is appropriate for predicting phenological events if
the temperature response of the phenological rate is linear over the
range of temperatures experienced but is inappropriate when the
event being simulated is nonlinear with temperature.

2. Days from Square to Flower

The reciprocal of square maturation period, the inverse of time from
square appearance to flower, was modeled as a function of temperature (Fig. 3). Daily progress of the square maturation period increased as temperature increased at the low end of the temperature
range; however, the developmental rate at temperatures above 27°C
did not increase. Developmental progress equations also projected to
zero development from squaring to flower at about 15°C. Our results
on rates of progress from square to flower formation were similar to
those reported by Hesketh and Low (1968) and Hesketh et al. (1972),

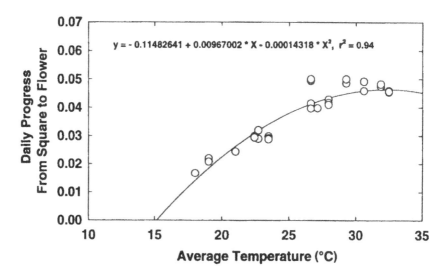

FIGURE 3 Daily progress of upland cotton from the appearance of first
square until first flower. First flower is predicted when the sum of the daily
progress values equals 1. (From K. R. Reddy et al., 1993a.)

except at low temperature where they had only limited data. Apparently changes in varieties during the past few decades have not influenced the sensitivity of cotton to temperature during the square maturation period. About 46 days were required at 20°C to develop flowers after squares were visible, but only 21 days at 30°C.

3. Boll Maturation Period and Boll Size

The time required from flower formation to a mature boll is also very temperature-dependent. The daily developmental rate concept was used in GOSSYM to calculate the boll maturation period. The daily progress was nearly linear throughout the temperature range tested (Fig. 4). Progress toward open bolls was only about one-half as fast at 20°C as at 30°C. About 40 days were required from anthesis to open boll at 30°C. Boll maturation period of these modern cotton cultivars was faster at low temperatures and slower at high temperatures than those reported by Hesketh and Low (1968) for the cultivars used in commercial cotton production several decades ago.

We did not see a decline in rate of boll development in temperatures above 27°C like that observed for rates of square formation; however, we did observe lower boll weights (data not shown) for

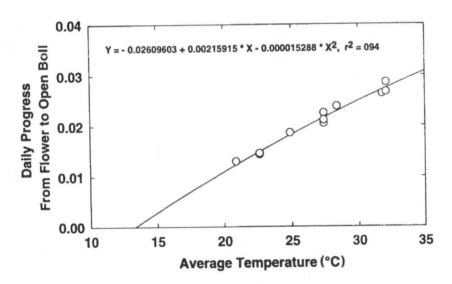

FIGURE 4   Daily progress of upland cotton from flowering to maturity as a function of temperature. Appearance of open boll is predicted when the sum of the daily progress values equals 1. (From K. R. Reddy et al., 1993a).

bolls produced at higher temperatures. Maximum boll weight of pima cotton was attained at 26°C with only 88% and 44% as much dry matter produced per boll when plants were grown at 20°C and 32°C, respectively (Hodges et al., 1993). Upland cotton averaged 6.0 g per boll on plants grown at 26°C, but bolls weighed 10% less on plants grown at 20°C and 32°C. Pima was considerably more sensitive to high temperatures than upland cotton.

4.   Nodes of Mainstem, Fruiting, and Vegetative Branches

The addition of mainstem and fruiting branch nodes are important aspects of cotton crop development, because these determine the number of leaves produced and thus canopy development and light interception. The reciprocal of days required to produce a node above the first fruiting branch on the mainstem and the reciprocal of days required to produce nodes on the fruiting branches themselves are presented as a function of temperature in Fig. 5. Mainstem nodes were produced at progressively faster rates as temperature

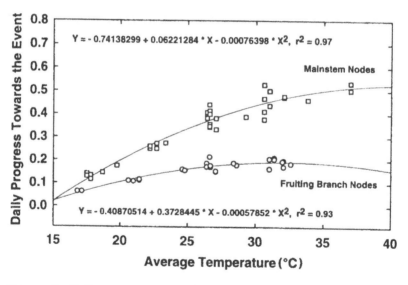

FIGURE 5   Daily progress to produce a mainstem or fruiting branch node since the previous node as a function of temperature. A node is defined as a discrete event by determining the time the three main veins can be seen on the unfolded subtending leaf. The next node (or leaf) is predicted when the sum of the daily progress values equals 1 since the previous node (or leaf) on that organ was formed. (From K. R. Reddy et al., 1997b.)

increased to 37°C. The rate of mainstem node addition increased more than rates of fruiting branch nodes increased as temperature increased. Production of leaves and nodes on the mainstream was more rapid at warmer temperatures than the production of fruiting branch leaves. Thus, growing temperature alters the architectural form of the plant.

Studies in which phenology of cotton is predicted by even simple temperature models (Jackson, 1991) need detailed organ development–temperature functions to accurately predict plant development and growth relationships. Progress toward the next node of both mainstem and fruiting branches decreased to zero as temperature decreased to 14.5°C. The number of days required to produce an additional mainstem node above the first fruiting branch was very stable at near-optimum temperature (Fig. 6). The first fruiting branch was produced at the sixth mainstem node. Only 2–3 days were required to produce any mainstem nodes in positions 6–17. There were no significant differences in time required to produce mainstem nodes due to atmospheric CO_2 concentrations, suggesting that carbon supply was not limiting vegetative development during this period.

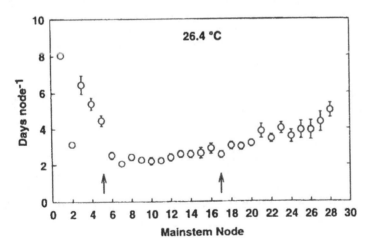

FIGURE 6 The number of days required for cotton plants to produce a node at different positions on the mainstem when grown at near-optimum temperature. The arrows are the positions at which the first fruiting branch and the node of the unfolding leaf at first flower were produced. (From K. R. Reddy et al., 1995b)

The time required to produce mainstem nodes above position 17 increased as the nodes were formed later. These nodes were formed while fruit was being produced at lower positions. These delays in node formation rate were probably due to limited carbohydrates. In GOSSYM, the carbohydrate supply/demand ratio (discussed in detail in Section II.E) was used to delay the potential developmental rates of various organs.

The time required to produce mainstem nodes prior to the first fruiting branch node was considerably longer than that required after a fruiting branch was produced. The reasons cotton plants require longer to produce prefruiting mainstem branch nodes are not known. We know that during early development a considerable fraction of the total dry matter produced was partitioned to root growth (Hodges et al., 1993). One could argue that such a genetically controlled partitioning coefficient early in the life of the plant might leave only limited energy available for shoot growth and thus delay leaf development. However, if leaf development was delayed due to insufficient carbohydrates, then growing plants in high CO_2 environments should alleviate the carbohydrate shortage and allow faster leaf developmental rates on the prefruiting nodes. That did not occur in either of our high CO_2 environments, so the reason for slower leaf development at the prefruiting nodes is still unexplained. In GOSSYM, the rate of prefruiting node formation was calculated to be slower than that of the fruiting branch nodes by applying a multiplication factor to the temperature rate function for fruiting branch node formation on the mainstem.

Vegetative branch production depends on the growth rate or carbohydrate status of the plant. In the field, we find vegetative branches produced on plants adjacent to positions where more light or nutrients are available. In controlled environments, vegetative branches are produced where growth is slow, as in low temperature or where atmospheric CO_2 is high. These results suggest that some unknown level of carbohydrates must be available for vegetative branches to be formed. Since we do not know that threshold, GOSSYM does not predict vegetative branches before the first square is produced. The model may then predict a vegetative branch if conditions are favorable.

5. Expansion Duration of Leaf and Internode

The duration of mainstem leaf expansion and of each mainstem internode elongation decreased gradually as temperature increased to about 30°C. The daily progress of leaf expansion and internode elon-

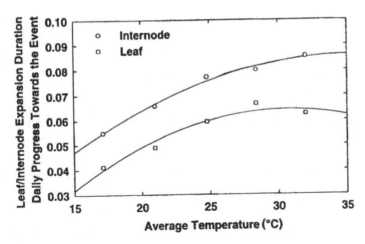

FIGURE 7   Daily progress toward leaf and internode expansion duration as a function of temperature. The model predicts that the organ is fully expanded when the sum of the daily progress values equals 1. (From K. R. Reddy et al., 1997b.)

gation duration is plotted against temperature in Fig. 7. Leaf petiole elongation occurred simultaneously with the lamina expansion (data not shown). About 21 days were required for leaves to fully expand at 20°C and only 15 days at 30°C, whereas internodes required about 16 days to expand at 20°C and only 12 days at 30°C. Leaf expansion duration at a particular temperature was similar regardless of the leaf position on the mainstem (K. R. Reddy et al., 1993b). Expansion durations of leaves and internodes were not available for the models of Baker et al. (1983), Jackson (1991), and Kiniry et al. (1991) but have since been incorporated into GOSSYM (K. R. Reddy et al., 1997a, 1997b).

B.   Photosynthesis

Light interception and therefore photosynthesis depends primarily on plant height and row spacing. The rationale for this is stated in Baker et al. (1978). Briefly, they found that leaf area index (LAI) was a poor predictor of canopy light interception, because internode lengths and leaf canopy density vary in cotton depending on whether the crop is grown under water-deficit conditions or in moist, well-watered conditions. Canopies produced in dry conditions produced shorter internodes and captured a lower percentage of the incoming radiation.

Therefore, they reasoned that the ratio of plant height to the distance between rows is a better index of ground cover and a better indicator of light interception in cotton. However, in a mature, defoliating canopy this relation does not remain true. They also empirically adjusted the light capture relationship from field data so that during maturation, when the LAI became less than 3.1, the light interception decreased linearly with LAI. Photosynthesis is estimated in GOSSYM on a canopy basis. The rationale for using a canopy model, which is relatively empirical in nature, includes several considerations for the alternatives. The basic physiology of photosynthesis is controlled at the leaf level; however, there is considerable difficulty in scaling up to a canopy level. The reasons for such difficulties include the problem of defining leaf environment. How much radiation strikes a leaf varies with position in the canopy and the angle to the sun. That angle is also modified by the heliotropic nature of cotton leaves. In addition, the age and nutritional status of individual leaves need to be taken into account, and, perhaps most important, the difficulty of validating a leaf element model requires data that are not readily available. For these reasons, Baker et al. (1983) chose to model photosynthesis on a canopy basis. Their aim was to provide the model each day of the growing season with an appropriate increment of dry matter produced that was a function of the plant age, size, physiological status, and environmental conditions and to distribute it to the growing points in the plant.

Canopy gross photosynthesis (P_g) responses to radiation are plotted and shown in Fig. 8. Gross photosynthesis is equal to net photosynthesis plus respiration. Respiration is a function of biomass and temperature. Canopy photosynthesis was measured under a range of temperature, vapor pressure deficits, and fertility conditions but with good soil moisture conditions. These individual factors significantly influenced photosynthesis, but none changed the estimate of photosynthesis by 5% over the conditions tested. Therefore, Baker et al. (1983) used the relationship in Fig. 8 to calculate hourly photosynthate yield and daily totals. Those values are used from daily total radiation to simulate daily potential P_g in GOSSYM. They did not adjust P_g for nitrogen status because in their tests the nitrogen range was not sufficient to cause differences.

Data more recently acquired but not incorporated into a commercially used version of GOSSYM show that nitrogen-deficient plants are also carbon-deficient plants because of the effect of nitrogen deficits on photosynthesis (K. R. Reddy et al., 1997a). A 25% decrease in photosynthesis occurred when leaf nitrogen declined to 20% of

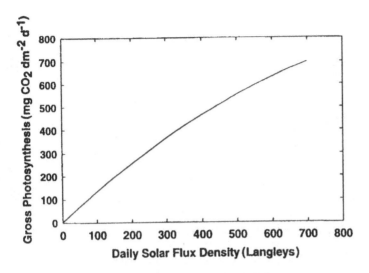

FIGURE 8 Daily canopy gross photosynthesis (net photosynthesis plus respiration) as a function of solar radiation. (From Baker et al., 1983.)

optimum nitrogen levels, and photosynthesis decreased to 50% of maximum when leaf nitrogen was reduced to 60% of its maximum value.

The effects of water deficits on photosynthesis are shown in Fig. 9. Cotton canopy photosynthesis declined as midday leaf water potentials decreased. It appears that little change occurs in canopy photosynthesis at leaf water potential values above about −1.5 to −1.7 MPa, although the trend line causes one to infer that the maximum values are nearer zero.

C. Expansion of Organs

The size and age of each organ on the plant is considered, and the potential growth rate is calculated as a function of temperature, assuming no shortage of photosynthesis or nitrogen. Potential growth is calculated for day and night periods separately in GOSSYM, using temperature and water deficits appropriate to those respective time periods.

1. Leaves

The potential leaf area expansion is estimated based on temperature (Fig. 10) (Marani et al., 1985; K. R. Reddy et al., 1995a, 1997b) and then reduced by amounts determined from estimates of leaf water

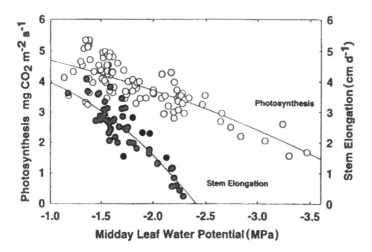

FIGURE 9 Photosynthesis and stem elongation as functions of midday leaf water potential of upland cotton. (From K. R. Reddy et al., 1997a).

potential and mepiquat chloride (PIX) concentration in the leaves (Fig. 10 and K. R. Reddy et al., 1995a). Maximum leaf growth rate occurs at 30°C, and zero expansion occurs at 12.5°C. Leaf expansion also decreased rapidly as temperatures increased above 30°C, even in well-watered plants. After the potential leaf growth was calculated based on temperature, leaf expansion was decreased by a factor based on the estimate of midday leaf water potential followed by a reduction factor based on the PIX concentration in the leaves. A more complete discussion of PIX effects and how they are modeled is described in K. R. Reddy et al. (1995a).

2. Stem Expansion

Potential plant height growth rates are based on the estimated expansion of the top three nodes. Controlled environmental data were not available for stem expansion rates at the time GOSSYM was first developed, so field data were used (Bruce and Romkens, 1965). In the model, stem expansion is based on the age of the top three nodes. Thus, simulated stem expansion is related to developmental rate, for which there were no reliable data from controlled environmental studies in the literature. The expansion rate is written as a function of temperature and of metabolite supply, which may be reduced by water deficits and PIX. A total carbohydrate demand is calculated for day and night periods separately, using appropriate temperature and

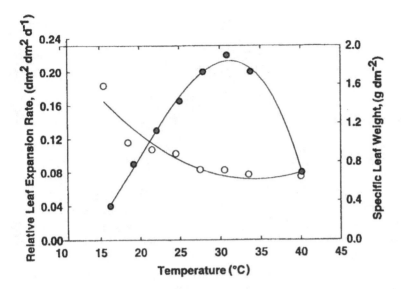

FIGURE 10   Relative leaf expansion rate and specific leaf weight of mature cotton leaves as functions of the temperature at which the leaves were grown. ● = Relative Leaf Expansion Rate; ○ = Specific Leaf Weight. (From Baker et al., 1983.)

water deficit for those time periods. Since data are now available, plant height can be calculated using the prevailing conditions present at the time each internode is elongating (K. R. Reddy et al., 1997a).

3.   Root Expansion

Root expansion is simulated as a function of soil temperature and biomass of roots in each of three age categories: (1) ≤5 days, (2) 5–15 days, and (3) >15 days. The root environment is estimated by keeping track of the physical and chemical properties of a hypothetical grid that is the width of the row spacing and 2 m deep. There are 20 × 40 cells in the grid in which the properties and conditions are initialized at the beginning of the season and updated daily. Further discussion of this is presented in Section III. A materials balance of nitrogen, water, and root mass is maintained for each cell. The concept of age dependency on water uptake and root growth is discussed by Graham et al. (1973). Direction of root growth from a parent cell to adjacent cells in the matrix is controlled by relative amounts of mechanical impedance in the various cells where growth may occur.

Potential root growth is reduced in each cell of the matrix where physical conditions or nitrogen supply is limiting. If growth is limited in certain cells due to unfavorable physical or chemical conditions, the carbohydrate that might have been used in those cells is allocated to other positions in the matrix. This causes root growth to be stimulated in those regions of the profile where conditions are most favorable.

D. Mass Accumulation

Biomass of each organ on the plant is estimated independently and then summed to determine total mass of the plant. To estimate potential growth, the size and age (since initiation) of each organ are used at ambient temperature with the assumption that metabolites are not limiting. A total carbohydrate demand is calculated as the sum of all the individual organ growth potential increments. We have since determined that a better estimate is based on canopy temperature rather than ambient air temperature (K. R. Reddy et al., 1997a). Canopy temperature is simulated from air temperature and leaf water potential, but that concept is not yet incorporated into the commercially used GOSSYM/COMAX. Total carbohydrate demand is compared with carbohydrate supply, and potential growth is adjusted proportionally to accommodate the available supply.

1. Leaves, Stems, Roots, and Bolls

Mass of leaves is estimated in GOSSYM by an adjustment of the leaf area with estimates of specific leaf weight. Specific leaf weight is less sensitive to temperature than leaf expansion (Fig. 10). Specific leaf weight decreased from about 1.6 g/dm^2 to less than 0.8 g/dm^2 as temperature increased from 15 to 40°C.

Potential stem mass accretion follows one rate for the first 32 days and a different rate after 32 days. Stem growth becomes proportional to stem biomass accumulation during the preceding 32 days. The rationale for that concept was based on an assumption that woody structure in stem tissues was not capable of dry matter accretion. Therefore, stem growth was proportional to the recently added dry matter.

A materials balance of water, nitrogen, and root biomass in three age categories is updated daily in each of the cells of the three-dimensional soil profile mentioned under "root expansion" (Section II.C.3) (Boone et al., 1995). Root growth is calculated in two steps. First, potential growth is estimated on the basis of soil temperature and root biomass for an age capable of growth (from 0 to 15 days old). Second, potential root growth is reduced in areas of the soil matrix with a shortage of nitrogen and/or higher penetration resistance. This makes additional carbohydrate available for root growth in grid positions with good nitrogen supplies and/or low penetration resistance. Third, based on the level of water stress in the plant and the ratio of root to stem plus leaf biomass, potential root growth may be increased. This effectively increases root growth in periods of drought if there is a small or negligible fruit load. Finally, the directions of root growth from a parent soil cell to adjacent cells in the root matrix is controlled by relative amounts of mechanical impedance in the various cells in which growth may occur and weighted for downward growth (geotropism). Actual root growth is then calculated as the product of potential growth multiplied by the carbohydrate supply/demand ratio, CSTRES. This partitions photosynthate to each organ on the plant purely on the basis of the contribution of that organ to the total carbohydrate demand.

Potential fruiting structure growth is calculated from temperature and adjusted downward under water-deficit conditions. The temperature responses of squares and boll growth are presented in Figs. 11 and 12.

E. Calculating Stress Effects

The model calculates the carbohydrate requirements for the various organs, then the subroutine NITRO is called. This subroutine esti-

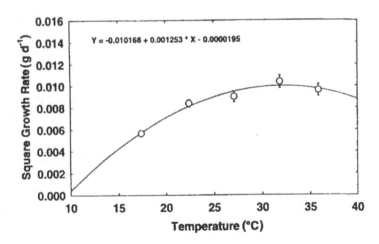

FIGURE 11   Mass accumulation of cotton flower buds (squares) as a function of the temperature at which the plants were grown. (From Reddy et al., 1997a.)

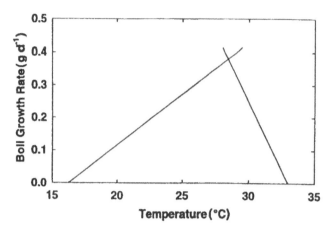

FIGURE 12   Mass accumulation rate of cotton bolls as a function of growing temperature. (From MacArthur et al., 1975.)

mates the nitrogen required for the assimilation of the carbon just estimated for all the organs. The nitrogen requirements are summed for the vegetative and fruiting structures. These sums are used to estimate the total nitrogen requirement (nitrogen demand) for each day.

The nitrogen supply is the nitrogen taken up that day plus any that may have been taken up but not used on previous days. Such a nitrogen "reserve" is assumed to be stored in the leaf. The daily supply/demand ratio is calculated to estimate the maximum fraction of the potential carbohydrate that can be assimilated in above-ground vegetation, roots, and fruiting organs considering the nitrogen limitations. These calculations allow the model to estimate organ growth taking into account the amount of nitrogen available. A conceptual diagram of the nitrogen supply/demand rationale is shown in Fig. 13. A more mechanistic nitrogen supply and stress model is needed.

A carbohydrate supply/demand ratio is calculated in a similar manner to the nitrogen supply/demand ratio (Fig. 14). In this case the carbon supply is estimated from photosynthesis, which is a function of the light intercepted by the canopy and leaf turgor. The C demand is estimated as a function of the total growth potential of all plant organs. That potential is controlled by turgor, age, and temperature of all the organs. Carbon stress is assumed to be the ratio of C supply to C demand. When C stress and N stress are equal to 1, the various organs are not limited by carbon or nitrogen. As the ratio of

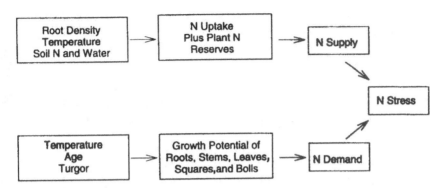

FIGURE 13   Factors that control nitrogen supply and nitrogen requirements of plants. When the supply/demand ratio is less than 1, plant growth is nitrogen-limited.

supply to demand for carbon or nitrogen (carbon or nitrogen stress) becomes less, growth is more limited by those respective nutritional deficits. In extreme stress situations or when there is a heavy fruit load on the plants, nutritional deficits cause abscission of squares or young bolls and delay morphogenesis. A conceptual diagram of nutritional stresses and their physiological effects is presented in Fig. 15. In GOSSYM there is no preference assumed in the allocation of photosynthates on the basis of boll age or proximity to the photosynthesis source.

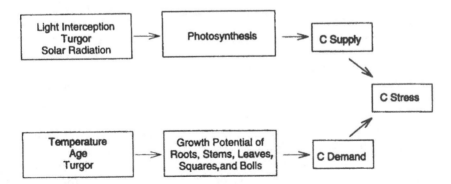

FIGURE 14   Factors that control available carbon. When the carbon supply/demand ratio is less than 1, plant growth is carbon-limited.

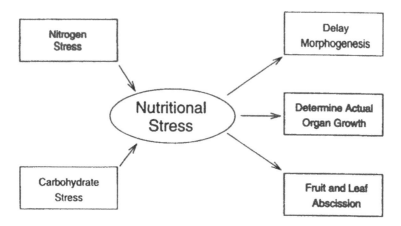

FIGURE 15 Relationships among carbon and nitrogen deficits and plant growth, development, or organ abscission.

III. SOIL-RELATED PROCESSES

A. Soil Water

The soil is treated as a slab 1 cm thick, 2 m deep, and extending from the center of one planted row to the next (Fig. 16). This mathematically convenient grid is superimposed over the natural soil horizons. The soil physical and hydrological properties are given by horizons. A complete description of the processes and how they are formulated and coded is given in Boone et al. (1995) and follows the program flowchart of Fig. 1. The commercially available GOSSYM/COMAX considers the plants to be at the upper corners of the soil slab, but a revised version (Theseira, 1994) places the plants in the middle of the slab in order to consider alternate furrow irrigation and alternate side of the row banding of fertilizer.

In the commercially available version of the simulator, rainfall and irrigation are treated as daily amounts. These are broken up into five increments during the daylight hours and one increment during the dark period. Water is simulated to move into each layer, starting at the surface and moving downward, as calculated by the Darcey fluxes (Hillel, 1980). Within each layer, water is moved from side to side by the same type of flux equations based on the soil water gradient and the soil water diffusivity. A revised version of the model uses the soil matrix potential gradient and the hydraulic conductivity to calculate these fluxes. Both versions have advantages and disad-

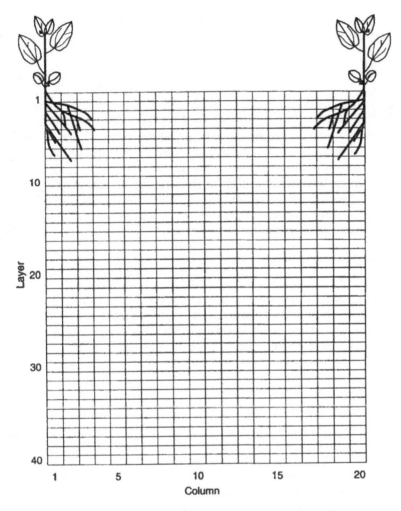

FIGURE 16   A hypothetical grid between plant rows that allows a description of soil physical conditions and growth and distribution of plant roots. GOS-SYM assumes that each cell is 5 cm × 5 cm.

vantages. Data needed for these calculations are, for each horizon, the depth of the horizon, the saturated hydraulic conductivity, the soil moisture release curve (from 0.01 to 15 bar), the bulk density, percent sand, and percent clay. These latter three properties are used in other calculations or model modules. Any soluble chemicals, primarily nitrate nitrogen, are assumed to move with the water fluxes. This mass movement is assumed to be much greater than diffusion.

B. Runoff

The runoff from rainfall or irrigation is based upon the Soil Conservation Service "curve number method" [USDA (1972); for details see Boone et al. (1995)]. The runoff is based on the soil water content at the soil surface for the preceding 5 days and the surface horizon sand and clay content. There is a variable in the soil hydrology file that can be changed if the user does not want runoff to be considered. A grower might select this option if the field has been landleveled, such as for rice. The runoff values are subtracted from the rainfall or irrigation amounts for each day, thus giving the "effective" rainfall value to be infiltrated into the soil.

C. Drainage

The lower layer of soil is considered to be a time-dependent water content boundary. Therefore, the drainage out of (or into) that layer depends on the hydraulic gradient associated with the water content. In GOSSYM, the lower boundary is usually 2 m deep, and therefore the water content does not change rapidly, especially under irrigated conditions.

D. Soil Mechanical Impedance

Soil mechanical impedance or penetration resistance helps determine the root growth in each horizon (Whisler et al. 1986; Boone et al., 1995). An equation for cotton root penetration resistance versus root growth was published by Taylor and Gardner (1963). It was found to cover a wide range of soil textures. A relationship for penetration resistance as a function of bulk density and soil water content was published by Campbell et al. (1974). Thus, the model requires bulk density as mentioned earlier. These relationships have been used in two studies (Whisler et al., 1982, 1993) and are discussed in Section V.B of this chapter.

E. Soil Temperature

In the commercially available version of GOSSYM/COMAX, an empirical model of soil temperature is used that is based on a regression of air temperature and measured soil temperature at four depths (Boone et al., 1995). The equations used in this module are based on the work of McWhorter and Brooks (1965). Daytime and nighttime temperatures are determined from average hourly temperatures and an algorithm of Stapleton et al. (undated). This routine does not account for variations in soil thermal properties or soil moisture. A revised version has been tested (Khorsandi, 1994; Whisler et al., 1986). For this version of the simulator, the percent sand and percent clay are needed for each soil horizon as this allows calculation of the soil thermal conductivity.

F. Evapotranspiration

Evapotranspiration is calculated using a modification of Ritchie's approach (Ritchie, 1972). The modifications involve using light interception instead of leaf area index and minimum daily temperature instead of relative humidity or dew point. These latter two properties are very difficult to measure in a weather station located near agricultural fields owing to dust. For complete details, see Whisler et al. (1986) and Boone et al. (1995). The evaporation amount is taken from the surface layer of soil cells and by soil water gradients from the lower cells on an iterative basis each day. The transpiration amount is taken from the rooted cells and weighted for root amount (mass) and root age in each cell containing roots. This causes moisture gradients among the cells with the lowest water content in the cells with the most active roots. Water is simulated to move from regions of high water content to drier regions by Darcey fluxes as described earlier (Section III.A).

G. Nitrogen Transformations

The nitrogen transformations in GOSSYM are calculated using a modified version of a model by Kafkafi et al. (1978). The modifications include mineralization rates from A. M. Briones (1988, personal communication). The mineralization rate, Q_{10}, is assumed to double with every 10°C rise in temperature, and the mineral content also varies with depth (Stanford and Smith, 1972). Mineralization from organic nitrogen to ammonium nitrogen decreases as the soil water potential decreases (becomes more negative), and so does the rate of nitrification of ammonium nitrogen to nitrate nitrogen (Miller and Johnson, 1964). This model simulates the changes from organic matter to ammonium and to nitrate forms of nitrogen. A fraction of the ammonium ion concentration in each rooted cell, as water is taken up by transpiration, is assumed to be extracted by the roots in active uptake. This model does not account for denitrification or the break-down of plant residues of the current crop.

IV. PUTTING A CROP SIMULATION MODEL AND EXPERT SYSTEM TOGETHER

One can readily see that processes involved in plant growth and development are complex. GOSSYM simulates only a few of the seemingly more important processes. However, there are many other processes that are critical to crop growth. These involve both plant and soil processes that appear important to simulating crop responses to real-world situations. As additional information becomes available, it should be incorporated into the model to allow crop simulation in a wider array of environmental conditions.

Unfortunately, model complexity requires that additional information about the crop or conditions be provided by the user. Although users typically want all the information possible about their crop, during busy times they may also be unwilling to provide that information or examine and interpret the details of the model simulations. We attempted to hide some of the complexity and simplify the operational aspects of running the model by providing an expert system that could aid the decision-making process.

A. The Expert System—COMAX

An expert system is a computer-based system that implements the expertise of one or more human experts in some well-defined domain. Expert systems are usually developed to make scarce human expertise available to a larger number of people. In the 1980s, scientists at the Crop Simulation Research Unit recognized that a major limitation to the wide use of GOSSYM as a crop management tool by growers was the lack of a user-friendly interface and the lack of knowledge of how to plan a set of simulation runs to support decision making. COMAX was developed by Lemmon (1986) to address both of these issues. The knowledge domain of COMAX is knowledge of how to use the GOSSYM model as a decision support tool for crop management; in effect, COMAX is an expert user of GOSSYM. The power of any expert system lies in the knowledge it contains. In this case, there are two primary types of knowledge embedded in the system. The first is a set of simulation scheduling and interpretation strategies that have been developed by both scientists and growers who have used GOSSYM to support crop management

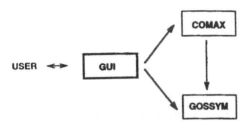

FIGURE 17 Major components of GOSSYM/COMAX. GUI represents a graphical user interface; COMAX represents an expert system that is inextricably linked to GOSSYM, a dynamic cotton crop model.

decisions for a number of years. The second is expertise in the area of cotton management. In the original version of COMAX, the term "COMAX" referred to both the user interface and the expert system component. The interface has been redesigned by Phillip Baulch (1993, personal communication) as a separate component, and the term "COMAX" is now used to refer only to the expert system component. Figure 17 shows the relationship of the three major GOSSYM/COMAX components: (1) the simulation model GOSSYM, (2) the expert system COMAX, and (3) the graphical user interface (GUI). The user interacts with the program through the GUI. The GUI can run the GOSSYM model directly, or the user can request a COMAX analysis through the GUI and the GUI will invoke COMAX, which in turn invokes GOSSYM.

1. The Task and General Approach of COMAX

The management decisions currently addressed by a COMAX analysis are irrigation, nitrogen fertilization, plant growth regulation, and harvest timing. Figure 18 illustrates how COMAX divides the growing season into three sections: (1) the portion of the growing season that has already elapsed and for which "actual" weather data are available, (2) a short period of time (at most 1 week) following the current date for which predicted weather data are available, and (3) the remainder of the growing season. Historical weather data are typically used as input for this third portion of the simulated season. Each historical weather file is called a weather scenario or future weather file. COMAX can run the simulation with up to three future weather files. When COMAX runs the GOSSYM simulation, it runs the simulation through the last day of actual weather, saves the values of all the GOSSYM state variables at the end of the simulated last day of actual weather, and then runs the simulation from this point to the end of the season with up to three different weather scenarios. COMAX uses the stored state variables to reinitialize the simulation to the state at the last day of actual weather before running each new weather scenario.

Management decisions that a COMAX user wants to consider are selected using the GUI. Prior to a COMAX analysis, all weather, soil, and cultural information are specified as for GOSSYM. Users are also required to provide additional information about crop management practices that are relevant to the task. For example, a user who wants COMAX to determine an irrigation schedule for the crop must provide information about how irrigation is delivered (drip, sprinkler, etc.), the maximum amount of each irrigation, the minimum time between irrigations, and the point in the growing season when the system should start to consider applying irrigations. COMAX assumes that any cultural inputs specified prior to the last day of actual weather have already been applied to the crop and cannot be changed. It develops a strategy for using the GOSSYM model to analyze the set of cultural management decisions specified by the user. This task includes planning a schedule of simulation runs, determining what the inputs should be for each simulation run, determining what data need to be collected from each simulation run, and determining what data analysis is needed at the end of each run. COMAX then runs the model one or more times using the strategy, modifies the strategy after each run as necessary, and evaluates the output of these GOSSYM runs for the user.

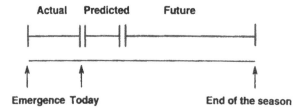

FIGURE 18   Weather files for a single weather scenario.

2. New Design

The original COMAX design had several disadvantages that have been addressed in the new design. These shortcomings were as follows:

1. The original COMAX was a one-dimensional tool that provided crop management advice on either fertilization or both fertilization and irrigation. The design of the tool made it difficult to enhance the knowledge base of the irrigation and fertilization advisors. This design also severely limited the types of analysis that could be done with the model.
2. The COMAX expert system ran the GOSSYM model as an independent program. This loose coupling of the two programs meant that COMAX did not have access to the status of the model until simulation of a season was complete. This limited the way COMAX could use the information generated by the model. As a result, for tasks where COMAX needed access to the status of the crop on a daily basis, the GOSSYM code was modified to include COMAX functionality.
3. Little provision was made for adapting the advisory system to the diversity of agronomic practices found across the Cotton Belt.
4. The one-dimensional design of COMAX effectively prevented the addition of other advisory components.

These shortcomings have been addressed in the new toolbox architecture of COMAX. The system has been redesigned as a set of cooperating experts (tools), each responsible for a different type of decision. The control unit of COMAX knows which tool should be invoked for each type of decision and integrates the activities and output of the tools to accomplish the overall goal. This new architecture has many advantages. First and foremost, the design is flexible and allows several different types of analysis. This also facilitates the addition of new advisory systems. For example, the current version of COMAX includes three different irrigation advisors—a short-term advisor that decides if the crop will need irrigation within the next week, a long-term advisor that develops an irrigation schedule for the remainder of the growing season, and a water conservation advisor that pays special attention to meeting the moisture demands of the crop with the minimum amount of irrigation. The new architecture also makes it much easier to enhance the knowledge base of the individual tools. This means that as scientists develop new expertise for using the GOSSYM model for decision making, this expertise can easily be incorporated into new versions of the system. The knowledge content of the advisors in the new COMAX has been substantially increased over that of the original version. The new toolbox architecture also makes it much easier to add new advisors. A new PIX advisor has recently been tested with the system with little difficulty.

In the new design, COMAX and GOSSYM are two separate modules of one program with clearly defined interfaces (Fig. 19). The simulation control unit and knowledge-based tools of COMAX have access to the state variables of GOSSYM. The simulation control unit can invoke the three major GOSSYM components—one that reads the input and initializes state variables, one that runs the model on a daily basis, and one that produces output describing the status of the model. This new design increases modularity while allowing im-

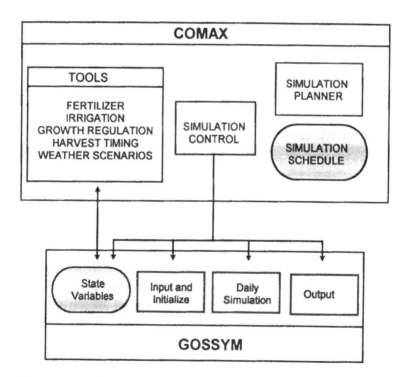

FIGURE 19   Interfaces between COMAX and GOSSYM.

proved interaction between the two modules. The new architecture also allows the COMAX simulation planner (described in Section IV.C) to construct an efficient schedule of simulation runs, where each run meets multiple decision support goals, thereby minimizing the number of simulation runs needed.

B. The COMAX Toolbox

COMAX is a toolbox of advisors that can be used together or independently. The COMAX tools have two purposes: (1) to improve the users' efficiency when using the GOSSYM simulation and (2) to automate the use of the GOSSYM model for making recommendations to manage the cotton crop. Two categories of tools have been developed to fulfill these purposes. "Scenario" tools are used primarily to automate running the simulation model and to summarize results. Scenario tools do not make recommendations but rather run the simulation under conditions specified by the user and summarize the results of the simulation. "Advisor" tools, on the other hand, use rules to determine how to plan a set of simulation runs and make recommendations for management decisions based on the simulation runs. Figure 19 shows the COMAX toolbox that is currently implemented. Any one of these tools can be selected for a particular COMAX analysis. In addition, one irrigation advisor (long-term or water conservation), the fertilization advisor, and the PIX scenario tool can be used in any combination.

Several factors were taken into account in the design of the mechanism for using multiple tools:

1. The number of simulation runs required for the analysis should be minimized.
2. Changing or enhancing the knowledge content of one tool in future versions should have minimal impact on the operation of the other tools.
3. It should be straightforward to add new tools to the toolbox and to integrate the activities of the new tools with activities of existing tools.

1. User-Specified Rules

Many users had requested changes to the original irrigation and fertilization tools that would allow the recommendations of the advisors to be adapted to cultural practices of individual growers. For example, the rules used by the original irrigation advisor were ill-suited for the high plains of Texas, where water resources are extremely limited. Two alternative solutions to this problem were considered. The first was the development of sets of "region-specific" rules based on the expertise of cotton specialists and growers from major cotton-growing regions. The second was the implementation of user-specified rules that would allow individual growers to customize the tools to their specific crop management practices. Although the use of regional rules has the advantage of allowing the system to make very specific recommendations, this alternative was rejected for the following reasons:

1. The number of "regions" is very large. Texas alone has five distinct cotton-growing regions.
2. There are large variations in crop management practices within each region that are due to a variety of factors—other crops being grown on the same farm, field history, equipment available to the grower, the economic status of the grower, and many more.
3. The use of a specific set of regional rules would limit the users' ability to use COMAX to experiment with the use of different cultural inputs and practices.

User-customized rules, on the other hand, allow the recommendations to be tailored to each grower's crop management practices. Since most growers have adapted extension service recommendations to their particular situation, it was considered preferable to give them a mechanism for customizing COMAX rules. Some aspects of the customization allow users to adapt the rules to their particular delivery capabilities, while others reflect the types of "rules of thumb" developed by growers and cotton specialists through years of experience. Several examples of the type of customization enabled by user-specified rules are illustrated on the Irrigation Advisor screen from the GUI in Fig. 20. The check boxes on the left require the user to select one of the three irrigation tools (short-term, long-term, or water conservation). Each of these tools is described in more detail in the examples of the following section. To customize the irrigation rules, the user specifies the maximum amount of irrigation that should be delivered for a single application, the minimum number of days between applications, and the application method (furrow, sprinkler, or drip). These allow the user to tailor the rules to his irrigation delivery capability. They also allow the user to play what-if games with different types and amounts of irrigation delivery. The starting and stopping criteria for irrigation allow the grower to specify heuristics for when irrigation should be initiated and terminated

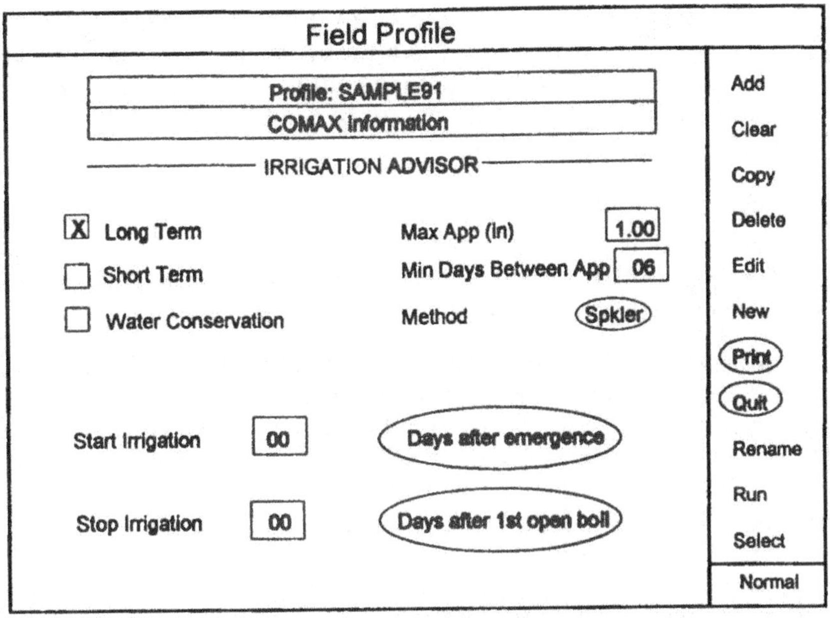

```
┌─────────────────────────────────────────────────────────┐
│                     Field Profile                        │
├──────────────────────────────────────────┬──────────────┤
│  ┌───────────────────────────────────┐    │     Add      │
│  │        Profile: SAMPLE91          │    │              │
│  │        COMAX Information          │    │    Clear     │
│  ─────────── IRRIGATION ADVISOR ──────────     │         │
│                                            │    Copy      │
│  [X] Long Term      Max App (in)   [1.00]  │   Delete     │
│                                            │              │
│  [ ] Short Term     Min Days Between App [06] │  Edit     │
│                                            │              │
│  [ ] Water Conservation  Method  (Spkler)  │    New       │
│                                            │   (Print)    │
│                                            │   (Quit)     │
│  Start Irrigation  [00]  (Days after emergence) │ Rename   │
│                                            │    Run       │
│  Stop Irrigation   [00]  (Days after 1st open boll) │ Select │
│                                            │   Normal     │
└────────────────────────────────────────────┴─────────────┘
```

FIGURE 20   Irrigation advisor screen for COMAX.

for a particular growing region or crop management strategy. The start criterion specifies the point in the growing season when the system should start considering the possibility of scheduling irrigation. Interviews with growers and cotton extension specialists from many different regions revealed that some people use a specific calendar date as a rule of thumb while others use a cotton plant development stage. The four possible choices for start criteria are (1) calendar date, (2) days after emergence, (3) days after first square, and (4) days after first bloom.

If the user selects Calendar Date from the roll bar on the GUI, then the format of the Start Irrigation box changes to accept a date. Otherwise, the user enters a number of days (as shown in Fig. 20). The Stop criterion for irrigation illustrates the wide range of types of heuristics that users were using to decide the point in the season at which irrigation should be terminated. The four possible choices for the Stop criterion are (1) calendar date, (2) days after first open boll, (3) percent open bolls, and (4) days before harvest. In regions of the Cotton Belt where early frosts are a problem, growers have often identified a specific calendar date beyond which they will not apply irrigation; that date is usually based on the average first frost date. In other regions, where boll rot is a particular problem, growers almost never irrigate after the first open boll. The customization options available for other advisors are discussed briefly in the following sections describing each advisor.

2.   Toolbox Examples

Each of the tools in the COMAX toolbox is described briefly in this section. Section IV.C describes how the simulation planner schedules a set of simulation runs to accomplish the tasks of each tool.

Irrigation   Three different irrigation advisors have been developed to accommodate the way different growers use the model for irrigation timing. These three irrigation advisors and their input requirements are shown in Fig. 20. All three advisor systems interact with GOSSYM on a daily basis. Each of these advisors examines the average soil water potential each day, and if it falls below −0.5 bar, an irrigation is scheduled.

The short-term irrigation advisor is meant to be used in those situations when the only decision the user is trying to make is whether the crop will need to be irrigated within the next 2 weeks. It cannot be used in conjunction with any other advisor. The short-term irrigation advisor runs the simulation for 2 weeks past the last day of actual weather and then informs the user if the GOSSYM model is predicting the need for an irrigation application within that time period. Users often use a weather scenario with all rainfall removed from the future weather file so they can determine when they will need to irrigate if there is no rainfall.

The long-term irrigation advisor develops an irrigation schedule for the remainder of the growing season using each weather scenario specified by the user. This advisor should be used in situations where the grower wants to know what type of irrigation schedule will be needed under different weather scenarios for the entire growing season. This is also the advisor that will most often be used in conjunction with the fertilization advisor and PIX scenario tool for an irrigated field.

The water conservation advisor is a modification of the long-term advisor that modifies the amount of irrigation recommended at each irrigation using the predicted evaporative demand of the crop. One simulation run is made in which irrigation is applied in the same manner as for the long-term advisor. A subsequent run is made that uses the evaporative demand from the first run to determine the amount of irrigation that should be used for each application. This advisor is most useful in situations where irrigation amounts tend to be quite large (in excess of 2 in.), the amount of water available for irrigation is limited, and the amount of water applied per irrigation can be varied.

Fertilizer The fertilizer advisor has been modified to accommodate more detailed information about the user's fertilizer delivery practices. A wide variety of fertilizer application amounts, materials, and schedules are used by cotton growers. Lemmon's (1986) original fertilizer advisor was mainly concerned with determining the amount of fertilizer to apply. The system had little knowledge of how to schedule fertilizer applications to not only meet the nitrogen requirements of the plant but also accommodate the application practices and capabilities of the grower. The new fertilizer advisor divides the growing season into four segments for purposes of scheduling nitrogen applications. The general markers for these segments are emergence, first square, first bloom, and late season, respectively. For each of these markers, the user is asked to specify the form of nitrogen that will be applied, the method of application, and the percentage of total nitrogen requirement that the grower will try to supply. A percentage of zero for a particular segment indicates that the user does not want to consider an application at this marker. COMAX makes a series of simulation runs to determine how much nitrogen should be applied to the crop and develops an application schedule that meets the percentages specified by the user as closely as possible.

Earlier versions of COMAX used two different strategies to determine how much nitrogen should be applied to the crop. The first version developed by Lemmon (1986) ran the simulation with no added nitrogen until a nitrogen stress was detected, applied a fixed amount of nitrogen a few days before the simulated stress was encountered, and then repeated the process until the crop did not encounter nitrogen stress. This strategy worked quite well when the amount of nitrogen that needed to be applied was relatively small, but it required a large number of simulation runs when a lot of nitrogen was needed to meet the demand. James Siefker (personal communication) modified COMAX to use a "demand" strategy rather than the previous "stress" strategy. In Siefker's strategy, an excessive amount of nitrogen was applied to the crop on the first simulation run, the actual demand of the crop was calculated from the amount of nitrogen actually used by the crop, and then this modified amount was applied on the second simulation run. Additional simulation runs were often necessary to refine the nitrogen amount. This strategy worked very well when a large amount of nitrogen was needed but required too many simulation runs when only a small amount of nitrogen was needed, such as late in the growing season. The new version of COMAX implements both of these strategies and uses a set of heuristics to select the most appropriate strategy. The user is asked to enter the typical yield for the crop, and this is used to compute a rough estimate of the total nitrogen requirement of the crop. The heuristics use this estimate, along with the amount of nitrogen already available for the crop (residual nitrogen and fertilizer that has already been applied), the point in the growing season, the availability of irrigation, and the cotton variety to select either the "demand" strategy or the "stress" strategy. These heuristics were developed on the basis of discussions with many experienced GOSSYM/COMAX users, both cotton growers and scientists. For both strategies, the goal of the advisor is to find an amount of nitrogen to apply that will allow only a slight nitrogen stress late in the growing season so there will be sufficient nitrogen to support plant growth and boll development but there will not be nitrogen available to support late season vegetative growth that will interfere with boll maturation and harvest. The time window used by COMAX is the 10 days prior to the date the crop reaches 80% of the maximum yield.

Plant Growth Regulators   The most common reasons for using plant growth regulators with cotton are to reduce the size of the plant structure while increasing yield. GOSSYM models the effect of PIX, the most widely used growth regulator. The PIX scenarios tool was developed to automate the process of making simulation runs that users typically make in order to decide if they should apply PIX. With each scenario, the tool makes one simulation run with no PIX applied and one run with PIX applied as specified by the user. A summary comparing the results of these runs is then presented. This tool does not currently determine when and how much PIX should be applied, but it does allow the user to experiment with the effects of different PIX tools. An experimental PIX advisor has been developed that uses an analytical model to analyze the sigmoidal growth curve of the plant and examines the growth rate during the early part of the growing season to determine how much PIX should be applied. It was very straightforward to integrate this new experimental advisor into the existing toolbox.

Harvest Timing   The harvest timing scenarios tool was implemented to automate the analysis that many users have been doing manually to determine when they should apply boll-opening and defoliation chemicals. At the point in the growing season when harvest timing decisions are being made all other crop management decisions are complete, so this is a stand-alone tool. When this tool is invoked, one simulation run is made with no harvest chemicals added, and up to three additional simulation runs are made with applications of boll-opening or defoliant materials applied as specified by the user. These applications are made at 40%, 60%, and 80% open bolls if each of these maturity points occurs after the last day of actual weather and before termination of the simulation.

Weather Scenarios   When the weather scenarios option is selected, COMAX will run GOSSYM with the specified weather scenarios and summarize the results of each run. This tool is useful if the user wishes to run the GOSSYM model with more than one weather scenario and a specified set of inputs. Using the weather scenarios option has several advantages over making a single GOSSYM run with each scenario. The user needs only specify one set of inputs for all three files; the use of a state file by COMAX results in a significant time savings in running the simulation, and a summary of all simulation runs is provided, thus facilitating analysis of the simulation results.

C.   The Simulation Planner

The heart of the new COMAX design is the simulation run planning component. After the user selects a set of COMAX tools for a particular analytical session and initiates a COMAX analysis, the COMAX simulation planner must schedule a set of simulation runs to achieve the goals of each advisor. This includes specifying how to modify input values, deciding which data values should be recorded during the simulation run, and determining how the results need to be analyzed. The heuristics used by the planner are based on the expertise of many GOSSYM users. The work of each of the advisors is called a *task*, and there are a set of available actions that can be used to achieve each of the tasks. The planner has a set of rules that it uses to develop an initial schedule of simulation runs and the actions that are to be executed before, during, and after each simulation run. These rules are very straightforward in cases where only one advisor has been selected by the user. When several advisors are selected for one analysis, the planner must try to achieve the goals of each advisor while using a minimum number of simulation runs. When planning a set of simulation runs with multiple tasks, the planner constructs a simulation run schedule using its knowledge of which actions are prerequisites for others, which actions can be carried out on the same simulation run, and how the actions interact. For example, the long-term irrigation advisor typically requires one simulation run, and the fertilizer advisor typically requires two simulation runs. If both of these advisors are requested for one analysis, the planner examines the actions required by both advisors and schedules the actions required by the irrigation advisor with those required by the fertilizer advisor on the first simulation run, thus reducing the number of required simulation runs by one. The second simulation run will include only actions needed by the fertilizer advisor but will incorporate recommendations produced by the irrigation advisor on the first simulation run.

The initial simulation plan produced by the planner schedules the simulation runs that are typically needed by the advisors selected by the user. A variety of circumstances can occur that may make it necessary to modify the schedule. For example, consider a fertilizer analysis where COMAX is using the stress strategy described above. If one simulation run is made and no nitrogen stress is encountered during the growing season, the second simulation run in the initial schedule stress will not be necessary. In contrast, under other growing conditions, more than two simulation runs may be necessary to fine-tune the amount of nitrogen fertilizer that should be added. The simulation planner must be able to dynamically change the simulation schedule to accommodate this type of situation. This requires that each advisor be able to notify the planner when and how the simulation schedule needs to be changed to accomplish its task and that COMAX knows how revisions of actions scheduled to accomplish one task affect the scheduling of actions for another task.

The dynamic planning module of COMAX allows the expert system to function well under a large variety of cultural and environmental conditions and to react in a reasonable manner to unexpected simulation results.

V. APPLICATIONS

A. On-Farm Applications

GOSSYM/COMAX has many applications for on-farm use (McKinion et al., 1989). A grower might be considering leasing or buying a new field or farm. That person can determine from the county soil survey what soils are there, but he would like to know what he might expect from his normal production practices on this new area. If the necessary soil files are in the GOSSYM/COMAX file (we have over 350 soils, mainly from the Cotton Belt, in that file), then he can make several simulations using his normal production methods and selecting different years of weather data. The results would indicate whether or not he might want to consider irrigation for that field if it was not irrigated, how much nitrogen he should apply preplant and/or as side-dressing, what varieties might be best for that specific soils, what row spacings to use, etc. We have found that on some soils cotton generally yields more than on other soils, some varieties yield more than others, some planting dates are more optimum than others for a particular climatic location, and some row spacings seem better most of the time under those same climatic conditions and soils (Wang and Whisler, 1994). Another exercise a grower might do is to test the timing and rate of chemical applications such as TEMIK, PIX, or PREP (a pesticide, plant growth regulator, and harvest aid chemical, respectively) by simulating the crop with these different management practices, with different weather and soil scenarios. These types of exercises, which can be done before the planting decisions are finalized, are often called strategic exercises or strategic decision making.

During the growing season, there are several other types of decisions to be made, and these are called tactical decisions. They might be, Should I irrigate tomorrow, or should I wait? Should I side-dress with nitrogen or wait and apply foliar nitrogen after an expected rain? Should I apply PIX today or wait? Should I apply PREP and a defoliant tomorrow or wait? All of these decisions are made on a field-by-field basis and might be made with precision agriculture, using variable rate applications on a soil-by-soil, variety-by-variety subset of a field. This assumes that the grower has access to up-to-date weather data, either from his own weather station or from one within close proximity to his field and has rainfall and/or irrigation amounts for the specific field. In the Midsouth, that proximity is around 5 miles. In the more arid regions it might be greater.

The simulator-assisted decisions result is greater economic returns as a consequence of better information about the crop provided by GOSSYM/COMAX. Scientists from Texas A&M University, at the request of the U.S. Extension Service, studied the impact of GOSSYM/COMAX on production decisions and practices. In that independent study, Ladewig and Thomas (1992) found that the average GOSSYM/COMAX user attributed $55/acre net return to the crop model when it was used in his cotton management. First-year users gained less, and more experienced users found it more helpful and required less time. More experienced users found more ways, sometimes unexpected by the developers, in which the information provided by the simulator could be useful.

B. Other Applications

There have been other uses made of GOSSYM/COMAX. In the mid-1980s, the model was used to study retrospectively, the so-called cotton yield decline (V. R. Reddy et al., 1987, 1989, 1990; Whisler et al., 1993). It was found that there was no one specific cause of the apparent cotton yield decline between 1964 and 1980. At different locations within the Cotton Belt there were apparently different conditions responsible for the yield decline. For example, in the Midsouth a change in the insect management strategies might have been partly responsible. In the southwest, such as in Phoenix, Arizona, increases in the atmospheric ozone level might have been partly responsible. On the lighter, sandier textured soils, soil compaction might have been partly responsible as tractor sizes increased.

In another study, the effects of erosion were analyzed (Whisler et al., 1986). One soil profile, 100 cm deep, was assumed to have a traffic pan 17–24 cm below the soil surface, and it was assumed that the soil surface soil was eroded by 5 or 10 cm. A relatively dry year (1980) and a relatively wet year (1982) were compared. For the dry year, 5 cm of erosion reduced the simulated yield by 9%, and 10 cm of erosion reduced it by 19%. For the wet year, the maximum simulated yield reduction was only 2%. For a 30 cm deep profile of the same soil, but where the traffic pan was re-formed each year at 17–24 cm below the soil surface, the reductions in yield were greater. The shallower soil reduced the predicted yield by 32% in a dry year, and increased erosion further reduced the yield another 10–20%. In a wet year the shallower soil reduced the simulated crop yields by only 14%, but more erosion further reduced the yields by 20–40%.

Whisler et al. (1982) studied the effects of simulated tillage and wheel traffic on cotton crop growth and yield. The soil compaction due to wheel traffic and subsequent loosening of the soil surface due to cultivation can change the root distribution patterns and water and nutrient extraction patterns, especially in lighter, sandier textured soils. Landivar et al. (1983) used the model to study whether a particular genotypic variation will be advantageous over other characteristics.

All of these applications of GOSSYM/COMAX, and there can be many more, are possible only because the model is mechanistic in its concepts of plant growth and plant–environment interactions. Also the erosion and tillage studies could be done in a meaningful and quantitative way only because the soil matrix is considered to be a three-dimensional grid.

VI. BRIDGING TECHNOLOGIES

Recently, scientific visualization system (SVS) technology has made tremendous strides with the advent of fast, inexpensive desktop computers with large amounts of memory and disk space, comprehensive graphics libraries, and hardware graphics accelerators. Combining SVS and crop simulators will provide the user with the capability of simulating crop development and yield for more crops and more complex mechanisms. A recent programming technique called object-oriented programming seems likely to provide a way to gain efficiency in developing crop models (Booch, 1991; Sequeira et al., 1991). It allows one to develop modules, on either biological or physical processes, that can be used in a unified, coherent architecture with a standardized file structure for data input and results. Using the features of object-oriented programming, some modules might be usable for models of more than one crop. Users with plug-and-play capability might then be able to evaluate alternative mechanisms for building models of complex systems. For example, different models of photosynthesis or evapotranspiration could be used, or entire systems may be substituted as alternative ways to calculate water and agrochemical movement in soil. The user will be able to scale the system from simple, computationally fast crop models with limited prediction capability, i.e., final yield, to complex, computationally expensive crop models that can be used for tactical farm management decision support and as physiological probes to further elucidate the mechanisms that govern plant responses to the environment. This is suggesting that the technology may be approaching a point where model developers may be able to select appropriate modules and assemble a crop model that is specific for one crop and some different combination of modules that is more appropriate for another crop. Considerable work needs to be done, however, before such generic models are achievable.

The object-oriented program system coupled with an SVS will aid scientists in using this tool. The SVS coupled with the simulation model and expert system technologies can be used to aid three-dimensional visualization of several processes in soil–plant–insect systems in a very realistic manner. The user will be able to visually relate model-calculated physiological processes with the three-dimensional plant representation. Through colorization, such effects as variations in plant temperature can be displayed to show the effect of water deficits on stomatal closure, with the resulting slowing of transpiration rate and diminishing evaporative cooling. Another example is a three-dimensional soil representation with roots, soil water, chemical fertilizer, and/or pesticide movement in both time and space.

With the addition of expert systems technology that can be used to hide system complexity from the user, the combined scientific visualization system will also provide a unique tool for academic training of both undergraduate and graduate students for departments that teach agronomy, soil science, botany, agricultural engineering, and environmental science. Students, through the use of an expert system that facilitates communication between the model and the user, will be able to easily access and exercise the visualization system and the generic plant modeling system. The expert system–SVS combination will provide the student with analytical capability so that soil, weather, and cultural practice effects are apparent in the context of the physical and biological processes involved. The student will be able to quickly gain insights into plant physiology, micrometeorology, soil physics, crop management, and many other associated areas of expertise.

Another area of rapid technological development that has a possible role in crop management is a combination of geographic information systems (GIS), global positioning systems (GPS), and intelligent implements (II) technologies for precision agriculture. These techniques may be integrated with the simulation model and expert system technologies to deliver dynamic, precision agriculture decision support systems that can respond to within-season problems and provide real advantages to productivity. The primary product of this research will be the maintenance and improvement of food and fiber production systems while minimizing erosion and the use of agrochemicals, thus lowering the negative environmental impacts of agricultural production.

Crop models are the only automated source of dynamic knowledge within computer technology, or any other known technology, that can respond to crop status, current meteorology, and management practices. Crop models can provide the essential linkages between the crop manager (decision maker) and the information made available with GIS, GPS, and II. The crop model is the only technique with which we are familiar that has a possibility of accepting, processing, and responding to the vast quantities of information provided by the GIS and GPS to deliver variable rates of input through intelligent implements. Appropriate responses to precision agriculture information will include recommendations on a soil type-by-soil type or situation basis within single management units using intelligent implements to deliver variable seeding and agrochemical rates. Crop models are the ideal source of dynamic information that can optimize crop production and minimize resource utilization while protecting the environment. Crop models also have the advantage that they allow producers to simulate the application of new technology, test it on their soils with historic weather data for each site, and develop the optimum management practices to use with the new technology.

VII. TECHNOLOGY TRANSFER

The availability of crop models and expert systems to farmers has added tremendously to the information available for aiding crop management decisions. The prospects of having spatial mapping of crop yields and automated control of farm machinery greatly enhance the information available and the possibility of tailoring management decisions to specific sites. Unfortunately, the reasons for yield differences will often not be obvious. Therefore, producers with these new technologies will need assistance to understand the reasons for variations in crop yields in a given field and all the diagnostic assistance available to explain the failure to attain the potential crop yields.

By agriculture's very nature, crops are produced in complex environments with varying weather, soils, cultural practices, and genetic resources. The introduction of any new technology into such a complex situation requires considerable time, experience, and understanding of the whole crop production system. The resources for introducing new materials or technologies are always expensive and are sometimes simply not available. Scarce resources limit the preparation and distribution of educational materials as well as real-world local demonstrations. Crop models may provide a tool to illustrate new technology. New concepts or cultural practices may be incorporated into the model. The upgraded model may be distributed with suggestions that the user test the concepts with his soil, previous weather, and other cultural practices. Those results should provide sufficient information to convince the user that he should or should not adopt the new technology. Upgrades or new software may be distributed over the Internet with minimal cost for support.

For the first time in history, scientists, purveyors of various forms of high technology computing and other electronic gadgets, and agriculturalists working at the farm production level have an opportunity to apply the best-known information regarding specific crop conditions. The information can be assessed on a quantitative basis and will also allow economic and social values to be assigned to that practice, in addition to its crop production benefits. The model can be used to study crop yield-limiting aspects of certain sites and thus render the other aspects of site-specific agriculture more valuable. It will be a major challenge for agricultural leadership to maintain public support for this research.

REFERENCES

Baker, D. N. 1980. Simulation for research and crop management. In: *World Soybean Research Conference* II. *Proceedings*. F. T. Corbin (Ed.). Boulder, CO: Westview, pp. 533–546.

Baker, D. N., and J. A. Landivar. 1991. The simulation of plant development in GOSSYM. In: *Predicting Crop Phenology*. T. Hodges (Ed.). Boca Raton, FL: CRC Press, pp. 153–170.

Baker, D. N., J. D. Hesketh, and R. E. C. Weaver. 1978. Crop architecture in relation to yield. In: *Crop Physiology*. U. S. Gupta (Ed.). New Delhi, India: Oxford and IBH, pp. 110–136.

Baker, D. N., J. R. Lambert, and J. M. McKinion. 1983. GOSSYM: A Simulator of Cotton Crop Growth and Yield. South Carolina Exp. Sta. Tech. Bull. 1089.

Booch, G. 1991. *Object Oriented Design*. New York: Benjamin/Cummings.

Boone, M. Y. L., D. O. Porter, and J. M. McKinion. 1995. RHIZOS 1991: A Simulator of Row Crop Rhizospheres. USDA, ARS Bull. 113. Washington, DC: Govt. Printing Office.

Bruce, R. R., and M. J. M. Romkens. 1965. Fruiting and growth characteristics of cotton in relation to soil moisture tension. *Agron. J.* 57:135–139.

Campbell, R. B., D. C. Reicosky, and C. W. Doty. 1974. Physical properties and tillage of paledults in the southeastern coastal plans. *J. Soil Water Conserv.* 29:220–224.

Graham, J., D. T. Clarkson, and J. Sanderson. 1973. Water uptake by roots of narrow and barley plants. *Ann. Rep. Letcombe Lab. Wantage*, p. 9–12.

Hesketh, J. D., and A. Low. 1968. Effect of temperature on components of yield and fibre quality of cotton varieties of diverse origin. *Cotton Grower Rev.* 45:243–257.

Hesketh, J. D., D. N. Baker, and W. G. Duncan. 1972. Simulation of growth and yield in cotton. II. Environmental control of morphogenesis. *Crop Sci.* 12:436–439.

Hillel, D. 1980. *Fundamentals of Soil Physics.* New York: Academic Press.

Hodges, H. F., K. R. Reddy, J. M. McKinion, and V. R. Reddy. 1993. Temperature Effects on Cotton. Miss. Agric. Forest Exp. Sta. Bull. 990.

Jackson, B. S. 1991. Simulating yield development using cotton model COTTAM. *Predicting Crop Phenology.* T. Hodges (Ed.). Boca Raton, FL: CRC Press, pp. 171–180.

Kafkafi, U., B. Bar-Yosef, and A. Hadas. 1978. Fertilization decision model— a synthesis of soil and plant parameters in a computerized program. *Soil Sci.* 125:261–268.

Khorsandi, F. 1994. Testing soil temperature models in GOSSYM. Ph.D. Dissertation, Mississippi State Univ.

Kiniry, J. R., W. D. Rosenthal, B. S. Jackson, and G. Hoogenboom. 1991. Predicting leaf development of crop plants. In: *Predicting Crop Phenology.* T. Hodges (Ed.). Boca Raton, FL: CRC Press, pp. 29–42.

Ladewig, H., and J. K. Thomas. 1992. A Follow-Up Evaluation of the GOSSYM-COMAX Program. College Station, TX: Texas Agric. Extension Service.

Landivar, J. A., D. N. Baker, and J. N. Jenkins. 1983. The application of GOSSYM to genetic feasibility studies. I. Analysis of fruit abscission and yield in okra-leaf cotton. *Crop. Sci.* 23:497–506.

Lemmon, H. E. 1986. COMAX: An expert system for cotton management. *Science* 233:29–33.

MacArthur, J. A., J. D. Hesketh, and D. N. Baker. 1975. Cotton. In: *Crop Physiology: Some Case Histories.* L. T. Evans (Ed.). London: Cambridge Univ. Press, pp. 297–395.

Marani, A., D. N. Baker, V. R. Reddy, and J. M. McKinion. 1985. Effect of water stress on canopy senescence and carbon exchange rates in cotton. *Crop Sci.* 25:798–802.

Mauney, J. R. 1984. Anatomy and morphology of cultivated cottons. In: *Cotton.* R. J. Kohe and C. F. Lewis (Eds.). Agron. Monograph No. 24. Madison, WI: ASA-CSSA-SSSA, pp. 59–80.

Mauney, J. R. 1986. Vegetative growth and development of fruiting sites. In: *Cotton Physiology.* J. R. Mauney and J. M. Stewart (Eds.). The Cotton Foundation, Memphis, TN: pp. 11–28.

McKinion, J. M., and H. F. Hodges. 1985. Automated system for measurement of evapotranspiration from closed environmental growth chambers. *Trans. ASAE* 28:1825–1828.

McKinion, J. M., D. N. Baker, F. D. Whisler, and J. R. Lambert. 1989. Application of GOSSYM/COMAX system to cotton crop management. *Agric. Syst.* 31:55–65.

McWhorter, J. C., and B. P. Brooks, Jr. 1965. Climatological and Solar Radiation Relationships. Miss. Agric. Exp. Sta. Bull. 715.

Miller, R. D., and D. D. Johnson. 1964. The effect of soil moisture tension on carbon dioxide evolution, nitrification, and nitrogen mineralization. *Soil Sci. Soc. Am.* 28:644–647.

Moraghan, B., J. D. Hesketh, and A. Low. 1968. The effects of temperature and photoperiod on earliness of floral initiation among strains of cotton. *Cotton Grower Rev. 45*:91–100.

Mutsaers, H. J. W. 1983. Leaf growth in cotton (*Gossypium hirsutum* L.). 1. Growth in area of main-stem and fruiting branch leaves. *Ann. Bot. 51*: 503–520.

Parsons, J. E., J. L. Dunlap, J. M. McKinion, C. J. Phene, and D. N. Baker. 1980. Microprocessor-based data acquisition and control software for plant growth chambers (SPAR system). *Trans. ASAE 23*:589–595.

Phene, C. J., D. N. Baker, J. R. Lambert, J. E. Parsons, and J. M. McKinion. 1978. SPAR—A Soil–plant–atmosphere research system. *Trans. ASAE 21*: 924–930.

Reddy, K. R., V. R. Reddy, and H. F. Hodges. 1992a. Temperature effects on early season cotton growth and development. *Agron. J. 84*:229–237.

Reddy, K. R., H. F. Hodges, J. M. McKinion, and G. W. Wall. 1992b. Temperature effects on pima cotton growth and development. *Agron. J. 84*:237–243.

Reddy, K. R., H. F. Hodges, and J. M. McKinion. 1993a. A temperature model for cotton phenology. *Biotronics 22*:47–59.

Reddy, K. R., H. F. Hodges, and J. M. McKinion. 1993b. Temperature effects on pima cotton leaf growth. *Agron. J. 85*:681–686.

Reddy, K. R., M. Y. L. Boone, A. R. Reddy, H. F. Hodges, S. Turner, and J. M. McKinion. 1995a. Developing and validating a model for a plant growth regulator. *Agron. J. 87*:1100–1105.

Reddy, K. R., H. F. Hodges, and J. M. McKinion. 1995b. Carbon dioxide and temperature effects on pima cotton development. *Agron. J. 87*:820–826.

Reddy, K. R., H. F. Hodges, and J. M. McKinion. 1997a. Crop modeling and applications: A cotton example. *Adv. Agron. 59*:225–290.

Reddy, K. R., H. F. Hodges, and J. M. McKinion. 1997b. Modeling temperature effects on cotton internode and leaf growth. *Crop Sci. 37*:503–509.

Reddy, V. R. 1995. Modeling ethephon and temperature interaction in cotton: The model. *Comput. Elec. Agric. 8*:227–236.

Reddy, V. R., D. N. Baker, F. D. Whisler, and R. E. Fye. 1987. Application of GOSSYM to yield decline in cotton. I. Systems analysis of effects of herbicides on growth, development, and yield. *Agron. J. 79*:42–47.

Reddy, V. R., D. N. Baker, and J. M. McKinion. 1989. Analysis of effects of atmospheric carbon dioxide and ozone on cotton yield trends. *J. Environ. Qual. 18*:427–432.

Reddy, V. R., D. N. Baker, F. D. Whisler, and J. M. McKinion. 1990. Analysis of the effects of herbicides on cotton yield trends. *Agric. Syst. 33*:347–359.

Reddy, V. R., D. N. Baker, and H. F. Hodges. 1991. Temperature effects on cotton canopy growth, photosynthesis, and respiration. *Agron. J. 83*:699–704.

Ritchie, J. T. 1972. Model for predicting ET from a row crop with incomplete cover. *Water Resources Res. 8*:1204–1213.

Sequeira, R. A., P. J. H. Sharpe, N. D. Stone, K. M. El-Zik, and M. E. Makela. 1991. Object oriented simulation: Plant growth and discrete organ to organ interaction. *Ecol. Modeling* 58:55–89.

Stanford, G., and S. J. Smith. 1972. Nitrogen mineralization potentials of soils. *Soil Sci. Soc. Am. Proc.* 36:465–472.

Stapleton, H. N., D. R. Buxton, F. L. Watson, D. J. Nolting. No date. Cotton: A Computer Simulation of Cotton Growth. Ariz. Agric. Exp. Sta. Tech. Bull. 206.

Taylor, H. M., and H. R. Gardner. 1963. Penetration of cotton seedling taproots as influenced by bulk density, moisture content and strength of soil. *Soil Sci.* 96:153–156.

Theseira, G. W. 1994. Development and testing of an amended GOSSYM model for simulation of asymmetric inputs about the crop row. Ph.D. Dissertation. Mississippi State Univ., Mississippi State, MS.

USDA, Soil Conservation Service. 1972. *National Engineering Handbook.* Section 4.

Wang, X. N., and F. D. Whisler. 1994. Analyses of the Effects of Weather Factors on Predicted Cotton Growth and Yield. MS Agric. Forest Exp. Sta. Bul. 1014.

Whisler, F. D., J. R. Lambert, and J. A. Landivar. 1982. Predicting tillage effects on cotton growth and yield. In: *Predicting Tillage Effects on Soil Physical Properties and Processes.* P. W. Unger, D. M. Van Doren, Jr., F. D. Whisler, and E. L. Skidmore (Eds.). ASA Special Publication #44 ASA-SSSA, Madison, WI: pp. 179–198.

Whisler, F. D., B. Acock, D. N. Baker, R. E. Fye, H. F. Hodges, J. R. Lambert, H. E. Lemmon, J. M. McKinion, and V. R. Reddy. 1986. Crop simulation models in agronomic systems. *Adv. Agron.* 40:141–208.

Whisler, F. D., V. R. Reddy, D. N. Baker, and J. M. McKinion. 1993. Analysis of the effects of soil compaction on cotton yield trends. *Agric. Syst.* 42:199–207.

9

Integrated Methods and Models for Deficit Irrigation Planning

Neil L. Lecler
University of Natal, Pietermaritzburg, South Africa

I. INTRODUCTION

In the design and planning of irrigation projects, most "water requirement" studies have focused on climatic factors that influence water use for maximum crop production. This approach may be losing its relevance. In many parts of the world the demands on limited water resources are of major concern and the available water resources are subject to increasing pollution. The net result is that water users, especially the larger users like irrigated agriculture, will face shortages and in order to survive will need to make more efficient use of diminishing and more costly water supplies. For example, in South Africa, a country that is about 90% of the way toward developing water resources infrastructure to accommodate the annual storage potential, more than half of the 70 state irrigation schemes had to cut farmers' water quotas by 50–100% in 1983 when the country nominally required only 40% of the annual storage potential (Green, 1984).

Irrigation systems and application strategies designed to achieve maximum crop yields in even the driest years need to be scrutinized and alternative ideas and methodologies investigated. In this regard, deficit irrigation is worth consideration.

Deficit irrigation is an optimizing strategy whereby net returns are maximized by reducing the amount of irrigation water applied to a crop to a level that results in some yield reduction caused by water stress.

The fundamental goal of deficit irrigation is to improve water use efficiency and maximize profits through, among other things, a reduction in capital and operating costs. A major challenge to the introduction of deficit irrigation strategies is the need to convince irrigation practitioners not only of their value but also of their practicality. Often, to improve irrigation management a practitioner may perceive the need to forsake other aspects of his or her management portfolio, which may not be feasible or desirable.

The aim of this chapter is to give the reader an overall perspective of deficit irrigation, from the basic concepts to some of the options, challenges, and tools that could be considered and integrated during the development of a viable deficit irrigation strategy. The use of simulation models is shown to be a key element in the development of such a strategy.

II. DEFICIT IRRIGATION CONCEPTS

Recognition of the following points is fundamental to an understanding of deficit irrigation:

1. The efficiency of irrigation water applications diminishes as the number and magnitude of the applications increases.
2. The application of irrigation water is costly.
3. The determination of an optimal irrigation strategy is dependent on whether a shortage of water or a shortage of land is the factor limiting production.

A. Efficiency

With most irrigation applications, water is lost owing to, among other things, spray evaporation, wind drift, and evaporation from the soil surface and potentially through surface runoff and deep percolation. The potential for losses through surface runoff and deep percolation increases as the number and magnitude of irrigation applications increases, and, in addition, the conjunctive use of rainfall is likely to

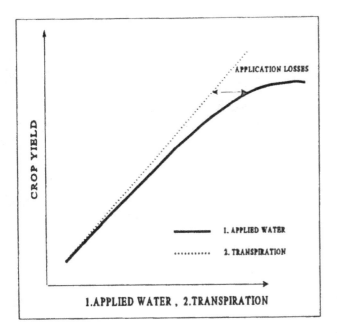

FIGURE 1 General form of a crop production function. (After English, 1990.)

diminish. The net result is that if a plot of crop yield versus actual transpiration and applied irrigation water is drawn, the yield versus applied irrigation water line will curve away from the yield versus transpiration line as shown in Fig. 1. The more generous the irrigation, the more modest the proportional contribution to crop water requirements (transpiration) is likely to be. This is one of the reasons the application of the large amounts of irrigation water need to maximize yield can be inefficient and is not always likely to result in maximum economic return.

B. Costs

Irrigation water application costs are related to actual costs of water, interest on capital equipment, energy, labor, management, and opportunity costs, especially if water is limited. When water as opposed to land is limited, the water saved by reducing irrigation applications per hectare may be used to irrigate additional land, possibly resulting in an increase in total farm income. The potential income from the irrigation of the additional land is an opportunity cost of water. For

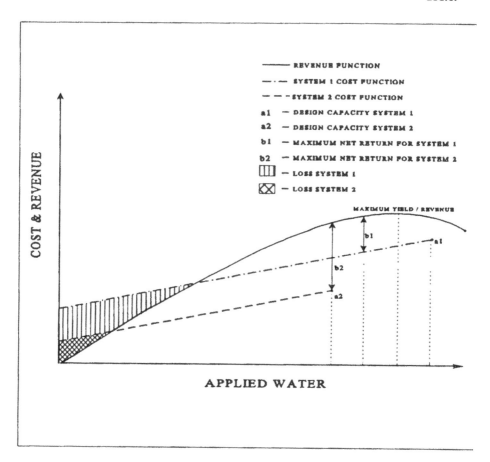

FIGURE 2   Revenue and cost functions. (After English, 1990.)

example, 40 mm of water could be applied to 10 ha, or 20 mm of water could be applied to 10 ha and the remaining 20 mm applied to a further 10 ha, thereby irrigating a total of 20 ha, which may result in an increase in total profit. If land is limited, the question reverts to what irrigation application amount results in the maximum difference between irrigation application costs and yield-related returns.

In Fig. 2, two cost functions and a revenue function are shown. The revenue function has the same shape as the yield versus applied water curve (see Fig. 1) because revenue is simply the product of yield and crop price. The lower limit of the cost functions represents fixed costs such as capital costs, crop insurance, fixed costs of irrigation, planting, tillage, chemical use, and harvesting. The slope of the cost functions represents the marginal variable costs of production, for example, pumping costs, water costs, and yield-related costs. The upper limit of the cost functions represents the maximum water delivery capacity of the two different irrigation systems. The maximum water delivery capacity of an irrigation system is a hardware limitation that can have significant cost and flexibility implications. For example, for the two systems shown in Fig. 2, the system with the smaller capacity, i.e., system 2, does not have sufficient capacity to irrigate for maximum crop yields; however, the implications of lower capital and operating costs are such that the net returns are nearly double those attainable with the larger system, i.e., system 1.

C. Optimal Strategy

Given that the relationships shown in Fig. 2 were known precisely, the selection of an optimal strategy for either water- or land-limiting circumstances would be relatively simple. The analytical framework for such an exercise has been well documented by English (1990). However, while it is fairly easy to determine the cost function, the following two questions remain:

1. Is water limiting, or is land limiting?
2. How can the information relating yield to irrigation applications be determined?

In order to answer these questions there is the need to determine (1) the quantity and timing of irrigation water supplies in relation to the quantity and timing of irrigation crop water demands and (2) the response of a crop (in terms of yield) to the supply constraints and the associated irrigation water applications.

III. CROP WATER REQUIREMENTS AND ASSOCIATED YIELDS

To develop a viable deficit irrigation strategy, information on the response of a crop to different irrigation watering regimes and environmental constraints is needed. This information can be determined from experimental trials or by using crop growth/yield simulation models. Some of the problems with experimental trials are that

1. Experimental trials are expensive and time-consuming, and therefore it is difficult to test a wide range of crops and water application and management strategies.

2. Experimental trials are location-specific, and it is difficult to extrapolate the results beyond the particular site and season of the experiment.
3. Usually only a few years of trial results are available for a specific location.
4. Many field trials are designed to determine only the maximum crop yield and do not include crop yield responses to various combinations of inputs.

Simulation models rely on experimental trials for their development and verification, but by conceptually representing a range of systems in the real world they provide a way of transferring knowledge from a measured or trial situation to an unmeasured situation where management decisions are needed (Branson et al., 1981). In areas where there is no information from experiments or this information is limited, models can be used to

1. Evaluate a wide range of design/management strategies relatively quickly and cheaply.
2. Evaluate different strategies over a wide range of uncertain conditions such as wet and dry years or seasons.
3. Integrate numerous and complex procedures and knowledge into a system that can be easily operated and used to aid decision making.

Crop yield simulation models range from simple formulas to complex physiologically based models. Each has limitations in terms of the availability of input data and information, model accuracy, and potential uses. Schulze et al. (1995a) describe the range as follows.

Simple crop yield models, for example, yield–mean annual precipitation (MAP) relations, have the advantage that they require only simple, readily obtainable input. However, the models' estimates are not always accurate because they are not representative of the physical system. These models should not be used to extrapolate yield estimates outside of the conditions for which they were developed, nor can they be applied for risk analysis.

Complex crop growth models, for example, the models in the DSSAT v3 (Tsuji et al., 1994) modeling system, are generally accurate predictors (although not necessarily so), but difficulties may be experienced in securing adequate data or information for various parameters. The development of complex models from the processes of analysis, assembly of data, and model construction and verification takes up costly resources in the form of skilled person-hours and computer time.

Models of intermediate complexity, for example, soil water budgeting and total evaporation–driven models, are sensitive to plant water stress and crop phenology and endeavor to capture the approximate response of crops to environmental conditions while sacrificing certain details of physiological processes such as photosynthesis and respiration. Models of this nature have been shown to be of sufficient accuracy for irrigation decision support applications (de Jager, 1994).

When it comes to the practical selection of an appropriate model, the overriding factor is the quantity and availability of input information. If all that is known about an area is the MAP, then the simple MAP–yield relation will likely prove more accurate than the most advanced mechanistic crop growth model, which has little usefulness without proper inputs. For deficit irrigation applications, models of at least intermediate complexity that endeavor to represent the response of the crop to water stress should be considered, and, ideally, if adequate input data are available, then the more complex models should be selected.

IV. IRRIGATION SCHEDULING STRATEGIES

The amount of irrigation water used to attain a certain crop yield is dependent on the irrigation scheduling strategy. For this reason it is important that different modes of irrigation scheduling be considered during the planning phase of an irrigation project. An efficient way of accomplishing this is through the use of simulation models.

Models of irrigation scheduling depend on, among other things, the irrigation system (i.e., equipment), the level of management (which is often overlooked during the design and planning phase), water availability, climatic conditions, and the type of crop and its stage of growth. Three modes of irrigation scheduling are discussed here, with emphasis on their potential for deficit irrigation.

A. Demand Mode Scheduling According to Soil Water Depletion Levels

When scheduling according to soil water depletion levels, water is required when the depletion of soil water in a defined zone (normally the active root zone) has reached a threshold level. For example, a threshold equivalent to a depletion of 50% of the plant available water (PAW) is often chosen when irrigating to avoid crop water stress. If this method of irrigation scheduling is used for deficit irrigation, the crop is allowed to deplete soil water during certain growth stages to a level that results in some stress and some reduction in yield.

The determination of the timing and magnitude of the depletion-to-stress-level events is not a simple exercise. In the planning stages of an irrigation project, multiple simulations of different depletion levels at different crop growth stages could provide a database for the selection of a suitable strategy. It is important when developing this database to also consider the actual physical characteristics of the irrigation systems that would be required to implement the simulated strategies. It is difficult to optimize a depletion level schedule and hardware requirements concurrently. This is an important consideration, as the level of flexibility built into an irrigation system during the design phase can have a major influence on capital equipment costs and therefore on the potential returns (see Fig. 2). To implement a given depletion level schedule, an irrigation system with a large capacity is often required even though the potential capacity is only used for relatively short but possibly crucial periods during a crop's growing season.

In practice, the logistics of moving irrigation hardware to maintain an optimum depletion level schedule over all sectors of a field is extremely difficult unless a solid set irrigation system is used. These difficulties are compounded when the water budget of different sectors of a field are unequally influenced by rainfall events. Many farmers find the management of a simple depletion level schedule overtaxing and would be doubtful participants in a more complex and varied approach. In addition, untimely irrigation equipment breakdowns and unforeseen delays that may occur in practice may result in severe crop water stress if irrigation applications are routinely "triggered" when more than 50% of PAW is depleted and there is no safety buffer of soil water.

When scheduling according to soil water depletion levels, the soil profile is normally recharged to the drained upper limit (DUL). An additional option for which depletion level scheduling to a planned deficit is implemented is worth consideration. The root zone soil profile is then deliberately recharged by irrigating to below the DUL when the soil water threshold is reached. The irrigation amount is therefore planned to leave a portion of the potential soil water store to be filled by the expected rainfall. One assumption in this mode of scheduling is that irrigation is supplementary to rainfall in areas where there is a high probability of rain falling between irrigation applications. Rainfall effectiveness may then be maximized and stormflow generation reduced (Furniss, 1988). The potential rainwater store for which irrigation applications are reduced can be varied according to local climatic conditions.

B. Irrigation with a Fixed Cycle and in Fixed amounts of Water Application

In this mode of irrigation either a preselected or otherwise predetermined amount of irrigation water is applied in a fixed cycle. The selected cycle length is assumed to continue throughout the growing season. Simple improvements to this strategy can considerably enhance the efficiency of irrigation applications. For example, the irrigation cycle could be stopped for a period of time after a rainfall event, and different cycle lengths could be used for different parts of the growing season. The length of time that the cycle is halted can be related to the magnitude of the rainfall event and the effective daily irrigation application according to the equation.

$$t = \frac{p}{i/c} \tag{1}$$

where

t = time that cycle is halted (days), $t \leqslant c$
p = precipitation (mm)
i = irrigation application per cycle (mm)
c = irrigation cycle time (days)

This mode of irrigation is commonly used in practice because it is easily managed. With the use of simulation models there is the potential to optimize cycle times and application amounts for specific crops and environments and thereby considerably improve irrigation efficiency (Furniss, 1988). This approach to irrigation scheduling is more suited to deficit irrigation than to full irrigation because of the efficiency considerations. Full irrigation has the potential for higher irrigation losses because more water is applied. As a deficit irrigation scheduling strategy, this approach to scheduling has the following advantages.

1. The strategy is relatively simple.
2. Because of this simplification, it is easy to select the optimal hardware specifications that coincide with irrigation management considerations and yield consequences by using computer simulation models.

3. Computer simulations are easily implemented as practical on-farm operational methods.

C. Irrigation with a Fixed Cycle and in Varying Amounts of Water Application

The fixed cycle/varied application irrigation scheduling strategy is particularly suited to hand-moved sprinkler or dragline irrigation systems. The irrigation hardware (e.g., the lateral pipes) is moved according to a fixed, regular schedule, thereby simplifying management. Irrigation applications take place as often as required, within the system limitations. The amount of irrigation water applied after each move or for each set of lateral positions is determined by the available potential soil water storage capacity, which can be either fully or partially refilled, depending on the limitations of the irrigation systems' capacity and the available supply of water. The amount of water applied is controlled by varying the pumping times associated with each set of lateral positions.

This is a very efficient scheduling strategy, and together with the fixed cycle/fixed amount scheduling strategy it can also be used to maximize the benefits of small irrigation systems. Irrigation water applications can be initiated prior to planting and together with applications early in the growing season can help to refill the potential soil water store when evapotranspiration requirements are low. This soil water is then available to supplement irrigation water applications later in the season when the increased evapotranspiration is in excess of what can be supplied by an irrigation system with a limited capacity.

As with demand mode scheduling, the option to leave a portion of the potential soil water store to be filled by probable rainfall could also be considered. The fixed cycle/varied application scheduling strategy holds good potential for the derivation and practical implementation of a highly efficient deficit irrigation strategy that can be managed relatively easily.

D. Integrating Crop Yields, Irrigation System Limitations, Scheduling Strategies, and Profit

To determine optimal irrigation strategies, it is necessary to quantify the yield and hence profit implications associated with various scheduling strategies and system limitations. Crop yield simulation models can be used for this. Simulations can be performed whereby the cost of applying irrigation water with a defined irrigation system using various possible timing and application constraints can be compared to the associated returns derived from the corresponding simulated crop yields together with crop price information. An example of the results that could be derived from such simulations is shown in Table 1. The information shown in Table 1 is hypothetical and is presented here for illustrative purposes, but similar information for specific environmental and economic conditions can be derived for a given site with the aid of simulation models (see, e.g., Lecler et al., 1994). Apart from the crop yield model inputs, most of the economic information necessary for such an exercise such as crop prices and irrigation system and energy costs can be obtained relatively easily from local sources. Information analogous to what is presented in Table 1 is highly dependent on the economic and environmental data and information used to generate it and can vary significantly. Therefore, the reader is encouraged to focus on the observations that have been drawn from an analysis of this or similar information rather than on the numbers themselves.

The following observations based on the hypothetical results shown in Table 1 are presented to illustrate how the analysis of crop yield model outputs can aid in the selection of optimal irrigation system and scheduling specifications for either water or land production constraints.

1. Increasing the irrigation system capacity above a threshold level, equivalent in this case to 60 mm of water applied every 7 days, did not improve yields. This can be deduced by comparison of the differences in crop yields and seasonal irrigation water use for systems 3 and 4. These differences are negligible. Therefore, even though system 4 had the capacity to apply 70 mm of water in 7 days, simulations could show that this larger potential capacity was not required. If system 4 was not scheduled correctly, the surplus system capacity may indeed be detrimental as a result of increased potential for overwatering.

2. Irrigation systems with capacities that did not permit at least a 60 mm irrigation application every 7 days could not maintain soil water at a stress-free level. This can be deduced by comparing yields: 8.5 t/ha and 7.6 t/ha for systems 2 and 1, respectively, compared to 9 t/ha for system 3. However, from an economic perspective, the reduction in income resulting from the reduced crop yield for system 2 is more

Table 1 Comparison of Economic Returns and Seasonal Water Applications for Different Irrigation System Limitations and a Fixed Cycle/Varied Application Irrigation Scheduling Strategy

Irrigation system	Maximum capacity (mm/7 days)	Seasonal irrigation application (mm)	Crop yield* (t/ha)	Net returns to land ($/ha) (a)	Irrigable area relative to system 4 (ha) (b)	Relative net returns if water is limited ($) (a) × (b)
1	20	350	7.6	56	1.440	80.64
2	40	450	8.5	65	1.120	72.80
3	60	500	9.0	60	1.008	60.48
4	70	504	9.0	57	1.000	57.00

*Non-water-stressed crop yield potential = 9.0 t/ha.

than compensated for relative to system 3, owing to reduced capital (i.e., as influenced by the smaller capacity) and operating costs (i.e., less water applied with a smaller irrigation system).

3. Using a fixed cycle/varied amount irrigation scheduling strategy should result in system 2 realizing maximum profits if land is limited. If water is limited, however, system 1 should potentially give the maximum returns. This is because the capacity limitations of system 1 allow for only a seasonal irrigation water application of 350 mm compared to the 450 mm that was applied using system 2. (Systems 3 and 4 applied 500 and 504 mm, respectively.) Therefore if system 1 were selected, the same volume of water as was used for 1 ha in system 4 could be spread over a relatively larger area (1.440 ha at 350 mm), which, depending on the economic information, could result in a relatively larger profit, as was the case for the information shown in Table 1.

After the selection of preliminary irrigation scheduling strategies and system design specifications, the next step is to determine the irrigation water supply constraints and an appropriate area to irrigate. This latter analysis will reveal whether land or water is the factor limiting production. After this procedure, it may be necessary to reassess the selected irrigation scheduling strategy.

V. IRRIGATION WATER SUPPLY

Water for irrigation applications may be supplied from a number of sources, including reservoirs, rivers, or a combination of rivers and water releases from upstream reservoirs, or it may be supplied via canal systems from areas remote to the irrigated fields. The area being irrigated may vary seasonally according to the crop(s) being grown and is dependent on the selected scheduling strategy and available water supply. An estimate of the available water supply is therefore needed in order to determine an optimum deficit irrigation strategy.

In some situations, irrigation water supply information is available from direct measurement at the desired site. Most often, however, the need for this information has resulted in the development and use of hydrological simulation models to simulate, among other things, streamflow. The simulation of irrigation water supply should allow for a large degree of flexibility so that complex real situations can be represented. The irrigation supply options in the ACRU model (Lecler et al., 1995a) are good examples of what is required and are discussed in this section. Reference should be made to Fig. 3, which depicts schematically various possible sources of water supply to an irrigation project.

A. Unlimited Water Supply

In the planning and design phase of an irrigation project the need may arise to determine irrigation water demands when there are no

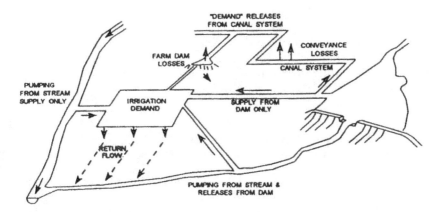

FIGURE 3 Simulating irrigation water supply—a schematic of various options. (After Lecler et al., 1995a.)

water resource limitations. For example, this is an option that could be used in assessing crop yields and seasonal water use under different deficit irrigation scheduling strategies. Then, once a strategy has been selected, the relationships between irrigation supply and demand and the area to be irrigated can be investigated.

B. Irrigation Water Supply from a Reservoir

In the ACRU model the streamflow from a watershed can be routed through reservoirs of various sizes. This enables the concurrent selection of an optimum reservoir capacity and area irrigated in relation to the hydrological characteristics of the watershed and the irrigation water demand associated with the selected irrigation strategy. Irrigation demands are simulated to take place in tandem with a reservoir yield analysis (Schulze et al., 1995b). All irrigation demands by the crop, plus all supply losses (conveyance, etc.), are then abstracted from the reservoir, subject to its containing water; i.e., when dead storage levels are reached in the reservoir, no more irrigation is supplied to the fields and the crops may suffer yield losses due to plant stress caused by water shortages.

C. Irrigation Water Supply Directly from a River

It is a common practice for irrigators to pump water directly out of a river onto their adjacent fields. When this option is selected in the ACRU model, only those irrigation water demands (crop plus losses) for which the stream has enough water at the point of abstraction can be satisfied. Daily streamflow downstream from that point is consequently reduced by the amount of the daily irrigation requirements that can be met. If the irrigation requirements exceed the available daily streamflow amount, then the irrigation applications are reduced accordingly and the crop may suffer water stress and yield reduction.

D. Irrigation Water Supply from Rivers in Combination with Releases of Water from Upstream Reservoirs

Numerous irrigation projects have evolved along river frontages, where, along a length of river, water is abstracted by a series of farmers by pumping. A consequence is that the farmers on the downstream end of the river often cannot irrigate to full demand because the upstream farmers have used up all the streamflow for their own irrigation demands. It is then often necessary to build a dam within the system so that stored water can be released to downstream users when the natural streamflow is not sufficient to supply their abstraction demands.

When this option is invoked in ACRU and the simulated streamflow on a given day fails to satisfy abstraction demands, extractions from the storage reservoir then operate on a first come, first served basis. The water requirements of the most upstream irrigation area are satisfied first, followed in a cascade by those areas with abstraction demands further downstream.

The options to simulate and integrate irrigation water supply and demand in the ACRU modeling system are commendable and make for a powerful irrigation planning tool. At present, intermediate level crop yield models for maize (corn), wheat, and sugarcane that can be integrated with various irrigation water supply and scheduling options are available. In order to provide integration with a wider range of more sophisticated crop growth models, water supply options similar to those used in the ACRU model are being developed for the DSSAT v3 (Tsuji et al., 1994) crop growth models. A reservoir yield routine (Schulze et al., 1995b) and irrigation application routines (Lecler et al., 1995b) have been added to a prototype version of DSSAT v3 (Lecler and Hoogenboom, 1995). From a user's perspective the only additional requirement for this prototype version of DSSAT is a file containing daily streamflow information and data, either recorded or simulated with, for example, the ACRU model.

VI. DISCUSSION AND CONCLUSIONS

The theoretical basis and analytical framework for deficit irrigation are well established. It has been difficult, however, to turn theory into practice. Difficulties arise in the determination of crop responses to various irrigation water applications and also in assessing the available water resources and associated implications. With advances made in crop growth/yield modeling and also in modeling watersheds using hydrological simulation models, the potential exists for irrigation design and planning procedures to be refined and for deficit-type irrigation strategies to be more easily developed.

As a note of caution, though, an advanced, efficient, but very complex deficit irrigation strategy that may seem the perfect solution during computer simulations may be of only limited use to a farmer making day-to-day irrigation decisions who places a premium on management considerations. The challenge is to optimize irrigation practice on the farm, not on a computer! The benefits of simple deficit irrigation strategies that can be easily managed at the farm level may outweigh an academically impressive but very complex "computer-derived" strategy that is very difficult to manage at the farm level.

The options to simulate irrigation water supply discussed in this chapter cover a wide range of probable occurrences. Their integration with irrigation water demand and crop growth models leads to a very powerful deficit irrigation planning tool and provides the facility to investigate in an integrative mode

1. Various irrigation application strategies, namely, different timing criteria and application limitations
2. Various irrigation water supply options, for example, from a river, a reservoir, or reservoirs of different capacities
3. Various areas under irrigation
4. The associated crop responses in terms of yield as simulated by the crop growth/yield modules, which is vital for economic optimization and the determination of an optimal irrigation strategy

REFERENCES

Branson, F. A., G. F. Gifford, K. G. Renard, and R. F. Hadley. 1981. *Rangeland Hydrology*. Dubuque, IA: Kendall/Hunt.

de Jager, J. M. 1994. Accuracy of vegetation evaporation formulae for estimating final wheat yield. *Water SA* 20(4):307–314.

English, M. 1990. Deficit irrigation. I: Analytical framework. *J. Irrigation Drainage Eng.* 116(3):399–412.

Furniss, P. W. 1988. Planning for effective irrigation water utilization. M.Sc.Eng. Dissertation. Department of Agricultural Engineering, University of Natal, Pietermaritzburg.

Green, G. 1984. A Green look at irrigation. *SA Water Bull.*, May, pp. 31–33.

Lecler, N. L., and G. Hoogenboom. 1995. Personal communication. Dept. Agric. Eng., University of Natal, Pietermaritzburg, and Dept. Biolo. Agric. Eng., University of Georgia, Griffin.

Lecler, N. L., G. A. Kiker, and R. E. S. Schulze. 1994. Integration of factors affecting irrigation system capacities. ASAE Paper No. 942028. St Joseph, MI: ASAE.

Lecler, N. L., R. E. Schulze, and W. J. George. 1995a. Irrigation water supply and return flows. In: *Hydrology and Agrohydrology: A Text to Accompany the ACRU 3.00 Agrohydrological Modelling System*. R. E. Schulze (Ed). Report TT69/95. Pretoria: Water Research Commission, pp. AT18-1–AT18-8.

Lecler, N. L., and R. E. Schulze. 1995b. Irrigation crop water demand. In: *Hydrology and Agrohydrology: A Text to Accompany the ACRU 3.00 Agrohy-*

drological Modelling System. R. E. Schulze (Ed). Report TT69/95. Pretoria: Water Research Commission, pp. AT17-1–AT17-16.

Schulze, R. E., F. B. Domleo, P. W. Furniss, and N. L. Lecler. 1995a. Crop yield estimation. In: *Hydrology and Agrohydrology: A Text to Accompany the ACRU 3.00 Agrohydrological Modelling System.* R. E. Schulze (Ed). Report TT69/95. Pretoria: Water Research Commission, pp. AT19-1–AT19-14.

Schulze, R. E., J. C. Smithers, N. L. Lecler, K. C. Tarboton, and E. J. Schmidt. 1995b. Reservoir yield analysis. In: *Hydrology and Agrohydrology: A Text to Accompany the ACRU 3.00 Agrohydrological Modelling System.* R. E. Schulze (Ed). Report TT69/95. Pretoria: Water Research Commission, pp. AT14-1–AT19-17.

Tsuji, G. Y., G. Uehara, and S. Balas. 1994. *DSSAT v3.* Honolulu, Hawaii: University of Hawaii.

10

GRAZE: A Beef-Forage Model of Selective Grazing

Otto J. Loewer, Jr.

College of Engineering, University of Arkansas, Fayetteville, Arkansas

I. INTRODUCTION

A. Objectives

The quantitative description of either a plant or animal system is indeed complex. Describing the plant/animal grazing interface presents an even greater challenge. The beef-forage grazing model described in this chapter is the result of nearly 20 years of effort by many individuals. Model developments are dynamic; that is, models change as information and research objectives change. With this in mind, the objectives of this material are to (1) present a brief history of the submodels contained in the GRAZE model and (2) conceptually describe the logic and mathematics of the model as related to quantifying biological processes.

B. Historical Perspective

The ancestor of the GRAZE model is the original Kentucky BEEF model (Loewer et al., 1980, 1981; Walker et al., 1977). BEEF was funded in 1976 by the National Science Foundation (NSF) to examine energy use in beef production systems. Four major submodels in BEEF work together to provide a daily analysis of the effects of changes in resources and management on the production of beef cattle. These four submodels relate to economics, energy, plant growth, and animal growth.

In 1978, further development of BEEF became one of the objectives of the North Central Regional Research Project "Forage Production and Utilization Systems as a Base for Beef and Dairy Production" (NC-114). One goal was to develop an improved stand-alone plant model that would directly simulate the impact of weather on a daily basis and also compute various measures of forage quality. This plant model was called GROWIT and is described by Smith and Loewer (1981). Another goal of NC-114 was to develop an improved stand-alone animal model capable of considering climate, breed characteristics, and some measure of forage quality in estimating beef animal growth, reproduction, and milk production. The animal model (called BEEF NC-114 or BABYBEEF) was created as an interactive model (Loewer et al., 1983a). Validation of the submodel for feedlot conditions is given by Brown (1982). It was the long-term goal that these plant and animal models would again be combined into a single grazing model. This was accomplished later on with the development of GRAZE by two other regional research projects.

The primary limitation of the plant model GROWIT was that it did not have direct linkage to an animal grazing model. Rather, GROWIT used various harvesting strategies to reflect grazing conditions and used a very simple algorithm to reflect forage utilization by grazing animals.

BEEF NC-114 had several limitations. Gain and maintenance were based on NRC (National Research Council, 1976) regression equations (Lofgreen and Garrett, 1968), and protein needs were determined through a similar formulation (Preston, 1966). These equations were developed for steers and heifers under feedlot conditions. BEEF NC-114 extrapolated these equations over the life of the animal. The degree to which error is introduced when going beyond the boundaries of the regressions is unknown; yet the NRC equations appeared to provide the best available estimates. Another limitation of BEEF NC-114 was that metabolizable energy (ME) and digestible protein (DP) were the measures of forage quality, and little allowance was made for minerals or changes in the animal body composition over time.

For the above reasons, the initial objective of the Southern Regional Research Project "Simulation of Forage-Beef Production in the Southern Region" (S-156), initiated in 1981, was to develop an improved beef-forage model that could more accurately test the effects of nutrition and environment on animal maintenance, growth, and production. This larger model was called GRAZE and was the emphasis of another Southern Regional Research Project S-221 "Development of Profitable Beef-Forage Production Systems for the Southern Region." At the end of this project in 1994, a user's guide and a collection of "validation" case studies for GRAZE version 2.3 were prepared (Loewer and Parsch, 1995; Parsch and Loewer, 1995). The GRAZE software package includes both a DOS version (Version 2.3, available since mid-1995) and a Windows version (Version 3.02, available since June 1996). Both versions are available for downloading free of charge from the Internet at the home page of the University of Florida Agricultural & Biological Engineering at the following Universal Resource Locator (URL) address:

http://www.agen.ufl.edu

C. GRAZE Overview

GRAZE is a dynamic simulation model that mathematically describes the daily interrelationships among plants and grazing beef animals over a grazing season (Fig. 1). Plant growth and development are

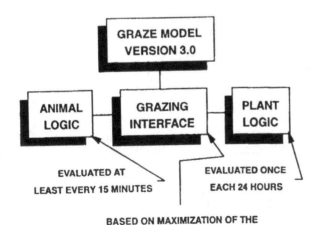

FIGURE 1 Overview of the major components of the GRAZE model.

simulated on a daily basis, and the growth and development of animals are computed at least every 15 min. The model is composed of three submodels: plants, beef animal, and plant/animal interface. The plant/animal interface simulates selective grazing and is based on the concept that an unbound grazing area may be conceptually divided into subareas or "partial fields" that are created during the grazing process (Fig. 2). These subareas may differ in forage quality and density. The animal then selects among these as desired based on the premise that grazing priority is established by an attempt to maximize the digestible dry matter intake rate. Subareas may be created or consolidated on the basis of the selective grazing process. Subareas are modeled as separate entities with regard to plant growth.

GRAZE represents a significant improvement over the BEEF model in that it more fully describes plant and animal physiology during the grazing process. However, GRAZE may be used to simulate forage growth without grazing, or beef animal growth in a feedlot situation. The plant submodel is based on the premise that all forages may be described collectively by a number of parameters. The animal submodel is somewhat unique in that it uses body composition as the key to defining physiological development of the animal. A study conducted in the early stages of this model's development

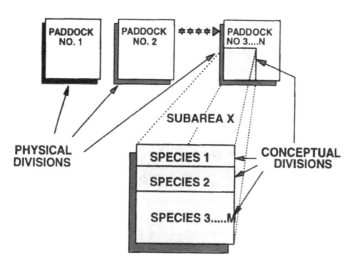

FIGURE 2   Physical and conceptual divisions within GRAZE.

indicated that the body composition approach closely followed National Research Council (NRC) predictions for average energy and animal inputs but provided deeper insights into distribution of growth under exceptionally high or low energy feed supplies (Loewer et al., 1982, 1983b). These results were further enhanced by comparing the protein deposition predictions of GRAZE to those of other researchers (Hintz et al., 1982).

D. Industrial Dynamics Approach to Modeling

The plant–animal ecosystem is composed of various systems that interact and depend upon each other. One key to understanding the response of animals to various internal and external stimuli is to use a flowchart that depicts the flow of various types of material and information with the associated feedback loops. The industrial dynamics (Forrester, 1968) approach to modeling provides a set of symbols that will be used in many of the figures to follow in describing some of the processes used in GRAZE. The list of symbols is given in Fig. 3.

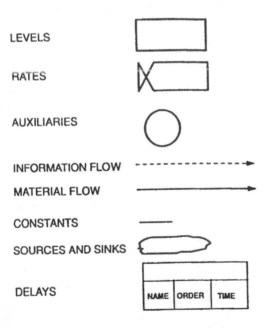

FIGURE 3   Industrial dynamics symbols.

1. *Levels.* The rectangles (or "levels") represent the quantity of material present at one point in time. The more mathematically inclined would refer to the "levels" as the resulting quantity obtained by integrating an input expression (a rate) over time. An example of a level is the quantity of fat contained in the animal.

2. *Rates.* The valve-type symbols indicate rates of flow per unit of time.

3. *Auxiliaries.* The circles are called "auxiliaries." Auxiliaries may be viewed as gauges that indicate a measure of a level. For example, the heat content of the animal would be a level, while the animal's temperature would be an auxiliary.

4. *Information flow.* Dashed lines indicate the flow of information, such as when to initiate eating.

5. *Material flow.* Solid lines indicate the flow of material such as nutrients.

6. *Constants.* Constants maintain their values over time and are indicated by a relatively short straight line over which a name appears.

7. *Line closed on itself.* The curved line closed on itself indicates a point beyond the boundary of the system to be considered. For example, the source of dry matter for feeding may be outside the area of interest.

8. *Delays.* Delays are symbolized by a box with three compartments and may be viewed as a type of surge basin where outflow is related to but not necessarily equal to inflow at a given point in time.

E. GRAZE Approach to Modeling

In describing GRAZE, it is important to recognize that we are all modelers of one type or another. The mathematical-logical concepts in GRAZE were developed in part by interdisciplinary teams of researchers, usually beginning with an ideal about how something functioned and ending with a quantitative description of the process. Most of these researchers had little or no formal training in modeling or the use of computers. Rather, their contributions were to provide a scientific understanding of the beef-forage ecosystem and a willingness to discuss and justify their concepts in a group setting. Thus, GRAZE represents a pooling of ideas and experience and reflects many types of models.

1. Mental Models

Each of us observes the world that surrounds us. We formulate how we believe it functions in terms of physical, social, and economic relationships. This type of model is called a mental model and is composed of images and words that we use in communicating our understanding to others. When we make an observation that cannot be explained by our existing mental model, we reformulate it, taking into account the added information. The process of model restructuring is called learning and has been an integral part of GRAZE development. As part of the regional projects, researchers would gather several times a year to discuss some facet of the beef-forage ecosystem, with the longer term goal being to quantify the relationships in question. Thus, GRAZE represents a type of quantitative consensus about the workings of the beef-forage ecosystem.

2. Physical Models

Mental models may be used to form physical models, the term "physical" being used in its broadest sense to include components associated with physics, biology, and chemistry. A physical model is an object or system made to scale, a simple example being a model airplane. Grazing trials, however, may also be classified as physical models in that they attempt to represent scaled-down grazing systems. Likewise, grazing trials have been used to confirm that the predictions given by GRAZE are within the limits of what has been observed in nature (Loewer and Parsch, 1995).

3. Mathematical Models

Mathematical models result when mental models are expressed in mathematical form, thus quantifying our description of the system in question. Mathematical models may be placed into categories: statistical, mechanistic, or simulation.

Statistical Models. Statistical models include regression models. Regression models use experimental observations to mathematically relate one or more input parameters with an associated output. An example of a statistical regression model is the net energy system that predicts average daily gain of beef cattle given dry matter intake and energy content (Lofgreen and Garret, 1968). However, GRAZE also contains a few equations that were based on statistical observations.

Mechanistic Models. Mechanistic models are based on a mathematical form that is believed to best represent the relationship between one or more input parameters and an associated output. Input terms in the mathematical expression may often be conceptual rather than physical. For example, a power function may be used to describe the weight–age growth relationship (Brown et al., 1976). Thus, for the equation

$$Y = a*X^b \tag{1}$$

the exponent b of the mathematical function has conceptual rather than physical significance. GRAZE contains many of these types of expressions that are used to define the intervals in a continuous manner between known points of relationships.

Simulation Models. Simulation models most often use computers for computational purposes and are often referred to as computer models. Computer models are collections of mathematical-logical expressions that relate inputs and outputs. These expressions may include both statistical and mechanistic relationships. Optimization models, such as linear programs, are sometimes referred to as static simulations. Models that predict relational changes that occur over time are called dynamic simulations. Dynamic simulation models offer the greatest potential for describing the complex relationships associated with grazing trials; GRAZE is included in this category of models.

F. Major U.S. Modeling Efforts in Beef that Paralleled GRAZE

The biological processes associated with an animal may reflect far different levels of aggregation. For example, the following modeling efforts, which originated in the 1960s and 1970s, represent different levels of methodology and aggregation in their approach to describing beef animal physiology. These models are all discussed in chapters of a book edited by Spreen and Laughlin (1986) and have been used as the basis for other modeling efforts.

1. California Net Energy Model

The California net energy system was mentioned earlier (Lofgreen and Garret, 1968). Essentially, this is a very simple static regression model that relates intake energy to beef animal gain. Its equations can be used as part of a dynamic simulation.

2. Texas A&M Model

The Texas A&M model (Sanders and Cartwright, 1979a, 1979b) describes beef animals in terms of a herd composed of various categories of animals. This model emphasizes the breeding aspects of herd management.

3. UC Davis Model

The University of California at Davis model (Baldwin and Black, 1979) examines the animal at the organ level. The collective function of the body organs describes the total function of an animal.

4. Kentucky BEEF Model

The Kentucky BEEF model is like the Texas A&M model in that it considers a herd of animals composed of various categories within the total. However, the emphasis of this model is the daily prediction and associated interaction of forage and animal performance. This model formed the basis for the GRAZE model, which came out of the same working groups. GRAZE emphasizes the internal functioning of an individual beef animal (but not to the extent of the UC Davis model) and its interaction with forage growth and availability.

II. MODELING PLANTS

A. An Overview of the Plant Submodel

The following discussion of the plant logic associated with GRAZE draws heavily on the work of Smith and Loewer (1981, 1983) and Smith et al. (1985).

1. Nonspecific Models

The plant modeling logic in GRAZE is nonspecific in that it is not restricted by site, crop, or management specificity and consequently can be used in different physiographic areas by inputting those parameters that relate site, crop, and management variables to the attributes being simulated (Fig. 4). The nonspecificity is accomplished by developing continuous mathematical-logical equations that describe the fundamental relationships among the variables that affect time-related changes in the attributes being simulated by the model. These mathematical-logical equations form the bases for the nonspecific model, as opposed to the algorithms of specific data that form the bases for specific models.

2. Genetics

As a nonspecific model, GRAZE identifies 25 variables that describe the genetic potential of the crop to be simulated (Parsch and Loewer,

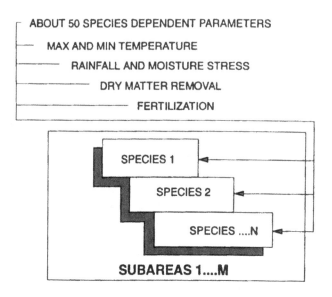

ABOUT 50 SPECIES DEPENDENT PARAMETERS

MAX AND MIN TEMPERATURE

RAINFALL AND MOISTURE STRESS

DRY MATTER REMOVAL

FERTILIZATION

SPECIES 1

SPECIES 2

SPECIES ....N

SUBAREAS 1....M

FIGURE 4   Factors affecting plant growth are determined for each species growing on each subarea once each 24 h.

1995). Thus, each crop is viewed as functioning in basically the same way, with differences among crops being a result of differences in the values of these variables.

3.   Time Step for Calculations

The time step used for determining crop status is 1 day. This is to be contrasted with the animal portion of GRAZE where the status of an animal is updated at least every 15 min.

4.   Exogenous Variables

The exogenous variables that directly influence plant growth in GRAZE are daily maximum and minimum temperature, leaf area, day length, photoperiod, water, and nutrients (N, P, K—nitrogen, phosphorus, and potassium).

5.   Endogenous Variables

There are many endogenous variables within the plant model of GRAZE, as will be discussed later. However, it is important to recognize that GRAZE simulates the availability of dry matter. Dry matter is described physiologically as collections of indigestible and potentially digestible portions of material that is further characterized as belonging to several physiological states (new, old, dead).

6. Daily Growth

Daily growth is defined initially by genetic potential as influenced by temperature, and it is assumed that radiation levels are adequately expressed by temperature levels over the day. Temperature, in turn, is defined continuously over the course of the day. The daily growth projection is modified downward by a series of multipliers in the range of 0–1. Multipliers are computed for leaf area, photoperiod, and rainfall. Growth may also be limited by the availability of nutrients from the soil.

B. Overview of Flows Within the Plant

The flows in the plant as modeled by GRAZE consist of movement of water and nutrients, changes in physiological states, and removal of dry matter through harvesting and grazing. All of these are discussed at length later. The following is a brief overview of these flows.

1. Energy

Energy flow is expressed in the plant portion of GRAZE by converting sunlight, as represented by day length and temperature, into plant growth. The ability of the plant to convert energy into dry matter may be limited by a number of other factors, including leaf area, photoperiod, water, and available nutrients. Also, stored sugars in the various physiological states of the plant contain stored energy that may be used for plant growth.

2. Water

GRAZE contains a water balance model that considers rainfall, runoff, percolation losses, and plant utilization. The availability of water has a direct impact on plant growth potential.

3. Nutrients

Nutrients (N, P, and K) are extracted from the soil by the plant. The availability of soil nutrients may limit plant growth. Nutrients may be replenished by a fertilizer application, and nitrogen may be leached from the soil. Currently, GRAZE does not consider recycling of nutrients from deposition of fecal material by grazing animals.

4. Physiological States

GRAZE places dry matter into one of three physiological states: new, old, or dead material. Dry matter is also divided into three cellular states (cell contents, potentially digestible cell wall, and indigestible cell wall), all of which are subject to change based on their physiological states. In general, physiological states flow from new to old to dead material with an accompanying change in cellular makeup.

C. Impact of Air Temperature on Growth

Crops are seasonal, as exemplified by the accepted characterization of perennial grasses as cool season and warm season grasses (Health et al., 1973). The predominant environmental attribute that distinguishes the different seasons is air temperature. Crops are seasonal because they are genetically adapted to grow in certain temperature regimes.

Plant physiologists (Salisbury and Ross, 1969) have documented three characteristic air temperatures as they relate to crop growth: (1) A minimum air temperature below which growth does not occur, (2) a maximum air temperature above which growth does not occur, and (3) an optimum air temperature at which the growth rate of each crop is optimum. These three characteristic air temperatures are combined with a fourth parameter, the maximum possible growth rate, to characterize the genetic potential of each crop as a function of temperature.

The continuous quadratic function used in GRAZE to relate growth rate to the three characteristic air temperatures is (Fig. 5)

$$dG/dH = A + B*TEMP + C*TEMP**2 \qquad\qquad (2)$$

where

dG/dH = plant growth rate, kg/ha per hour
TEMP = air temperature (°C)

The constants A, B, and C can be determined by applying the four crop parameters mentioned previously. This differential equation could not be solved to obtain accumulated growth (G) as a function of time (H), because the presence of air temperature (TEMP) makes it nonhomogeneous. A function was needed to relate air temperature (TEMP) to time (H) so that growth rate (dG/dH) could be expressed as a function of time (H).

Growth depends upon photosynthesis, which occurs on a diurnal period, so it seemed logical to express air temperature for each

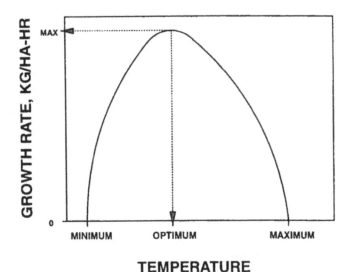

FIGURE 5   Growth rate of plant as influenced by temperature.

diurnal period. Penrod (1960) and Smith et al. (1968) did considerable research on the diurnal variation in air and soil temperatures. In the time period from sunup to solar noon to sundown, daily air temperature follows a function of the form

$$\text{TEMP} = A + B*H + C*H**2 \tag{3}$$

where

   A, B, C = constants
      H = time in hours
   TEMP = temperature (°C)

From sundown to sunup, the function is of the form

$$\text{TEMP} = D*e**(-E*H) \tag{4}$$

where D and E are constants.

   The constants $A, B, C, D,$ and E can be determined with four environmental parameters: (1) minimum daily air temperature, which occurs at sunup; (2) maximum daily air temperature, which occurs at solar noon; (3) the mean daily air temperature, which occurs at sundown; and (4) day length (hours from sunup to sundown), which can be calculated as a function of latitude (Fig. 6).

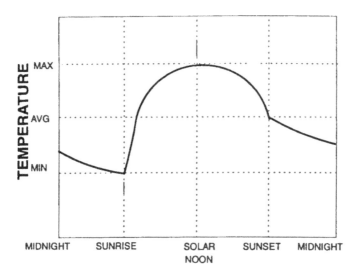

TIME OF DAY

FIGURE 6   Temperature distribution based on maximum and minimum daily temperatures.

Equation (3) can be substituted for temperature TEMP in Eq. (2). Equation (2) can then be solved for the accumulated growth (G) that will occur in one diurnal period. This accumulated growth for each day is the daily growth rate in kilograms per hectare per day based upon the air temperature regime (Fig. 7).

D. Latitude and Day Length

Day length (XL) is determined by latitude for each day of the year (JULIAN) using the following mathematical and logical expressions. For a range of JULIAN values, an intermediate term DELTA is defined as follows.

For JULIAN values less than or equal to 41:

$$DELTA = 0.2073 * JULIAN - 23.5 \tag{5}$$

For JULIAN values between 42 and 108:

$$DELTA = 0.3788 * JULIAN - 30.91 \tag{6}$$

For JULIAN values between 109 and 240:

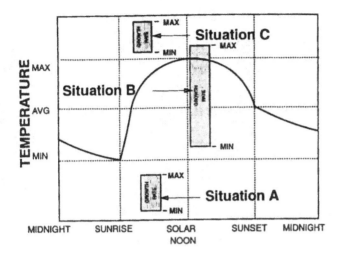

 Range of daily temperature

Situation A: Temperature too low for growth

Situation B: Temperature allows some growth

Situation C: Temperature too high for growth

FIGURE 7   Situations reflecting crop growth and the daily temperature range. Situation A: Temperature too low for growth. Situation B: Temperature allows some growth. Situation C: Temperature too high for growth. Shaded rectangles indicate range of daily temperature.

$$DELTA = (0.4122*(JULIAN - 108)$$
$$- 0.0032*(JULIAN - 108)**2) + 10.0 \qquad (7)$$

For JULIAN values between 241 and 309:

$$DELTA = 0.3676*JULIAN + 98.62 \qquad (8)$$

For JULIAN values greater or equal to 310:

$$DELTA = -0.1518*(JULIAN - 309) - 15.0 \qquad (9)$$

Given the intermediate value of DELTA, day length (XL) is determined as follows:

$$XL = 0.1333*ARCOSE(- (SIN(0.01745*XLAT)$$
$$*SIN(0.01745*DELTA))/(COS(0.01745*XLAT)$$
$$*COS(0.01745*DELTA)))*57.31 \tag{10}$$

where XLAT is the latitude of the location, in degrees.

Sunrise (SUNRIS) and sunset (SUNSET) are computed as

$$SUNRIS = 12.0 - XL/2.0 \tag{11}$$

and

$$SUNSET = 12.0 + XL/2.0 \tag{12}$$

E. Impact of Leaf Area on Growth

The daily growth rate as a function of air temperature cannot be achieved unless there is enough leaf area to intercept sufficient solar energy to support photosynthesis. Functions were developed to relate the proportion of the daily growth rate (based on air temperature) that can be achieved to the quantity of accumulated dry matter (sum of daily growth rates minus the quantity harvested) (Fig. 8). These functions are of the forms

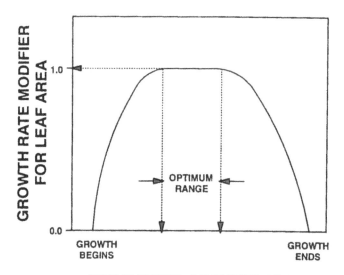

DRY MATTER ACCUMULATION

FIGURE 8   Daily plant growth as influenced by leaf area.

$$LAF = A + B*ADM + C*ADM**2 \tag{13}$$

$$LAF = 1.0 \tag{14}$$

and

$$LAF = D + E*ADM + F*ADM**2 \tag{15}$$

where

> LAF = a leaf area factor that is the proportion of the daily growth rate that will occur, $0 \le LAF \le 1.0$
> ADM = accumulated dry matter, which is the sum of daily growth rates minus the quantity harvested

The constants A, B, C, D, E, and F can be determined with four crop parameters: (1) proportion of the daily growth rate that can be achieved when the crop first emerges from seed or other organs that initiate growth before photosynthesis is active; (2) minimum quantity of accumulated dry matter that provides enough leaf area to support the full daily growth rate in an optimum temperature regime; (3) maximum quantity of accumulated dry matter that provides enough leaf area to support the full daily growth rate in an optimum temperature regime (leaf senescence and shading begin to reduce the effective leaf area); and (4) maximum quantity of accumulated dry matter when growth terminates.

Equation (13) is used when the accumulated dry matter value is greater than or equal to zero and less than or equal to the minimum quantity of accumulated dry matter that provides enough leaf area to support the full daily growth rate.

Equation (14) is used when the accumulated dry matter is between the minimum and maximum quantities of accumulated dry matter that provide enough leaf area to support the full daily growth rate.

Equation (15) is used when the accumulated dry matter is greater than or equal to the maximum quantity of accumulated dry matter that provides enough leaf area to support the full daily growth rate, and less than or equal to the maximum quantity of dry matter when growth terminates.

Equations (13)–(15) are continuous equations that, in effect, force the crop growth to follow a sigmoidal relationship on a physiological time basis.

F. Photoperiod

Holt et al. (1975) describe a photoperiod effect on the growth rate of perennial crops as a physiological phenomenon of storing carbohydrates in the lower portion of the crops after the summer solstice when the day length (hours from sunup to sundown) is decreasing. The stored carbohydrates are used to regenerate growth after a period of dormancy. The function used by Holt et al. (1975) was adopted for use in the GRAZE plant model. This function is of the form (Fig. 9).

$$PF = 1.0 + [\{(C - 1.0)/(B - A)\}*(DL - A)] \tag{16}$$

where

> PF = a photoperiod factor that reduces the daily growth rate as a consequence of a portion of the carbohydrates that are produced by photosynthesis being stored instead of being used for growth ($C \le DL \le 1.0$)
> A = day length (after the summer solstice) when the crop begins to store carbohydrates (h)
> B = day length when the portion of carbohydrates going into storage reaches a constant value (h)

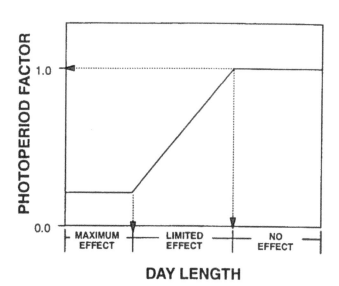

FIGURE 9   Plant growth per day as influenced by day length.

C = maximum effect of photoperiod as a proportion of the optimum growth rate (0.0 ≤ C ≤ 1.0)

DL = day length in hours (B ≤ DL ≤ A)

As day length continues to decrease,

$$PF = C \tag{17}$$

until the winter solstice is reached, and then

$$PF = 1.0 \tag{18}$$

A, B, and C in Eqs. (16) and (17) are crop parameters that reflect the genetic characteristics of each crop with respect to storing carbohydrates to regenerate growth after a period of dormancy. The photoperiod factor is calculated each day of simulation and is multiplied by the rainfall factor (discussed later), the optimum growth rate [Eq. (2)], and the three leaf area parameters [Eqs. (13)–(15)] to revise these parameters to reflect an altered growth regime.

G. Utilization of Water

The water stress factor calculated in GRAZE ranges in value from 0.0 to 1.0 and is a direct multiplier of potential plant growth (Fig. 10). For the most part, GRAZE does not "borrow" logic from other mod-

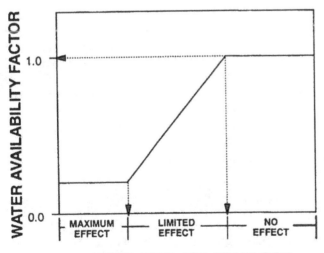

FIGURE 10   Plant growth per day as influenced by soil water.

els, the exception being that GRAZE uses water balance logic from CERES-Maize (Jones and Kiniry, 1986) and specifically subroutines SOILRI and a modification of WATBAL (called WATBL2) (Costello et al., 1988). A new subroutine, NEWAT1, replaces the original GRAZE Version 1.0 subroutine PRECIP.

In GRAZE, the water balance logic in WATBL2 calculates a daily water stress factor as a function of daily rainfall, average daily radiation, and various soil and plant parameters. The following explanation of WATBL2 (WATBAL) is a condensation from Jones and Kiniry (1986).

1. Water Stress Logic

Conceptually, the subroutine WATBAL receives values of daily rainfall and calculates runoff according to the U.S. Soil Conservation Service (SCS) curve number method. If precipitation is greater than a computed minimum, runoff occurs. Potential infiltration equals the difference between precipitation and runoff. Potential and actual soil evaporation and potential and actual evapotranspiration are calculated on the basis of specific soil, plant, and weather conditions. Water flow between soil layers is determined. Root growth and associated potential root water uptake from the soil is calculated. The actual water uptake is computed, and the soil water content of each layer is updated.

Two soil water deficit factors are calculated. The less sensitive factor is used to affect photosynthesis and is the ratio of the potential root water uptake and transpiration. The more sensitive factor affects cell expansion and is a weighted ratio of the same two factors of potential root water uptake and transpiration. In CERES-Maize, plant growth and leaf senescence associated with water stress are reduced according to the water stress indices calculated in the original subroutine WATBAL. The less sensitive of the two water deficit factors is used in GRAZE as the water stress factor. The function of the water stress factor is to reduce optimum plant growth in proportion to the available water.

2. TWATBL2: Modification to WATBAL

Although GRAZE uses the water balance logic from CERES-Maize (Jones and Kiniry, 1986), there are some modifications. These include a substituted method of calculation for potential evapotranspiration, an abbreviated determination of accumulated degree-days, and estimations of initial input values.

In CERES-Maize, potential evapotranspiration is a function of solar radiation, air temperature, and an integrated crop and soil albedo. To maintain the universal nature of GRAZE that makes the model useful and adaptable to more than one specific region, the calculation of potential evapotranspiration in WATBL2 is made according to an equation reported by Hargreaves and Samani (1982). This equation calculates potential evapotranspiration as a function of day of the year, latitude, and daily air temperatures and was selected, in part, because these values are available at most recording weather stations.

Extraterrestrial radiation (RA) is calculated in subroutine NEWAT1 as a function of day of the year and latitude, in accordance with the *Handbook of Meteorology* (Berry et al., 1945). The coefficient of temperature difference (KT) is a function of the average relative humidity of the day. The relative humidity is also calculated in NEWAT1, according to the ASAE Standard D271.2 (Hahn and Rosentreter, 1987), as a function of the minimum and average daily temperatures with the assumption that the dew point temperature is the minimum daily temperature.

Accumulation of growing degree days, DTT (°C), is calculated in the CERES-Maize subroutine PHENOL (Jones and Kiniry, 1986) as a function of mean air temperature and base temperature. Base temperature varies according to plant stage of development and is adjusted for extreme temperatures. This subroutine also performs other calculations and therefore requires values for parameters not needed by WATBL2. To reduce complexity and the need to estimate these extraneous values, the determination of DTT is made in WATBL2 following the basic guidelines set in subroutine PHENOL. DTT is equal to zero if the average daily temperature is less than the base temperature of 8°C or if the maximum daily temperature exceeds 34°C. If the daily temperature is within these bounds, DTT is determined as the difference between the average daily temperature and the base temperature.

Only subroutines SOILRI and WATBAL were extracted from CERES-Maize; hence, several parameters in WATBL2 are estimated as constants or functions. The weight of roots per unit area of soil is a function of several factors and was determined to be irrelevant to the water stress calculation in GRAZE and so is input as unity. The leaf area index is estimated as a function of dry matter yield and a factor that varies from spring to summer found from growth chamber trials and actual measurements. Maximum daily root water uptake per unit root length is estimated as a ratio of the maximum evapotranspiration and an average root length per unit of land surface area. The initial root length per unit of soil volume is expressed as an exponential function of soil layer depth. The root length density rate of decline is an estimated, plant-dependent input variable.

3. Model Input Requirements

The WATBL2 water balance routine in GRAZE requires input parameters describing soil, plant, and water conditions. Many of the factors used by subroutine WATBL2 must be estimated because of a lack of measurement or scientific description. Input variables are listed and described by Parsch and Loewer (1995). Soil water input parameters for each paddock include

> Number of days from planting to emergency
> Root length per gram root weight (cm/g)
> Maximum increase in root depth per degree-day accumulated if there is no water stress (cm/degree-day)
> Weight of roots (g/m^2)
> Maximum daily root water uptake per unit root length (cm^3/cm root)
> Leaf area index factor for summer growth
> Leaf area index factor for spring growth
> Soil Conservation Service curve number for calculating runoff
> Number of soil layers (maximum of 10)
> Bare soil albedo
> Soil water constant for calculating drainage rate
> First, second, and third coefficients for root water uptake equation
> Upper limit of stage 1 soil evaporation, mm
> Depth of each layer, cm
> Drained upper limit of soil water, decimal fraction
> Drained lower limit of soil water, decimal fraction
> Field saturated soil water content, decimal fraction
> Initial soil moisture content, decimal fraction
> Weighting factor for soil depth to determine new growth distribution

Water inputs include daily values of precipitation, extraterrestrial radiation, and maximum and minimum air temperatures.

H. Plant Composition and Physiological State

In the plant portion of GRAZE, dry matter is considered collectively rather than subdivided into subcategories such as stems, leaves, and

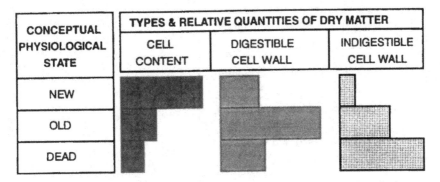

CONCEPTUAL PHYSIOLOGICAL STATE	TYPES & RELATIVE QUANTITIES OF DRY MATTER		
	CELL CONTENT	DIGESTIBLE CELL WALL	INDIGESTIBLE CELL WALL
NEW			
OLD			
DEAD			

FIGURE 11   Dry matter types and physiological states for each species on each subarea created by the GRAZE simulation model.

roots. Dry matter, however, is defined in terms of both its cellular makeup and its physiological state (Fig. 11).

1.   Cellular Composition

All plant dry matter is further described as falling into one of the following cellular states.

    1.   Cell content material
    2.   Potentially digestible cell wall material
    3.   Indigestible cell wall material

The following plant parameters describe cellular nutrient and carbohydrate composition for each plant species in GRAZE (Fig. 12).

    1.   Fraction of the cell content that is nitrogen
    2.   Fraction of the cell content that is phosphorus
    3.   Fraction of the cell content that is potassium
    4.   Fraction of the crop material in the new and old partitions that can be stored as nonstructural carbohydrates.

2.   Physiological State

There are three physiological states of the dry matter (Fig. 11) in GRAZE, referred to as (1) new material, (2) old material, and (3) dead material. Material in each of these physiological states ages according to a physiological time scale, with new plant material eventually becoming old plant material, old material eventually becoming dead material, and dead material eventually disappearing. The following plant parameters for each plant species in GRAZE describes the

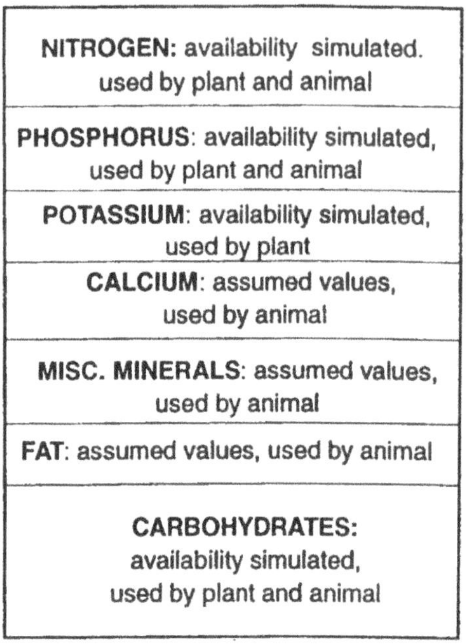

NITROGEN: availability simulated. used by plant and animal
PHOSPHORUS: availability simulated, used by plant and animal
POTASSIUM: availability simulated, used by plant
CALCIUM: assumed values, used by animal
MISC. MINERALS: assumed values, used by animal
FAT: assumed values, used by animal
CARBOHYDRATES: availability simulated, used by plant and animal

FIGURE 12   Divisions of cell content and potentially digestible cell wall as simulated and utilized by GRAZE.

change in cellular composition as the plant material changes in physiological age.

Cell wall fraction of new growth that is 1 day old
Cell wall fraction when new growth is changing to old growth
Cell wall fraction when old growth is changing to the dead partition
Cell wall fraction of dead crop material
Half-life of dead crop material, days
Half-life of nonstructural carbohydrates stored in the dead crop material, days
Fraction of the crop material in the new and old partitions that can be stored as nonstructural carbohydrates
Fraction of the maximum quantity of stored dry matter that is unavailable for harvest but can be removed from reserve to support crop growth

The following plant parameters define the rate of flow of dry matter from one physiological state to another:

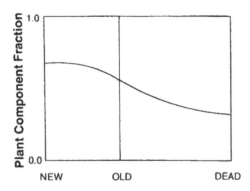

Physiological Age of Material, days

FIGURE 13 Conceptual relationships between cellular composition and physiological age.

Maximum physiological age of the initial quantity of dry matter in the new partition, which is analogous to the life span in days of a stated day length (this may be viewed as the time required for a crop leaf to reach its mature size).

Maximum physiological age of the initial quantity of dry matter in the old partition, which is analogous to the age when leaf senescence is essentially complete.

Changes in physiological state should be viewed as a continuum from new material to old material, with each physiological state also changing the ratio of cellular composition. In GRAZE, a range of functions are used to "connect" the defined plant parameters, including linear, parabolic, negative exponential, and, under extreme conditions, step function relationships. Figure 13 depicts the relationships between cellular composition and physiological age.

3. Physiological Aging in Plants

Physiological aging in GRAZE relates physiological age to chronological age. Physiological age advances at the same rate as chronological age when (1) day length is exactly the same as that defined as a plant parameter for the context in which physiological age was specified or (2) the temperature remains above freezing.

In GRAZE, physiological age advances in direct proportion to the ratio of day length to the reference day length provided as a plant parameter (Fig. 14). Each of the physiological states (new, old, and

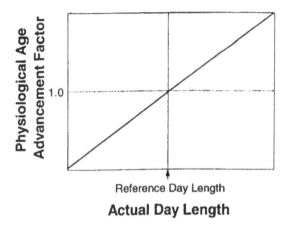

FIGURE 14 Physiological age advances in direct proportion to the ratio of day length to the reference day length provided as a plant parameter.

dead material) is represented by a physiological age that is a weighted average of the ages of all the material contained within that state. A "forcing function" (ratio of physiological age of old material to the physiological age span of old material) is used to reflect a gradual change in the photosynthetic activity of old material and does not instantaneously "dump" the old material into the dead material classification. "Dumping," however, does occur if temperatures reach critical limits. For example, when freezing occurs, all new and old material is reclassified as dead material.

I. Availability of Soil Nutrients

The plant model in GRAZE simulates growth based in part on the availability of (1) soil nitrogen (N), phosphorus (P), and potassium (K) and (2) stored sugars in the plant. Concentrations of other nutrients are estimated as used by the animal but do not directly influence plant growth.

The soil nutrient mathematical-logical relationships in the current version of GRAZE (Version 3.02) are relatively simplistic (Fig. 15). It is assumed that the root structure of the plant will be in close enough proximity to soil nutrients to remove them if needed. The function of the roots in the water logic of GRAZE, as noted above, is considerably more complex in this regard.

There is no limitation to the rate at which a nutrient can be extracted from the soil by the plant, regardless of the soil concentra-

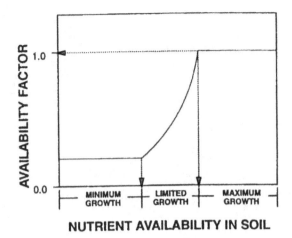

FIGURE 15 Plant growth per day as influenced by soil nutrients.

tion of that nutrient, except when soil nutrient levels are relatively low. For this situation, the potential rate of removal is reduced until a specified removal rate is reached. Nutrient extraction from the soil is modeled using the following soil parameters:

1. Fraction of nitrogen in the soil that is unavailable for immediate crop growth
2. Fraction of phosphorus in the soil that is unavailable for immediate crop growth
3. Fraction of potassium in the soil that is unavailable for immediate crop growth

Initially the amount of a given nutrient present in the soil is specified. Fertilizer applications of this nutrient to the soil adds to the total amount available to the plant. In DOS Version 2.3, nutrients are assumed to be instantly available after application. In GRAZE for Windows Version 3.02, the user may specify several forms with associated time delays for which the applied nutrient may become available to the plant (instantaneous, half-life, ramp, and exponential delay functions). A soil nutrient may be either transferred to the plant or, in the case of nitrogen, lost by leaching through the soil (Fig. 16) as modeled using as the soil input the half-life of the nitrogen-leaching process in days.

Although GRAZE computes the nutrient composition of the animal's fecal matter, this nutrient source is not added back to the soil.

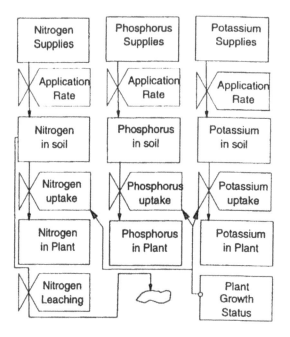

FIGURE 16 Soil nutrients may be either transferred to the plant or, as with nitrogen, lost by leaching through the soil.

Likewise, no attempt is made to reduce the effective grazing area as a function of fecal deposition. Future versions of GRAZE will address nutrient recycling and associated issues such as water quality.

J. Nutrient Demand by the Plant

The overriding assumption in GRAZE with regard to nutrient uptake is that the concentrations of N, P, and K in the cell content of new growth are constant. Thus, availability in the soil of any one of these nutrients may limit the growth of the plant. GRAZE uses the following equations to determine the amount of each nutrient required to satisfy plant growth given that there may be other limiting factors such as water or leaf area.

$$RNC = XN*(1.0 - CWALL1)*C \tag{19}$$

$$RPC = XP*(1.0 - CWALL1)*C \tag{20}$$

$$RKC = XK*(1.0 - CWALL1)*C \tag{21}$$

where

> RNC, RPC, RKC = amounts of N, P, K, respectively, required for
> growth, kg/ha
> XN, XP, XK = fraction of the cell content that is N, P, or K,
> respectively
> CWALL1 = fraction of the plant growth that is cell wall
> material
> C = projected plant growth, kg/ha

GRAZE assumes that there exist minimum amounts of N, P, and K in the soil that are always available to the plant. The soil input parameters that define these values are

1. Minimum amount of nitrogen always in the soil (kg/ha)
2. Minimum amount of phosphorus always in the soil (kg/ha)
3. Minimum amount of potassium always in the soil (kg/ha)

K. Potential Digestibility of the Plant Material

Digestibility is a plant "quality" factor in grazing systems. In GRAZE, all cell content material is assumed to be digested by the animal. Similarly, by definition the animal cannot digest the indigestible portion of the cell wall. The remaining material, called potentially digestible cell wall, may be digested if the retention time in the animal is long enough. This logic is discussed in more detail in Section III.

Plant parameter inputs that are used to determine potential digestibility of cell wall material are

1. Fraction of digestible cell wall in very young growth
2. Fraction of senescent digestible cell wall
3. Minimum fraction of digestible cell wall in very old crop material

Figure 17 depicts how potential digestibility changes over the physiological age and state of the plant. Potential digestibility does not directly influence crop growth. However, it does influence the rate at which dry matter is removed by grazing and thus will indirectly affect plant growth.

L. Species Competition

Plant species competition is modeled in GRAZE version 3.02 (Windows version) but not in Version 2.3 (DOS version), which was presented by Parsch and Loewer (1995) and Loewer and Parsch (1995).

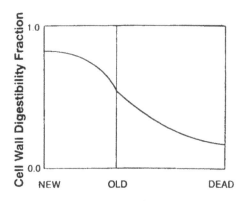

FIGURE 17 Conceptual changes in potential digestibility of cell wall as impacted by physiological age and state of the plant.

In GRAZE 3.02, many different species can be specified as growing on the same grazing paddock, with each species effectively occupying a stated fraction of the land area (Fig. 18). Assumptions are that all species are uniformly distributed and that the proportion of land area occupied by each species does not change even though the relative quantities of dry matter may. This logic is closely related to that presented by Smith et al. (1981).

A detailed discussion of subareas or partial fields (refers to the same thing) is given in Section IV. For the moment, a partial field should be viewed as a conceptual division of the paddock that is created by the GRAZE selective grazing logic. A paddock may have up to 25 partial fields. Partial fields are not bounded by physical barriers such as fences and are all available for grazing in priority order based upon forage quality and availability. At the beginning of each simulated day, GRAZE computes the plant growth for each species on each partial field in much the same way that it would if only one species were present. However, the forage available is multipled by the fraction of the paddock area occupied by the species as specified initially, thus effectively altering the effects of leaf area index as a function of dry matter accumulation. In addition, nutrients and water are removed from the soil each day in a sequential manner rather than by computing their removal simultaneously. Most of the time, sequential removal will not make any difference, and if it does the effects will not generally last for more than one simulated day at a time.

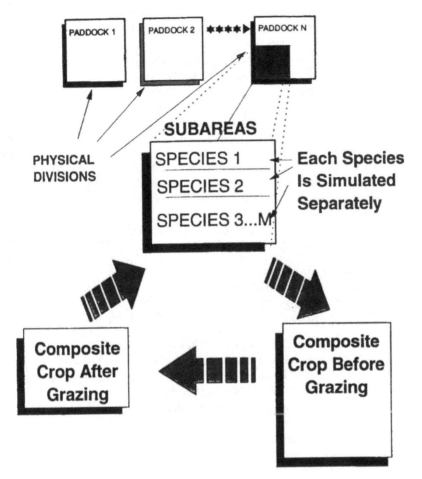

FIGURE 18 Competing species in GRAZE are simulated separately but aggregated for grazing.

In terms of grazing, the GRAZE model creates a "composite" forage composed of weighted averages of all plant attributes that influence animal selection of forage and response to plant quality. It is assumed that if grazing occurs, forage is consumed in direct proportion to the amounts of forage available from each species in the mix. GRAZE then reduces this proportional amount from each plant species on each partial field, recomputes the individual species characteristics, and independently simulates plant growth for each species. The process then repeats itself.

The selective grazing logic with species competition can be used to evaluate different mixes of conventional forages. This logic can also be used to determine the impact of undesirable plants (weeds) assuming that their plant parameters are defined.

III. MODELING BEEF ANIMALS

A. An Overview of the Animal Submodel

The beef animal portion of the GRAZE model (Fig. 1) uses engineering principles to describe biological processes. Although each animal genotype is somewhat unique, there are overriding principles that govern them all.

The biological processes of the beef animal in the GRAZE model may be quantified using

1. Body composition
2. Heat transfer and thermodynamics
3. Intake and digestion
4. Energy and nutrient priorities

A simplified flowchart of the animal portion of GRAZE is shown in Fig. 19. Two categories of factors impact the animal submodel: (1) the availability of feed nutrients and energy and (2) environment (temperature, relative humidity, radiation).

The beef animal, as described mathematically and logically by GRAZE, has survival (or maintenance) as its highest priority. This is expressed in the animal by its attempt to maintain a nearly constant body temperature. After survival, fetal development, body growth, lactation, and nutrient or fat stores follow in priority.

Under practically all circumstances, the animal loses heat to the environment. Likewise, heat is constantly being produced as a result of food metabolism and cell respiration. Growth and/or production can occur only if the intake of energy through digestion is greater than heat energy losses to the environment. The digestive system disposes of intake energy in one of three ways: Energy may be purged from the body in the form of excrement or gas, it may enter the body's pool of heat (thermal) energy, or it may enter the body's pool of chemical energy.

Chemical energy may be converted into tissue, milk, a fetus, fat in excess of that normally associated with tissue, or heat energy. The conversion to heat energy is associated either with the inefficiency of producing tissue, a fetus, or fat or with activity such as shivering.

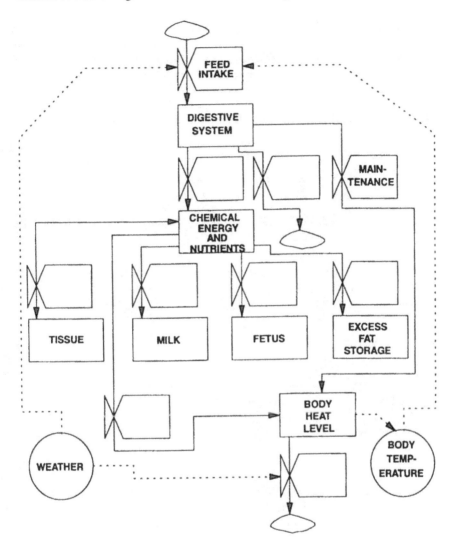

FIGURE 19 Overview of animal system.

The model contains a priority for distribution of chemical energy. Chemical energy may be stored in the form of tissue or "excess" fat (the term "excess fat" is discussed in detail later). These stores may be converted back into chemical or thermal energy forms if there is the need to increase body temperature or maintain the fetus growth rate, respectively. "Excess" fat can also be converted into the chemical energy needed to synthesize tissue or produce milk. The chemical energy level may trigger the initiation or termination of eating. A high body temperature also may terminate eating. Nutrient distribution is handled similarly.

B. Overview of the Types of Flows Within the Animal

Before examining the details of how the animal's biological processes can be described quantitatively, it is important to gain a broad perspective of the overriding principles associated with these processes.

1.   Laws of Conservation

Energy. The law of conservation of energy states that energy can be neither created nor destroyed, However, it can be changed from one form to another.

Matter. Matter is the material of which substances are made and on which energy acts. For purposes of the discussion of animal physiology, matter is not destroyed but may be converted into different forms.

2.   Dry Matter

Dry matter refers to material that does not contain any water. In terms of the animal, water may be viewed as an essential component of the body. However, water contains no energy that can be used to convert animal feed dry matter into another form nor is any energy derived directly from water when it is lost from the tissue of the animal body. In GRAZE, it is assumed that water is always available to the animal.

3.   Nutrients

Nutrients are categories of feed dry matter that are consumed by the animal and then converted into another form as part of the animal body tissue.

4.   Energy

Energy is used to convert the feed dry matter that is consumed into matter that makes up the body of the animal. In terms of the animal, energy is generally viewed as being either chemical or thermal.

Chemical. Chemical energy is contained within the material consumed or in the body of the animal.

Thermal. When energy is converted from one form to another, there is some loss of efficiency, which means that part of the chemical energy is converted to thermal energy. In converting matter from one form to another, biological processes use chemical energy and, as a consequence, result in some thermal energy being generated.

5.   Feed Dry Matter Composition

The intake of "feed" dry matter provides the material and energy required to "build" (grow), maintain, and operate the animal body. An animal is literally what it eats. While the GRAZE plant submodel simulates the presence of nitrogen, phosphorus, potassium, and stored sugars (carbohydrates), the animal submodel also considers several other categories. In GRAZE Version 2.3, some of the values were assumed to be constant percentages of grazed (as opposed to fed) forages as noted below. GRAZE Version 3.02 (Windows version) considers these components to be part of the plant parameter inputs.

Plant Cells.   In GRAZE, plant material is divided into cell content material and cell wall material (Fig. 11).

Cell Content: Cell content is the material contained within a cell. It is generally considered totally digestible in a relatively short period of time, less than 24 h. The cell content contains the soluble cell minerals (not silica), soluble protein, nonprotein nitrogen, organic acids, sugars, soluble carbohydrates, starch, and pectin.

Cell Wall: Cell wall material may be categorized as

1. Potentially digestible cell wall: material (cellulose, hemicellulose) that can be digested in the animal body given sufficient time. However, some of this material would be expected to exit the body before being totally digested.
2. Indigestible cell wall: material that will not be digested by the animal regardless of the time that the material spends in the digestive tract. Indigestible material is composed of lignin, cutin, silica, tannins, essential oils, and polyphenols.

Dry Matter Components.   The feed dry matter is composed of minerals, protein and nonprotein nitrogen, fat, and carbohydrates.

Minerals: Although at some level all minerals are important, calcium and phosphorus are the key minerals considered by the GRAZE animal submodel. Minerals do not contain energy for purposes of animal growth or maintenance. GRAZE Version 2.3 (DOS version) assumes that the cell content of grazed material is 1.7% calcium and 27.13% miscellaneous minerals not including calcium and phosphorus. In GRAZE for Windows (Version 3.02), these parameters are assigned as part of the input data file. The potentially digestible cell wall material is assumed to contain no calcium or miscellaneous minerals.

Protein Nitrogen: Plant protein compounds contain nitrogen. On average, multiplying the percentage of nitrogen, as provided by the plant submodel, by 6.25 gives the percentage of protein nitrogen in the material. This form of protein also has value as an energy source.

Nonprotein Nitrogen: Generally, nonprotein nitrogen is not a significant factor in plant material. However, ruminants like beef animals may use nonprotein nitrogen such as that contained in urea fertilizer. GRAZE assumes that there is no nonprotein nitrogen in the cell content or potentially digestible cell wall of grazed forage.

Fat: Fat (lipids) represents the largest concentration of energy per unit of weight in plant material. GRAZE Version 2.3 assumes that the fat content percentages of cell content and potentially digestible cell wall material in grazed forage are 7.37% and 0%, respectively. In GRAZE for Windows (Version 3.02), these parameters are assigned as part of the input data file.

Carbohydrates: Carbohydrates are the primary provider of energy in most plant material. In the GRAZE animal submodel, all material in the cell content or potentially digestible cell wall that has not been placed into one of the above categories is assumed to be carbohydrate.

Methods of Estimating Plant Dry Matter Partitions.   The rate at which nutrients and energy are made available for animal growth and maintenance is a function of the rate of digestibility. One way of estimating digestibility is through a series of laboratory procedures, referred to as the detergent system, that result in quantifying the neutral detergent fiber and acid detergent factor (Fig. 20) (Van Soest, 1982).

Fluent detergent fiber (NDF).   In simplest terms, NDF is the dry matter that is left over after the plant material is exposed to a neutral (pH 7.0) compound. The dry matter that is dissolved and hence removed during the process is considered to be the cell content. Mathematically,

$$\text{Cell content, } \% = 100 - \text{NDF, }\% \tag{22}$$

Acid detergent fiber (ADF).   Acid detergent fiber is what remains after the neutral detergent fiber material has been further exposed to an acid solution. ADF material constitutes the indigestible cell wall material. Thus, by process of elimination, the

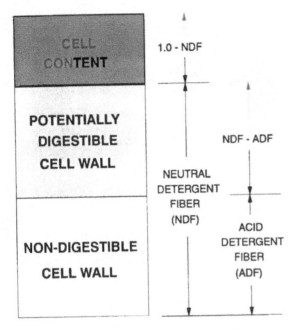

FIGURE 20   Divisions of dry matter intake.

digestible cell wall material is the difference between the NDF and ADF components, Mathematically,

$$\text{Digestible cell wall, } \% = 100 - (100 - \text{NDF, }\%) - \text{ADF, }\%$$

$$= \text{NDF, }\% - \text{ADF, }\% \qquad (23)$$

6. Energy Flow Terminology Concepts

The animal will always maintain an energy balance, even if it kills it to do so! Thus, the flow of energy may be described in GRAZE as a series of processes that each require some energy (Fig. 21).

Gross Energy.   Gross energy is the amount of heat that is released when a substance is completely oxidized in a bomb calorimeter containing 25–30 atm of oxygen. A similar term is "heat of combustion." For example, Baxter and Rook (1953) determined the energy values of fat and protein to be 39.2 and 23.8 MJ/kg, respectively, using bomb calorimeter tests. In GRAZE, the consumed feed represents the gross energy available for the animal's biological processes.

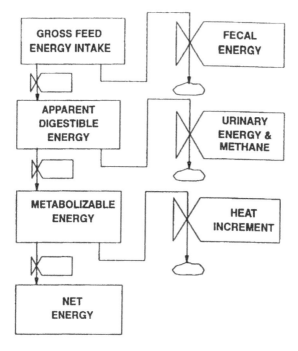

FIGURE 21   A model of animal utilization of energy supplied by feed.

Fecal Energy.   Fecal energy is the gross energy of the feces. In GRAZE, it consists primarily of the energy content of the undigested cell wall material.

Apparent Digestible Energy.   Digestible energy is the difference between gross energy and fecal energy. The modifier "apparent" appears because the term does not consider losses from gas or urine. GRAZE computes the energy generated as a result of digestion and adds this to the body heat pool.

Gaseous Products of Digestion.   Gaseous products of digestion include the combustible gases produced in the digestive tract by the fermentation of the ration. GRAZE computes the production of methane, which is by far the dominant gas produced and represents a significant loss of energy only in ruminants such as beef cattle.

Urinary Energy.   Urinary energy is the gross energy of the urine, it includes the energy content of the nonoxidized portion of the absorbed nutrients and the energy contained in the endogenous (body) fraction of the urine. GRAZE computes an energy loss associated with conversion of the various types of body stores including maintenance.

Metabolizable Energy.   Metabolizable energy is the energy that remains after subtracting from the gross energy intake the fecal energy, energy in the gaseous products of digestion, and urinary energy. Essentially, GRAZE computes metabolizable energy, from which it then proceeds to determine the extent to which other biological processes occur.

Heat Increment.   Heat increment is the increase in heat production following consumption of food when the animal is in a thermally neutral environment. It consists of increased heats of fermentation and of nutrient metabolism. GRAZE considers heat increment in that the inefficiencies of digestion are converted to thermal energy and added to the body heat energy pool.

Net Energy.   In GRAZE, net energy is the energy available above that required for maintenance that can be used to satisfy the needs of other biological processes such as growth, lactation, and gestation. Physical work could also be considered a user of net energy. GRAZE uses the concept of net energy to define the partitioning of functions above maintenance (such as growth, lactation, and pregnancy).

7. Summary

Many of the concepts just discussed are "mental models" that form the basis for mathematical models that are needed to quantify biological processes. The key concept to remember in GRAZE is that there is conservation of energy and mass as the feed supply moves through the animal. At the same time, energy and mass are being converted to different forms by biological and chemical processes. The GRAZE model attempts to relate these processes in a mechanistic manner reflective of the actual biological processes.

C. Body Composition

1. Introduction

The modeling of growth in biological systems relates changes in mass with respect to time as influenced by nutritional and physical environments. In animal systems (as in other biological systems), the nutritional environment (that is, the intake and utilization of feed inputs) is dependent in part on previously accumulated growth. One of the most difficult tasks in modeling animal growth is to mathematically describe the quality and quantity of intake as the animal matures, given an unlimited availability of feed. The concept of "physiological age" may be used effectively in developing more accurate models. A discussion of a particular version of this concept, as applied to beef animals, follows.

The objectives of this section are to

1. Present a conceptual and mathematical framework for (a) defining physiological age, (b) describing the growth process as a function of input energy and nutrients, and (c) examining interactions among body components as related to growth
2. Discuss the laboratory measurements needed to make full use of this approach and present some general observations about its effectiveness in modeling beef animals

2. Physiological Age and Weight

The term "physiological age" is used to denote some measure of maturity (i.e., physiological weight) that is more descriptive than chronological age. In modeling animal growth, a common practice is to develop a "weight–age" relationship that expresses the expected weight of a given genotype at any age (Fig. 22) (Brown et al., 1976; Fitzhugh, 1976; Fitzhugh and Taylor, 1971; Nelsen et al., 1982; Oltjen et al., 1986a, 1986b; Taylor and Fitzhugh, 1971).

Conventional Weight–Age Assumptions. The inherent assumptions of the conventional weight–age relationship are that

1. Weight is a satisfactory indicator of maturity.
2. Chronological age is a satisfactory measure of physiological age.
3. Weight and chronological age are directly related.

Let's examine these assumptions in more detail.

Weight is certainly an indicator of maturity in beef animals in that heavier animals are generally more mature. However, "excessively fat" animals may not reach sexual maturity any sooner than lighter weight animals of the same age and genotype. Likewise, thin animals may not reach sexual maturity as quickly as heavier animals. Thus, the observation that weight and chronological age are directly related may not be sufficient to infer that chronological age is a satisfactory measure of physiological age, especially for animals fed

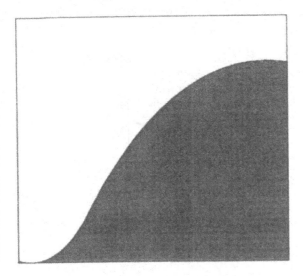

CHRONOLOGICAL AGE

FIGURE 22   Relationship between chronological age and expected animal weight.

widely differing levels of energy and nutrients. Thus, if growth is to be accurately modeled over a broad range of diets, a more satisfactory set of assumptions must be made.

Definitions. The process of more precisely describing the relationship between physiological age and weight begins with definitions, the term "weight" always referring to empty body weight. For purposes of this discussion,

1. Physiological weight is the minimum weight necessary for a given level of physiological development.
2. Physiological age is the minimum chronological age required to achieve a given physiological weight. Thus, chronological age will never be less than physiological age (Fig. 23).
3. "Empty body weight" refers to the cumulative weight of animal tissue and does not include material in various states of digestion, including associated water. Empty body weight is less than live (full) body weight and never less than physiological weight (Fig. 24).

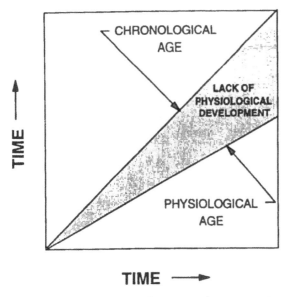

FIGURE 23   Relationship between physiological and chronological age.

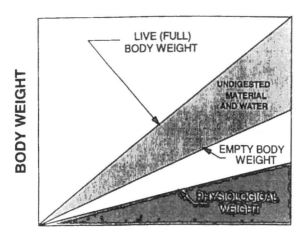

FIGURE 24   Relationship between empty body, live (full) body, and physiological body weight.

PHYSIOLOGICAL COMPONENT

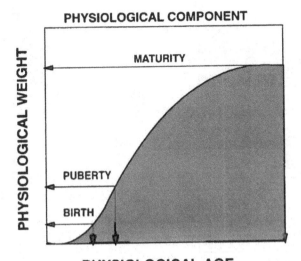

FIGURE 25   Key physiological identifiers.

4. The chronological weight–age relationship shown in Fig. 22 was replaced in GRAZE by the physiological weight–age relationship shown in Fig. 25.

Categories of Composition.   Animal growth may be expressed in terms of body composition. GRAZE assumes that all animal tissue may be divided into four categories: (1) water, (2) nitrogen in the form of protein, (3) fat, and (4) essential minerals.

Essential minerals are grouped together for purposes of illustration. In developing a model, minerals could be subdivided into as many groupings as desired. In GRAZE, calcium and phosphorus are grouped as separate minerals, and a category called "miscellaneous minerals" contains all the rest.

Assumptions about Composition.   Given the above categories, the following assumptions are made:

1. There is an independent physiological weight–age relationship for each tissue component that may be expressed mathematically as a segmented sigmoid curve extending from conception to maturity (Fig. 25).
2. Each unit of physiological growth will contain each of these components in a predictable ratio that changes simultaneously with physiological age (Fig. 26).

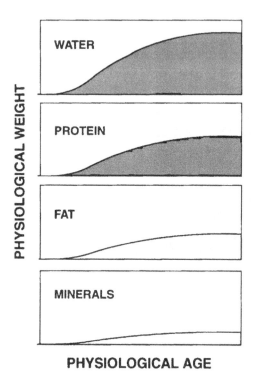

FIGURE 26   Physiological weight–age relationships.

3. Water cannot be stored as part of tissue in excess of its phys-
    iological maximum for a given physiological age, thus mak-
    ing empty body water content the most accurate measure of
    physiological age.
4. Nitrogen and essential minerals may be stored in excess of
    their physiological maxima at a given physiological age but
    not in quantities greater than constant percentages of their
    physiological maxima (Fig. 27).
5. Fat storage is not limited by a physiological maximum.

Each of these assumptions is addressed independently.

3. Mathematical Formulation

Each tissue component (water, nitrogen, fat, essential minerals) may
be described by a segmented sigmoid curve, the Gaussian integral
best fitting the research data (Loewer et al., 1983a, 1983b). Mathe-

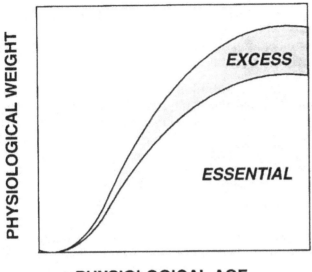

FIGURE 27 Conceptualization of the division of a body component.

matically, the following Gaussian relationship is integrated to obtain
the physiological weight of each tissue component.

$$RGAIN = RGAINM*e^{[ln(RGAINC/RGAINM)*(XNOW\ TCON)**2/TCON**2]}$$

(24)

where

> RGAIN = potential change in weight in a particular compo-
> nent (kg/day)
>
> RGAINM = maximum potential rate of gain of a particular com-
> ponent (kg/day)
>
> XNOW = physiological age (days)
> If the animal has not yet reached the inflection point
> of the sigmoid curve, i.e., where maximum daily
> rate of gain occurs,
>
> RGAINC = the growth rate of that component at conception
> (kg/day)
>
> TCON = number of physiological days from conception to
> the inflection point
> If the animal is past the physiological age where
> maximum daily rate of gain is obtained

RGAINC = growth rate of that component at the defined age of
maturity

TCON = number of physiological days from the inflection
point to maturity

The above mathematical formulation of the sigmoid curve is not
symmetrical around the inflection point (Fig. 28). An example of laboratory data for beef cattle fitting the segmented sigmoid is given by
Moulton (1923). Further discussion of the mathematics may be found
in Bridges et al. (1985), who used swine as the animal genotype.

4. Independence and Dependence in Components

Each component is assumed to follow its own physiological weight–
age relationship independently of the others. However, all components have physiological age as a common measure of comparison
(Fig. 26). Thus, when physiological age is known, the instantaneous
rate of physiological change of any component can be computed. As
physiological age changes, the relative ratios among components
change, but always in a mathematically predictable manner for a

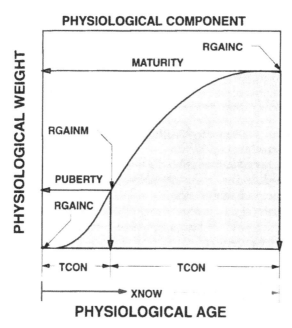

FIGURE 28 Growth curve parameters.

given point in physiological time. A somewhat similar approach is used by Whittemore (1976) and Whittemore and Fawcett (1976).

Water. Every empty body component but water is assumed to have the potential for storage. Therefore, of the component categories, empty body water weight is assumed to be the best measure of physiological age.

Protein and Essential Minerals. In the GRAZE approach to modeling, nitrogen and essential minerals all have two weight components. The first is "physiological" weight, and the second is referred to as "excess" weight (Fig. 27). Both weight classifications represent stores of their respective component, but they differ in priority of utilization. Excess material is always used prior to physiological material. Physiological material cannot exceed the quantities defined by physiological age. The maximum quantity of excess material may not exceed a fixed percentage of physiological material. For example, excess can be no greater than $y\%$ of the physiological component x regardless of physiological age. Thus, maximum excess weight is directly linked to physiological age. The body excretes through urine the nitrogen or minerals in excess of the maximum allowable levels.

Fat. As with nitrogen and essential minerals, fat is further classified as being either "physiological" or "excess." Unlike nitrogen and essential minerals, however, there is no upper limit on the quantity of fat that may be stored. In essence, excess fat is the storage mechanism for all the digested energy in the body that has not been used to synthesize other body components.

5. Physiological Growth and Utilization of Components

Limitations to Physiological Growth. In GRAZE, physiological growth is limited by (1) genetic potential or (2) availability of any single component in the form of either (a) digested energy or nutrients from feed (Figs. 29 and 30) or (b) stored reserves in the form of "excess" material. In effect, the lack of any input for purposes of physiological growth will cause physiological growth to cease in all components. However, excess growth may occur at the same time that a component is experiencing negative or zero physiological growth. In addition, the body has several demands for energy and nutrients that have priority over growth.

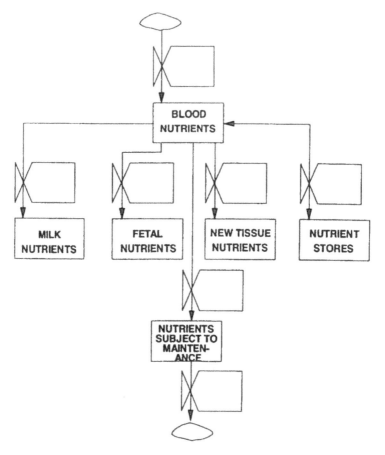

FIGURE 29   Nutrient pathways.

Priorities for Biological Processes.   The following priority of biological processes is assumed in GRAZE (Fig. 31):

1. Maintenance
2. Fetal development
3. Physiological growth
4. Milk production
5. Excess nutrient or fat stores

For purposes of illustration, assume that the animal type is beef steers, thereby eliminating the fetal development and milk production priorities. Each component experiences a certain amount of

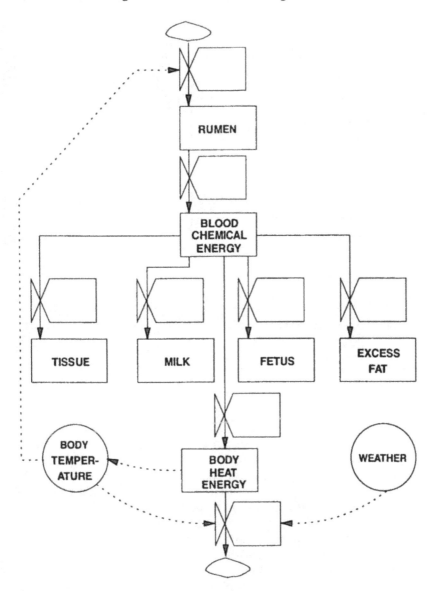

FIGURE 30   Blood chemical energy pathways.

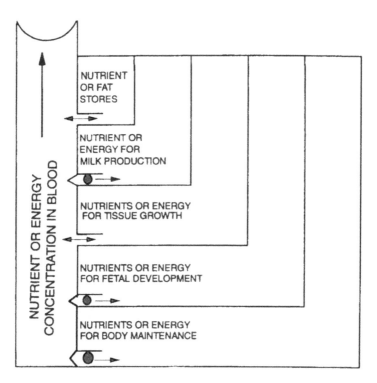

FIGURE 31 Nutrient and energy priority.

"turnover" maintenance, so that input replacement and associated energy are needed before physiological growth can occur. Likewise, "negative" physiological growth can occur if the daily sum of the input component and the percentage of excess stores that can be removed is not sufficient to sustain existing component levels. If negative physiological growth occurs in one component, it occurs in all. A younger physiological age is calculated based on the physiological weight–age relationship, and a reclassification is made in all components concerning physiological and excess material. To illustrate these concepts, a series of examples follow, beginning with a single component.

The weight–age relationship for a single component is shown in Fig. 32. Physiological growth is limited by genetic potential to the maximum gain that can be experienced in one physiological day. The maximum physiological gain is dependent on existing physiological age. The following definitions are used:

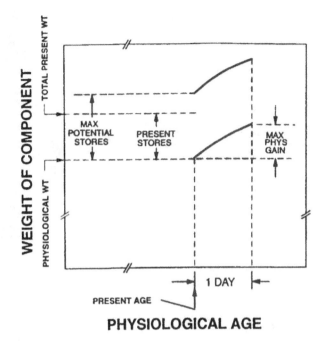

FIGURE 32 Conceptual physiological weight–age relationship for a single component.

Ec	Energy content per unit of weight of the component
Wa	Current physiological age
Wa'	Projected change in physiological age (no more than 1 day)
Wg'	Fraction of physiological growth potential that will occur on a given day
Wp'	Maximum daily physiological weight gain
Wp	Current physiological weight
Ws	Current level of excess stores for a single component
Wsmax	Maximum allowable level of excess storage for the component, equal to a constant percentage of Wp (the percentage may range from zero in the case of water to infinity for fat)
Wtot	Total weight of the animal

Abbreviated descriptions are shown in Figs. 32–37 rather than the above terms. However, these symbols are used extensively in the up-

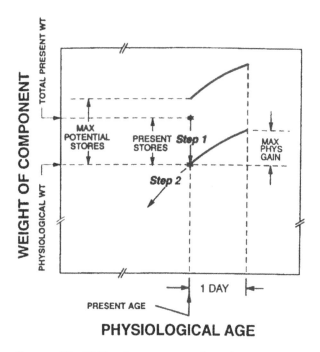

PHYSIOLOGICAL AGE

FIGURE 33   Utilization process when there is insufficient energy to satisfy maintenance needs.

coming discussion. In predicting daily changes in physiological and excess weight, the following situations must be addressed.

Situation 1.   There is sufficient available energy to satisfy maintenance needs regardless of the quantity of component available for increases in physiological and excess weight.

In Situation 1 (Fig. 33), no physiological growth will occur because all energy from digestion or conversion of tissue is used for maintenance. The body will always maintain an energy balance. If the conversion of the physiological component to energy is required to meet energy maintenance needs, the animal will first use the excess material (Ws) as is shown in step 1 and then, if necessary, become physiologically "younger" in the model as depicted by step 2. All other components must adjust accordingly so as to maintain the same physiological age. Should step 2 occur, the distribution of physiological and excess weight for all components will have to be recalculated based on the younger age. Potentially, this could result in some of a component being excreted through the urine because the maximum

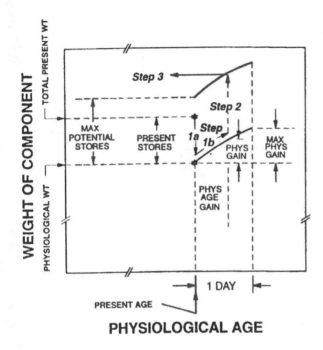

FIGURE 34 Utilization process when there is insufficient energy for maximum gain. Step 1a for fat or protein only. Steps 2 and 3 for excess nonenergy component.

excess storage level Wsmax could be exceeded. If material cannot be converted fast enough to meet maintenance needs, the animal will die. For example, if animals are placed in a sufficiently cold environment, they will die regardless of how much "excess" fat is present in their bodies.

Situation 2. There is sufficient energy to satisfy maintenance but not enough to satisfy all of the physiological growth potential regardless of the quantity of the component available for increases in physiological and excess weight.

In Situation 2 (Fig. 34), the total energy above maintenance needed to satisfy the physiological growth potential is the sum of the products of Wp' and Ec (energy component per unit of weight) for all components. The ratio of available energy above maintenance to the total energy above maintenance needed for maximum physiological growth is the fraction of physiological growth potential that will occur on a given day (Wg', step 1b). Again, all components will have

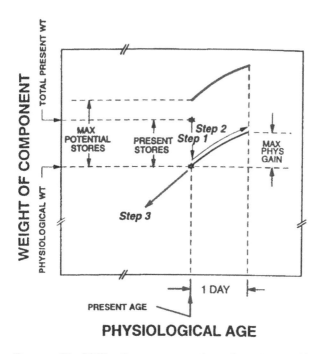

FIGURE 35 Utilization process when there is insufficient availability of a nonenergy component to meet maintenance needs. (Step 2, digestion only; step 3 total.)

the same change in physiological age (Wa'). Excess weight will be adjusted accordingly, with extra material going into excess weight within the storage limits of the particular component (step 2). If the storage limits are reached, material above the limit will be excreted (step 3). If the component is being used to provide part of the energy for physiological growth, excess material may be reclassified as physiological material in some situations (step 1a).

Situation 3. There is an insufficient quantity of the component available to satisfy maintenance needs regardless of the quantity of energy available for maintenance and tissue synthesis.

In Situation 3 (Fig. 35), energy is not a limiting factor. Limitations in daily component "availability" may be confined to only digestion or could include both digestion and the percentage of excess stores that may be removed each day (this is described mathematically later). If component availability includes only digestion, excess material may be in sufficient supply that maintenance is satisfied and

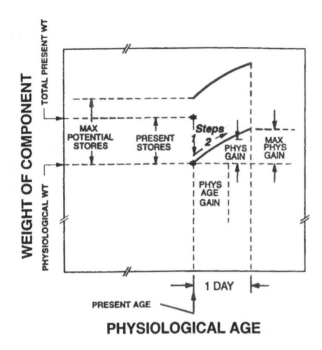

PHYSIOLOGICAL AGE

FIGURE 36   Utilization process when there is sufficient energy for maximum gain with another component greater than maintenance but less than maximum.

part of the excess is converted to physiological material up to the limits of either the quantity of excess material or the physiological growth potential (steps 1 and 2). In this situation, it is possible that maximum physiological growth will be achieved through the conversion of excess material. If, however, the lack of component "availability" includes both digestion and excess material, physiological age will be reduced to the degree necessary to achieve a maintenance balance (steps 1 and 3). It is possible that physiological age will be reduced even though there are excess stores in that only a fraction of the excess may be converted each day. The distribution between excess and physiological material will be adjusted accordingly for all other components, given the reduced physiological age level.

Situation 4.   There is a sufficient quantity of component available to satisfy maintenance needs but not enough to satisfy all of the physiological growth potential regardless of the quantity of energy available for maintenance and tissue synthesis.

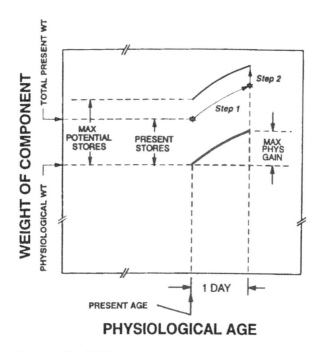

FIGURE 37 Utilization process when there is sufficient energy and other components to obtain maximum physiological gain with the possibility for storage.

In Situation 4 (Fig. 36), the ratio of amount of component available above maintenance (from either digestion or "excess") to the maximum daily physiological growth potential Wp' is the fraction of physiological growth that will be obtained. The situation in which excess material is used is shown in step 1. Otherwise, only step 2 occurs. In either case, the changes in physiological weight gain and age are depicted by Wg' and Wa', respectively, and physiological age will advance no further than that of the limiting component.

Situation 5. There are sufficient quantities of both energy and component to exceed both maintenance and physiological growth potential.

In Situation 5 (Fig. 37), in which neither energy nor components are limiting, maximum physiological growth potential is obtained (step 1). Material that remains after physiological growth is satisfied is converted to excess up to the maximum limit allowed, Wsmax (step 2).

6. Energy and Nutrient Utilization

The situations just described apply to all components. However, components differ in their levels of excess storage and in the energy required for maintaining and synthesizing tissue. Given the component categories of water, fat, nitrogen (in the form of protein), and essential minerals (grouped into a single category), consider how these factors interact.

Let Em be the body's daily demand for maintenance energy for all components, and Ed the energy available each day from digestion. The only sources of body energy are fat and nitrogen (in the form of protein). The energy above maintenance required for maximum physiological growth (Emaxpg) is

$$\text{Emaxpg} = \text{Wpn'*Epn} + \text{Wf'*Ef} \tag{25}$$

where

> Wpn', Wf' = maximum physiological weight change potential for nitrogen (as protein) and fat, respectively
>
> Epn, Ef = energy content per unit of weight for nitrogen (as protein) and fat, respectively

The total energy available to meet maintenance and growth needs in the living animal (Eta) is

$$\text{Eta} = \text{Ed} + \text{Rens*Wen*Epn} + \text{Wpn*Epn}$$
$$+ \text{Refs*Wef*Ef} + \text{Wpf*Ef} \tag{26}$$

where

> Ed = energy available from digestion
>
> Rens, Refs = maximum daily decimal percentages of excess nitrogen (as protein) and fat, respectively, that may be removed form excess material
>
> Wen, Wef = excess weights of nitrogen (as protein) and fat, respectively
>
> Wpn, Wpf = weights of physiological nitrogen (as protein) and fat, respectively

Let Eag be the total energy available for physiological growth and additions of excess material. Then

$$\text{Eag} = \text{Ed} + \text{Rens*Wen*Epn} + \text{Refs*Wef*Ef} \tag{27}$$

Let Eex be the total energy available from excess protein nitrogen and fat for physiological growth. Then

$$\text{Eex} = \text{Rens*Wen*Epn} + \text{Refs*Wef*Ef} \tag{28}$$

If Eag is less than Em, physiological age will decrease with accompanying losses in physiological weight in all components to the extent necessary to maintain an energy balance (Em − Eag) (see Situation 1 and Fig. 33). If Eag is greater than Em but less than Emaxpg, then

$$PA' = (Eag - Em)/Emaxpg \qquad (29)$$

where PA' is the change in physiological age per chronological day (see Situation 2 and Fig. 34). If Eag − Em is greater than Emaxpg and other components are not limiting, then

$$PA' = 1 \qquad (30)$$

(see Situation 5 and Fig. 37). By definition, PA' cannot exceed 1; that is, changes in physiological age cannot be greater than changes in chronological age. Energy supply in excess of Emaxpg is converted to excess nitrogen protein until (1) nitrogen becomes limiting, (2) maximum nitrogen storage is obtained, or (3) energy becomes limiting. In the first two cases, the remaining energy is converted to excess fat, there being no limit as to excess fat storage.

In the case of nitrogen or essential minerals, the model assumes that a constant percentage of the physiological weight is excreted, this quantity being referred to as the maintenance level. The energy required for maintenance may be specified as a function of body composition or total body metabolic weight, the latter being used in our current modeling efforts. Situations 3, 4, and 5 apply as shown in Figs. 35–37.

In summary, this conceptual approach to modeling limits physiological age and physiological growth either to the most limiting body component or to maximum genetic potential. "Excess" growth may occur in some components even when there is negative physiological growth.

7. Utilization of Laboratory Data

Laboratory data must be used to mathematically describe the concepts discussed above. We have drawn heavily on body composition studies conducted at the University of Missouri as reported by Trowbridge et al. (1918), Moulton et al. (1921, 1922a, 1922b, 1923), Moulton (1923), and Ellenberger et al. (1950). Although the studies are old, they are complete with regard to data and represent major efforts in determining body composition. The overall goal of these researchers was to determine changes in the body composition of beef steers (Hereford-Shorthorn) from birth to maturity under differing levels of nutrition. Sequential techniques were used to determine the composition of gain over time. Chemical analyses were conducted on both the total body and selected cuts. Chemical constituents were categorized as being water, protein, fat, ash, and phosphorus. The animals were divided into three groups according to nutrition. Group I was fed all it could eat of a ration composed of corn, oats, linseed, and alfalfa. Group II "was fed to secure maximum growth without the storage of excess fat." Group III was fed to levels so as "to have the growth distinctly retarded."

Given these feeding levels, the modeling group hypothesized that the Group I animals advanced at their physiological maximum while the Group III animals were sufficiently limited by nutrient and energy intake as not to have any excess fat stores. The Group I data were used to determine the physiological weight–age relationship for water, protein, and essential minerals. It was assumed that there was no storage at any of these components. This assumption is in conflict with our earlier statements concerning protein and essential minerals. However, the amount of error is believed to be relatively small because the body does not store large quantities of excess protein and the total amount of minerals stored in the body is relatively small compared to other components. By plotting water, protein, and essential mineral accumulations over time, it is possible to estimate the factors needed in Eq. (24) for each of these categories. However, fat distributions between physiological and excess amounts cannot be determined directly using Group I animals. Rather, an additional set of assumptions must be made using Group III animals.

In Group III, the composition of gain included fat even though animal intake was severely restricted. This occurrence confirmed our earlier assumptions concerning composition of gain as a function of physiological age. If there is no excess fat on these animals, all fat deposits are physiological. The physiological age of the Group III animals may be determined by comparing their water content to that of the Group I animals using Group I as the standard, thereby giving the physiological weight–age relationship for fat. The terms computed for entry into Eq. (24) as a result of this analysis of laboratory data are given by Parsch and Loewer (1995) including extrapolated values for several other beef genotypes. Equation (24) was integrated for each component, and, when compared to observed values, the mathematical calculations and the laboratory data were in close agreement.

8. Observations

The modeling approach discussed above incorporates physical as well as nutritional environments. In using this model over a range of conditions, we have made several observations concerning the performance and limitations of our assumptions concerning physiological development (Watson and Wells, 1985).

The mathematics of the physiological weight–age relationship provide a sigmoid curve extending from conception to maturity. This relationship holds quite well for the fetus during its gestation period if the weight of the placenta is not included. By using this approach to modeling and assuming no component stores, it is possible to determine the added nutrient and energy demand placed on the pregnant cow if the composition of the placenta is known.

It is possible in the model for an animal to be gaining weight for several days while in a negative energy balance. Trowbridge et al. (1918) observed a similar occurrence in their study of animals fed maintenance and submaintenance diets. The situation for which this occurs in the model is when the animal has excess fat stores and is fed a submaintenance diet with regard to energy but with sufficient nitrogen and essential minerals for growth (see Situation 1). The excess fat stores are converted to energy to synthesize protein resulting in physiological growth at the expense of excess fat. The increase in physiological age is accompanied by increases in physiological levels of all components including water (we assume water to be nonlimiting). Although the weight of excess fat decreases, the weight of the physiological gains in protein, water, and essential minerals more than offset the loss of fat. The same type of occurrence results in the model when animals are deprived of essential minerals while having sufficient stores with physiological growth continuing until stores are depleted

Validation and testing of this type of model using field experiments require many assumptions, especially in terms of body composition, because these measurements are rarely (if ever) made. In making its initial estimate of body composition, the model requires animal age and empty body weight. Empty body weight is estimated from live body weight values. Given these inputs, the model integrates the physiological weight gains for each body component, comparing the summed weight to the initial estimate of empty body weight and the physiological age to the initial chronological age. One of two possibilities will first occur. Either the physiological age will equal the state chronological age or the physiological weight will reach or exceed the stated empty body weight. Should physiological age first equal the stated chronological age, the stated weight in excess of the computed weight will be classified as excess fat. If the physiological weight first meets or exceeds the stated weight, all the weight is classified as physiological. In either case the physiological age cannot exceed the stated chronological age.

The model computes empty body capacity for feed in terms of weight rather than volume, establishing capacity as a function of physiological weight rather than of total weight. Conceptually, two genetically identical animals of the same physiological age but differing weights would consume the same quantity of dry matter, other factors being the same. The model demonstrates compensatory gain by recognizing the differences in feed quality, feed quantity, body composition, and physiological age among genetically similar animals. In most instances in which similar animals are of the same physiological age but of different weights, the model will predict that the lighter animal will have the higher daily weight gain because if its energy allocation for maintenance will be less. However, if there is not sufficient energy for the animals to obtain maximum physiological growth, the heavier animal may be able to offset the gain differences associated with energy for maintenance by mobilizing excess energy stores (mainly fat) for physiological growth. For similar animals of the same weight but different physiological ages, the general case is that the physiologically younger animal will gain at the faster rate because a greater percentage of its gain is in the form of protein and water rather than in fat as long as both animals are past the inflection point of the physiological weight–age curve. This animal also has greater excess energy stores to convert to physiological growth if energy intake is limiting. From this discussion, it is evident that there are many subtleties associated with GRAZE in regard to compensatory gain. GRAZE indicates that the highest levels of compensatory gain are in animals that are just past the inflection point of the physiological weight–age curve, have no excess stores, and are fed a diet sufficiently high in energy and nutrients as to gain at the maximum physiological rate while adding excess stores of all components.

When there is a severe shortage of energy or nutrient(s) (Situations 1 and 3), the animal will become physiologically "younger," always maintaining an energy and nutrient balance. "Backward" integration is used when the animal becomes physiologically younger. Currently, the model does not limit the animal's ability to regress along the physiological weight–age curve. Realistically, at some point the process becomes irreversible and death occurs.

9. Concluding Comments

A conceptual model for relating physiological weight and age has been presented. The approach is based on dividing the components of the body into categories in order to determine growth composition and efficiency. Each category is further divided into physiological and excess material. The maximum levels of excess material vary from zero to an unlimited percentage of physiological weight depending on the particular category. This method has been demonstrated in other studies to work well in simulating beef cattle growth but should perform equally well with other types of animals. In fact, it has been modified for use in a swine model (NCPIG) as part of the U.S. regional research projects NC:179 (Modeling Responses of Growing Swine to Environmental and Nutritional Conditions) and its the companion project that followed, NC-204 (Bridges et al., 1992a, 1992b; Usry et al., 1992; Usry, 1989).

D. Heat Transfer and Thermodynamics

1. Overview of Thermal Energy Gains and Losses

If the quantity of heat contained within the animal's body is known, and if a representative specific heat value can be computed, body temperature can be determined. The quantity of heat stored within the body can be expressed as

BODY HEAT present = BODY HEAT past

+ DT*(HEAT FLOW in − HEAT FLOW out)   (31)

where DT = time increment and HEAT FLOW is measured in heat units per unit of time.

The flow of heat in GRAZE is shown in a diagram and simplified flowchart, Figs. 38 and 39. All of the heat loss functions may also result in heat gain under some environmental conditions, although

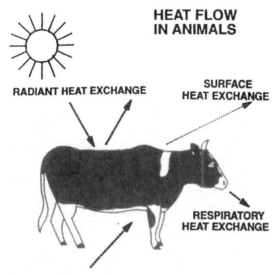

HEAT FLOW IN ANIMALS

RADIANT HEAT EXCHANGE

SURFACE HEAT EXCHANGE

RESPIRATORY HEAT EXCHANGE

INTERNAL HEAT GAIN
a. Maintenance
b. Fermentation
c. Pregnancy
d. Milk

FIGURE 38   Heat flow in an animal.

respiratory latent heat exchanges almost always result in heat losses. The remaining rates in Fig. 39 cause heat to be added to the body unless they are switched off completely. The direction, magnitude, and status (on or off) depends in part on the external environment defined by ambient weather conditions.

2.   The "External Drivers"—Weather Input

Temperature, relative humidity, and solar radiation are all referenced on a continuous basis when computing heat transfer and thermodynamic factors. All factors are used in each day of the GRAZE simulation model. Given that daily maximum and minimum temperature values are used in the GRAZE model, the following logic is employed.

Temperature.   Temperature is determined in the same way in both the animal and plant portions of GRAZE. The inputs required to determine temperature at any time during the day are (1) Julian

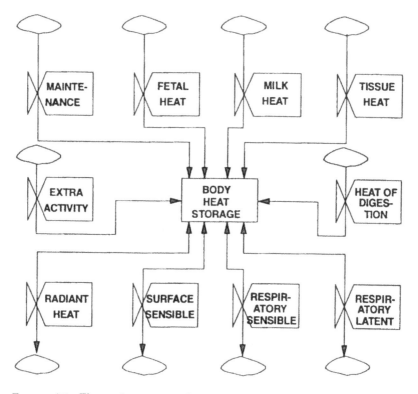

FIGURE 39   Thermal energy pathways.

date (day of the year), (2) latitude, (3) minimum daily temperature, (4) maximum daily temperature, (5) time lag for minimum temperature after sunrise, (6) time lag for maximum temperature after solar noon, and (7) coefficient that controls temperature decrease at night (Fig. 6). Note that factors 5 and 6 were not mentioned in the discussion on plant growth but may be used to "shift" the temperature curve over the day if desired.

The model computes sunrise, sunset, and day length, using latitude and the Julian data. Temperatures over the day are computed, using relationships developed by Parton and Logan (1981).

Relative Humidity.   Relative humidity is needed only in the animal portion of GRAZE. Relative humidity is usually near 100% when temperature is at a minimum. This assumption removes the need to enter relative humidity values from weather records. This can be especially important when these values are not provided, as is

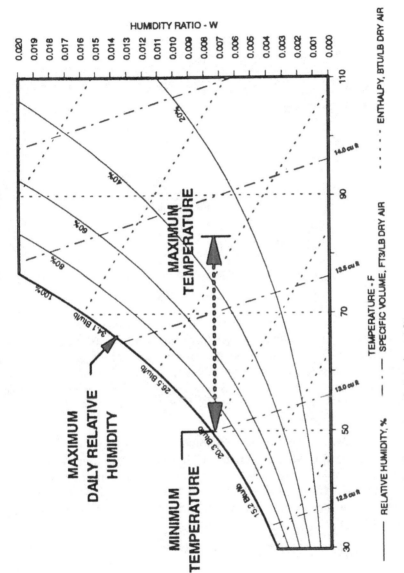

FIGURE 40 Daily change in relative humidity.

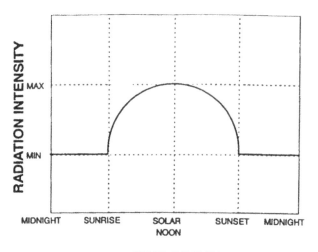

TIME OF DAY

FIGURE 41   Radiation distribution based on projections of the daily maximum.

often the case. Using the above assumption, GRAZE assumes that the absolute humidity can then be determined by entering its psychrometric chart subroutine with the existing temperature and the absolute humidity value (Fig. 40). A maximum daily relative humidity other than 100% also may be specified by the model user; the default value currently used is 90%.

Solar Radiation.   Although solar radiation is not used directly in the plant submodel, it can be a significant consideration in the animal submodel. Solar radiation is entered as the maximum received per hour, which is assumed to occur at solar noon (Fig. 41). The amount of radiation received from sunrise to sunset is assumed to follow a parabolic function of the form

$$R = Rmax - (4.0*Rmax/DAYL**2)*(144 - TIME + TIME**2)$$

$$(32)$$

where

R = radiation received, kcal/h
$Rmax$ = maximum radiation received, kcal/h
$DAYL$ = day length, h
$TIME$ = time of day, h

The following sections examine how the model uses these environmental factors.

3. Resistance to Heat Flow

Internal. The internal resistance to heat flow from the body, RI, was cited by Webster (1974) as

$$RI = (115.0 + 0.387*WT + 2.07*T)/1000 \qquad (33)$$

where

RI = internal resistance, C m^2/W
WT = live animal weight, kg
T = ambient temperature, °C

Webster states that Eq. (33) is valid for temperatures below 0°C. Note that the resistance to heat flow decreases with a decrease in temperature, reflecting an attempt by the animal to maintain the temperature of its outer extremities. However, one also would expect the resistance to heat flow to decrease at higher temperatures when the animal is experiencing heat stress and is trying to void itself of as much heat as possible. The model reflects this situation by assuming a linear decrease in internal resistance to heat flow from the value computed at 0°C to a zero value at body temperature, but never to be less than a stated minimum value. In effect, the assumed relationship between internal resistance to heat flow and ambient temperature is as shown in Fig. 42.

External. The external resistance to heat flow, i.e., heat flow through the air coat of the animal, is given in Eq. (34) (Webster, 1974) (Fig. 43):

$$RE = (118.0 + 132.0*HL - 16.4*HL**2)/1000 \qquad (34)$$

where RE = external resistance, C m^2/W, and HL = hair coat length, cm.

The model contains two input parameters that describe the hair coat of the animal. They are (1) minimum effective hair length, cm, and (2) maximum effective hair length, cm.

As assumption in the model is that hair length can be correlated with photoperiod alone. Presently, in the model, temperature and nutrition do not affect hair length. For beef cattle, the winter coat begins shedding 12 weeks after the shortest day of the year (Yeates, 1995), or Julian day 75. About 3 months (90 days) is required to fully shed or fully grow a winter coat (Miller and Berry, 1970). Growth of the

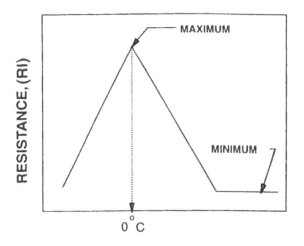

AMBIENT TEMPERATURE

FIGURE 42   Internal resistance to heat flow.

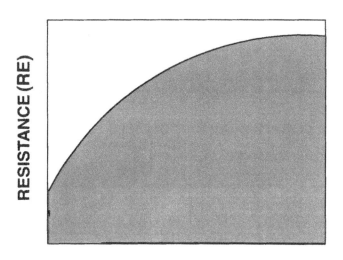

HAIR LENGTH (CM)

FIGURE 43   Relationship between hair coat length and resistance to heat flow.

winter coat begins 12 weeks after the longest day of the year (Yeates, 1955), or Julian day 256. Linear growth and shedding rates are assumed (Fig. 44). Hair lengths were 1.12–2.02 cm in one group of exposed cattle (Webster et al., 1970), with another report implying a 1.0–2.0 cm range (Webster, 1974).

Combined. The combination of Eqs. (33) and (34) is used to compute heat loss potential per hour per square meter of surface area (U),

$$U = (1/RE + 1/RI)*0.86 \text{ kcal/W} \cdot h \tag{35}$$

where U = heat loss potential, $\text{kcal}/(\text{C} \cdot \text{m}^2 \cdot \text{h})$.

4. Body Surface Area

The surface of the beef animal, SA, is given by Webster (1974) as

$$SA = 0.09*WT**0.67 \tag{36}$$

where SA = surface area, m^2, and WT = live weight of the animal, kg.

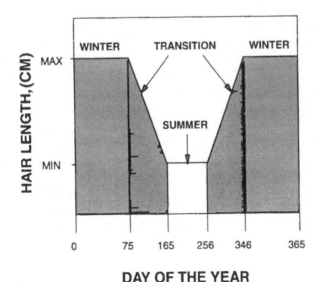

FIGURE 44   Hair coat growth as related to day of the year.

5. Sensible Losses from Body Surface

Sensible losses refer to heat exchange associated with differences in temperature. The sensible losses per day through conduction of heat from the surface of the animal (Fig. 45) becomes

$$\text{HTLSUR} = \text{U} * \text{SA} * (\text{RECTLT} - \text{AMBT}) \tag{37}$$

where

HTLSUR = heat losses from the surface, kcal/h
RECTLT = core temperature of the animal, °C
AMBT = ambient temperature, °C

6. Latent Losses from Body Surface

Latent heat exchange is associated with a change of state; that is, for animals, it is the heat exchange associated with evaporation of water.

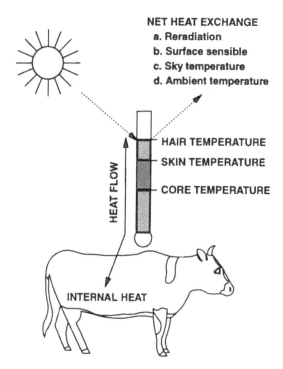

NET HEAT EXCHANGE
a. Reradiation
b. Surface sensible
c. Sky temperature
d. Ambient temperature

HAIR TEMPERATURE
SKIN TEMPERATURE
CORE TEMPERATURE

HEAT FLOW

INTERNAL HEAT

FIGURE 45   Surface heat flow in an animal.

Animals may be classified as "sweaters" or "nonsweaters." Beef animals are classified as "nonsweaters" in that they do not lose significant amounts of moisture through their skin surface. Thus, GRAZE considers the latent heat loss from the surface to be zero, whereas the heat loss would need to be considered for the "sweater" category.

7. Respiratory Heat Losses (Sensible and Latent)

Heat loss through respiration includes both sensible and latent losses. When the animal takes in air at a given temperature and relative humidity, heat from the body warms the air in the lungs to near body temperature, resulting in sensible heat losses for the animal. At the same time, the warmed air in the lungs approaches saturation following an adiabatic process. This removal of heat from the body through the removal of moisture is referred to as "latent" heat loss. (By definition, latent heat loss is the heat transfer associated with the removal of moisture.)

The animal can increase its respiration rate above some minimum level up to some maximum value. Respiration is highly correlated with surface area. Minimum latent heat loss is given by Baxter and Wainman (1964) as 17 $W \cdot m^2$, which is about equivalent to the latent losses resulting from a volumetric respiratory airflow rate of 0.084 m^3/min per square meter of surface area. The maximum volumetric respiratory airflow rate was given by Guyton (1947) as 0.114 m^3/min per square meter of surface area.

The GRAZE model allows the model user to specify other respiratory airflow rates if so desired. Likewise, the degree to which ambient air will be warmed and saturated with moisture also may be altered for both the lower and upper respiratory rates. The ambient air conditions are specified in terms of temperature and relative humidity. The sensible and latent heat losses can then be computed using standard psychrometric relationships (Figs. 46 and 47). The computations are carried out in the following order.

Minimum Respiratory Sensible Heat Losses

$$RHTSMN = RESMIN*PSYSMN*SA*C \qquad (38)$$

where

  RHTSMN = respiratory sensible heat losses associated with the
        minimum respiratory rate, kcal/h
  RESMIN = minimum respiratory area, m^3 $(min \cdot m^2)$

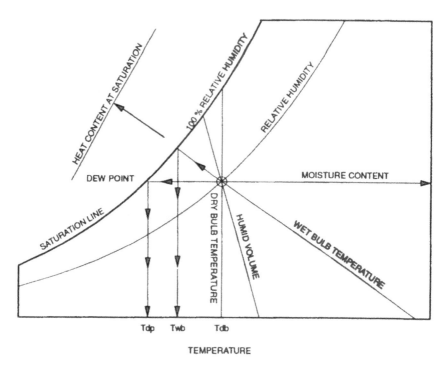

FIGURE 46   Psychrometric chart.

PSYSMN = sensible enthalpy difference between incoming and
exiting air as calculated through psychrometric relationships, kcal/m³ air
SA = surface area, m²
C = conversion constant equal to 60 min/h

Maximum Respiratory Sensible Heat Losses

$$RHTSMX = RESMAX*PSYMX*SA*C \tag{39}$$

where

RHTSMX = respiratory sensible heat losses associated with the
maximum respiratory rate, kcal/h
RESMAX = maximum respiratory rate, m³/(min·m²)
PSYSMX = sensible enthalpy difference between incoming and
existing air as calculated through psychrometric relationships, kcal/m³ air

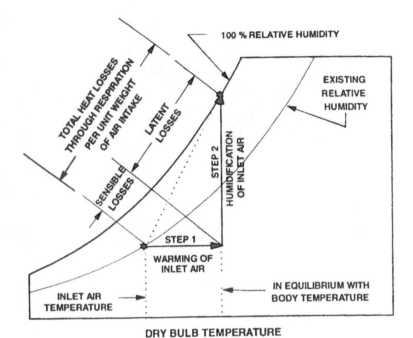

FIGURE 47 Use of psychrometric chart to compute latent and sensible heat loss during respiration.

Minimum Respiratory Latent Heat Losses

$$RHTLMN = RESMIN*PSYLMN*SA*C \qquad (40)$$

where

> RHTLMN = respiratory latent heat losses associated with the minimum respiratory rate, kcal/h
>
> PSYLMN = latent enthalpy difference between incoming and exiting air as calculated through psychrometric relationships, kcal/m³ air

Maximum Respiratory Latent Heat Losses

$$RHTLMX = RESMAX*PSYLMX*SA*C \qquad (41)$$

where

RHTLMN = respiratory latent heat losses associated with the minimum respiratory rate, kcal/h

PSYLMX = latent enthalpy difference between incoming and exiting air as calculated through psychrometric relationships, kcal/m³ air

Minimum Total Heat Loss

$$HTLSMN = HTLSUR + RHTSMN + RHTLMN \tag{42}$$

where HTLSMN = the minimum possible heat losses that the animal can experience, kcal/h.

Maximum Total Heat Loss

$$HTLSMX = HTLSUR + RHTSMX + RHTLMX \tag{43}$$

where HTLSMX = the maximum possible heat losses that the animal can experience, kcal/h.

8.   Zone of Thermal Neutrality

In GRAZE, the difference between HTLSMN and HTLSMX, i.e., the lower and upper bounds of heat losses as controlled by the respiration rate, defines the animal's zone of thermal neutrality (Fig. 48). Outside these lower and upper boundaries, the animal can maintain normal body temperature by regulating activity and production functions to maintain a thermal balance.

9.   Heat Production

Basal Metabolism.   To survive, a warm-blooded animal must generate heat to maintain its body temperature within a relatively narrow range. The functionings of the body, such as circulating blood, breathing, and digestion, require the expenditure of energy, with heat being a by-product. The minimum heat production level associated with maintaining life is referred to as basal metabolism, and the relationship, given by Crampton and Harris (1969), is

$$BASL = 70*ACTWT**0.75 \tag{44}$$

where BASL = heat production associated with basal metabolism, kcal/day, and ACTWT = empty body weight of animal, kg.

Maintenance and Work.   *Metabolic Weight Basis*: If the animal is active, the total heat produced includes some allowance for the

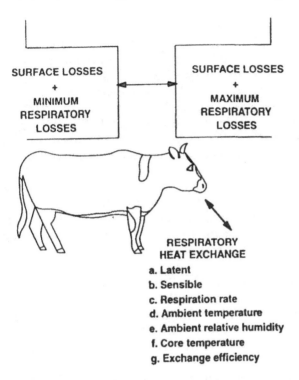

FIGURE 48   Zone of thermal neutrality.

work associated with this activity, and the relationship used (Lof-green and Garrett, 1968) is

$$MAINT = 77*ACTWT**0.75 \qquad (45)$$

where MAINT = heat production associated with basal metabolism and moderate activity, kcal/day. Certainly, this level of energy would increase if the animal were being used as a power source for some type of activity such as plowing.

Body Composition Basis: In GRAZE, some consideration has been given to making maintenance a direct function of body composition. That is, differing energy levels are required to maintain fat and protein. The resulting equation is

$$MAINT = Epm*P + Efm*F + Emm*M + Ewm*W \qquad (46)$$

where P, F, M, and W are quantities of protein, fat, minerals, and water, respectively, contained in the body, and Epm, Efm, Emm, and

Ewm are energy required for maintaining a unit of protein, fat, minerals, and water, respectively. At present, however, Eq. (45) is being used. All maintenance energy is added to the body heat energy pool.

Synthesis of Tissue and Excess Fat. When tissue (tissue is defined as physiologically essential material) or excess fat is formed, there is an inefficiency that results in part of the pool of body chemical energy being converted to heat energy. The model adds this heat energy source to the total body heat energy pool. The process also occurs if tissue or excess fat is converted to chemical energy.

Growth of Fetus. In GRAZE, the growing fetus and its associated body tissue are treated the same as other tissue insofar as efficiency of gain is concerned. The difference is that the body will not pull fetal energy back into the chemical energy pool.

Lactation. As with tissue and fat, thermal energy is produced when chemical energy is converted to milk (Moe and Flatt, 1969). All milk that is produced is assumed to exit the body, and therefore the model does not allow milk to be converted back into chemical energy for reconversion to body tissue or to meet other energy demands.

Digestion. The digestion system involves the fermentation of feed, as discussed in detail later. A by-product of fermentation is heat (thermal) energy, and this is added to the pool of body heat energy. The model assumes fixed heat losses per unit of digested material associated with each basic category of feed ingredient. The total heat generated is the sum of the individual category losses.

Solar Radiation. Heat Gain: In daytime, the animal may receive direct energy from the sun in the form of radiation. However, only a fraction of the animal's surface will be exposed. It would be logical for the animal to increase exposure during colder weather and decrease it in the summer, but no attempt has been made to model this occurrence. Instead, the model user specifies the fraction of the animal's surface area that is exposed to radiation. The quantity of heat received is expressed by

ENRDHR = [RMAX − (4.0*RMAX/DAYL**2)

*(144.0 − 24.0*TIME + TIME**2)]

*0.01433*60.0*SA*PCANEX/100.0   (47)

where

ENRDHR = radiation received, kcal/h
RMAX = radiation at solar noon, kcal/h
DAYL = day length, h
TIME = time of day, h
SA = surface area, m^2
PCANEX = percent of animal exposed to radiation

Heat Loss: The animal may reradiate heat energy to the sky at night. Equations (48) and (49) are used to simulate this situation for the skin and hair coat temperatures.

$$Tskin = (Tair + RE/RI*Tcore)/(1.0 + RE/RI) \qquad (48)$$

$$Thair = (Tskin - Tair)/4.0 + Tair + 273.0 \qquad (49)$$

where

Tair = temperature of the air, °C
Tcore = body core temperature, °C
RE = external resistance to heat flow [see Eq. (34)]
RI = internal resistance to heat flow [see Eq. (33)

The effective sky temperature (K) is then calculated as

$$Tsky = 0.0552*(Tair + 273.0)**1.5 \qquad (50)$$

Emittance of the hair coat is assumed to be 1.0, in which case the reradiation of the animal to the sky becomes

$$ENRDHR = 1.0*SA*SF*SBC*(Tsky**4 - Tair**4)*0.1443*60.0 \qquad (51)$$

where

ENRDHR = radiation losses from the animal, kcal/h
SA = surface area, m^2
SF = fraction of surface area exposed
SBC = Stefan–Boltzmann constant, 5.6697E-8

The major weaknesses in this logic is not the basic heat transfer mechanisms but rather the computation of the fraction of surface area that is exposed. Future logic may consider a maximum and minimum possible exposure and allow the animal to use solar radiation or reradiation to its best advantage insofar as body temperature is concerned. Presently, the model user has the option to bypass the solar radiation logic.

Activity. If the outflow of heat from the body is greater than the animal's ability to minimize heat loss [Eq. (42)], then body temperature will begin to fall. This will trigger activity that will result in the conversion of fat to heat energy, followed by conversion of tissue if necessary. The initiation of eating also will occur, but indirectly. More will be said about this later. If body temperature continues to fall, death will occur.

Summary. Environmental factors (i.e., temperature, relative humidity, and solar radiation) establish the short-term heat losses from the animal. Wind, rain, and snow effects may be added to the model at a later date if desired. In GRAZE, the animal may alter its chemical and physical systems using such options as regulating blood flow, respiration, shivering, and eating to maintain a constant body temperature.

E. Intake and Digestion

1. Introduction

The key to accurately predicting animal performance in terms of weight gain is to be able to accurately predict feed intake. Remember that tissue gain or loss occurs after other demands for energy are met. Thus, small changes in energy input can greatly impact weight gains and losses.

In GRAZE, beef animal intake is regulated by one or more of the following: (1) physical fill, (2) chemical energy levels in the blood, (3) body temperature, and (4) nighttime regulation. The rate of digestion influences each of these factors, and vice versa, although some of the effects are indirect. The quality and content of the feed in the rumen at a point in time establishes the instantaneous rate of digestion. In this section, the logic regarding intake, digestion, and feedback loops associated with environment and feed quality is examined. A simplified flowchart of the intake–digestion system is shown in Fig. 49.

2. Dry Matter Composition

The dry matter consumed by the animal is broadly categorized as being digestible cell content (CC), potential digestible cell wall (DCW), or indigestible cell wall (UCW) material (Fig. 11). The digestible material is further divided into protein, nonprotein nitrogen, minerals, fat, and carbohydrates. The animal is assumed to digest all of the digestible cell content. The amount of potential digestible ma-

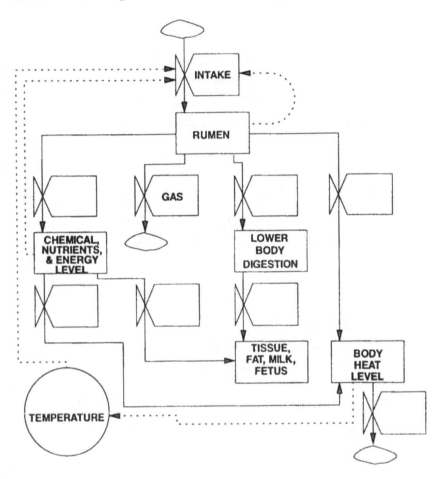

FIGURE 49   Digestive processes.

terial that is actually digested is dependent upon the time that the material resides in the animal's body.

In GRAZE, the user may specify a feed source to simulate a feedlot situation or the animal may obtain its feed supply through grazing. In addition, pastures may be supplemented with external feed sources, the assumption being that the external sources will be consumed first, regardless of quality. If an external feed source is used, the following inputs must be specified. These inputs are computed internally in a grazing situation.

PCCELC = percent of ration that is composed of cell content material

PCNCEC = percent of protein nitrogen in cell contents

PCNPNC = percent of nonprotein nitrogen in cell contents

PCFCEC = percent of fat in cell contents

PCCACC = percent of calcium in cell contents

PCPHCC = percent of phosphorus in cell contents

PCMNCC = percent of other minerals in cell contents

PCDPMW = percent of the cell wall material that is digestible

PCNCEW = percent of protein nitrogen in the cell wall material

PCNPNW = percent of nonprotein nitrogen in the cell wall material

PCFCEW = percent of fat in the cell wall material

PCCCEW = percent of calcium in the cell wall material

PCPHCEW = percent of phosphorus in the cell wall material

PCMNCW = percent of other minerals in the cell wall material

The cell content portions of the dry matter are assumed to be completely digestible. The potentially digestible cell wall material is composed in the remaining dry matter:

$$PCCELW = ((100.00 - PCCELC)*PCDPMW)/100.0 \qquad (52)$$

where PCCELW is the percent potentially digestible cell wall in the dry matter.

The equations used to determine the nutritive composition of the cell are

$$CCPKG = (PCNCEC/100.0)*6.25 \qquad (53a)$$

$$CCNPKG = (PCNPNC/100.0) \qquad (53b)$$

$$CCFTKG = (PCFCEC/100.0) \qquad (53c)$$

$$CCCAKG = (PCCACC/100.0) \qquad (53d)$$

$$CCPHKG = (PCPHCC/100.0) \qquad (53e)$$

$$CCMNKG = (PCMNCC/100.0) \qquad (53f)$$

where CCPKG, CCNPKG, CCFTKG, CCCAKG, CCPHKG, and CCMNKG are the portions of a kilogram of cell content that are protein, nonprotein nitrogen, fat, calcium, phosphorus, and other minerals, respectively.

The remaining portion of the cell content is designated as carbohydrates:

$$CCCRKG = 1.0 - CCPKG - CCNPKG - CCFTKG$$
$$- CCCAKG - CCPHKG - CCCMKG \qquad (54)$$

where CCCRKG is the portion of a kilogram of cell content that is carbohydrates.

The procedure for apportioning the potentially digestible cell wall material is similar to that for cell content:

$$CWPKG = (PCNCEW/100.0){*}6.25 \tag{55a}$$

$$CWNPKG = (PCNPNW/100.0) \tag{55b}$$

$$CWFTKG = (PCFCEW/100.0) \tag{55c}$$

$$CWCAKG = (PCCCEW/100.0) \tag{55d}$$

$$CWPHKG = (PCPHCW/100.0) \tag{55e}$$

$$CWMNKG = (PCMNCW/100.0) \tag{55f}$$

$$CWCRKG = 1.0 - CWPKG - CWNPKG - CWFTKG$$
$$- CWCAKG - CWPHKG - CWMNKG \tag{55g}$$

where RKG, CWPKG, CWNPKG, CWFTKG, CWCAKG, CWPHKG, and CWMNKG are the portions of a kilogram of cell wall material that are carbohydrate, protein, nonprotein, fat, calcium, phosphorus, and other minerals, respectively.

From the above information, the nutrient composition of each kilogram of dry matter entering the animal is known.

Each feed component is evaluated for energy content using proximate analysis where

> PAPRE = energy content of protein, kcal/kg
> PAEEE = energy content of fat (ether extract), kcal/kg
> PANFEE = energy content of carbohydrates (nitrogen-free extract), kcal/kg
> PAFIBE = energy content of fiber (fiber extract), kcal/kg

By knowing the energy composition of the dry matter intake, the energy available to the animal may be determined if the digestive rates are known. The gross energy values are computed as

$$GEKGCC = CCPKG{*}PAPRE + CCFTKG{*}PAEEE$$
$$+ CCCRKG{*}PANFEE \tag{56}$$

where GEFGCC is the gross energy per kilogram of cell contents, kcal,

$$GEKGCW = CWPKG{*}PAPRE + CWFTKG{*}PAEEE$$
$$+ CWCRKG{*}(PANFEE + PAFIBE)/2.0 \tag{57}$$

where GEKGCW is the gross energy per kg of potentially digestible cell wall, kcal, and

$$GEDM = (PCCELC/100)*GEKGCC + (100 - PCCELC)$$
$$*(PCDDMW/100)*GEKGCW \qquad (58)$$

where GEDM is the gross energy of the dry matter that is available to the animal in terms of digestion, kcal. Note that GEDM does not include the energy contained in the indigestible portion of the cell wall.

3. Rate of Dry Matter Intake

A number of factors affect the rate of dry matter intake.

Physiological Weight. Dry matter intake rate increases with increases in physiological weight, a measure of maturity.

Quality of Feed Source. Dry matter intake rate increases with the quality of the feed source. However, defining quality quantitatively is difficult. Generally, quality increases with increases in the proportion of cell content and protein.

Quantity of Feed Source Available per Unit Area. In grazing situations, dry matter intake rate increases to a maximum with increases in the availability of the feed source per unit land area (Fig. 50). That is, the animal can consume forage at a faster rate if it does not have to significantly change locations to obtain its food supply.

Methods Used to Predict Intake Rate. In quantitatively defining intake rate, the following are among the approaches that may be taken.

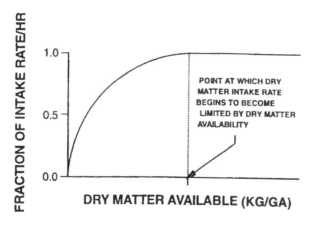

FIGURE 50   Influence of dry matter density on animal intake rate.

Percent of Body Weight.   Oftentimes, intake is expressed as a percentage of the animals body weight. In beef cattle, this percentage generally ranges from 2 to 3% per day. Depending upon the degree of accuracy required, it may be necessary to make some allowances for physiological maturity compared to actual weight if the animal is overweight for its age.

Counting Number and/or Size of Bites.   Another approach is to assume that the animal will take a given number of bites of a certain size over a given time interval.

Estimating Digestibility.   Estimates of digestion can be obtained for different feed sources. If the amount of fecal material deposited per day is known, daily intake can then be estimated.

Mechanistic Approaches.   In GRAZE, the intake feed source is placed into three "quality" categories based on cell content and physiological age of the material. Feeds in each category may be consumed at a rate that depends on the maturity of the animal. The intake rate is determined based on the weighted average of the three categories of material that is consumed after accounting for forage density, total availability, and whether the animal may eat as influenced by physical fill, chemostatic limits, thermostatic limits, or rest requirements.

4. Rate of Digestion

The rate of digestion in the model is dependent upon the relative distribution of intake into cell content and potentially digestible cell wall material. Each material is subject to a third-order exponential delay as it enters the rumen. This delay reflects a lag time in digestion associated with the population increases (or declines) in rumen bacteria, followed by a decay similar to a half-life function. The resulting digestion rate is similar to that given by Van Soest (1982) and includes the effects of rumination. Van Soest states that there is a lag in obtaining the maximum digestion rate partially because new material is being introduced into the rumen and partially because of the natural lag in the microflora population associated with this entry of a new food source. GRAZE reflects both of these situations. Inputs to the model are

HAFLCC and HAFLCW = digestion rate (%/h) of cell content and cell wall material, respectively, at optimum rumen bacterial levels

DELCC and DELCW = time required (h) for the rumen microflora population level to reach its optimum level for newly introduced inputs of cell content and potentially digestible cell wall material, respectively.

In GRAZE, potential digestibility of cell wall material is determined from the plant submodel.

The effective rate of digestion is shown in Fig. 51 for a batch load situation (never experienced by the animal) and a typical eating situation. The rates of disappearance for cell content and cell wall material are defined as characteristics of the feed.

5. Rumen Passage

The inputs to the model for this section are

DMIHR = dry matter intake rate, kg/h
RCFAC = maximum dry matter rumen capacity (a feed characteristic), percent of physiological weight

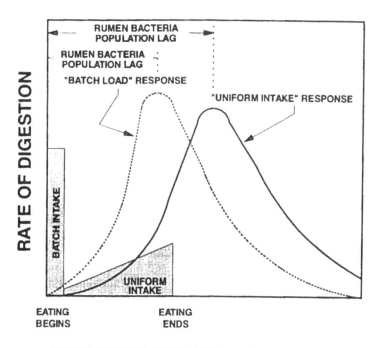

FIGURE 51 Digestion rate response to batch load and uniform dry matter intake.

PTRMAX = maximum rumen pass-through rate, percent of rumen capacity per hour

As dry matter enters, it is placed into (1) cell content (assumed
to be 100% digestible), (2) potentially digestible cell wall, or (3) indigestible cell wall. The quantities of cell content and potentially digestible cell wall material are either digested in the rumen (each following a modified half-life digestive rate) or passed through the
rumen. Cell content that passes through the rumen is digested later
in the body. Potentially digestible cell wall material that passes
through the rumen before being digested exits the body. Indigestible
cell wall can only pass through the rumen and be excreted.

The pass-through rate function is a symmetric sigmoid curve
based on the integral of a parabola (Fig. 52). The flow rate is zero
when the rumen is empty and equals

(PTRMAX/100)*RCFAC*physiological weight

when the rumen is totally full. The pass-through material is in pro-

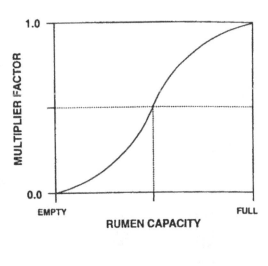

FIGURE 52 Dry matter in rumen; pass-through relationship. Pass-through
(kg/h) = multiplier factor × maximum pass-through rate (kg/h).

portion to the three categories of material that may be present in the rumen (Fig. 53).

Ellis (1978) reported that GI tract fill was found to vary from his value of 2.224 kg DM (dry matter) per 100 kg body weight to a high literature value of 2.998 kg DM per 100 kg. On an empty body weight basis, assuming empty body weight to be 90% of live body weight, these values translate to 2.471 and 3.331 kgDM/kg empty body weight, respectively. The value used in the model is the mean of the two values, 2.9 kg DM/kg empty body weight. However, this value is also defined as a characteristic of a particular supplemental feed-stuff to account for differences in dry matter density.

6. Digestive Energy Losses

Each component of the dry matter is given a heat energy loss per unit digested. The terms are

ELDPCC, ELDFCC, ELDCCC, ELDMCC, ELDCAC, ELDHPC, and ELDNPC = the thermal energy losses associated with the digestion of 1 kg of cell content protein, fat, carbohydrates, other minerals, calcium, phosphorus, and nonprotein nitrogen, respectively.

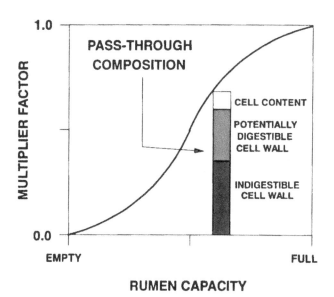

FIGURE 53 Composition of dry matter that passes through the rumen is proportional to its components.

ELDPCW, ELDFCW, ELDCCW, ELDMCW, ELDCAW, ELDPHW, and ELDNPW = the thermal energy losses associated with the digestion of 1 kg of cell wall protein, fat, carbohydrates, other minerals, calcium, phosphorus, and nonprotein nitrogen, respectively.

The total digestive losses per kg of dry matter component become

$$DLCCKG = CCPKG*ELDPCC + CCFTKG*ELDFCC + CCCRKG$$

$$*ELDCCC + CCMNKG*ELDMCC + CCCAKG$$

$$*ELDCAC + CCPHKG*ELDPHC + CCNPKG*ELDNPC \qquad (59)$$

where DLCCKG is the digestive energy losses per kilogram of cell contents digested, and

$$DLCWKG = CWPKG*ELDPCW + CWFTKG*ELDGCW$$

$$+ CWCRKG*ELDCCW + CWMNKG*ELDMCW$$

$$+ CWCAKG*ELDCAW + CWPHKG*ELDPHW$$

$$+ CWNPKG*ELDNPW \qquad (60)$$

where DLCWKG is the digestive energy losses per kilogram of cell wall digested. The digestive losses may be used to calculate the heat increment losses of the animal once the digestive rates are determined.

It is difficult to determine experimentally the portion of digested energy that is lost as fermentation heat. Therefore, theoretical analyses were used to obtain the fermentation-associated heat losses for protein, carbohydrate, and fat using, in part, the theoretical heats of combustion (Lange and Forkew, 1952).

Hershberger and Hartsook (1969) reported experimental values for fermentation heat loss of 5.5 and 5.1% of the heat of combustion of alfalfa hay fermented. Also cited is work by Marston (1948) in which 6% of the combustible energy of the fermentation products was lost as heat where cellulose was the substrate. A theoretical analysis for carbohydrate by Wolin (1960) was used to calculate a heat loss value of 6.4% of substrate heat content.

Reaction equations were formulated, and efficiency calculations were made. The equation used by Wolin (1960) for carbohydrate is

$$57.5(C_6H_{12}O_6) = 65 \text{ acetate} + 20 \text{ propionate} + 15 \text{ butyrate}$$

$$+ 60 \text{ } CO_2 + 35 \text{ } CH_4 + 25 \text{ } H_2O \qquad (61)$$

From this, Hershberger and Hartsook (1969) calculated a heat loss of 6.4% of the heat content of carbohydrate fermented. It was assumed that protein, when broken down to amino acids, takes on 8% water of hydration. Therefore, the energy content of amino acids is

$$5686 \text{ kcal/kg} * 0.92 = 5231 \text{ kcal/kg} \qquad (62)$$

A second assumption is that the ratios of volatile fatty acids (VFAs) produced are 67% acetate, 21% propionate, and 12% butyrate. The reaction equation given by Baldwin et al. (1977) for amino acids is

$$1.0 \text{ mol amino acid} = 0.43 \text{ mol pyruvate} + 0.54 \text{ mol ATP}$$
$$+ 1.07 \text{ mol NH}_3 + 0.03 \text{ mol H}_2\text{S} + 0.30 \text{ mol BCFA}$$
$$+ 0.21 \text{ mol acetate} + 0.89 \text{ mol H}_2 + 0.6 \text{ mol CO}_2 \qquad (63)$$

Using this equation, Webster (1979) calculated a heat loss of 3.4% of protein fermented.

Limited data for lipid fermentation heat losses are available because the total amount of dietary lipid fermented is relatively insignificant. A 1% heat loss value was assumed.

In summary, the inefficiency values associated with fermentation are

```
Carbohydrate   6.4%
     Protein   3.4%
         Fat   1.0%
```

Similarly, the heat loss values are

```
Carbohydrate   6.4% × 4179 kcal/kg = 267 kcal/kg
     Protein   3.4% × 5686 kcal/kg = 193 kcal/kg
         Fat   1.0% × 9367 kcal/kg = 94 kcal/kg
```

An assumption is made that the heat loss values are the same whether the material is from cell content or cell wall.

Energy losses also are associated with the addition of tissue components. Moe and Flatt (1969) reported an efficiency of 75% for the addition of body fat from metabolizable energy. Pullar and Webster (1977) reported an energy requirement of 53 kJ/g (12,661 kcal/kg) of fat or protein tissue added. Therefore, the efficiency should be

Efficiency (%) = (energy content of fat*100%)/

(energy required for fat addition)

= (9367 kcal/kg*100%)/12,661 kcal/kg = 74%

$$(64)$$

This agrees quite well with the Moe and Flatt value. The energy loss for addition of fat is

$(100 - 74)/100*12,661 = 3293 \text{ kcal/kg}$ $\qquad (65)$

Therefore, the efficiency of protein added from metabolizable energy is

Efficiency (%) for Protein Addition

= (5686 kcal/kg DM protein*100%)/12,661 kcal/kg of

DM protein added = 44.9% $\qquad (66)$

The energy loss for addition of protein may be computed as either

$(100 - 44.9)/100*12,661 = 6975 \text{ kcal/kg}$ $\qquad (67a)$

or

$(12{,}661 \text{ kcal/kg} - 5686 \text{ kcal/kg}) = 6975 \text{ kcal/kg}$ $\qquad (67b)$

Energy losses associated with the removal of tissue must also be considered. The energy associated with the removal of nitrogen is 7.9 kcal/gN removed (Crampton and Harris, 1969). This nitrogen compound then contains 1.264 as much energy as is normally associated with protein. The energy loss is estimated to be

$(1.264 - 1.0)*5686 \text{ kcal/kg of protein}$

= 1501.1 kcal/kg of protein removed $\qquad (68)$

The efficiency of fat removal is estimated to be 99% because very little heat loss is associated with fat removal. The energy loss associated with removal of fat is

$0.01*9367 \text{ kcal/kg of fat} = 93.67 \text{ kcal/kg}$ $\qquad (69)$

7. Regulation of Intake

Physical Regulation. GRAZE integrates the sum of all rates into and out of the rumen (Fig. 49). Once the animal initiates eating, the process continues until the animal is filled (Fig. 54), unless other limits (discussed later) are imposed first. All this time, digestion and pass-through empty the rumen. Emptying the rumen will not trigger

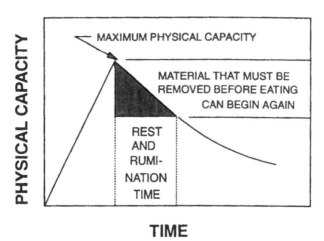

FIGURE 54   Physical fill control mechanism.

eating directly but rather indirectly through reduction of the body (blood) chemical energy level. However, once the rumen has been filled, eating will not be reinitiated until the physical fill level is reduced to a lower level (presently defaulted to 95% of capacity) irrespective of chemical regulation. Effectively, this allows the animal time for rest and rumination after eating.

Chemical Regulation. The blood chemical energy level must be sufficient to satisfy at minimum the broad maintenance needs of the body. As food is digested, the blood chemical energy level rises in direct relation to the digestive rate as was shown by Simkins et al. (1965a, 1965b) and Montgomery and Baumgardt (1965a, 1965b). At some point there is enough blood chemical energy to satisfy all energy requirements and initiate storage of energy in the form of excess fat. The rate at which excess fat is formed is proportional to the blood energy level above that required for other activities.

The model uses a maximum hourly rate of fat addition as a basis for determination of the energy that goes to stored (or "excess") fat each time step if excess chemical energy exists in the pool. The minimum value for this rate was determined from data compiled by Moulton et al. (1922a, 1922b). Two groups of steers that were part of the Moulton tests were classified as Group II and Group I. Group II steers were fed so as to give "maximum growth without permitting the layering on of much fat." Group I steers were fed all they would eat, allowing the maximum fat addition. Gain in body fat over time was compared for both groups with the essential fat curve used in the model that was developed from Moulton data on starvation. The animals in Group II followed the model's essential fat curve fairly well. The animals fed for maximum possible gain showed much more fat addition. It was hypothesized that the maximum rate of fat addition was proportional to the amount of essential fat present in the body. This allows the rate to increase as the animal increases in physiological age until maturity. At maturity, the maximum possible rate should be constant. The actual maximum rate may be much larger because of environment or other factors, such as increased maintenance requirements or restricted feed availability, and because excess fat may be deposited during only part of the day due to fluctuations in the chemical energy level of the blood.

The data show that from approximately 34 to 48 months of age the rate is a constant, which is assumed to be the least maximum daily rate of fat addition. This rate is calculated as

$$((285 \text{ kg} - 169 \text{ kg}) * 12 \text{ mo} * 1 \text{ yr}) / ((48 \text{ mo} - 34 \text{ mo})$$

$$* 1 \text{ yr} * 365 \text{ days}) = 0.272 \text{ kg/day or} = 0.60 \text{ lb/day} \tag{70}$$

In terms of energy use, this becomes

$$0.272 \text{ kg/day} * 9367 \text{ kcal/kg} = 2548 \text{ kcal/day or } 106 \text{ kcal/hr}$$

$$\tag{71}$$

This value is for an essential fat level of 68.8 kg, which transforms the least maximum fat energy addition rate on a per-unit-of-essential-fat basis to

$$(106 \text{ kcal/hr})/68.8 \text{ kg} = 1.54 \text{ kcal/hr per kg of essential fat}$$

$$\tag{72}$$

Eating is terminated in the model if the blood energy level is sufficiently above that required to synthesize maximum excess fat (Fig. 55). Eating will not be initiated again until the sum of the chemical energy removal rates, acting over time, has depleted the blood chemical energy level to the point where excess fat must be reconverted to chemical energy to meet demand. The process then continues in a cyclic manner.

If the maintenance requirements are sufficiently increased because of adverse weather, the blood energy pool will deplete more rapidly, leading to more frequent eating and subsequently to an increase in daily dry matter intake. If the animal eats slowly, the upper chemical level will not be reached as quickly. Of course, if the animal

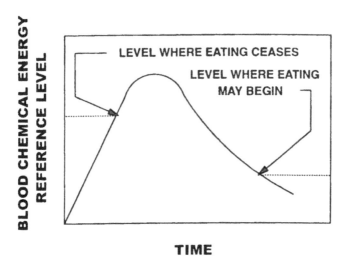

FIGURE 55   Chemostatic control mechanism.

is consuming low energy feed, physical fill rather than chemical limits will regulate intake.

The chemostatic theory of intake control relies upon some measure of energy level or concentration in the blood to serve as a control signal. The present model uses blood energy concentration above a base level as a control signal. The base level used is the energy content of the blood at which desired growth and production functions are completely satisfied and fat is being added at the maximum possible rate.

The weight of the blood must be known in order to compute the energy concentration in the blood. It is hypothesized that blood weight is related to the amount of fat-free empty body mass and the amount of fat in the body. Using data from Moulton et al. (1922a, 1922b), a regression was performed of blood weight on fat-free empty body weight and fat weight. Ten observations were from animals that were fed for maximum growth with little excess fat, and 10 were from animals that were fed all they would eat. The other observations were from animals that were fed at or below maintenance. The age of the animals varied from 3 months to 4 years. The resulting regression equation for blood weight is

$$BW = 0.05707127{*}WTFTFR + 0.00635873{*}FTWT \qquad (73)$$

where

> BW = blood weight, kg
> WTFTFR = fat-free empty body weight, kg
> FTWT = weight of total fat, kg

The R^2 value for the equation based on these data is 0.996. The values predicted agree fairly well with data from Marcilese et al. (1966) for Herefords of 5.12 L/100 kg of live body weight assuming a specific gravity of blood of 1.05. Using Eq. (73), an animal with 20.10 kg of fat would have 4.69 kg blood/kg empty body weight. Converting to live weight using a factor of 1.14 gives 5.35 kg blood/100 kg live weight, which is near the previously mentioned value. Paine (1971) used a value of 7.7 kg/100 kg live weight, with provision for some adjustments. Haxton et al. (1974) showed that calves with a mean age of 9 weeks had blood/weight ratios of 11.6 kg/100 kg live weight. The regression equation, therefore, provides a reliable estimate of blood weight, except perhaps in the case of very young animals.

At present, the energy concentration that initiates eating is set at the base of zero energy level. The energy concentration above that level is computed by dividing the excess energy in the blood by the total kilograms of blood as calculated from Eq. (73). At some energy concentration above the base, the animal will stop eating. Eating will resume when the energy concentration in the blood returns to the base level. Paine (1971) used a single value to initiate and stop eating and based his values on the acetate energy concentration only. A total energy concentration value comparable to Paine's acetate value of 5 kcal/kg would be 8.77 kcal/kg. However, the GRAZE model value is for energy concentration above a base level, which is therefore not directly comparable. Currently, GRAZE uses a single value near the base for both initiating and terminating eating. The use of a single value sometimes contributes to relatively rapid cycles of on–off eating activity, which merits further review in later versions of GRAZE.

Thermal Regulation. The body heat pool determines body temperature. Once the maximum rate of heat removal has been reached [Eq. (43)], the body temperature may begin to rise. If body temperature becomes sufficiently high (presently 39°C), the model assumes that the animal will stop eating to prevent additional heat load (Fig. 56). In the model, low temperatures may trigger eating indirectly by increasing maintenance and thereby reducing the body chemical energy level more rapidly.

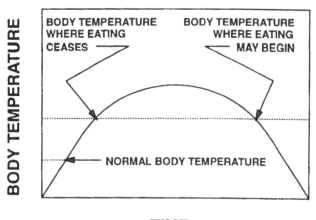

FIGURE 56 Thermostatic control mechanism.

Nighttime Regulation of Intake. There is a time during the evening when no eating occurs (Castle et al., 1950). The model uses two inputs to regulate nighttime eating:

TMEATE = military time that intake ends
TMEATB = military time that eating may begin

When TMEATE is reached, the animal ceases to eat irrespective of fill, chemical, or thermal conditions (Fig. 57). Eating may begin only when TMEATB is reached. However, fill, chemical, and thermal regulation may override TMEATB.

Summary. Dry matter intake and digestion involve a number of feedback mechanisms and governing regulations. In GRAZE, eating may be initiated only by a sufficiently low blood energy level. However, eating will cease if body temperature rises sufficiently above normal, if blood energy levels are sufficiently high, if sleeping occurs, or if physical fill is reached. These concepts are illustrated simultaneously in Fig. 58.

F.   Other Topics

GRAZE has been used primarily to evaluate beef animals' grazing forages. It does contain other logic that can be used to evaluate other types of grazing systems. These are presented below for informational

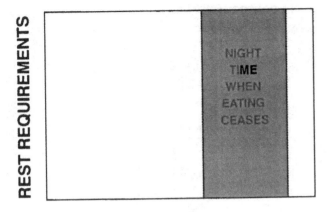

TIME

FIGURE 57   Nighttime rest regulation.

purposes, recognizing that much needs to be done in some of these areas to build confidence in the predicted output.

1. Reproduction

The GRAZE model has the following logic to determine if conception can occur.

1. The animal must be female (engineers and computer programmers have to consider all the possible logic situations!).
2. The animal must be past the point of puberty as defined by the animal's physiological age.
3. A minimum postpartum time must have passed since the animal last gave birth.
4. The animal must contain a minimum percentage of stored fat and each stored nutrient before conception can occur.

The fourth condition helps to explain why two similar females of the same weight may not both conceive if one is gaining and the other is losing weight. Effectively, the stored levels of energy and/or nutrients may be significantly different in the two animals, reflecting a body composition difference. This concept is still in the verification phase.

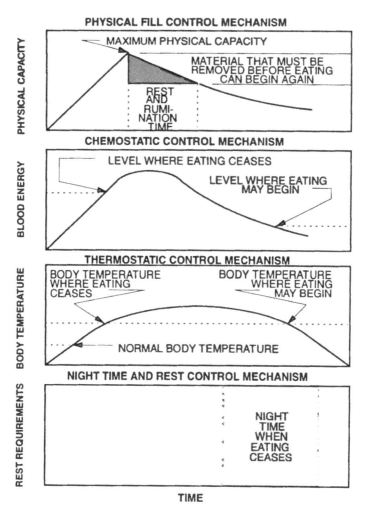

FIGURE 58   Controls for regulating intake.

2.   Fetal Development

The fetus develops along the physiological growth curves discussed earlier. However, different input parameters may be used to show effects of cross-breeding. In addition, body composition components of the fetus are aggregated into a single growth curve rather than simulating each component separately.

Current logic has development of the fetus second only to body maintenance in priority for energy and body nutrients. Maintenance for the fetus is calculated separately from that for its mother. Should there not be sufficient energy and nutrients for fetal maintenance, abortion will occur. Physiological development may be restricted under some low energy and low nutrient conditions. Current logic has the fetus being born when it reaches a certain chronological rather than physiological age. Usually, the two ages would be the same. Future additions to the logic may include establishing the composition of the placenta over the gestation period to have a firmer base for energy and nutrient utilization.

3. Milk Production

Milk production has important implications for animal performance and reproduction. A "good milker" may have more trouble getting bred. However, her calf may perform better than others in the herd, other things being equal.

Milk quantity and composition change over the lactation cycle. Typically, milk production increases for about 1 month, begins a steady decline, and finally reaches a steady plateau. For discussion purposes, these stages of production will be referred to as phase 1, phase 2, and phase 3 (Fig. 59).

A review of the literature indicates that the percentage of fat in the milk generally increases over time. Protein behaves similarly but to a lesser degree. Mineral content remains reasonably constant. As milk quantity drops, total energy in the milk increases, but not enough to give a constant daily level of milk energy.

Although GRAZE does contain logic related to milk production, this logic has not been sufficiently refined for it to be used. The following discussion should be considered as background information from which further enhancement of GRAZE will be directed. Current GRAZE logic concerning milk production is based on the assumption that either the genetic potential, demand for milk, or one milk input ingredient will be limiting. Genetic potential is supplied by the model user as the maximum kilograms of dry matter that can be produced per day. Demand for milk may be from a calf or a milking machine. The latter is assumed for discussion so as to state the logic without having to consider the increase in demand associated with a growing calf.

Input ingredients to milk include energy, protein (nitrogen), calcium, phosphorus, and other minerals. The degree to which each of these components is available is dependent on the outflow rates for

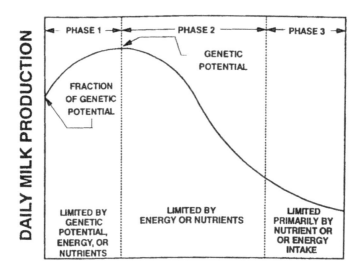

TIME

FIGURE 59 Conceptual lactation curve under constant milk demand with no thermal stress.

milk production. These rates are functions of the amount of stored material and the energy and nutrient priority discussed previously. For example, excess fat will not be available to provide for milk production until maintenance, fetal, and growth energy demands are satisfied. Thus, a period of cold weather could decrease milk production if energy were the limiting factor. A simplified logic flowchart is shown in Fig. 60.

In the GRAZE logic, milk is assumed to be composed of minimum percentages of protein (nitrogen), calcium, phosphorus, other minerals, fat, and a maximum percentage of lactose. Protein cannot exceed a specified maximum percentage. Calcium, phosphorus, and other minerals are considered to be in the milk in nearly constant quantities. Fat and protein content may increase, and lactose may decrease, as described later.

The absolute amounts of the milk components are influenced by the genetic potential of the animal and the rate at which this potential is reached after lactation begins. The number of days from the initiation of lactation to when maximum genetic potential is reached is specified, as is the percentage of maximum genetic potential when

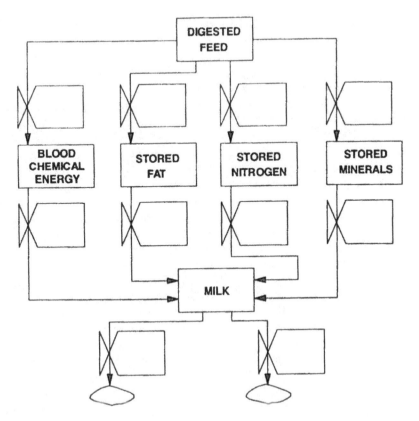

FIGURE 60   Milk production.

lactation begins. These two points are connected by a parabolic function similar to that shown in phase 1 of the lactation curve. Thus, GRAZE effectively assumes that reaching "genetic potential" may be limited by factors other than milk components until maximum production is reached. Milk production cannot exceed the genetic potential and may be limited to lesser milk output by demand for milk or the availability of milk components.

GRAZE determines the daily milk output by first "looping" through each simulated day every DT time step to maintain a current value of nutrient stores and blood concentrations of nitrogen and energy. DT will typically be 15 min. The milk logic begins by establishing a proposed milk composition based on minimum percentage levels of calcium, phosphorus, other minerals, fat, and protein (nitrogen) and maximum and minimum percentage levels of lactose on the basis of dry matter per kilogram. The amount of energy per kilogram of milk is computed using proximate analysis. The genetic potential is established and is expressed as a production rate in kilograms per hour. (The genetic potential may be reduced to a lower production level if demand for milk by a calf is less than potential.) This rate is converted to the maximum potential genetic energy and nutrient flow rates associated with milk production. The model evaluates the existing potential flow of energy and nutrients to milk based on storage and blood concentration levels at a point in simulated time. If existing nutrient and energy flows are sufficient to meet the genetic (or demand) production levels, milk is produced at this level, and nutrient and energy storage and concentration levels are reduced accordingly.

If energy and any other nutrient are limiting, the following logic applies. The model identifies the limiting factor and computes the kilograms of milk that can be produced with this limit. The model then alters the milk proportions of lactose, fat, and protein above the minimum composition levels. Energy available for adding more fat or protein is defined as the energy input difference between the maximum and minimum levels of lactose in the milk. This extra energy will first be used to increase the protein content, the degree being dependent on the nitrogen concentration in the blood. Any energy remaining after protein is added is used to manufacture fat. In effect, any time the animal is under sufficient stress that genetic factors do not limit production, the proportions of milk protein and fat increase and lactose decreases.

GRAZE does not consider any effects of disease or physical damage to the cows. Thus, the cow may provide milk indefinitely in phase 3 as long as energy and nutrients are available. However, if the cow produces less than a user-stated amount for three consecutive days, the model will stop lactation.

Much remains to be done in predicting levels of lactation, and the above logic has not been adequately tested. However, in concept it is safe to say that energy is a major consideration when attempting to describe the lactation curve.

4. Castration

GRAZE recognizes the animal categories of bulls, cows, and steers. The program contains growth curve characteristics that describe the physiological development of each.

When a bull is castrated, the model immediately considers that the animal will follow the steer growth parameters for future compositional growth. The existing body composition obviously does not change. (Such may not be the case with animal disposition, but that's beyond the scope of our modeling efforts!) After the model computes future changes in composition, the "steer" rates are added to the old "bull" levels. The earlier castration occurs, the closer the male animal resembles a steer insofar as physiological age for puberty is reached.

5. Internal and External Parasites

The GRAZE animal model does not attempt to simulate internal or external parasites directly. However, the effects of parasites may be demonstrated by increasing the maintenance levels for energy and nutrients (primarily protein) over designated time periods. The increase in energy maintenance is reflective of increased activity associated with movement of the animal to avoid insect contact. Another interpretation is that increases in maintenance for both energy and nutrients reflect the removal of body energy and nutrients in the form of blood.

IV. MODELING IN THE INTERFACE BETWEEN PLANTS AND GRAZING BEEF ANIMALS

A brief overview is presented to tie together the three major submodels of GRAZE (Figs. 1 and 2).

A. Plant Submodel Review

GRAZE simulates plant growth and composition on a daily basis given maximum and minimum daily temperatures, rainfall, forage removal rates, fertilization data, and numerous crop parameters. The computational interval is once per day. The above-ground plant material is partitioned into the categories new, old, and dead. For grazing situations, an "unavailable for grazing" category is added. The "new" material has a physiological age that is less than the number of days required for a leaf to reach its mature size. The "old" material is experiencing senescence. The "dead" material is derived from the "old" material after growth and senescence have been completed. "Unavailable" material refers to a minimum quantity of above-ground dry matter that is sufficiently sparse that cattle are not able to graze it. Each of the plant categories (i.e., new, old, dead, and unavailable material) is further partitioned into cell content, potential digestible cell wall, or indigestible cell wall. Nitrogen, phosphorus, and potassium contents within each partition are also computed.

B. Animal Submodel Review

GRAZE uses a division of energy flow to describe animal growth, intake, and response to environment (temperature, relative humidity, and solar radiation). Ingested plant material is partitioned into the categories of cell content, potentially digestible cell wall, and indigestible cell wall material. Cell wall and indigestible cell wall material have different rates of digestion within GRAZE. In the model, digestion logic computes the flows of chemical energy, thermal energy, and nutrients that are used by the animal for maintenance, growth, and production (fetal growth and lactation). GRAZE describes the animal in terms of chemical components and physical parameters. Chemical body composition is in the form of protein, water, fat, and minerals. Physical parameters include surface area and respiration (maximum and minimum air exchange rates). Chemical composition and physical parameters change as the animal changes in physiological development.

Intake of feed is regulated within the model by the following relationships:

1. Physical fill of the rumen.   If intake ceases because of physical fill, a portion of the material must be emptied from the rumen before the animal can resume eating regardless of other control factors.
2. Upper blood energy chemostatic limits.   Relatively lower chemostatic limits must be reached before eating can resume regardless of other control factors.
3. Internal body temperature.   Before the animal can resume grazing, its temperature must fall below a defined upper limit.
4. Time of day.   Animals do not eat during certain designated hours during the evening, regardless of the status of the other control mechanisms.
5. The computational interval for the animal model.   The grazing status of the animal remains constant over the computational interval. This interval is an input parameter that may be varied (typically, a 15 min interval is used).

C. An Overview of the Plant/Animal Interface

One of the most difficult aspects of modeling beef-forage systems is in accurately predicting intake. Reasons include the dynamic aspects of both plant and animal growth, influences of environment, definition of plant and animal maturity, and the difficulty of obtaining an accurate measure of forage quality. GRAZE is a dynamic simulation model that relates beef animal performance to environment, management practices, and selective grazing of pastures. Daily computations of pasture quality and availability are made. Measures of the animal's weight (physiological and total) and age (physiological and chronological) are computed at 15 min intervals or less, and body composition, efficiency of growth, and feed utilization are determined. Rotational grazing may be evaluated, with animal movement being based on time of grazing and/or availability of dry matter. During a simulated day, each pasture may be further divided into a variable number of subareas that reflect herbage quality and mass (kg/ha) differences associated with selective grazing. The number and size of the subareas vary with environment and forage removal and reflect conceptual rather than distinct physical boundaries within the pasture. Selection by the animal of the order in which these areas may be grazed is based on the premise that the animal will attempt to maximize its digestible dry matter intake rate. The following is a discussion about how this is accomplished.

D. GRAZE Interface

The GRAZE model contains selective grazing logic that interfaces the plant and animal submodels. The major divisions of the GRAZE interface logic relate to plant quality and the conceptual division of a pasture into partial grazing areas. The grazing animal selects from among the partial grazing areas based on differences in forage quality and density (kg/ha).

1. Selective Grazing Logic as Related to Plant Quality

Conceptually, the first assumption of major importance to the GRAZE selective grazing logic is that the beef animal will always attempt to maximize its ingestion rate of digestible dry matter. That is, the GRAZE decision algorithm used during active grazing maximizes the animal's ingestion of digestible dry matter. Active grazing occurs in the model when physical fill, chemostatic, thermostatic, or restricted evening hour controls are not in effect. The digestible dry matter intake rate is influenced by the physiological age of the animal, plant quality, and plant availability. The physiological weight–age status of the animal determines its maximum potential dry matter intake rate for a given quality of material. Similar observations have been reported by Thiessen et al. (1984) and Taylor et al. (1986). Furthermore, the faster dry matter is digested, the greater the potential for intake as long as chemostatic, thermostatic, and restricted grazing time controls are not in effect. For example, Holloway et al. (1979) observed that the daily dry matter intake of lactating beef cows increased with increases in forage quality. The higher the ratio of cell content to cell wall material, the greater the dry matter digestion rate given the same percentage of indigestible dry matter. In the partitions of dry matter, new material generally contains a higher percentage of cell content material. Accordingly, in GRAZE it is assumed that for a given forage the dry matter intake rate for a diet of entirely new material would be higher than for a diet composed entirely of old material. Similarly, the animal would be expected to ingest old material at a faster rate than dead material. In essence, "palatability" and "quality" are defined in GRAZE to be functions of the new, old, and dead plant partitions. An input to the model is the rate (kg/h) at which the animal would independently consume each of these plant categories. By definition, "new," "old," and "dead" do not represent different physical components of the plant but physiological states. Hence, GRAZE assumes that each bite the animal takes is in direct proportion to the availability of new, old, and dead material. The rate at which material is ingested is a weighted average of the portion of the plant category and its potential intake rate. A somewhat similar approach was used in a modeling effort by Johnson and Parsons (1985) in that the independent weights of four structural plant components were computed. The contribution of each of the structural components to the leaf area index was used to develop their algorithm for modeling selective grazing.

2. Selective Grazing Logic as Related to Forage Availability

The maximum potential intake rate may be reduced by forage availability (kg/ha). The premise is that at some point dry matter availability will begin to limit intake rate (kg/h) independent of forage quality. The functional form used in GRAZE is shown in Fig. 50. This relationship is similar to that reported by the National Research Council (1987) and discussed by Minson (1983) and Hodgson (1977). The following equations are used to determine maximum potential intake rate:

Maximum potential intake rate (kg/h)

= (maximum intake associated with quality for a given
forage, kg/h)*(forage availability factor)   (74)

where

Maximum intake associated with quality for a given
forage (kg/h)

= (% of available dry matter that is "new"/100)

*(maximum potential intake rate of "new"
material based on physiological weight-age
of animal (kg/h))

+ (% of available dry matter that is "old"/100)

*(maximum potential intake rate of "old" material
based on physiological weight-age of animal
(kg/h))

+ (% of available dry matter that is "dead"/100)

*(maximum potential intake rate of "dead"
material based on physiological weight-age of
animal (kg/h)) (75)

and

$$FAF = 2*A/D - A**2/D**2 \qquad (76)$$

where

FAF = forage availability factor that serves as the fractional multiplier of the potential dry matter intake rate as influenced
by dry matter availability less than $D(0 \leq FAF \leq 1.0)$.
D = available dry matter mass level below which the dry matter intake rate will begin to become limiting, kg/ha
A = available dry matter mass for grazing, kg/ha

If $A \geq D$, then FAF will equal 1.0. FAF will equal 0.0 when $A = 0$.
However, A is defined as the available dry matter on the field. Some
minimum dry matter level is present even after the animals have
removed as much dry matter as is physically possible.

For purposes of example, suppose the proportions of new, old,
and dead material are 40%, 35%, and 25%, respectively. Similarly,
assume that the associated independent intake rates for new, old, and
dead materials are 4.0, 2.5, and 1.0 kg/h, respectively, and that the
forage availability factor (FAF) limits intake to 50% of potential. Using
the above equations,

Maximum digestible dry matter intake rate (kg/h)

= [(40/100)*4.0 kg/h + (35/100)*2.5 kg/h

   + (25/100)*1.0 kg/h]*0.5

= 1.36 kg/h

3. Selective Grazing Logic as Related to Partial Fields

The second assumption of major importance to the selective grazing logic in GRAZE is that the area available for grazing may be divided into subareas (also referred to as partial fields) for purposes of selective grazing by the animal. The partial fields are conceptual in nature, are not separated by physical boundaries, and do not necessarily represent physically distinguishable areas within the total available pasture. The sum of the areas of the partial fields is equal to the total area available for grazing on a given day. In GRAZE, the partial fields are involved in the decision algorithm used by the grazing animal in maximizing its digestible dry matter intake rate. Each day, the size and number of partial fields are reevaluated based on plant growth, plant composition, and the quantity of dry matter removed by the animals the previous day. At the simulated time of midnight, the plant model updates the growth and composition of each partial field. GRAZE computes the potential digestible dry matter intake rate for each partial field using Eqs. (74)–(76) and ranks the partial fields in the order in which they may be grazed if there is sufficient demand for dry matter by the animal. Lower ranked partial fields on a particular day may not be grazed.

After the partial fields are ranked, the model calculates the quantity of dry matter that must be removed from the highest ranked field before the animal would be equally satisfied with the second ranked partial field. The dry matter that may be removed from the highest ranked partial field is computed in two parts consisting of the available material both above and below the point where forage mass (kg/ha) begins to restrict intake (Fig. 50). Within GRAZE, the animal may remove all the dry matter above the limiting dry matter mass point (expressed as kg/ha) without changing its digestible dry matter intake rate. Thus, step 1 in partial field 1 (Fig. 61) will be the first material removed, followed by step 2 until availability limits intake to the point at which a move to partial field 2 would be equally satisfying. The reference to a partial field number refers to its relative priority in terms of maximization of digestible dry matter intake rather than a permanent field designation. The order of partial field

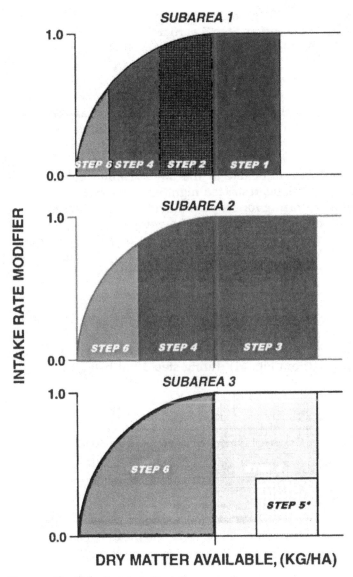

FIGURE 61 Selective grazing of composite crop.

grazing is subject to change each day dependent on plant growth, availability, and composition. Step 3 partial field 2 follows in priority. However, the digestible dry matter intake rates of partial fields 1 and 2 are equal at the dry matter liming point of partial field 2 and are both equal to or superior to partial field 3. Therefore, GRAZE computes the quantity of material that may be removed when simultaneously grazing partial fields 1 and 2 (step 4) before the animal would be equally satisfied, in terms of maximizing its digestible dry matter intake rate, to be grazing partial field 3. The above logic continues until either all the partial fields are of the same quality or the animal ceases to graze because of one of the internal limits (physical fill, chemostatic, thermostatic, restricted nighttime grazing).

The order of grazing for six partial fields is shown in Table 1. If all the partial fields are grazed until the animal is equally satisfied with any of them in terms of maximizing its digestible dry matter intake rate, GRAZE has the animal consume forage uniformly and simultaneously from all partial fields in proportion to the total dry matter remaining.

If the animal reaches one of the internal limits and ceases to graze in a partial field before the field reaches the same desirability status as all the partial fields grazed previously, another partial field may be created. For example, assume that the animal ceases to graze on partial field 3 (from Fig. 61) during step 5 but before dry matter

Table 1 Grazing Priority for Partial Fields as Defined by GRAZE

Partial No.	Grazing Priority Order of Two-Part Steps[a]					
	1	2	3	4	5	6
1	1a → (1)	1b → (2)	1b +	1b +	1b +	1b +
2		2a → (3)	2b → (4)	2b +	2b +	2b +
3			3a → (5)	3b → (6)	3b +	3b +
4				4a → (7)	4b → (8)	4b +
5					5a → (9)	5b → (10)
6						6a → (11)

[a]Grazing priority designates the steps of selective grazing in two parts ("a" and "b" except for partial field 1). The first number in each cell indicates the partial field to be grazed in terms of maximizing digestible dry matter intake. For each partial field, "a" and "b" refer to the portion of dry matter above and below, respectively, the dry matter level that begins to limit intake. The numbers in parentheses indicate the order of grazing. The "+" connects those partial fields in the same column that are being grazed simultaneously.

availability begins to become limiting. Likewise, assume that the animal grazes to point 5*. Given these assumptions, GRAZE assumes that step 5* did, in fact, reach the intake-limiting point but on partial field 3a rather than 3b. The areas of partial fields 3a and 3b are adjusted with regard to relative area so as to appear as shown in Fig. 62. At midnight, partial field 3 (3a and 3b) becomes 3 and 4 with separate calls to the crop growth model.

An input to GRAZE is the number of allowable partial fields within the major field that are available for grazing. This number is

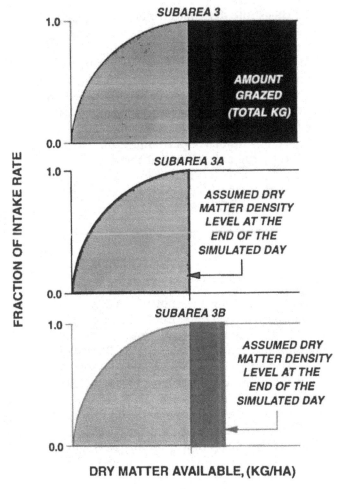

FIGURE 62 Creation of two subareas from one.

divided into the grazing area per animal to obtain a minimum partial field size. If a new projected partial field is not sufficiently large, the last partial field to be grazed retains its identity.

4. Summary of Selective Grazing Logic in GRAZE

In summary, the selective grazing logic in GRAZE serves the following functions:

> It calls the plant submodel at midnight of each simulated day for each partial field and updates plant growth and composition.
> It computes the digestible dry matter intake rate (kg/h) associated with the grazing animal for each partial field.
> It determines the order in which each partial field is subject to grazing based on its forage quality and availability.
> It calls the animal submodel as required over the day and moves the animal systematically from the highest ranked partial field to the lowest ranked as necessary so as to maximize the animal's digestible dry matter intake rate.
> It adjusts the size and number of partial fields based on the dry matter removed by the animal during the day.
> It controls the reading and writing of input and output information, respectively, and the scheduling of management practices such as fertilizer applications and moving of animals to other fields.

V. MODEL STRUCTURE, INPUT, AND OUTPUT

GRAZE is a Fortran program that does not use any simulation language. It contains approximately 17,000 lines of computer code and can be executed on a microcomputer (minimum configuration requirements are 386 machine, math coprocessor, 4 MB of RAM, and 4 MB of disk storage). The DOS release of GRAZE is Version 2.3, which is described fully by a user guide (Parsch and Loewer, 1995) and a case study publication (Loewer and Parsch, 1995). GRAZE for Windows, Version 3.02, was first released to the public in June 1996. Major enhancements of the Windows version allow GRAZE to simulate competition among plant species (Loewer et al., 1989), gives greater flexibility in examining fertilizer applications, provide direct access to the definition and default values of input parameters, and allow for direct creation of output graphs that can be used to compare different scenarios. GRAZE for Windows is the version that will be supported in the future.

Input to GRAZE is accomplished through the use of two files, one related to weather and the other containing the scenario that defines an individual system to be evaluated. GRAZE may generate several output files as specified by the model user. Maximum frequency of the output is 1 day. Each output file (except for the one "tracing" file) can be imported into a spreadsheet, where it can be manipulated at the discretion of the model user. In this situation, comparisons among systems, graphic displays, statistical analysis, etc. are the responsibility of the user, and these kinds of operations can usually be done within the spreadsheet. In GRAZE for Windows, Version 3.02, graphical displays may be generated, printed, and copied within the program. Again, details concerning execution and utilization of GRAZE may be found in the GRAZE *User Guide* (Parsch, 1995).

VI. SUMMARY

The beef-forage selective grazing model GRAZE had its beginnings in the late 1970s as part of the Kentucky BEEF modeling effort. GRAZE incorporates physiologically based plant and beef animal submodels developed by agricultural scientists from 25 states working as part of three regional research projects: S-156, NC-114, and S-221. The effects of environment on both plants and animals are considered, as is animal selectivity of pasture. The selectivity portion of the model is based on the premise that the beef animal attempts to maximize its digestible dry matter intake rate. Output from the model relates to plant growth, plant utilization, and animal performance. GRAZE Version 2.3 (DOS version) was released to the public in April 1995. GRAZE Version 3.02 (Windows version) was released to the public in June 1996. Both versions may be downloaded free of charge via the World Wide Web (URL address: http:\\www.agen.ufl.edu). Supporting publications (user guide and case study) with diskettes for Version 2.3 may be obtained from Agricultural Publications, AGRI 110, University of Arkansas, Fayetteville, AR 72701. There is a $4.00 handling charge.

Efforts to expand GRAZE by adding to its utility and credibility are continuing. Interested parties are encouraged to contact the author.

REFERENCES

Baldwin, R. L., and J. L. Black. 1979. Simulation of the effects of nutritional and physiological status on the growth of mammalian tissues: Description and evaluation of a computer program. Tech. Paper No. 6. CSIRO Animal Research Laboratory.

Baldwin, R. L., L. J. Koong, and M. J. Wyatt. 1977. The formation and utilization of fermentation end products. In: Microbial Ecology of the Gut. R. J. J. Clarke and T. Bauchop (Eds.). London: Academic Press, pp. 347–391.

Berry, F. A., E. Bollay, and N. R. Beers. 1945. *Handbook of Meteorology*. New York: McGraw-Hill.

Blaxter, K. L., and J. A. F. Rook. 1953. The heat of combustion of the tissues of cattle in relation to their chemical composition. *Br. J. Nutrition* 7:83–91.

Blaxter, K. L., and F. W. Wainman. 1964. *J. Agric. Sci. Camb.* 62:207.

Bridges, T. C., L. W. Turner, E. M. Smith, T. S. Stahly, and O. J. Loewer. 1985. A mathematical procedure for estimating animal growth and body composition. ASAE Paper No. 85-4508. St. Joseph, MI: ASAE.

Bridges, T. C., L. W. Turner, J. L. Usry, and J. A. Neinaber. 1992a. Modeling the physiological growth of swine, Part 1: Model logic and growth concepts. *Trans. ASAE* 35(3):1019–1028.

Bridges, T. C., L. W. Turner, J. L. Usry, and J. A. Neinaber. 1992b. Modeling the physiological growth of swine, Part 2: Validation of the model logic and growth concepts. *Trans. ASAE* 35(3):1029–1033.

Brown, B. 1982. *Computer Simulation of Plant and Animal Growth*. MP 43. Nebraska Agricultural Experiment Station, p. 5

Brown, J. E., H. A. Fitzhugh, Jr., and T. C. Cartwright. 1976. A comparison of nonlinear models for describing weight-age relationships in cattle. *J. Anim. Sci.* 2(4):810–818.

Castle, M. E., A. S. Foot, and R. J. Halley. 1950. Some observations of the behavior of daily cattle with particular reference to grazing. *J. Dairy Res.* 17(3):215–228.

Costello, J. L., H. D. Scott, O. J. Loewer, C. P. West, and L. D. Parsch. 1988. Incorporation of water-balance logic into a selective grazing model. ASAE Paper No. 88-4064. St. Joseph, MI: ASAE.

Crampton, E. W., and L. E. Harris. 1969. *Applied Animal Nutrition*. 2nd ed. San Francisco: Freeman.

Ellenberger, H. B., V. A. Newlander, and C. H. Jones. 1950. Composition of the Bodies of Dairy Cattle. Agric. Exp. Sta. Bull. 558, Burlington, VT: University of Vermont and State Agricultural College, July.

Ellis, W. C. 1978. Determinants of grazed forage intake and digestibility. *J. Dairy Sci.* 61:1828–1840.

Fitzhugh, H. A., Jr. 1976. Analysis of growth curves and strategies for altering their shape. *J. Anim. Sci.* 42(4):1036–1051.

Fitzhugh, H. A., Jr., and C. S. Taylor. 1971. Genetic analysis of degree of maturity. *J. Anim. Sci.* 33(4):717–725.

Forrester, J. W. 1968 *Principles of Systems*. Cambridge, MA: Wright-Allen.

Guyton, A. C. 1947. Analysis of respiratory patterns in laboratory animals. *Am. J. Physiol. 150:*78.

Hahn, R. H., and E. E. Rosentreter (Eds.). 1987. *ASAE Standards 1987.* St. Joseph, MI: ASAE.

Hargreaves, G. H., and Z. A. Samani. 1982. Estimating potential evapotranspiration. *J. Irrig. Drainage Div., Proc. ASCE 180*(IR3):225–230.

Haxton, J. A., M. D. Schneider, and M. P. Kaye. 1974. Blood volume of the male Holstein-Fresian calf. *Am. J. Vet. Res. 35:*835–837.

Heath, M. E., D. S. Metcalfe, and R. F. Barnes. 1973. *Forages,* 3rd ed. Ames, IA: The Iowa State University Press.

Hershberger, T. V., and E. W. Hartsook. 1969. In vitro rumen fermentation of alfalfa hay. Carbon dioxide, methane, VFA and heat production. *J. Anim. Sci. 30:*257–261.

Hintz, R. L., F. N. Owens, and O. J. Loewer. 1982. Prediction of protein deposition and gain in beef cattle. Orlando, FL: Southern Section of the American Society of Animal Science, Feb. 7–10, Abstract 17.

Hodgson, J. 1977. Factors limiting herbage intake by the grazing animal. Proceedings of the International Meeting on Animal Production from Temperate Grassland, Irish Grassland, and Animal Production Association, p. 70–75.

Holloway, J. W., W. T. Butts, J. D. Beaty, J. T. Hopper, and N. S. Hall. 1979. Forage intake and performance of lactation beef cows grazing high or low quality pastures. *J. Anim. Sci. 48*(3):692–700.

Holt, D. A., R. J. Bula, G. E. Miles, M. M. Schreiber, and R. M. Peart. 1975. Environmental Physiology, Modeling, and Simulation of Alfalfa Growth. I. Conceptual Development of SIMED. Purdue Agric. Exper. Sta. Res. Bull. 907.

Johnson, I. R., and A. J. Parsons. 1985. A theoretical analysis of grass growth under grazing. *J. Theor. Biol. 112:*345–367.

Jones, C. A., and J. R. Kiniry. 1986. *CERES-Maize: A Simulation Model of Maize Growth and Development.* College Station, TX: Texas A&M Univ. Press.

Lange, A. L., and G. M. Forkew. 1952. *Handbook of Chemistry.* Sandusky, OH: Handbook Publishers.

Loewer, O. J., and L. D. Parsch (Eds.). 1995. GRAZE Beef-Forage Simulation Model: Case Studies. Southern Cooperative Series Bull. 381B. Fayetteville, AR: University of Arkansas.

Loewer, O. J., E. M. Smith, G. Benock, N. Gay, T. C. Bridges, and L. G. Wells. 1980. Dynamic simulation of animal growth and reproduction. *Trans. ASAE 23*(1):131–138.

Loewer, O. J., E. M. Smith, G. Benock, T. C. Bridges, L. G. Wells, N. Gay, S. Burgess, L. Springate, and D. Debertin. 1981. A simulation model for assessing alternative strategies of beef production with land, energy and economic constraints. *Trans. ASCE 24*(1):164–173.

Loewer, O. J., E. M. Smith, K. L. Taul, L. W. Turner, and N. Gay. 1982. A body composition model for predicting beef animal growth. ASAE Paper No. 82-4555. St. Joseph, MI: ASAE.

Loewer, O. J., E. M. Smith, N. Gay, and R. Fehr. 1983a. Incorporation of environment and feed quality into a net energy model for beef cattle. *Agric. Syst.* 11:67–94.

Loewer, O. J., E. M. Smith, K. L. Taul, L. W. Turner, and N. Gay. 1983b. A body composition model for predicting beef animal growth. *Agric. Syst.* 10:245–250.

Loewer, O. J., C. P. West, L. D. Parsch, and H. D. Scott. 1989. Modeling plant species competition in conjunction with selective grazing. Proceedings of the XVI International Grassland Congress, Nice, France. Vol. II, pp. 1347–1348.

Lofgreen, G.P., and W. N. Garrett. 1968. A system for expressing net energy requirements. *J. Anim. Sci.* 27:793–806.

Marcilese, N. A., H. D. Figueiras, R. M. Valsechi, and H. R. Camberos. 1966. Blood volumes and body: Venous hematocrit ratio in cattle. *Cornell Vet.* 56(1):142–150.

Marsten, H. R. 1948. The fermentation of cellulose in vitro by organisms from the rumen of sheep. *Biochem. J.* 42:564.

Miller, J. A., and I. L. Berry. 1970. Contribution of hairshedding to the loss of insecticidal deposits from cattle. *J. Econ. Entomol.* 101-63(4):1338–1339.

Minson, D. J. 1983. Forage quality: Assessing the plant/animal complex. Proceedings of the 14th International Grasslands Congress, pp. 23–29.

Moe, P. W., and W. P. Flatt. 1969. Net energy value of feed stuff for lactation. *J. Dairy Sci.* 6:928. (Abstract.)

Montgomery, M. J., and B. R. Baumgardt. 1965a. Regulation of food intake in ruminants. Pelleted rations varying in energy concentration. *J. Dairy Sci.* 48:569–574.

Montgomery, M. J., and B. R. Baumgardt. 1965b. Regulation of food intake in ruminants. 2. Rations varying in energy concentration and physical form. *J. Dairy Sci.* 48:1623–1628.

Moulton, C. R. 1923. Growth of the Herford-Shorthorn Steer. Univ. Missouri Agri. Exper. Sta. Bull. 62, pp. 11–25.

Moulton, C. R., P. F. Trowbridge, and L. D. Haigh. 1921. Studies in Animal Nutrition. I. Changes in Form and Weight on Different Planes of Nutrition. Univ. Missouri Agric. Exper. Sta. Res. Bull. 43.

Moulton, C. R., P. F. Trowbridge, and L. D. Haigh. 1922a. Studies in Animal Nutrition. II. Changes in Proportions of Carcass and on Different Planes of Nutrition. Univ. Missouri Agric. Exper. Sta. Res. Bull. 54, 76 p.

Moulton, C. R., P. F. Trowbridge, and L. D. Haigh. 1922b. Studies in Animal Nutrition. III. Changes in Chemical Composition on Different Planes of Nutrition. Univ. Missouri Agric. Exper. Sta. Res. Bull. 55

Moulton, C. R., P. F. Trowbridge, and L. D. Haigh. 1923. Studies in Animal Nutrition. V. Changes in the Composition of the Mature Dairy Cow While Fattening. Univ. Missouri Agric. Exper. Sta. Res. Bull. 61.

National Research Council. 1976. *Nutrient Requirements of Domestic Animals.* No. 4. *Nutrient Requirements of Beef Cattle.* 5th rev. ed. Washington, DC: National Academy of Sciences—National Research Council.

National Research Council. 1987. *Beef Cattle. Predicting Feed Intake of Food-Processing Animals*. Washington, DC: National Academy Press, Chapter 6.

Nelsen, T. C., C. R. Long, and T. C. Cartwright. 1982. Post-inflection growth in straightbred and crossbred cattle. I. Heterosis for weight, height and maturing rate. II. Relationships among weight, height and pubertal characters. *J. Anim. Sci.* 55(2):280–292, 293–304.

Oltjen, J. W., A. C. Bywater, and R. L. Baldwin. 1986a. Development of a dynamic model of beef cattle growth and composition. *J. Anim. Sci.* 62: 86–97.

Oltjen, J. W., A. C. Bywater, and R. L. Baldwin. 1986b. Evaluation of a model of beef cattle growth. *J. Anim. Sci.* 62:98–108.

Paine, M. D. 1971. Mathematical modeling of energy metabolism in beef animals. Ph.D. Dissertation. Stillwater, OK: Oklahoma State University.

Parsch, L. D., and O. J. Loewer (Eds.). 1995. GRAZE Beef-Forage Simulation Model: User's Guide. Southern Cooperative Ser. Bull. 381A. Fayetteville, AR: University of Arkansas.

Parton, W. J., and J. A. Logan. 1981. A model for diurnal variation in soil and air temperature. *Agric. Meteorol.* 23:205–216.

Penrod, E. B. 1960. Variation of Soil Temperature at Lexington, Kentucky. Univ. Kentucky Eng. Exper. Sta. Bull. 57. Lexington, KY: University of Kentucky.

Preston, R. L. 1966. Protein requirements for growing-finishing cattle and lambs. *J. Nutr.* 90:157.

Pullar, J. D., and A. J. F. Webster. 1977. The energy cost of fat and protein deposition in the rat. *Br. J. Nutr.* 37:355–363.

Salisbury, F. B., and C. Ross. 1969. *Plant Physiology*. Belmont, CA: Wadsworth.

Sanders, J. O., and T. C. Cartwright. 1979a. A general cattle production systems model. I. Structure of the model. *Agric. Syst.* 4:217.

Sanders, J. O., and T. C. Cartwright. 1979b. A general cattle production systems model. II. Procedures used for simulating animal performance. *Agric. Syst.* 4:289.

Simkins, K. L., Jr., J. W. Suttle, and B. R. Baumgardt. 1965a. Regulation of food intake in ruminants. 3. Variation in blood and rumen metabolites in relation to food intake. *J. Dairy Sci.* 48:1629–1634.

Simkins, K. L., Jr., J. W. Suttle, and B. R. Baumgardt. 1965b. Regulation of food intake in ruminants. 4. Effect of acetate, propionate, butyrate and glucose on voluntary food intake in dairy cattle. *J. Dairy Sci.* 48:1635–1642.

Smith, E. M., and O. J. Loewer. 1981. A nonspecific crop growth model. ASAE Paper No. 81-4013. St. Joseph, MI: ASAE.

Smith, E. M., and O. J. Loewer. 1983. Mathematical logic to simulate the growth of two perennial grasses. *Trans. ASAE* 26(3):878–883.

Smith, E. M., T. H. Taylor, and L. Brown. 1968. Interpretation of diurnal variation in soil temperatures. *Trans. ASAE* 11(2):195–197.

Smith, E. M., L. S. Ewen, and O. J. Loewer. 1981. Growth of fescue and red clover as influenced by environment and interspecific competition. ASAE Paper No. 81-4020. St. Joseph, MI: ASAE.

Smith, E. M., L. M. Tharel, M. A. Brown, G. W. Burton, C. T. Dougherty, S. L. Fales, V. H. Watson, and G. A. Pederson. 1985. The plant-growth component. In: *Simulation of Forage and Beef Production in the Southern Region*. V. H. Watson and C. M. Wells, Jr. (Eds.). Southern Cooperative Ser. Bull. 308. Mississippi State:MS: Mississippi State University.

Spreen, T. H., and D. H. Laughlin (Eds.). 1986. *Simulation of Beef Cattle Production Systems and Its Use In Economic Analysis*. Boulder, CO: Westview Press.

Taylor, St. C. S., and H. A. Fitzhugh, Jr. 1971. Genetic relationships between mature weight and time taken to mature within a breed. *J. Anim. Sci. 33*(4): 726–731.

Taylor, St., C. S., A. J. Moore, and R. B. Thiessen. 1986. Voluntary food intake in relation to body weight among British breeds of cattle. Animal Production, Vol. 42:11–18.

Thiessen, R. B., Eva Hnizdo, D. A. G Maxwell, D. Gibson, and St. C. S. Taylor. 1984. Multibreed comparisons of British cattle. Variation in body weight, growth rate, and food intake. Animal Production, Vol. 38:323–340.

Trowbridge, P. F., C. R. Moulton, and L. D. Haigh. 1918. Effect of Limited Food on Growth of Beef Animals. Univ. Missouri Agric. Res. Bull. 28.

Usry, J. L. 1989. Simulating the digesta movement through the gastrointestinal tract of the growing pig. Unpublished Ph.D. Thesis. Agricultural Engineering Department, University of Kentucky, Lexington.

Usry, J. L., L. W. Turner, T. C. Bridges, and J. A. Neinaber. 1992. Modeling the physiological growth of swine, Part 3: Heat production and interaction with the environment. Model logic and growth concepts. *Trans. ASAE 35*(3):1035–1042.

Van Soest, P. 1982. *Nutritional Ecology of the Ruminant*. Corvallis, OR: O&B Books.

Walker, J. N., O. J. Loewer, G. Benock, N. Gay, E. M. Smith, S. Burgess, L. G. Wells, T. C. Bridges, L. Springate, J. Boling, G. Bradford, and D. Debertin. 1977. *Production of Beef with Minimum Grain and Fossil Energy Inputs*. Vols. I, II, and III. Report of NSF. Lexington, KY: University of Kentucky.

Watson, V. H., and C. M. Wells, Jr. 1985. The animal component. In: *Simulation of Forage and Beef Production in the Southern Region*. V. H. Watson and C. M. Wells, Jr. (Eds.). Southern Cooperative Ser. Bull. 308. Mississippi State University.

Webster, D. 1974. Heat loss from cattle with particular emphasis on the effects of cold. In: *Heat Loss from Animals and Man*. J. L. Monteith and L. E. Mount (Eds.). University of Nottingham, Chapter 10.

Webster, A. J. F. 1979. Energy costs of digestion and metabolism in the gut. In: *Digestive Physiology and Metabolism in Ruminants*. Ruckebusch and Thivend (Eds.). Westport, CT: AVI, pp. 469–484.

Webster, A. J. F., J. Chlumecky, and B. A. Young. 1970. Effects of cold environment on the energy exchanges of young beef cattle. *Can. J. Anim. Sci. 50*:89–100.

Whittemore, C. T. 1976. A study of growth responses to nutrient inputs by modeling. *Proc. Nutr. Soc.* 35:383–391.

Whittemore, C. T., and R. H. Fawcett. 1976. Theoretical aspects of a flexible model to simulate protein and lipid growth in pigs. *Animal Prod.* 22:87–96.

Wolin, M. J. 1960. A theoretical rumen fermentation balance. *J. Dairy Sci.* 433:1452–1459.

Yeates, N. J. M. 1955. Photoperiodicity in cattle. *Aust. J. of Agric. Res.* 2:645–663.

11

Dynamical Systems Models and Their Application to Optimizing Grazing Management

Simon J. R. Woodward

AgResearch Ruakura, Hamilton, New Zealand

I. INTRODUCTION

The purpose of studying and modeling grazed pasture systems is to improve the use of farmland for agricultural production. Improvement may take the form of increased profit per hectare, reduced pollution or erosion, pest management, or adoption of sustainable practices.

From a human point of view, agricultural grazing involves a farmer managing a simple ecosystem, consisting of interacting climate, soil, plant, and animal components (Fig. 1), for certain social and economic purposes. Experimentation in an agricultural context is focused on understanding the pastoral ecosystem and identifying management strategies that improve farm productivity and sustainability. A dynamical systems model is a mathematical description of a system, and mathematical modeling similarly focuses on understanding the system and identifying opportunities for better management.

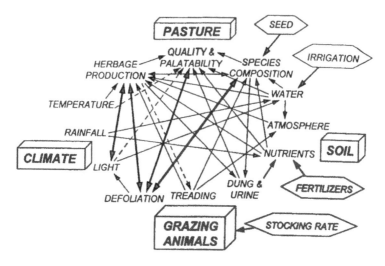

FIGURE 1 Interacting components in a grazed pasture system. [Redrawn from Snaydon (1981).]

One approach to modeling farm systems is to make measurements of the variables at intervals through time. A mathematical curve can then be fitted to these time series data. Provided the other variables can be provided, such a statistical regression model can be used to predict future trends of the system based on observations of past trends.

Unlike traditional regression models of farm system behavior, dynamical systems are constructed from relationships based on component experiments. The relationships are described mathematically and may be nonlinear, and the model allows the prediction of the evolution of the interacting system variables forward through time, similar to computer simulation modeling. However, dynamical systems models are amenable to a variety of useful mathematical analyses and optimization techniques so that the long-term stability of the modeled agricultural ecosystems may be completely understood and so that the model may explicitly focus on optimizing farm management.

For optimization and management applications, simpler models are more effective. Not only does additional complexity make parameterization difficult and reduce the chance that simple management principles may be identified, but a complex model may also be prone to instability (May, 1971) or chaotic behavior (Solé and Bascompte, 1995). Nevertheless, it is perhaps unlikely that chaos will be observed in grazing systems, as such systems are heavily "damped" (May, 1971).

This chapter considers only intensive systems where the number of grazing animals is completely controlled by the farmer. When we discuss optimization of such a grazing system, the farmer in practice has relatively few variables that can be controlled. The main ones are stocking rate, fertilizer and water application, herbage conservation and feeding-out, and the timing of animal events such as lambing, purchases, and sales. These are called the "control variables" at the farmer's disposal. Optimal management involves the optimal control of these inputs.

In discussing a range of grazing applications, I introduce the basic techniques of dynamical systems analysis and provide references to useful books that explain the mathematical techniques. It is hoped that the models and analyses presented here will prove thought-provoking and that the relative strengths of mathematical modeling compared to both static regression models and dynamic computer simulations will become evident.

We first work through the process of formulating a prototype model describing the dynamic interactions between pasture mass and liveweight gain by grazing animals. Second, we review mathematical models that have been used to analyze agricultural grazing systems management. Third, in Section IV we return to the pasture mass–liveweight gain model to illustrate how a qualitative analysis of the system stability is carried out through phase plane analysis. Fourth, we consider an example of applying optimal control theory to maximize annual production in a grazing system with a seasonal growth rate function. Finally I sketch out a simple model for predicting the effects of stocking density on pasture quality through time and discuss possible approaches to modeling selective grazing by animals.

II. A COUPLED PLANT–ANIMAL MODEL

The grazed ecosystem, though simple by comparison to natural ecosystems, is nevertheless enormously complex from a mathematical point of view. To provide useful management recommendations, simplification is necessary. This often involves assumptions about the system—for example, that the soil effects on herbage production are known in advance (as the farmer follows a prescribed fertilizer and irrigation regime) and that the effects of climate may be adequately represented through annual average light, temperature, and rainfall curves. This leaves us with the simplified system, consisting of the

FIGURE 2   The primary components of a grazing model.

animal, the pasture, and the grazing interface (Fig. 2). Indeed, many models have focused on the pasture component alone, assuming that changes in grazing pressure per hectare are small.

Ecologists have been studying natural plant–animal ecosystem dynamics for many years (e.g., Caughley and Lawton, 1981; Crawley, 1983; Stocker and Walters, 1984; May, 1971) where the animal populations may change in response to forage availability. However, in an intensively managed farming application, animal numbers are tightly controlled. Nevertheless, dynamical systems models are still useful.

Consider the situation of growing cattle grazing on a seasonally growing pasture. The objective of beef finishing operations, for example, is to maximize the animals' liveweight gain over a limited growing season and thus their market value. Optimal management includes choosing the right stocking rate so that animals are well fed while pasture supply is maintained throughout the grazing period.

A valuable skill in dynamical modeling is the ability to rapidly construct a prototype model of the system that contains all the basic structures even if some of the finer details have not been firmly established. The process of analyzing a prototype model sheds significant light on what kind of model will ultimately be needed to provide answers to the management issues and what biological knowledge will be essential for the realism of those answers.

The situation of growing cattle requires a model of pasture availability and animal weight gain that includes the effects of changing stocking rate. We will work through the construction of a prototype model for this system.

A.   Modeling Pasture Mass

For the purposes of beef finishing, it would be reasonable to assume that pasture mass is the main variable of interest and that pasture quality and composition are of secondary importance. Pasture growth studies such as those of Brougham (1956) showed that the logistic function offers a reasonable empirical approximation to the dependence of pasture growth on leaf area. Logistic pasture growth has

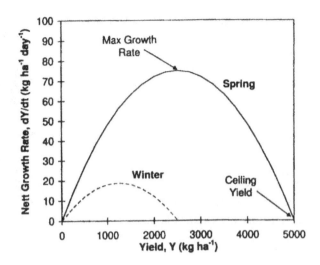

FIGURE 3   A sketch of the logistic growth curve for spring and winter pastures.

been used in many modeling studies (e.g., Morley, 1968; Thornley, 1990; Woodward et al., 1995).

We use a simple but elegant seasonal pasture growth model based on the logistic growth curve where both maximum growth rate and ceiling yield vary seasonally (Fig. 3). We model the rate of pasture accumulation [dY/dt, kg/ha·day] in the absence of grazing as

$$\frac{dY}{dt} = a(t)Y - bY^2 \tag{1}$$

where Y (kg/ha) is the pasture biomass at any point in time. The first term on the right represents the net balance between the rate of new growth and the rate of senescence and depends linearly on the pasture mass according to the seasonal parameter $a(t)$. The second term represents a damping of growth due to self-shading within the sward canopy. This term does not vary seasonally but is nonlinear. The first and second terms taken together comprise a logistic growth curve with ceiling yield

$$Y_{max} = \frac{a(t)}{b} \tag{2}$$

at any season and maximum growth rate of

$$g_{max} = \frac{a(t)^2}{4b} \tag{3}$$

Therefore,

$$a(t) = 4 \frac{g_{max}}{Y_{max}}, \qquad b = 4 \frac{g_{max}}{Y_{max}^2} \tag{4}$$

We have economized nicely by achieving seasonal growth rate and seasonal maximum yield using a single seasonal parameter, $a(t)$. In early spring, for example, we might expect Y_{max} = 5000 kg/ha, g_{max} = 75 kg/ha·day), which implies that $a(t)$ = 0.06 day ^{1} and b = 1.2 × 10 ^{5} ha/(kg·day).

B. Modeling Intake

The pasture model above gives the rate of pasture accumulation at any given pasture yield in the absence of grazing animals. How does grazing affect the rate of change of pasture yield?

The potential intake of a growing animal is approximately proportional to its body weight. However, when pasture mass is limiting, actual intake is determined by a curvilinear relationship as shown in Fig. 4. Therefore we may express the intake rate I of an animal as depending on both liveweight and pasture availability:

$$I = L \times \frac{rY}{c + Y} \tag{5}$$

Here, I is the daily dry matter intake rate of an animal in kilograms of dry matter per animal per day and L is its liveweight in kilograms per animal. The second term on the right, $rY/(c + Y)$, is a Michaelis–Menten response to pasture availability. A Michaelis–Menten function has the correct shape (as in Fig. 4), is easy to handle algebraically, and has been shown to be appropriate in several studies of grazing theory (Thornley et al., 1994; Spalinger and Hobbs, 1992; Ungar and Noy-Meir, 1988; Woodward, in press). In Eq. (5), r is the animal's potential pasture dry matter (DM) intake rate per kilogram of liveweight (LW) of the animal [expressed as kgDM/(animal·day·kgLW)] and c (kgDM/ha) is the pasture mass at which intake is half of the potential. For a beef animal, r might be around 0.035 and c, 1100 kg/ha.

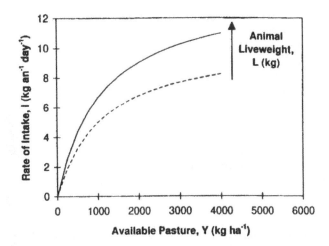

FIGURE 4 The dependence of intake on animal liveweight (L) and available pasture (Y). The dotted line shows the intake of a light animal, the solid line that of a heavy animal.

This intake model can be attached to the pasture growth model [Eq. (1)] to give a prediction of the accumulation (or disappearance) of pasture mass under grazing:

$$\frac{dY}{dt} = [a(t)Y - bY^2] - n(t) \times L\frac{rY}{c + Y} \qquad (6)$$

| Pasture accumulation | Pasture growth | Stocking rate | Intake |

This mathematical differential equation expresses the dependence of rate of pasture accumulation (dY/dt) on the component processes of pasture growth and intake, which themselves depend upon the variables Y and L, the pasture mass and the animal liveweight at a point in time. The ability to express the relationships and couplings between the interacting components of a system in differential equation form is the basis for the dynamical systems modeling methodology.

C. Modeling Liveweight Gain

Animal growth modeling attempts to predict the response of animal size to different feeding strategies. A simple model states that feed energy is used either for maintenance of body tissue or for deposition of new tissue. This model is equivalent to a linear approximation to

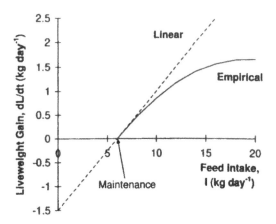

FIGURE 5   The dependence of animal liveweight gain on feed intake. The solid curve is the usual empirical relationship. The dotted line represents a linear approximation.

the standard liveweight gain response curve (Fig. 5) (e.g., Sheath and Rattray, 1985). This may be expressed mathematically as

$$ I \quad = \quad v\frac{dL}{dt} \quad + \quad mL \qquad\qquad (7) $$

| Feed intake | Body growth | Body maintenance |

Here, v (kgDM/kgLW) is the conversion efficiency of feed into liveweight, and m [kgDM/(kgLW·day)] is the feed maintenance requirement per kilogram of liveweight. Typical values for these parameters are $v = 4$ and $m = 0.02$.

Equation (7) is a first-order differential equation relating dL/dt, the animal's rate of liveweight gain, to feed intake I. We have already modeled feed intake from pasture using Eq. (5). Substituting this into the liveweight equation Eq. (7), and rearranging gives a prediction of liveweight gain for a grazing animal:

$$ \frac{dL}{dt} \quad = \quad \frac{1}{v}L\frac{rY}{c+Y} \quad - \quad \frac{m}{v}L \qquad\qquad (8) $$

| Liveweight gain | Feed intake | Body maintenance |

This gives an equation for the rate of liveweight gain of an animal of weight L grazing on a pasture of mass Y.

Table 1 Variables and Typical Parameter Values in Early Spring for Beef Grazers for the Yield–Liveweight Model

Parameter	Description
Y, kg/ha	Pasture yield
I, kgDM/(animal·day)	Daily dry matter intake per animal
n, animal/ha	Stocking density
L, kg/animal	Animal liveweight
Y_{max} = 5000 kg/ha	Maximum yield
g_{max} = 75 kg/(ha·day)	Maximum pasture growth rate
a = 0.06 kg/(kg·day)	Seasonal relative pasture growth rate
b = 1.2 × 10^{-5} ha/(kg·day)	Damping of pasture growth
r = 0.035 kg/(kg·day)	Potential intake per kilogram liveweight
c = 1100 kg/ha	Yield at which intake is half-maximum
v = 4 kg/kg	Feed conversion efficiency
m = 0.02 kg/(kg·day)	Feed maintenance requirement

D. A Coupled Yield–Liveweight Model

Dynamical systems models allow us to study the coupled evolution of system variables when their rates of change are interrelated. Equations (6) and (8) together form a coupled two-dimensional dynamical system for the rates of change of pasture yield Y and animal liveweight L at a given stocking density $n(t)$. The model parameters are summarized in Table 1. We return to this model in Section IV when we use it as an example to illustrate the methods of "phase plane" analysis.

E. "Simulating" with a Dynamic Systems Model

Using our yield–liveweight model, it is straightforward to simulate the system forward in time from some given initial conditions. An excellent introduction to simulation (or "integration") of dynamic systems is given by Press et al. (1989), who discuss the pitfalls of the Euler method used by almost all computer simulation models, and who also provide ready-made computer routines for several better numerical methods.

III. APPLICATION OF SIMPLE BIOMASS GRAZING MODELS

Even very simple models can be used to give useful grazing management information. There have been a number of grazing models

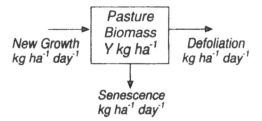

FIGURE 6   Flow diagram for simple biomass models of pasture dynamics.

published that take the form of a single differential equation with a single state variable, pasture biomass or cover. These models consider that for the purposes of making management decisions, the pasture is adequately characterized by its biomass alone (see Fig. 6). The pasture model presented in the previous section [Eq. (6)] is a biomass model of this form. Examples of management information derived from some simple models are reviewed in this section.

A.   Calculating Grazing Duration Using the McCall et al. Model

Perhaps the two simplest published grazing models are those of McCall et al. (1986) and Bircham and Sheath (1986). These models were designed to assist in feed rationing and studies of rates of disappearance under grazing down in a rotational grazing system. Both use a linear intake function (Fig. 7), which is reasonable under grazing down provided pasture mass is not too high (e.g., Laca et al., 1992). The Bircham and Sheath (1986) model also incorporates a constant daily rate of growth, g_0 kg/(ha·day) (see Appendix).

The McCall et al. model predicts that the change in pasture yield over time is

$$Y(t) = Y(0)e^{-nkt} \tag{9}$$

where $Y(0)$ is the pasture yield at time 0. This is not a remarkable equation in itself. What is remarkable rather is that this equation claims to be a dynamic model of pasture disappearance in a grazed pasture and as such offers *prediction* of herbage flow. This prediction may be tested statistically by comparison to the results of a grazing

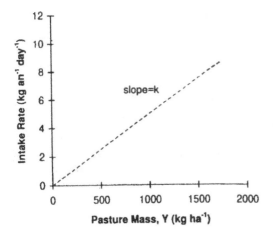

FIGURE 7   Linear intake relationship used by McCall et al. (1986), Bircham and Sheath (1986), Woodward et al. (1993a, 1993b, 1995), and Woodward and Wake (1994).

experiment. This then may lead to the modification or rejection of the model.

Methods of solution of differential equations are generally difficult. Some simple methods are presented in the Appendix. For more complicated equations or systems the reader is directed to undergraduate calculus texts such as Zill (1989). However, in most models of biological systems we should not expect to be able to calculate an exact solution of the dynamic system. Dynamic systems theory grew up around this eventuality, and most of the methods described in this paper do not require an analytical solution.

Nevertheless, if the solution of the system can be calculated, this is clearly a useful circumstance, and the equation can be used directly to deduce the consequences of the model. We offer a simple example using the solution to the McCall et al. model of pasture disappearance under grazing. This is the kind of question that the model was designed to answer.

Question: At what time does per cow intake fall below $I = 4$ kgDM/(animal·day) if the pregrazing mass is $Y(0) = 2000$ kgDM/ha, the stocking density is $n = 30$ cows/ha, and the relative intake rate of cows is $k = 0.005$ ha/(animal·day)?

The model assumes that per cow intake is given by

$$I = kY(t) = kY(0)e^{-knt} \tag{10}$$

We can rearrange this equation to make time t the subject:

$$t = -\frac{1}{kn} \ln \frac{I}{kY(0)}$$

$$= -\frac{1}{0.005 \times 30} \ln \frac{4}{0.005 \times 2000} = 6 \text{ days} \tag{11}$$

This is a much more satisfactory method than calculations based simply upon "herbage allowance," as it takes into account the change in intake rate as pasture mass declines. The simple model of McCall et al. (1986) provides simple management prescriptions appropriate to its intent.

B. Optimal Scheduling of a Rotational Grazing System Using Morley's Model

Rather than treat senescence explicitly (Fig. 6), most biomass models have modeled net pasture production. One such is the model of a regrowing pasture presented by Morley (1968):

$$\frac{dY}{dt} = aY\left(1 - \frac{Y}{Y_{max}}\right) \tag{12}$$

where Y_{max} is the ceiling yield, as before. This function is the same "logistic growth equation" that we met in Fig. 3 and is based on field data of ryegrass pasture growth gathered by Brougham (1956).

If a paddock of a rotational grazing system has a postgrazing mass of $Y(0)$ kg/ha when the animals are removed at time $t = 0$, then solving Eq. (12) (using separation of variables, see Appendix) gives the accumulation of regrowth at some later time t as

$$Y(t) = \frac{Y_{max}Y(0)}{Y(0) + [Y_{max} - Y(0)]e^{-at}} \tag{13}$$

This is the typical sigmoid pasture growth curve (Fig. 8). Assuming negligible pasture growth during the grazing phase for the paddock, Morley wished to calculate the time at which the average rate of pasture regrowth was maximized after the animals were removed from the paddock (Fig. 8). That is,

$$\text{Find } t \text{ that maximizes } \frac{Y(t) - Y(0)}{t} \tag{14}$$

Mathematically, to maximize some quantity we must set the derivative equal to zero, as follows:

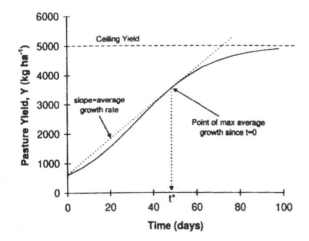

FIGURE 8 Typical S-shaped growth curve for a pasture regrowing after grazing in a rotational grazing system. The dotted line is the steepest line that touches the growth curve. The point where it touches is the point of maximum average regrowth rate. [Redrawn from Parsons and Penning (1988).]

$$\frac{d}{dt}\left(\frac{Y(t) - Y(0)}{t}\right)$$

$$= \frac{d}{dt}\left(\frac{1}{t}\left(\frac{Y_{max}Y(0)}{Y(0) + [Y_{max} - Y(0)]e^{-at}}\right) - \frac{1}{t}Y(0)\right) = 0 \qquad (15)$$

With a bit of work we can deduce from this that the average pasture growth rate is maximized if the animals are introduced into the paddock at time t satisfying the relation

$$Y(0) + [Y_{max} - Y(0)]e^{-at} = \frac{1}{2}Y_{max}(1 - at)$$

$$+ \frac{1}{2}[Y_{max}^2(1 - at)^2 + 4atY_{max}Y(0)]^{1/2} \qquad (16)$$

This is not soluble in an explicit form but can easily be solved numerically. Taking an initial pasture (postgrazing mass) of $Y(0) = 600$ kg/ha and the parameters in Table 1, the optimal resting time is $t^* = 48$ days.

This was an elegant way of using a simple logistic model of pasture growth to make recommendations about the length of resting in a rotational grazing system. Once again, a simple model was appropriate to answering the management problem.

C. The Pasture Stability Analyses of Noy-Meir

The work of ecologist Noy-Meir (1975, 1976, 1978a, 1978b) in modeling grazing systems is well known. His seminal paper in 1975 analyzed grazing models of the form

$$\frac{dY}{dt} = g(Y) - nI(Y) \tag{17}$$

where $g(Y)$ is the net instantaneous rate of growth of new herbage (minus losses due to senescence) depending on the photosynthetic biomass Y, and $I(Y)$ is the rate of intake of a single animal when offered herbage Y. The stocking density is n.

The relative merits of a number of explicit functional forms for $g(Y)$ and $I(Y)$ are discussed in Noy-Meir (1978b), but the 1975 paper carries out a simple graphical stability analysis (see Fig. 9) for a generic model [Eq. (17)]. Noy-Meir shows that for some particular forms of the growth and intake functions, there might be two stable steady-state values of pasture mass Y for a given stocking density—one at high pasture mass and one at low pasture mass.

A steady state is a point (in this case a value of pasture yield Y) at which all of the system fluxes are in balance, hence the system

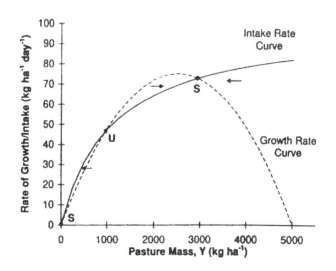

FIGURE 9 The balance of growth and intake rates in Noy-Meir's (1975) model for a given stocking rate. Steady states are shown as filled circles labeled S (stable) or U (unstable). Arrows indicate the accumulation or loss of pasture mass.

"state" is "steady." In the case of Eq. (17), for example, this would mean that the rate of herbage growth is equal to the rate of removal by grazing. In Fig. 9, therefore, the steady states occur at the intersections of the growth and intake rate function curves. At intermediate stocking densities, there are three such intersections. The leftmost, $Y = 0$, is a stable (attracting) steady state, because nearby states move toward it (i.e., the pasture goes to extinction, shown by an arrow in Fig. 9). There is another stable steady state where the declining growth rate intersects the intake rate at the higher pasture mass, labeled S. And in between these two is a third intersection, U, which is an unstable steady state. This point is very interesting. Above this point, herbage growth rate outstrips intake and the pasture mass increases to the higher stable steady state over time. Below this point, intake is greater than growth rate, and the animals graze the pasture to extinction. So this intermediate "unstable" steady state is a "watershed." The system moves away from it toward one of two steady stable states. This feature is termed "discontinuous stability." The system ends up in one of the two stable states after a long time, depending on the initial conditions; the set of initial conditions that lead to a particular stable steady state are called the "basin of stability" of that steady state. We discuss steady states in more detail in Section IV.

D. Bifurcation Diagram Analysis

The existence of three distinct steady states depends on the stocking density being within a critical range. A useful tool for examining this feature is called a bifurcation diagram. With this diagram we can examine how the steady states of a system change as we change one of the system parameters. Let's examine a model of the form of Eq. (17) where the herbage growth follows the logistic growth equation of Morley (1968) and the animal intake follows a Michaelis–Menten saturation curve (e.g., Noy-Meir, 1978b):

$$\frac{dY}{dt} = aY\left(1 - \frac{Y}{Y_{\max}}\right) - n\frac{KY}{c + Y} \tag{18}$$

where K is the potential rate of intake of herbage by one animal at a given liveweight, and the other parameters are familiar to us from Eq. (6). For a 300 kg cow, for example, we might have $K = 10.5$ kg/(animal·day).

Steady-state values of pasture mass occur when $dY/dt = 0$, i.e., either when $Y = 0$ or when

$$a \left(1 - \frac{Y}{Y_{max}}\right) = \frac{nK}{c + Y} \tag{19}$$

This is a quadratic equation with solutions at

$$Y = \frac{1}{2} (Y_{max} - c) \pm \frac{1}{2} \left[(Y_{max} - c)^2 - 4Y_{max} \left(\frac{nK}{a} - c\right) \right]^{1/2} \tag{20}$$

This has one feasible solution when $n \le ac/K$ and two real positive solutions when

$$\frac{a}{K} c < n < \frac{a}{K} \left(\frac{(Y_{max} - c)^2}{4Y_{max}} + c\right) \tag{21}$$

Henceforth, we denote the upper bound as n^*. This is the range of stocking densities within which we expect to observe discontinuous stability. Taking parameters again from Table 1 and using $K = 10.5$, the critical range is

$$\frac{0.06}{10.5} 1100 < n < \frac{0.06}{10.5} \left(\frac{(5000 - 1100)^2}{4 \times 5000} + 1100\right) \tag{22}$$

or

$$6.3 < n < 10.6 \text{ animals/ha} \tag{23}$$

which is higher than usual for continuously grazed cattle, because parameters for early spring have been used. At this time of year we would expect pasture growth to outstrip removal by grazing, and so pasture yield will normally move toward a high steady state.

This information can be summarized on a bifurcation diagram (Fig. 10). This diagram shows the positions (in this case the pasture mass yield) of the steady states in relation to some control parameter (in this case stocking density n). In this diagram, the steady states appear as lines. Stable steady states are often drawn as a solid line, and unstable steady states as a dotted line. Arrows are placed to show the evolution of the system with time. The bifurcation diagram is interpreted as follows. For low stocking density n, assuming that Y is initially greater than 0, the system moves toward the high steady state. As n increases we get to a critical stocking density, $n = ac/K$, beyond which if initial mass is low the pasture will be grazed to extinction, otherwise herbage will increase toward the high steady state. This region has discontinuous stability. As stocking density n increases further and passes n^*, the discontinuous stability disappears and the herbage mass drops inexorably toward extinction.

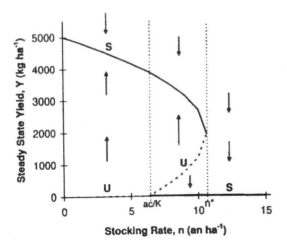

FIGURE 10 Bifurcation diagram showing the steady-state pasture masses at a range of stocking rates. The curve shows the position of the steady states, which are labeled S (stable, heavy line) or U (unstable, dotted line). n^* is the critical stocking rate. (The horizontal axis is a branch of steady states, stable when $n > ac/K$.)

These sudden changes that occur when n passes ac/K and n^* are called catastrophes. The bifurcation diagram shows how a small change in a system parameter, in this case n, can result in a sudden and dramatic change in the stability of the system.

This rich field of stability analysis offers powerful tools for the analysis of agricultural systems. Although there is little evidence from field trials that discontinuous stability exists in real grazed pasture, largely because of intervention by the farmer [but see the excellent study of Morley (1966)], it does provide us with a means for assessing the risk in a system, predicting which management practices are unsustainable, and determining how we may restore stability. In the above example, if the system is headed for extinction due to a high stocking density, the pasture may be restored to the high mass region by a period of time in which the animals are removed to allow the pasture to recover. The optimal region of the graph to operate in is where pasture growth rate is maximized, so stocking density is just below n^* and pasture mass is at the higher stable steady state. Clearly, though, this is also a risky situation to be in, because a small change in one of the system parameters could lead to a crash. This is the problem with "maximum sustainable yield" calculations done by fisheries economists (see Clark, 1990).

E. Models for Optimizing Rotational Grazing

Woodward and coworkers (Woodward et al., 1993a, 1993b, 1995; Woodward and Wake, 1994) have used simple biomass models similar to those above to explore opportunities for optimizing rotational grazing. In Woodward et al. (1993a), they considered growth and grazing in a two-field subsystem of a rotational grazing system. The rate of pasture production was assumed to be proportional to the herbage biomass with relative rate of growth a (per day), so that

Disappearance = production − grazing

or

$$\frac{dY}{dt} = aY - n_i k Y \tag{24}$$

where n_i is the stocking density in paddock i, that is, the total number of animals divided by the area of the paddock being grazed. This is a reasonable model over a short time period at low to medium biomass. If the animals sequentially graze two paddocks of areas h_1 and h_2 and initial pasture masses $Y_1(0)$ and $Y_2(0)$, respectively, then the total intake of the animals over T days and the total pasture remaining in the two paddocks at time T were calculated to be, respectively,

$$C(T) = \frac{kY_1(0)}{a - n_1 k} \left(e^{(a - n_1 k)t_1} - 1\right) + \frac{kY_2(0)e^{at_1}}{a - n_2 k} \left(e^{(a - n_2 k)(T - t_1)} - 1\right) \tag{25}$$

and

$$Y_1(T) + Y_2(T) = h_1 Y_1(0)e^{(a - n_1 k)t_1}e^{a(T - t_1)} + h_2 Y_2(0)e^{at_1}e^{(a - n_2 k)(T - t_1)} \tag{26}$$

where t_1 is the "swapover" time at which the animals were moved from paddock 1 to paddock 2.

In modeling rotational grazing, complications arise because the stocking density in any one paddock changes whenever animals are introduced or removed. This means that the dynamics are discontinuous with time, greatly complicating the analysis, as the untidiness of Eqs. (25) and (26) illustrates.

Woodward et al.'s (1993a, 1993b) analysis showed that herbage conservation $[Y_1(T) + Y_2(T)]$ would be maximized if the animals grazed only one of the paddocks during the T days. This was moderated when a constraint on the animals' minimum daily intake was applied. The intake over T days, $C(T)$, was maximized with respect to t_1 when

$$\frac{\partial C(T)}{\partial t_1} = 0 \tag{27}$$

This was found to be the case when t_1 satisfied the equation

$$(a - n_2 k) \frac{Y_1(0)}{Y_2(0)} e^{-n_1 k t_1} = a - n_2 k e^{(a - n_2 k)(T - t_1)} \tag{28}$$

Table 2 gives a numerical example, including the calculated value of the optimal swapover time, t_1.

Woodward et al.'s (1993a, 1993b) analysis did not reveal any significant new opportunity for utilization of pasture in a rotational grazing system but showed that intake was likely to be similar under continuous and rotational systems for realistic stocking rates. However, rotational systems were confirmed to offer greater scope for herbage conservation.

Woodward et al. (1995) then extended the analysis to explore opportunities presented by altering the sequence in which paddocks were grazed in a multipaddock rotational grazing system. Morley's model was used for pasture growth, and a linear grazing function (Fig. 7) was included:

$$\frac{dY}{dt} = aY \left(1 - \frac{Y}{Y_{max}} \right) - nkY \tag{29}$$

Systems of 100 ha divided into 5, 10, or 20 paddocks were compared, and an algorithm was presented for finding the optimal sequence for

Table 2 Parameter Values for Woodward et al.'s (1993a) Two-Field Grazing Problem

Parameter	Description
$n_1, n_2 = 120$ animals/ha	Stocking density in paddocks 1 and 2
$a = 0.06$ day^{-1}	Relative pasture growth rate
$h_1, h_2 = 0.25$ ha	Areas of paddocks 1 and 2
$Y_1(0), Y_2(0) = 1500$ kg/ha	Initial pasture mass in paddocks 1 and 2
$k = 0.005$ ha/(animal·day)	Relative intake rate per animal
$T = 10$ days	Total time in the two fields
$t_{1,opt} = 6.1$ days	Optimal swapover time
$C(T)_{opt} = 31.0$ kg/animal	Optimal 10-day intake per animal

grazing the paddocks for a total of 60 days, where animals were moved every 1, 2, or 4 days. These scenarios were also compared to continuously grazing the same 100 ha area for 60 days.

The results of this "simulated experiment" confirmed that stocking rate is the primary variable that determines animal performance on a given area of land and that management variables such as number of paddocks and rate of rotation are of secondary importance. A simple cyclic rotational strategy on an initial "wedged" distribution of pasture in the paddocks was shown to be optimal for maximizing intake in most circumstances. Most rotational systems are operated in this manner in practice. Animals should always be moved to the paddock with the most feed.

This study also confirmed that rotational grazing is more suited to feed rationing applications, which indeed is its primary use, and that no increase in animal intake can be expected from rotational systems over continuous grazing.

F. Woodward and Wake's Delayed Senescence Model

It has been suggested that rotational grazing takes advantage of the lag between new growth and senescence in a regrowing pasture, allowing greater utilization of green leaf. Woodward and Wake (1994) tested this claim theoretically by explicitly including a lag of τ days between the production and senescence of plant tissue in a linear model.

If animals graze all leaves equally, then the proportion of leaf tissue "born" at time $t - \tau$ that remains ungrazed at time t is

$$\exp\left[-\int_{t-\tau}^{t} \frac{nI(s)}{Y(s)}\, ds \right] \tag{30}$$

where $nI(s)$ is the rate of removal of pasture by the herd at time s. Using a linear model [Eq. (29)], $I(s) = kY(s)$. The rate of senescence at time t is then the rate of growth at time $t - \tau$ multiplied by this survival function [Eq. (30)], and a linear model for the overall pasture dynamics is

$$\frac{dY}{dt} = \underbrace{aY}_{\text{Growth}} \quad \underbrace{- n(t)kY}_{\text{Grazing}} \quad \underbrace{- aY(t - \tau)\exp\left[-\int_{t-\tau}^{t} n(s)k\, ds \right]}_{\text{Senescence}} \tag{31}$$

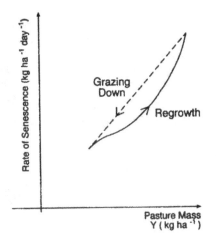

FIGURE 11   Dependence of rate of senescence on pasture mass under grazing down (dotted line) and under regrowth (solid curve).

With this model, it is not possible to write a simple expression for the prediction of pasture yield, $Y(t)$, although an approximate prediction may be calculated [see Woodward and Wake (1994)]. Treating senescence explicitly in this way, although biologically attractive, causes difficulties when we come to performing mathematical analyses. The model [Eq. (31)] predicts that senescence in a grazed pasture is approximately proportional to herbage biomass, as measured by Bircham and Hodgson (1983) for continuously grazed systems (Woodward and Wake, 1994). In the case of regrowth, the relative rate of senescence is reduced due to the effect of the delay (Fig. 11) (Woodward and Wake, 1994), as expected by Johnson and Parsons (1985). This may be expected to be significant in a rotationally grazed pasture. However, the comparison in Woodward (1993) of Eq. (31) to its equivalent nondelay form estimated that the differences would be very slight in practice. Therefore, treating senescence as being proportional to herbage mass may be an adequate model for most applications, including rotational grazing systems.

IV.   SKETCHING A PHASE PLANE

Although expressing a model as a mathematical system allows one to use powerful methods of integrating, as mentioned in Section II.E, a far more important property of dynamic systems is the ability to analyze the system dynamics without explicitly solving the equations. For a two-dimensional system such as the pasture–liveweight model constructed in Section II [Eqs. (6) and (8)], sketching a phase plane is an insight-generating first step toward understanding the structure and stability of the system dynamics.

A phase plane is a graph that shows how two variables, in this case pasture yield and animal liveweight, evolve together over time. The phase plane also gives insight into the stability of the system, that is, what state the system ends up in after a long time. The phase plane cannot tell us, however, how *fast* the system evolves. Strictly, phase plane methods are applicable only to systems whose parameters do not vary with time. In the case of the yield–liveweight model, the biological system is heavily influenced by the changes in pasture growth rate $a(t)$ and stocking density $n(t)$ during the year. Nevertheless, over a short period of time these parameters may be considered to be approximately constant, and in any case there is still a lot to be learned by analyzing the phase plane of the system. Therefore, we will consider $a(t)$ and $n(t)$ to be constant and write $a(t) = a$ and $n(t) = n$.

Fig. 12 shows a phase plane sketch for the yield–liveweight system described above. The phase plane and associated analysis help us to understand the qualitative interactions between pasture mass and liveweight and so can assist us in grazing management decisions. We first discuss the important features of the diagram and then delve into more detail about how a phase plane is constructed.

The dotted lines in Fig. 12 represent the system states (L, Y) where liveweight L or pasture mass Y are not changing. These lines are called isoclines and are calculated directly from the model equations; their position depends on the model parameters (in Fig. 12 the horizontal and vertical axes are also isoclines). These isoclines enable us to sketch the trajectories, which are shown as arrows in Fig. 12. The trajectories show us which direction the system will evolve in from any initial state.

Where the isoclines intersect we have a steady state. At this point both liveweight and pasture mass are in equilibrium. Each steady state has a certain stability: It may be "stable" or "unstable." A stable steady state is one that "attracts" nearby trajectories. This means that *all* trajectories near the steady state lead toward that steady state. In the long term, a system will usually evolve toward a stable steady state.

One of the isoclines indicated in Fig. 12 is the maintenance isocline, a horizontal line below which animals lose weight. Assuming

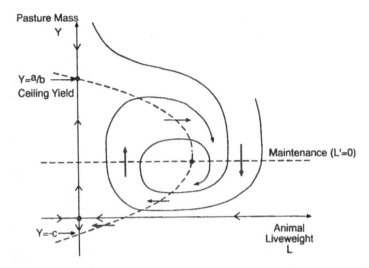

FIGURE 12   Phase plane diagram for the yield–liveweight model. Arrows indicate direction of system evolution. The three filled circles are steady-state points.

that we wish to feed the animals more than their maintenance requirement, we notice from the phase plane that this is not possible to maintain in the long term. Initially, when animals are small, it is possible to sustain above-maintenance feeding from pasture. However, as animals increase in weight, their intake rate begins to outstrip pasture production, until at some point the pasture barely provides maintenance for them. In order to maintain animal growth, the stocking density must then be reduced; this has the effect of stretching the parabolic isocline in Fig. 12 out to the right so that the current state (L, Y) is in the region of the graph where pasture mass and animal liveweight are both increasing.

Since the maintenance isocline is a horizontal line, a second prediction is that animals of all sizes require the same pasture mass for maintenance. Naturally this depends on the exact functions we have chosen, in particular the linear liveweight model [Eq. (7)]. Nevertheless, it is probably not far from the truth. Although larger animals eat more feed, their requirements are also greater, so the pasture mass on which they achieve maintenance may not be much different from that providing maintenance to a smaller animal.

Before we proceed with the details of how the phase plane is constructed, it will be helpful to restate the state equations of our system [Eqs. (6) and (8)], since these will be the basis of our analysis:

$$\frac{dY}{dt} = aY - bY^2 - nL\frac{rY}{c + Y} \tag{32}$$

$$\frac{dL}{dt} = \frac{1}{v}L\frac{rY}{c + Y} - \frac{m}{v}L \tag{33}$$

We now discuss methods for sketching the isoclines and trajectories and for determining the stablity of the steady-state points.

A. Draw the Isoclines and Calculate the Steady-State Positions

The isoclines are curves that show where the trajectories (curved arrows in Fig. 12) turn around, that is, where they are either vertical or horizontal. This occurs when either Y or L is not changing, that is, when $dY/dt = 0$ or $dL/dt = 0$.

First we consider all the points where $dL/dt = 0$, that is, the L' $= 0$ isoclines. From Eq. (33), we see that $dL/dt = 0$ when

$$0 = \frac{1}{v}L\left(\frac{rY}{c + Y} - m\right) \tag{34}$$

Equation (34) is satisfied either when $L = 0$ (no liveweight) or when Y is equal to (using basic algebra)

$$Y_m = \frac{mc}{r - m} \tag{35}$$

(Y_m is the pasture yield level at which the animals' intake is equal to their maintenance requirement.) The value of Y_m calculated with the parameters in Table 1 would be, for example,

$$Y_m = \frac{0.02 \times 1100}{0.035 - 0.02} = 1467 \text{ kg/ha} \tag{36}$$

So the $L' = 0$ isoclines consist of the line $L = 0$ (the Y axis) and the horizontal line $Y = Y_m$ (Fig. 12). At every point on these isoclines, the liveweight of the animals is not changing, so the system is evolving vertically in the phase plane. This is shown by small vertical arrows across the isoclines.

When growth rate slows to the point that the maximum yield $Y_{max} = a/b$ is less than that required for maintenance Y_m, this means that pasture availability is so low that animals of any liveweight will lose weight at any stocking rate. In such a case the farmer will be forced to supplement the animals' feed. Using the parameters from Table 1, this could occur if the relative growth rate a fell below

$$a \leq 1.2 \times 10^{-5} \times 1467 = 0.018 \text{ kg}/(\text{kg}\cdot\text{day}) \tag{37}$$

Second, we calculate the position of the $Y' = 0$ isoclines, where pasture mass is stationary. From Eq. (32), $dY/dt = 0$ when

$$0 = aY - bY^2 - nL\frac{rY}{c + Y} \tag{38}$$

Equation (38) is satisfied either when $Y = 0$ (no pasture) or when

$$L = \frac{1}{nr}(a - bY)(c + Y), \qquad Y \neq -c \tag{39}$$

Equation (39) is the equation of a parabola passing through the Y axis at $Y = a/b$ (ceiling yield) and $Y = -c$ (note, however, that $Y = -c$ is a singular value, and neither L' nor Y' is defined). We also plot these $Y' = 0$ isoclines in the phase plane (Fig. 12). At every state on these curves (except at $Y = -c$), $Y' = 0$, so the herbage biomass is stationary and the system state (i.e., liveweight L) is moving horizontally, as shown by the small arrows.

Aside from $(0, -c)$, the $L' = 0$ and $Y' = 0$ isoclines cross at three points, indicated by filled circles in Fig. 12. At these points, both pasture mass and liveweight are stationary. These are called the steady states of the system and are vitally important. It is the steady states that determine the stability of the whole system.

The three steady states of this system lie at the points $(L, Y) = (0, 0)$, $(0, a/b)$, and (L_s, Y_m), where the value of L_s is calculated by substituting $Y = Y_m$ [Eq. (35)] into Eq. (39):

$$L_s = \frac{1}{nr}\left(a - b\frac{mc}{r - m}\right)\left(c + \frac{mc}{r - m}\right) \tag{40}$$

Taking the parameters in Table 1 and the value of Y_m calculated in Eq. (36), we get

$$L_s = \frac{1}{0.035n}(0.06 - 1.2 \times 10^{-5} \times 1467)(1100 + 1467) = \frac{3109}{n} \tag{41}$$

So, for example, if $n = 5$ cows/ha, the liveweight L_s that would maintain the pasture at $Y = Y_m$ would be $3109/5 = 621$ kg. This high value reflects the high value of the parameter a for spring growth and indicates that at this time of year even very large animals can gain weight at a high stocking rate.

B. Add Direction Field Arrows and Sketch Trajectories

The directions of the small arrows (the "flow") on each isocline are found by looking at the sign $(+/-)$ of the other equation. Equation (34) shows us that $dL/dt > 0$ when $L > 0$ and $Y > Y_m$. Everywhere else, $dL/dt \leq 0$. This means that the direction of the trajectories (or the "direction of flow") is to the right in this region and to the left everywhere else. Similarly, from Eq. (38) we see that $dY/dt > 0$ when $Y > 0$ and L is to the left of the parabolic isocline. To the right of the parabola, $dY/dt < 0$. We can now draw in the direction of flow across or along the isoclines.

This information is summarized in Fig. 13. In fact, the directions of many of the arrows can be deduced from common sense. For example, it is obvious that when liveweight is zero, pasture mass will increase toward maximum yield and that when there is no pasture, animal liveweight will decline to zero.

It is a fundamental property of dynamical systems that the trajectories of the system (the curves that join up the arrows) can never cross except at a steady state. This is a very important property to keep in mind when sketching trajectories on phase planes. Along

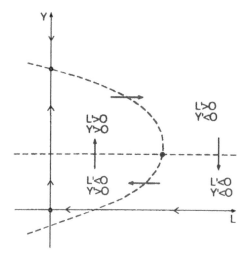

FIGURE 13 Phase plane diagram for the yield–liveweight model, showing the direction of flow in each region of the plane and on the isoclines.

with this is the property that every trajectory either (1) is a closed loop or (2) starts somewhere and ends somewhere. The "somewhere" may be (1) a steady state, (2) at infinity, or (3) in the vicinity of a closed loop.

Keeping these properties in mind, the full trajectories (lines of flow) can now be roughly sketched in, because the isoclines are the only places where the trajectories can be horizontal or vertical. In many cases this means that the isoclines fully determine the phase plane. In Fig. 12 the trajectories have been sketched in the positive ($L > 0$, $Y > 0$) part of the graph.

A numerical method to determine the direction of flow at any point is simply to calculate the values of dY/dt and dL/dt at many points and draw these as small (dY/dt, dL/dt) vectors. For example, if dY/dt and dL/dt are both positive at a particular point, then the direction vector is upward and to the right at that point. This is called plotting the "direction field" of the system. Some examples are given by Edelstein-Keshet (1987).

One important consequence of Fig. 13 is that the farmer must keep the system in the region of the graph where $Y > Y_m$ if the animals are to gain liveweight. Since the position of the parabolic $Y' = 0$ isocline depends on stocking density n [Eq. (39)], we can maintain growth if the stocking density is [from Eq. (40)]

$$n < \frac{1}{Lr} (a - bY_m)(c + Y_m) \tag{42}$$

(which is $n < 3109/L$ for the example parameter values), so that the parabolic isocline is always to the right of the system state on the phase plane. A liveweight of $L = 300$ kg would therefore require a stocking density of less than 10.3 animals/ha to maintain growth at this time of year.

C. Determining the Stability of the Steady States

If a steady state has at least one trajectory leading away from it, it is "unstable." So although system states near an unstable steady state may initially move toward that equilibrium, eventually they will move away. From Fig. 12, we can see that the two steady states on the Y axis have at least one trajectory leading away from them; both are unstable.

The stability of the remaining steady state at (L_s, Y_m) is more difficult to determine. From the sketch, it seems that the trajectories spiral clockwise around the steady state, perhaps spiralling inward.

But do they in fact spiral inward, do they spiral outward, or do they spiral toward a closed loop (a "limit cycle," Fig. 14)? The answer to this question may depend on the parameters (a, b, c, k, m, v) of the system.

The usual method for determining the stability of a point such as this is to study its eigenvalues. Eigenvalues are complex numbers that describe the dynamics of a linear system near a steady state. The basic rule is this: If the real parts of *all* the eigenvalues are negative, then the steady state is stable. But if even one of the eigenvalues has a positive real part, then the steady state is unstable. In addition, if any of the eigenvalues are complex, this indicates spiralling behavior near the steady state. An introduction to the method is given in Edelstein-Keshet (1987).

Eigenvalue analysis is a technical aspect of mathematical system analysis, but it is worthwhile to present briefly the results for the (L_s, Y_m) steady state of this yield–liveweight system, both to give a flavor of the technique and because several interesting insights come out. An attempt is made to focus on the biological implications and downplay the mathematical detail. I used the computer algebra software Maple V to assist with this analysis (Char et al., 1992).

The first step is to calculate the eigenvalues of the system linearized at the steady-state point. If we denote the right-hand side of

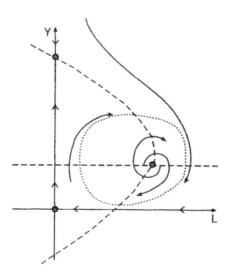

FIGURE 14 Phase plane diagram for the pasture yield–liveweight model in the case where there is a stable limit cycle (closed loop).

the Y state equation [Eq. (32)] as f_Y and the right-hand side of the L equation [Eq. (33)] as f_L, then the eigenvalues e_1, e_2 are the solutions of the matrix determinant equation

$$
\det \begin{pmatrix} \dfrac{\partial f_L}{\partial L}(L_s, Y_m) - e & \dfrac{\partial f_L}{\partial Y}(L_s, Y_m) \\[2ex] \dfrac{\partial f_Y}{\partial L}(L_s, Y_m) & \dfrac{\partial f_Y}{\partial Y}(L_s, Y_m) - e \end{pmatrix} = 0 \tag{43}
$$

where the partial derivatives are evaluated at the steady-state point (L_s, Y_m). Solving Eq. (43) gives the values of the eigenvalues at this steady state as

$$
e_1, e_2 = \frac{1}{2}\left(\frac{m}{r}\right)a - \frac{1}{2}\left(\frac{bmc(r + m)}{r(r - m)}\right)
$$
$$
\pm \frac{1}{2}\left(\frac{1}{r(r - m)}\right)\left(\frac{m}{v}\right)^{1/2} \times (\text{discriminant})^{1/2} \tag{44}
$$

where

$$
\text{Discriminant} = mv(r^2 - m^2)a^2 + [2mvbc(m^2 - r^2)
$$
$$
- 4r(r - m)^3]a + mvb^2c^2(r + m)^2 + 4bmcr(r - m)^2 \tag{45}
$$

Dynamical systems theory states that for the steady state to be stable, the real parts of both eigenvalues must be negative. In addition, when the discriminant of Eq. (44) is negative, both of the eigenvalues are complex and their real parts are equal. Then the steady state is the center of a spiral; this occurs when a is in the range (B_1, B_2), where

$$
B_1, B_2 = \frac{bc(r + m)}{r - m} + \frac{2r(r - m)}{mv} \pm \frac{2r}{mv}[(r - m)^2 + mvbc]^{1/2} \tag{46}
$$

Using the parameter values from Table 1, this range is

$$
0.030 < a < 0.093 \tag{47}
$$

In fact, $a = 0.06$ lies within this range, so we expect a spiral point. This would be observed as out-of-phase oscillations in pasture yield and animal liveweight, so that both Y and L will fluctuate, with a lag between the phase when the pasture is increasing and the phase when the liveweight is increasing (Fig. 15). If the oscillations are

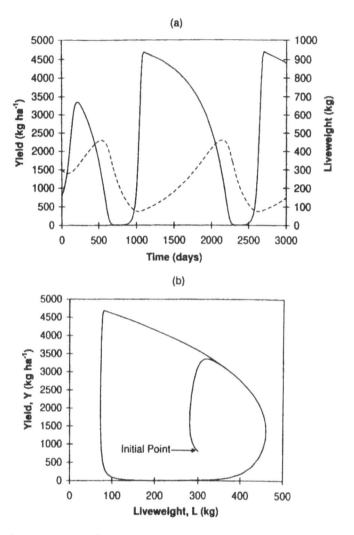

FIGURE 15 (a) Out-of-phase oscillations in yield (solid curve) and liveweight (dotted curve) when the steady state is a spiral point (in this case an unstable spiral point). (b) The same oscillations plotted on the phase plane.

damped, this indicates a stable spiral; if they are increasing in magnitude, an unstable spiral.

When a is in (B_1, B_2), the real part of the eigenvalues is given by the first two terms only in Eq. (44) because the value of the square root is imaginary. The real part is zero when

$$\frac{1}{2}\left(\frac{m}{r}\right)a - \frac{1}{2}\left(\frac{bmc(r + m)}{r(r - m)}\right) = 0 \tag{48}$$

That is,

$$a = \frac{bc(r + m)}{r - m} \tag{49}$$

This value of a (0.048 for the parameters in Table 1) is the watershed between a stable spiral point and an unstable spiral point. The real parts of the eigenvalues are plotted in Fig. 16, and we see that the steady state is unstable when $a > 0.048$. In this case the trajectories spiral outward toward a limit cycle (Fig. 14). At lower values of a, they spiral inward (as shown in Fig. 12).

When the eigenvalues are outside the range specified in Eq. (46), the eigenvalues are real, the spiralling behavior disappears, and the steady state is a node. The critical value for stability of this node occurs when

$$a = \frac{bcm}{r - m} \tag{50}$$

at which point $e_1 = 0$ and $e_2 < 0$. Below this value ($a < 0.018$) the

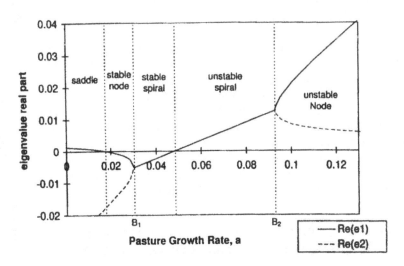

FIGURE 16 The eigenvalues e_1, e_2 of the yield–liveweight system at the steady-state point (L_*, Y_m) plotted against the pasture growth rate parameter a. Dotted lines indicate changes in stability. — Re(e1), --- Re(e2).

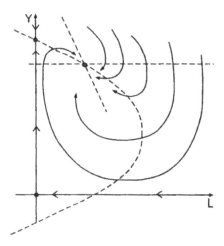

FIGURE 17   Phase plane diagram for the yield–liveweight model in the case when the steady state at (L_s, Y_m) is a stable mode.

steady state is a saddle point [and hence unstable (Edelstein-Keshet, 1987)]. In any case, we have already noted in Section IV.A that this situation ($Y_m > Y_{max}$) is biologically less interesting.

When a lies between the Eq. (50) value and B_1 [Eq. (46)] (i.e., $0.018 < a < 0.030$), the steady state is a stable node. The system does not oscillate on the way to the steady state (Fig. 17). When $a > B_2$ we also have an unstable node. The sketch of the phase plane in this case is left to the reader (this will include a limit cycle, as in Fig. 14).

The behavior of the eigenvalues and thus the steady-state point is summarized in Fig. 16. Although eigenvalue anaysis is technical, it does highlight critical sensitivities of the system, and we can see how one parameter (we have chosen growth rate a) affects the stability of the whole system.

At this point we leave our extended analysis of the yield–liveweight model and go on to look at an example of using a dynamic grazing model to calculate an optimal management strategy.

V.   OPTIMAL STOCKING ON A SEASONAL PATTERN OF PRODUCTION: USING OPTIMAL CONTROL THEORY

Agricultural systems are biological systems under intensive management. The farmer exerts a significant degree of control over the system. In a grazing system, the farmer is able to change the stocking rate, apply fertilizer or water, mow the pasture, sow seed, feed out supplements, and so on. All of these are managed controls.

Many agricultural problems involve identifying better control strategies. How should a farmer change the stocking density or grazing management to maximize animal growth? How much fertilizer should be applied, and when, to maximize pasture response? How much excess pasture should be conserved, and when should it be fed out? Optimal control theory allows us to take a dynamic systems model of a system, define some quantity that we wish to maximize, and then calculate the optimal management strategy to maximize this quantity. In the language of control, the quantity to be maximized is called the objective function, the management variable is called the control variable, and the system dynamics equations are called the state equations.

Control theory is well established in physics and economics and is equally useful for biological applications. An excellent introductory text to the application of optimal control methods in bioeconomic management is *Mathematical Bioeconomics* by Colin Clark (Clark, 1990), which gives simple methods and examples for calculating the optimal harvesting strategies for fisheries, which are biological systems not unlike agricultural systems. Here, we work through a simple problem in grazing management as an example of how the methodology works.

We take the example of calculating the optimal stocking rate (adjusted daily) to maximize production per hectare (measured as animal intake) over a year.

A. Formulation

The first requirement is a model of the system. The pasture model from our yield–liveweight system is suitable [Eq. (6)], although here we assume (for simplicity) that liveweight is constant with time, as might approximately be the case in a dairy farm, and so the parameter K is the potential daily intake rate *per animal*.

$$\frac{dY}{dt} = a(t)Y - bY^2 - n(t)\frac{KY}{c + Y} \tag{51}$$

It is clear that growth rate $a(t)$ and stocking rate $n(t)$ may change with time. Our objective is to maximize the total amount that animals eat in one year, which we assume is equivalent to maximizing the annual production per hectare. At any one time the intake rate per hectare is $n(t)KY/(c + Y)$, so the objective function J is

$$J = \int_0^{365} n(t)\, \frac{KY(t)}{c + Y(t)}\, dt \tag{52}$$

which we want to maximize by choosing the stocking rate $n(t)$. An additional constraint is that the stocking rate must be greater than or equal to zero: $n(t) \geq 0$.

B.   The Hamiltonian

The method of optimal control is as follows. We form a special function H called the Hamiltonian of the problem. This is done by appending the state equation to the objective function using a Lagrange multiplier, $\lambda(t)$. $\lambda(t)$ is called the adjoint variable.

$$H = n(t)\, \frac{KY}{c + Y} + \lambda(t) \left(a(t)Y - bY^2 - n(t)\, \frac{KY}{c + Y} \right) \tag{53}$$

The optimal stocking rate $n(t)$ is found when either the partial derivative of H with respect to $n(t)$ is zero or $n(t)$ itself is zero. In addition, the partial derivative of H with respect to $Y(t)$, the state variable, must also be zero. These partial derivatives are

$$\frac{\partial H}{\partial n} = \frac{KY}{c + Y} + \lambda \left(- \frac{KY}{c + Y} \right) = 0, \qquad n(t) > 0 \tag{54}$$

$$\frac{\partial H}{\partial Y} = n(t)\, \frac{Kc}{(c + Y)^2} + \lambda \left(a(t) - 2bY - n(t)\, \frac{Kc}{(c + Y)^2} \right) = 0 \tag{55}$$

When $n(t) > 0$, we find from Eq. (54) that $\lambda = 1$. Substituting $\lambda = 1$ into Eq. (55) and solving then tells us that

$$Y_{\text{opt}} = \frac{a(t)}{2b}, \qquad n(t) > 0 \tag{56}$$

This is the optimal pasture level to maximize animal intake. It follows the seasonal growth rate pattern, $a(t)$.

This particular problem turns out to be very simple. To maximize the amount that the animals eat per hectare, we should maintain the pasture at that level where growth rate is maximized. For a logistic growth function such as here, $Y_{\text{opt}} = Y_{\text{max}}/2$. The result that production is maximized when pasture growth rate is maximum has also been observed by Parsons and Johnson (1986), Morley (1968), and Chen and Wang (1988). Chen and Wang's (1988) paper used the more complex pasture model of Johnson and Parsons (1985) to determine the optimal stocking policy for a limited growing season.

Readers are referred to it as an example of optimal control theory being applied to a complex model.

C. Optimal Stocking Rate

What is the stocking rate required to maintain this optimal pasture level [Eq. (56)]? The pasture mass is determined by the dynamics in Eq. (51). Substituting the optimal pasture level Y_{opt} [Eq. (56)] into the pasture dynamics equation, Eq. (51) gives

$$\frac{dY_{opt}}{dt} = a(t)Y_{opt} - bY_{opt}^2 - n(t)\frac{KY_{opt}}{c + Y_{opt}}$$

$$\Rightarrow \frac{a'(t)}{2b} = a(t)\frac{a(t)}{2b} - b\left(\frac{a(t)}{2b}\right)^2 - n(t)\frac{K[a(t)/2b]}{c + a(t)/2b}, \qquad n(t) > 0$$

$$(57)$$

We rearrange this to find the stocking rate, $n(t)$, that holds pasture mass at the optimal level for maximum production:

$$n(t) = \frac{a(t)^2/4b - a'(t)/2b}{[Ka(t)/2b]/[c + a(t)/2b]}, \qquad n(t) > 0 \qquad (58)$$

Notice that the optimal stocking rate depends on $a'(t)$, which is the rate of change of pasture growth rate with time. This is because we need to anticipate changes in pasture growth in order to adjust stocking rate in advance to allow the pasture to remain at the optimal level.

Expressing $a(t)$ as a sine function facilitates a numerical example. Figure 18 shows a typical course of Y_{opt} and the optimal $n(t)$ calculated from Eq. (58)

It is possible that Eq. (58) might suggest that the stocking rate $n(t)$ should be negative. This occurs when pasture growth rate is increasing rapidly. In this case we take $n(t) = 0$, so that Eq. (58) is not needed and we do not need to find the optimal stocking rate–it is zero. It is then more difficult, however, to calculate the optimal pasture level. The model recommends removing all of the animals to allow pasture mass to grow to the optimal level, which itself is increasing. The animals are reintroduced when the pasture level achieves the optimal level recommended by Eq. (56). So in practice we cannot achieve the optimal pasture mass at all points in time, because we cannot have $n(t) < 0$. We are forced to accept a suboptimal pasture level during those periods when growth rate is increasing rapidly.

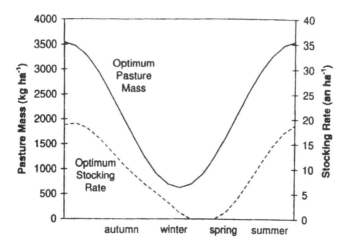

FIGURE 18 Optimal pasture mass for maximum growth, and optimal stocking density to hold pasture as near to this level as possible. In spring, the actual pasture may be suboptimal for a time.

The recommendation is that when pasture growth rates are increasing, for instance in the spring flush, one should withdraw animals for a short time so that the pasture can get to its maximum growth rate level. At this point, animals are reintroduced. Farming practice in early spring often follows this line—farmers hold back from increasing grazing pressure in the early spring in order to allow the pasture mass to accumulate.

However, this recommendation alerts us to some deficiencies in the model. First, there is no consideration of pasture quality. In the late spring we need to graze hard in order to maintain high green/dead ratios in the sward. Second, the model has no limitation to the rate at which animals may be added to or removed from the system. In a real farm this will be limited. Further constraints could be introduced to reflect the number of animals available to the farmer.

VI. A GREEN–DEAD MODEL FOR OPTIMIZING PASTURE QUALITY

Up to this point we have considered pasture models that express pasture using a single variable, pasture biomass Y. However, milk production in New Zealand dairy systems is adversely affected by summer–autumn buildup of dead leaf in the pasture sward. A biomass model is inadequate in this context, as it cannot address issues of pasture quality.

A. Mechanistic Modeling of Pasture Rate of Growth

A more appropriate model considers the coupled accumulation of green and dead leaf mass in a grazed pasture under the influence of selective grazing (Fig. 19). Define G as the herbage mass (kg/ha) of green leaf and stem in a ryegrass–white clover pasture, and D as the above-ground mass (kg/ha) of senescent leaf and stem in the pasture. The mixture of green leaf and dead leaf is in constant flux, with new leaves continually being formed and old ones dying.

There are four major processes at work: decay, senescence, new growth, and intake (Fig. 19). The experiment of Yates (1982) showed that the rate of dead matter disappearance (which is largely due to the action of earthworms and microorganisms) is approximately proportional to the amount of dead matter in the sward. So we can model the rate of decay of dead leaf D as some proportion δ of D per day. Similarly, evidence from Bircham and Hodgson (1983) in swards continuously grazed by sheep and the subsequent analyses by Woodward (1993) and Woodward and Wake (1994) showed that the rate of senescence of green pasture components may also be modeled as being linearly related to the amount of pasture, especially in the case of continuous grazing management. So we take the rate of senescence of green matter G into dead matter D as being some proportion σ of G per day.

The rate of new growth of green leaf is proportional to the rate of carbon assimilation by photosynthesis, which is in turn proportional to the quantity of light intercepted by the leaves. The amount of light intercepted is calculated from Beer's law as a Mitscherlich dependence on leaf area (Fig. 20) (Davidson and Philip, 1958; Johnson et al., 1989). The leaf area is roughly proportional to the leaf mass, G (e.g., Parsons et al., 1994). Since dead leaf in the sward tends to be

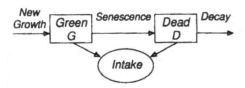

FIGURE 19 Flow diagram of material in a pasture model with green and dead matter components.

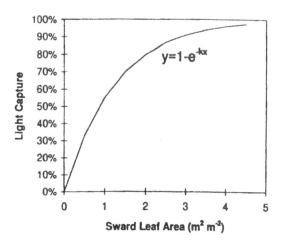

FIGURE 20   Efficiency of capture of light by a sward canopy.

located in the lower sward strata at most times of the year, its effect in intercepting light may be assumed to be negligible, and the rate of new growth of green leaf may be directly related to green leaf mass by a Mitscherlich curve. A more sophisticated model might attempt to include the effect of light interception by dead matter in the sward.

We approximate this relationship by a Michaelis–Menten curve (for ease of analysis):

$$\text{Rate of new growth} = \frac{g_{max}G}{q + G} \text{ kgDM}/(\text{ha}\cdot\text{day}) \tag{59}$$

where g_{max} is the maximum rate of new growth, which may vary seasonally with changes in available light, temperature, and moisture, and q is the herbage mass at which 50% of incident light is captured by the sward. The net rate of green leaf accumulation dG/dt in the absence of grazing is then modeled as

$$\underset{\text{Net accumulation}}{\frac{dG}{dt}} = \underset{\text{Growth}}{\frac{g_{max}G}{q + G}} - \underset{\text{Senescence}}{\sigma G} \tag{60}$$

where senescence is now treated explicitly. Setting $dG/dt = 0$ in Eq. (60) shows that G has a ceiling yield when $G = g_{max}/\sigma - q$. We have a similar differential equation for the rate of change of dead mass, dD/dt:

$$\frac{dD}{dt} = \sigma G - \delta D \qquad\qquad (61)$$

Net change in D Senescence   Decay

Equations (60) and (61) taken together form a two-dimensional dynamical system modeling the fluxes of green and dead leaf mass in regrowing pasture in the absence of grazing animals.

B. Daily Intake

To analyze the effect of selective removal of green and dead matter by animals on the green–dead balance of the sward, we need to predict the daily intake of green leaf mass I_g and dead leaf mass I_d as functions of the sward variables. Daily removal of pasture by grazing animals is the sum of a number of processes, from the rate and size of individual bites up to the length of time animals spend on the activities of grazing, ruminating, and idling each day. During this process the animal is continuously making a complex hierarchy of decisions that determine what activity it will perform and where it will graze (Fig. 21). These decisions are influenced by pasture composition and distribution and also by animal variables (Hodgson, 1985) such as physiological state (e.g., lactation), gut fill, nutrient and energy balance, body temperature, and group size.

Every bite that an animal ingests must be processed by mastication and rumination. We may suppose that the time required for this processing is proportional to the mass and quality of the ingested material (Laca et al., 1994; Penning et al., 1991). Rook et al. (1994)

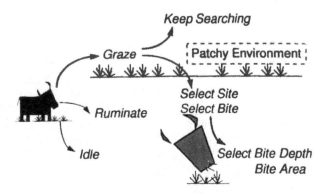

FIGURE 21   The hierarchy of decisions faced by a grazing animal.

have shown that animals need to spend a relatively fixed amount of time each day in nongrazing activities, although this is modified by an animal's physiological state and feeding history (e.g., whether the animal has been fasted).

Based on these ideas, Woodward (in press) constructed the following daily time budget for an animal's activities:

$$N[t_p + S(t_m + t_r)] + t_i = 1440 \text{ min} \tag{62}$$

where 1440 is the number of minutes in 24 h, N is the number of bites taken in a day, t_p is the time required for prehension of each bite (min/bite), S is the bite mass (kg/bite), t_m and t_r are the times required to masticate and ruminate one unit of material, respectively (min/kg), and t_i is the time spent by the animal each day in "idling" (min/day), i.e., performing nongrazing activities. t_m and t_r can be expected to change with the quality of the forage ingested. Values for these time parameters (t_m, t_r, t_i) for cattle and sheep are available from experimental studies (Woodward, in press). This approach is similar to that used by Spalinger and Hobbs (1992) and others to calculate feed intake rate.

Equation (62) constrains the number of bites N an animal takes in a day. Bite size S is constrained by the pasture availability and animal potential intake rate. Since daily intake is the product of number of bites per day and bite mass, Eq. (62) then enables us to calculate daily intake I (kg/day) as a function of bite size:

$$I = NS = \frac{1440 - t_i}{t_p + S(t_m + t_r)} S \tag{63}$$

This gives a Michaelis–Menten relationship between daily intake and bite size that agrees with experimental and other theoretical observations (Spalinger and Hobbs, 1992; Black and Kenney, 1984). Previous relationships have been based purely on experimental data, but this model [Eq. (63)] is mechanistic, based on the processing time (prehension, mastication, and rumination) for ingested material.

If the composition of each bite consisted of green and dead components, so that the total bite mass S was the sum of S_g and S_d, the time taken to process each bite would then be

$$t_p + S_g(t_{m,g} + t_{r,g}) + S_d(t_{m,d} + t_{r,d}) \tag{64}$$

and the total number of bites per day [constrained by Eq. (62)] would be modified accordingly. Then we would have the daily intake of green leaf mass, I_g, as

$$I_g = NS_g = \frac{1440 - t_i}{t_p + t_g S_g + t_d S_d} S_g \tag{65}$$

and a similar expression for the daily intake of dead leaf mass, I_d where t_g and t_d are the bracketed expressions from Eq. (64), that is, the total processing time (mastication and rumination) per kilogram of green or dead leaf eaten, respectively.

C. Selection of Bites

The challenge is to predict bite size and composition from the sward variables. This is complicated owing to one significant problem: herbage is not uniformly distributed in a pasture, and the mechanisms animals use to select which patch they will graze are still poorly understood. Animals use the patchy distribution of pasture mass and composition to their advantage by selecting a diet of higher quality than the average quality of the pasture and in doing so they create further patchiness. It has been suggested that within the constraints of their speed and size, animals forage to obtain an optimal diet from their environment (Newman et al, 1994). It may be that this theory of "optimal foraging" will provide the way forward. However, there are still many obstacles, not the least of which is the lack of suitable statistics for characterizing the spatial distribution of plant material in a pasture. The question of how one predicts the size, rate, and composition of an animal's bites when it is selectively grazing in an environment with a nonuniform spatial distribution of pasture mass and components is still very much open.

Animals prefer green mass, and the more green mass present, the more they will eat. However, increasing levels of dead mass will depress green intake, as more time is spent in discrimination. Similarly, the more dead leaf in the sward there is, the more will be eaten (assuming a principally green, vegetative sward). But increasing levels of green mass result in less dead matter being eaten as animals can more easily select the green leaf in the sward surface layers.

A prototypical approach might be to consider that the average bite size of each component is proportional to the availability (herbage mass) of that component, i.e.,

$$S_g = \alpha_g G, \qquad S_d = \alpha_d D \tag{66}$$

where α_g and α_d are constants (ha/bite). Such linear relationships may be derived from encounter and rejection arguments of selective foraging as described in Parsons et al. (1994). If $Y = G + D$ is the total pasture yield and $P_s = D/Y$ is the proportion of dead matter in the sward, then the total bite size S and proportion of dead mass in the diet P_d are predicted to be

$$S = S_g + S_d = \alpha_g G + \alpha_d D = [\alpha_g(1 - P_s) + \alpha_d P_s]Y \tag{67a}$$

and

$$P_d = \frac{S_d}{S} = \frac{\alpha_d D}{\alpha_g G + \alpha_d D} = \frac{\alpha_d P_s Y}{\alpha_g(1 - P_s)Y + \alpha_d P_s Y}$$

$$= \frac{[\alpha_d/(\alpha_g - \alpha_d)]P_s}{[\alpha_g/(\alpha_g - \alpha_d)] - P_s} \tag{67b}$$

This simple model gives predictions of total bite size and composition consistent with the experimental observations of Laca et al. (1992) that total bite size increases linearly with herbage mass (or height) and of Rattray and Clark (1984) and Parsons and Penning (1993) that animals select a diet of higher quality than the average quality of the pasture (Fig. 22).

If for convenience we define $t_t = 1440 - t_i$ as the total nonidling time in Eq. (65), then substituting the component bite sizes from Eq. (66) into Eq. (65) gives predictions for the daily intake of green mass I_g and dead mass I_d, respectively,

$$I_g = \frac{t_t \alpha_g G}{t_p + t_g \alpha_g G + t_d \alpha_d D} \tag{68}$$

and

$$I_d = \frac{t_t \alpha_d D}{t_p + t_g \alpha_g G + t_d \alpha_d D} \tag{69}$$

These are simple formulae for the daily intake of green and dead leaf mass by a selectively grazing animal and may be appended to the green–dead dynamics equations [(60) and (61)] in order to analyze the equilibrium green/dead ratio in a grazed sward.

D. Green/Dead Ratio of a Grazed Sward in Equilibrium

We now have a simple model of green and dead pasture fluxes in a pasture grazed at a stocking density of n animals per hectare:

$$\frac{dG}{dt} = \frac{g_{max}G}{q + G} - \sigma G - n\frac{t_t \alpha_g G}{t_p + t_g \alpha_g G + t_d \alpha_d D} \tag{70}$$

$$\frac{dD}{dt} = \sigma G - \delta D - n\frac{t_t \alpha_d D}{t_p + t_g \alpha_g G + t_d \alpha_d D} \tag{71}$$

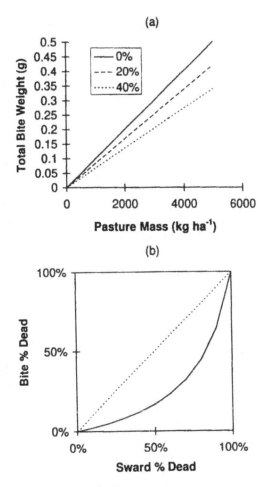

FIGURE 22 Predictions of (a) total bite size and (b) composition when the bite size of each component is proportional to the availability of that component.

We proceed with a basic phase plane analysis of this prototype model for the purpose of illustrating what information might be available from a pasture composition model of this type over a simple biomass model. Typical values of the parameters in this model are given in Table 3.

Under continuous grazing at a given stocking density n, we expect a sward to reach an equilibrium balance between intake and

Table 3 Typical Parameter and Variable Values in Early Spring for the Green–Dead Pasture Model Grazed by Steers

Parameter	Description
$g_{max} = 180$ kg/(ha·day)	Maximum pasture rate of new growth
$q = 1000$ kg/ha	Pasture mass at which 50% of light is captured
$\sigma = 0.03$ kg/(kg·day)	Relative rate of senescence
$t_p = 0.02$ min/bite	Prehension time for cattle
$t_{m,g}$, $t_{m,d} = 9.5$ min/kg	Mastication time for green or dead leaf mass
$r_{t,g} = 24.5$ min/kg	Rumination time for green leaf mass
$t_{r,d} = 35$ min/kg	Rumination time for dead leaf mass
$t_g = 34$ min/kg	Total processing time for green leaf mass
$t_d = 44.5$ min/kg	Total processing time for dead leaf mass
$t_i = 650$ min	Daily idling time for cattle
N, bites/day	Number of bites per day for cattle
S, kg/bite	Total bite size
$t_t = 790$ min	Daily nonidling time
$\alpha_g = 1 \times 10^{-7}$ ha/bite	Bite size constant for green leaf mass
$\alpha_d = 2 \times 10^{-8}$ ha/bite	Bite size constant for dead leaf mass
I_g, kg/(animal·day)	Daily intake of green mass
I_d, kg/(animal·day)	Daily intake of dead mass
$\delta = 0.05$ kg/(kg·day)	Relative rate of litter decay

growth. The green/dead ratio at this equilibrium may well be influenced by the stocking density chosen, so a farmer may be able to choose a stocking density that gives an improvement in pasture quality.

The green/dead ratio at equilibrium can be studied by using phase plane analysis (recall Section IV). However, the isoclines for this system are complicated. There is one isocline along the straight line $G = 0$, where there is no green leaf and the amount of dead leaf decays away to zero. The other isoclines are hyperbolas. Nevertheless, we can plot these curves numerically for the parameters given in Table 3. Figure 23 shows the isoclines of the GD phase plane for several stocking densities. The $D' = 0$ isocline, although a hyperbola, is approximately straight in the feasible portion of the phase plane (where $G \geq 0$ and $D \geq 0$). It intersects the D axis at $(0, 0)$, where there is a steady state. The slope of the $D' = 0$ isocline is approximately $G = (\delta/\sigma)D$ and is largely unaffected by stocking density.

The hyperbolic $G' = 0$ isocline is a curve whose position changes markedly as stocking density increases. When n is small, the isocline

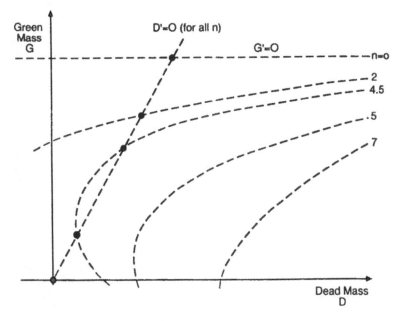

FIGURE 23 Isoclines of the green–dead pasture model. The diagonal line shows the $D' = 0$ isocline (which does not change much with stocking density n), and the dotted curves show the $G' = 0$ isocline for the various stocking rates. The horizontal axis is also a $G' = 0$ isocline.

is a gentle curve asymptotic to the line $G = g_{max}/\sigma - q$ (i.e., the ceiling yield of green), and there is a single stable steady state where this curve intersects the $D' = 0$ isocline at $(D, G) = (D^*, G^*)$, where G^* is near maximum yield and D^* is approximately $\sigma G^*/\delta$. The steady state at (0,0) is unstable.

As n increases, an additional steady state may appear (see the isoclines for $n = 4.5$ in Fig. 23). In this case there is discontinuous stability (see Sections III.C and III.D), with stable steady states at (0, 0), and at (D^*, G^*) and an unstable steady state in between. Which of these two states the pasture ends up in depends on the initial values of D and G.

At higher stocking densities still (illustrated by the isoclines for $n = 5$ and $n = 7$ in Fig. 23), only one steady state remains in the feasible region of the phase plane, a stable steady state at (0, 0), and the pasture will be grazed to extinction. Figure 24 summarizes the stability of the green–dead system for a range of stocking densities.

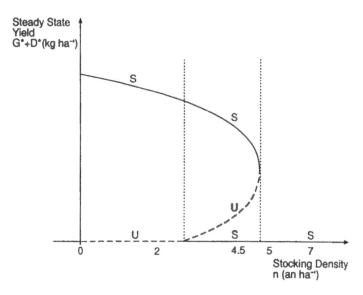

FIGURE 24   Bifurcation diagram showing the steady states of the green–dead pasture model for a range of stocking densities. The curve shows the position of the steady states, which are labeled S (stable, heavy line) or U (unstable, dotted line). The horizontal axis is also a branch of steady states.

Figure 24 is identical in structure with Fig. 10 for Noy-Meir's (1975) analysis of pasture stability using a simple biomass model. The implication of this is that a simple biomass model is likely to be adequate for the purpose of studying grazed pasture stability.

In terms of pasture quality, however, our green–dead model has yielded an additional result. We noticed from Fig. 23 that the steady states will always lie on the $D' = 0$ isocline, which has a slope of approximately δ/σ and passes through (0, 0). Therefore, at equilibrium we expect a green/dead ratio that is approximately δ/σ (5/3 for the parameters in Table 3). Since the slope of the line changes very little with stocking density, this implies that there is limited scope for improvements in pasture quality through manipulating stocking rate alone.

A possible extension might be to use the green–dead model [Eqs. (70) and (71)] to identify stocking density strategies that maximize pasture quality in the short term. However, optimal control of such multidimensional systems is in general difficult (Clark, 1990).

VI. CONCLUSION: REGRESSION, SIMULATION, OR DYNAMICAL SYSTEMS?

There is no one universal methodology that is suited to all problems. The art of applied modeling is the ability to synthesize the empirical knowledge and insight of experimental scientists into an appropriate model so that the applied problem can be answered. Dynamical systems models may be oversimplified for the purposes of coordinating diverse research efforts studying the detailed components of a complex system; however, when simple insights or robust management recommendations are required, simplicity is a virtue, and dynamical systems models come into their own—they allow a thorough analysis of the system's stability and are ideal for the purpose of calculating optimal management strategies.

The modeling of problems in grazing management is still very much a new frontier for the application of dynamical systems, and there is vast scope for further progress in this field. Despite its long history, the agriculture industry faces as many problems and opportunities today as ever in the past, and mathematical modeling is another tool in the scientist's workshop for addressing these issues. Formulating models that are both realistic enough to give meaningful answers to our questions and simple and robust enough to give reliable answers requires partnership between agricultural scientists and mathematicians. Likewise, interpretation of results is a biological and economic issue as much as it is a mathematical one.

The intention of this review has been to survey the variety of existing work in dynamical systems modeling of agricultural systems and to suggest some directions for the next step forward. Two of these have been the coupling of simple animal growth functions with pasture models and the expansion of pasture models to address pasture quality issues. An additional lack is a simple model to predict the development of animal body composition during growth.

In addition to a range of grazing applications, I have tried to introduce the basic techniques of dynamical systems analysis. I have also tried to provide references to useful books that explain the mathematical techniques. It is hoped that the models and analyses discussed in this chapter have proved thought-provoking and that the relative strengths of mathematical modeling compared to both static regression models and dynamic computer simulations are evident.

ACKNOWLEDGMENTS

Many thanks to Dr. David G. McCall and Professor Graeme C. Wake for their helpful comments and suggestions.

APPENDIX: METHODS FOR SOLVING SIMPLE DIFFERENTIAL EQUATIONS

Finding the function Y from a differential equation containing Y and its derivatives is a mysterious art. However, the study of methods for "solving" differential equations (and systems of differential equations) is actually a well-established branch of mathematics. It is worth mentioning here a few of the simpler methods for first-order ordinary differential equations like those we have been using. For more details and more sophisticated methods, the reader is directed to Zill (1989).

Two methods are very useful and straightforward to illustrate. These are the method of separable equations and the method of integrating factors. A third technique, substitution of variables, is useful in a wide range of problems and so is worth mentioning here.

A1. Separable Equations

If the equation is of the form

$$\frac{dY}{dt} = f(Y)a(t) \tag{A1}$$

where f and a are known functions of the dependent variable Y and the independent variable t, respectively, then the equation is said to be separable. For example, the so-called monomolecular equation (Fig. A1) is

$$\frac{dY}{dt} = a(t)\left(1 - \frac{Y}{Y_{max}}\right) \tag{A2}$$

where $a(t)$ is the time-varying potential growth rate. Note that it is often not necessary to fix the form of the function $a(t)$ in advance.

The method is to rearrange the equation so that all of the Y's appear on the left and all of the t's on the right. For this example, we divide by $f(Y)$ and multiply by dt as follows:

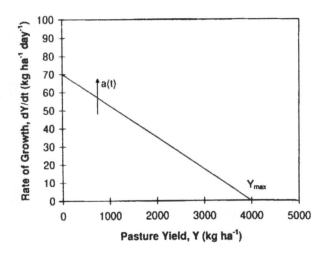

FIGURE A1 The monomolecular growth function.

$$\frac{dY}{1 - Y/Y_{max}} = a(t)dt \tag{A3}$$

Then we attempt to integrate both sides (this is the crunch of whether this method will work).

$$\int \frac{dY}{1 - Y/Y_{max}} = \int a(t)dt \tag{A4}$$

Remember that when we integrate this we must include an unknown constant (the constant of integration), say C, on one side.

$$-Y_{max} \ln\left(1 - \frac{Y}{Y_{max}}\right) + C = \int_0^t a(t)dt \tag{A5}$$

The value of this constant is determined by substituting the known initial conditions [$t = 0$, $Y(t) = Y(0)$] into the solution:

$$-Y_{max} \ln\left(1 - \frac{Y(0)}{Y_{max}}\right) + C = 0 \tag{A6}$$

This gives us the value of C in terms of the other parameters. This value is substituted into Eq. (A5), and the solution is rearranged (if possible) for $Y(t)$. After some algebra we obtain the solution

$$Y(t) = Y_{max} - [Y_{max} - Y(0)] \exp\left[-\frac{1}{Y_{max}} \int_0^t a(t)dt\right] \tag{A7}$$

A2. Integrating Factors

A second method is useful when we have equations of the form

$$\frac{dY}{dt} = a(t)Y + g(t) \tag{A8}$$

where a and g are known parameters (that may vary with time if desired). For example, consider Bircham and Sheath's (1986) model for disappearance of pasture under grazing:

$$\frac{dY}{dt} = g_0 - knY \tag{A9}$$

where g_0 is a constant pasture growth rate. The method is first to move the term involving Y to the left:

$$\frac{dY}{dt} + knY = g_0 \tag{A10}$$

Now we construct an integrating factor IF using the coefficient (kn) of Y on the left-hand side:

$$\text{IF} = \exp(knt) \tag{A11}$$

This strange formula is chosen because the derivative of IF is now

$$\frac{d\text{IF}}{dt} = kn \times \text{IF} \tag{A12}$$

Note: The general formula for the integrating factor for Eq. (A8) is

$$\text{IF} = \exp\left[-\int a(t)dt \right] \tag{A13}$$

So we multiply both sides of Eq. (A10) by this integrating factor and then collapse the left-hand side (using the reverse of the product rule of differentiation):

$$\frac{dY}{dt} \times \text{IF} + knY \times \text{IF} = g_0 \times \text{IF} \tag{A14a}$$

$$\frac{d}{dt}(Y \times \text{IF}) = g_0 \times \text{IF} \tag{A14b}$$

$$\frac{d}{dt}(Y \exp(knt)) = g_0 \exp(knt) \tag{A14c}$$

Now we integrate both sides, remembering the constant of integration, C.

$$Y \exp(knt) + C = \int_0^t g_0 \exp(kns)ds = \frac{g_0}{kn} [\exp(knt) - 1] \qquad \text{(A15)}$$

Substitute in the initial conditions to find the value of C. In this case assume that our initial conditions are $Y = Y(0)$ when $t = 0$.

$$Y(0) + C = 0 \qquad \text{(A16)}$$

C is then eliminated from Eq. (A15), and both sides are divided by the integrating factor to give

$$Y(t) = Y(0)\exp(-knt) + \frac{g_0}{kn} [1 - \exp(-knt)] \qquad \text{(A17)}$$

This is the solution to the Bircham and Sheath model.

A3. Substitution of Variables

A third "trick" is also very useful. It involves changing the variables. A particularly nice example is a form of the Ricatti equation (Birkhoff and Rota, 1989):

$$\frac{dY}{dt} = a(t)Y + b(t)Y^2 \qquad \text{(A18)}$$

This is a first-order nonlinear equation. As it stands, it is not solvable by either of the two methods mentioned above. It is not separable and is not first-order linear (and therefore not permeable to the integrating factor method). However, we may try a substitution. One that works for this equation is to substitute for the reciprocal of Y; i.e., try

$$Y = \frac{1}{Z} \Rightarrow \frac{dY}{dt} = \frac{-1}{Z^2} \frac{dZ}{dt} \qquad \text{(A19)}$$

We substitute these into the differential equation, which then becomes

$$\frac{-1}{Z^2} \frac{dZ}{dt} = a(t) \frac{1}{Z} + b(t) \frac{1}{Z^2}$$

$$\Rightarrow \frac{dZ}{dt} = -a(t)Z - b(t) \qquad \text{(A20)}$$

This is now a first-order linear equation and may be solved for $Z(t)$ by the method of integrating factors. The reciprocal of the solution $Z(t)$ is the solution of the original equation, $Y(t)$. For example, this substitution could be used to find the solution of the seasonal model in Eq. (1).

Choosing the right substitution is more a matter of art than science. Common substitutions are logarithms (ln Z), trigonometric functions (sin kZ, cos kZ, tan kZ), exponentials (e^{kZ}), and power functions (Z^n), where k and n are constants.

REFERENCES

Bircham, J. S., and J. Hodgson. 1983. The influence of sward condition on rates of herbage growth and senescence in mixed swards under continuous stocking management. *Grass Forage Sci.* 38:323–331.

Bircham, J. S., and G. W. Sheath. 1986. Pasture utilization in hill country: 2. A general model describing pasture mass and intake under sheep and cattle grazing. *N. Z. J. Agric. Res.* 29:639–648.

Birkhoff, G., and G. C. Rota. 1989. *Ordinary Differential Equations.* 4th ed. New York: Wiley.

Black, J. L., and P. A. Kenney. 1984. Factors affecting diet selection by sheep: 2. Height and density of pasture. *Aust. J. Agric. Res.* 35:565–578.

Brougham, R. W. 1956. The rate of growth of short-rotation ryegrass pastures in the late autumn, winter, and early spring. *N. Z. J. Sci. Technol.* A38:78–87.

Caughley, G., and J. H. Lawton. 1981. Plant herbivore systems. In: *Theoretical Ecology: Principles and Applications.* 2nd ed. R. M. May (Ed.). Blackwell, pp. 132–166.

Char, B. W., K. O. Geddes, G. H. Gonnet, B. L. Leong, M. B. Monagan, and S. M. Watt. 1992. *First Leaves: A Tutorial Introduction to Maple V.* New York: Springer-Verlag.

Chen, J. L., and Q. Wang. 1988. A theoretical analysis of the potential productivity of ryegrass under grazing. *J. Theor. Biol.* 133:371–383.

Clark, C. W. 1990. *Mathematical Bioeconomics: The Optimal Management of Renewable Resources.* New York: Wiley-Interscience.

Crawley, M. J. 1983. *Herbivory: The Dynamics of Plant–Animal Interactions.* London: Blackwell.

Davidson, J. L., and J. R. Philip. 1958. Light and pasture growth. Proceedings of UNESCO Symposium "Climatology and Microclimatology," Canberra, 1956. *Arid Zone Res.* 11:181–187.

Edelstein-Keshet, L. 1987. *Mathematical Models in Biology.* New York: McGraw-Hill.

Hodgson, J. 1985. The control of herbage intake in the grazing ruminant. *Proc. Nutri. Soc.* 44:339–346.

Johnson, I. R., and A. J. Parsons. 1985. A theoretical analysis of grass growth under grazing. *J. Theor. Biol.* 112:345–367.

Johnson, I. R., A. J. Parsons, and M. M. Ludlow. 1989. Modelling photosynthesis in monocultures and mixtures. *Aust. J. Plant Physiol.* 16:501–516.

Laca, E. A., E. D. Ungar, N. Seligman, and M. W. Demment. 1992. Effects of sward height and bulk density on bite dimensions of cattle grazing homogeneous swards. *Grass Forage Sci.* 47:91–102.

Laca, E. A., E. D. Ungar, and M. W. Demment. 1994. Mechanisms of handling time and intake rate of a large mammalian herbivore. *Appl. Anim. Behav. Sci.* 39:3–19.

McCall, D. G., R. J. Townsley, J. S. Bircham, and G. W. Sheath. 1986. The interdependence of animal intake, pre- and post grazing pasture mass and stocking density. *Proc. N. Z. Grassland Assoc.* 47:255–261.

May, R. M. 1971. Stability in model ecosystems. *Proc. Ecol. Soc. Aust.* 6:17–56.

Morley, F. H. W. 1966. Stability and productivity of pastures. *Proc. N. Z. Soc. Anim. Prod.* 26:8–21.

Morley, F. H. W. 1968. Pasture growth curves and grazing management. *Aust. J. Exp. Agric. Anim. Husbandry* 8:40–45.

Newman, J. A., A. J. Parsons, J. H. M. Thornley, P. D. Penning, and J. R. Krebs. 1994. Optimal diet selection by a generalist grazing herbivore. *Funct. Ecol.* 9:255–268.

Noy-Meir, I. 1975. Stability of grazing systems: An application of predator–prey graphs. *J. Ecol.* 63:459–481.

Noy-Meir, I. 1976. Rotational grazing in a continuously growing pasture: A simple model. *Agric. Syst.* 1:87–112.

Noy-Meir, I. 1978a. Grazing and production in seasonal pastures: Analysis of a simple model. *J. Appl. Ecol.* 15:809–835.

Noy-Meir, I. 1978b. Stability in simple grazing models: Effects of explicit functions. *J. Theor. Biol.* 71:347–380.

Parsons, A. J., and I. R. Johnson. 1986. The physiology of grass growth under grazing. In: *Grazing*. J. Frame (Ed.). British Grasslands Society, Reading, UK, pp. 3–13.

Parsons, A. J., and P. D. Penning. 1988. The effect of the duration of regrowth on photosynthesis, leaf death and the average rate of growth in a rotationally grazed sward. *Grass Forage Sci.* 43:15–27.

Parsons, A. J., and P. D. Penning. 1993. Plant/animal interactions. In: Institute of Grassland and Environmental Research (IGER) Annual Report 1993, pp. 93–100.

Parsons, A. J., J. H. M. Thornley, J. Newman, and P. D. Penning. 1994. A mechanistic model of some physical determinants of intake rate and diet selection in a two-species temperate grassland sward. *Funct. Ecol.* 8:187–204.

Penning, P. D., A. J. Parsons, R. J. Orr, and T. T. Treacher. 1991. Intake and behavior responses by sheep to changes in sward characteristics under continuous stocking. *Grass Forage Sci. 46*:15–28.

Press, W. H., B. P. Flannery, S. A. Teukolsky, and W. T. Vetterling. 1989. *Numerical Recipes in Pascal: The Art of Scientific Computing*. New York: Cambridge Univ. Press.

Rattray, P. V., and D. A. Clark. 1984. Factors affecting the intake of pasture. *N. Z. Agric. Sci. 18*(3):141–146.

Rook, A. J., C. A. Huckle, and P. D. Penning. 1994. Effects of sward height and concentrate supplementation on the ingestive behavior of spring-calving dairy cows grazing grass-clover swards. *Appl. Anim. Behav. Sci. 40*: 101–112.

Sheath, G. W., and P. V. Rattray. 1985. Influence of pasture quantity and quality on intake and production of sheep. Proceedings of the XV International Grassland Congress, Kyoto, pp. 1131–1133.

Snaydon, R. W. 1981. The ecology of grazed pastures. In: *Grazing Animals*. F. H. W. Morley (Ed.). New York: Elsevier, pp. 13–31.

Solé, R. V., and J. Bascompte. 1995. Measuring chaos from spatial information. *J. Theor. Biol. 175*:139–147.

Spalinger, D. E., and N. T. Hobbs. 1992. Mechanisms of foraging in mammalian herbivores: New models of functional responses. *Am. Nat. 140*(2): 325–348.

Stocker, M., and C. J. Walters. 1984. Dynamics of a vegetation–ungulate system and its optimal exploitation. *Ecol. Model. 25*:151–165.

Thornley, J. H. M. 1990. A new formulation of the logistic growth equation and its application to leaf area growth. *Ann. Bot. 66*:309–311.

Thornley, J. H. M., A. J. Parsons, J. Newman, and P. D. Penning. 1994. A cost–benefit model of grazing intake and diet selection in a two-species temperate grassland sward. *Funct. Ecol. 8*:5–16.

Ungar, E. D., and I. Noy-Meir. 1988. Herbage intake in relation to availability and sward structure: Grazing processes and optimal foraging. *J. Appl. Ecol. 25*:1045–1062.

Woodward, S. J. R. 1993. A dynamical systems model for optimizing rotational grazing. Ph.D. Thesis. Massey University, Palmerston North, New Zealand.

Woodward, S. J. R. (in press) Formulae for predicting animals' daily intake of pasture and grazing time from bite weight and composition. *Livestock Production Science*, accepted.

Woodward, S. J. R., and G. C. Wake. 1994. A differential-delay model of pasture accumulation and loss in controlled grazing systems. *Math. Biosci. 121*:37–60.

Woodward, S. J. R., G. C. Wake, A. B. Pleasants, and D. G. McCall. 1993a. A simple model for optimizing rotational grazing. *Agric. Syst. 41*:123–155.

Woodward, S. J. R., G. C. Wake, D. G. McCall, and A. B. Pleasants. 1993b. Use of a simple model of continuous and rotational grazing to compare herbage consumption. Proceedings of the XVII International Grassland Congress, 8–21 Feb. 1993, Palmerston North, New Zealand, pp. 788–789.

Woodward, S. J. R., G. C. Wake, and D. G. McCall. 1995. Optimal grazing of a multi-paddock system using a discrete time model. *Agric. Syst.* 48:119–139.

Yates, J. J. 1982. A technique for estimating rate of disappearance of dead herbage in pasture. *Grass Forage Sci.* 37:249–252.

Zill, D. G. 1989. *A First Course in Differential Equations with Applications.* 4th ed. Boston: PWS-Kent.

12

Modeling and Simulation in Applied Livestock Production Science

Jan Tind Sørensen

Danish Institute of Animal Science, Tjele, Denmark

I. INTRODUCTION

Systems-oriented research approaches have been given increased emphasis in publicly funded agricultural research. The underlying message seems to be that research should consider the detrimental implications (side effects and/or delayed effects) of research innovations that were formerly often overlooked.

Systems-oriented research is a challenge for the applied livestock production scientist. The animal scientist was trained to do experimental research and has typically not studied systems theory and model development. Although there is a history of modeling biological and physiological systems in animal science (Mertens, 1977; Baldwin and Hanigan, 1990), the technique has had limited use at higher levels of aggregation such as applied livestock production science.

The purpose of this chapter is to give an overview of modeling as a systems research method and to describe how computer simulation models can be used in livestock production research and development.

II. MODELING AS A SYSTEMS RESEARCH METHOD

Systems theory was introduced by the biologist Ludvig von Berta-lanffy in the 1940s (Bertalanffy, 1951). He found that different systems (plants, animals, societies, and ecosystems) could be described in general terms as "general systems theories" (Bertalanffy, 1973). Being holistic and general at the same time presents a paradox. As one of the "fathers" of general systems theory (Boulding, 1956) said, "All that we can say about practically everything is almost nothing." The relevance of systems theory is closely linked to the word "model." Systems theory may become clearer if we say as Le Moigne (1989) did that "systems theory is the theory of modelling."

A system is a part of reality (physical or organizational) that is separable from the environment. What is inside a conceptual border is part of the system, and what is outside the border is the environment. Most systems experience exchanges (for example, energy or information) with the environment; these exchanges are called inputs and outputs. When we want to describe systems or simply to communicate our thoughts concerning systems, we create a model of the system. A model is a simplification of the system and expresses the system from a certain perspective and with a certain purpose. A more detailed discussion of model definition in livestock production is given by Spedding (1988).

An important property of a system is that it can be defined within a hierarchy of systems (Rountree, 1977). A livestock farm as a hierarchy of systems is described in Fig. 1. The farm may include different "flocks" of animals (dairy cows, goats, swine, etc.) and different crops. Each of these subsystems consists of its own subsystems. The dairy herd, for example, consists of individual animals. In each of the animals, organs can be identified, each composed of tissue, cells, and organelles. Each level in the hierarchy has its own characteristic rules of input/output transformations.

$n+1$	The livestock farm
n	The dairy herd
$n-1$	The cow
$n-2$	An organ
$n-3$	A tissue

FIGURE 1   Example of a hierarchy of systems in a livestock farming system.

To understand a system it is often desirable to study the behavior of its subsystems and the relationships among them. The number of possible relationships or interactions between subsystems increases exponentially as you move down in the hierarchy (Weinberg, 1975). Although this means that a full description of a system by subsystems necessitates reductions, a certain system cannot typically be reduced to a simple function of subsystems. Some characteristics of a system may disappear when it is divided into subsystems. These systems characteristics were pointed out by the philosopher Mario Bunge and described as emergent properties (Agassi and Cohen, 1982). An emergent property differs from an interaction between subsystems. An emergent property cannot be linked to specific subsystems and their relations. Theoretically, a system with emergent properties cannot be described by subsystems without reductions.

The word "analyze" originated from the Greek for "to split." Traditional science is based on the analytical approach of gaining knowledge by splitting a problem into parts. A problem is reduced to hypotheses suitable for testing in a designed experiment. The word "systems" is also of Greek origin and derives from the opposite of "analyze," namely, "to aggregate." In systems-oriented research, knowledge is obtained by aggregation and not by splitting. Systems-oriented research can be conducted in various ways but will in general be based on the following steps:

1. A problem is perceived.
2. The system of concern is identified conceptually.
3. A model of the system is developed.

Step 1 implies identification of a task or problem to be solved or explained through research. Systems identification (step 2) means that the system relevant to the problem is described in terms of a conceptual model. The system is separated from the environment by a border, and the communications between the system and the environment are described. The conceptual model defines and summarizes the perspective and the purpose of the research activity. However, to study the behavior of the system, it is necessary to make a physical or symbolic model (step 3).

III. A CONCEPTUAL MODEL OF A LIVESTOCK FARM

The livestock production scientist needs to describe the system of concern by means of a conceptual model. The system description depends on the problem that is to be solved by the scientist.

It is necessary for the scientist to know how an animal is managed in order to understand the production of the animal. Since livestock management is conducted at herd level, the system of concern is often the livestock herd. Herd management may be difficult to separate from land management. If this is the case, the system of concern might be the livestock farm. Research results may suggest a certain general change in management. This change should be evaluated at the level of the system of concern. Conceptual modeling of a livestock farm is discussed by Sørensen and Kristensen (1992). The following description focuses on the livestock herd as the system of concern.

A conceptual model of the livestock herd needs to describe the relationship between production (transformation of input to output) and management (control of the transformation process). A relevant relationship between production and management for a livestock farm is found in a cybernetic system (Sørensen and Kristensen, 1992). A cybernetic system consists of an effector (a controlled system) and a sensor acting on the effector (a controlling system). A model of a dairy herd as a cybernetic system is shown in Fig. 2. The herd as a system is defined by two subsystems: a production system and a management system. The production system consists of all the animals in the herd. The production from the herd is its output, and the input to the production system is characterized as consisting of controllable and uncontrollable factors. Uncontrollable factors include the climate, legislative regulations, and price changes. The management

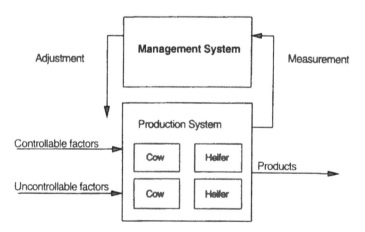

FIGURE 2   A dairy herd as a cybernetic system.

system is a feedback of information from the output and the state of the production used to control the controllable input. The state of the herd and its production are measured and compared with a production target. These comparisons lead to adjustments of the controllable factors such as feed input and disease treatment.

IV. A COMPUTER MODEL OF A LIVESTOCK HERD

It is often beneficial to develop a symbolic model or a physical model of the system of concern in order to study the system behavior. The most common symbolic model is a set of mathematical equations that constitute a mathematical model.

One classification of mathematical models distinguishes between simulation models and mathematical programming models (Dent and Blackie, 1979). A mathematical programming model includes an algorithm for optimization. The most commonly used types are linear programming models and dynamic programming models. The linear programming technique refers to a certain type of algorithm for optimization and is a very widespread technique. A comprehensive description of this technique in relation to application in agriculture is given by France and Thornley (1984). Dynamic programming refers to numerical methods for the solution of sequential decision problems. An obvious sequential decision problem is that of replacement. Dynamic programming applied to the livestock replacement problem is described in detail by Kristensen (1993). A mathematical programming model finds the combination of input that yields an "optimum" output. The purpose of a mathematical programming model is to answer What to do? questions. Mathematical programming models are typically designed for decision support or to describe optimum allocation of production input. Optimization of culling and reproduction strategies is an area in which mathematical programming models have been made such as the dairy cattle models of Kristensen (1987) and Van Arendonk (1985) and the swine model of Huirne et al. (1988).

A simulation model is a "freestyle" model in the sense that the model does not include an optimization algorithm. The purpose of a simulation model is to answer "what if?" questions, and this type of model is therefore often used for research purposes. Simulation models can be used for what to do questions by setting up simulation experiments exploring a certain response surface. The following description is confined to simulation models.

Simulation models can be classified by the following three dichotomies (France and Thornley, 1984):

1. Static versus dynamic
2. Deterministic versus stochastic
3. Empirical versus mechanistic

The difference between a static and a dynamic model is that a dynamic model includes time as a driving variable. In such cases the state of a system at time $t + 1$ is a function of the state at time t. A dynamic model is often represented by a loop. The feedback of information in Fig. 2 illustrates a dynamic loop. If the measurement refers to time t, the adjustment refers to time $t + 1$. The fact that a feedback loop describes a time change is often overlooked. A livestock herd model as illustrated here mimics a self-recruiting group of female animals over time and will therefore by dynamic as indicated in Fig. 2.

The difference between a deterministic model and a stochastic model is that input variables in a stochastic model are not described solely by their mean values but also by their probability distributions. The probability distributions will typically affect the mathematical transformation of input to output made in the model, and a deterministic model ignoring these distributions would not produce mathematically correct results. In a stochastic model a change in the variance of one or more input variables can change the mean of one or more important output variables, whereas the result in a deterministic model remains the same.

A deterministic livestock herd model uses classes of animals as the simulation unit to deal with discrete events such as conception, sex of offspring, death, and culling. A class of animals contains, for example, 10.7 animals at a certain time. A much-used deterministic cattle herd simulation model is one formulated at Texas A&M University of Sanders and Cartwright (1979).

A certain type of stochastic models are called probabilistic or Markov chain models. In a probabilistic livestock herd model, the classes of animals constitute a transition matrix expressing a probability distribution of animal movement from a specific class to all other classes in the next step. Probabilistic models follow the so-called Markovian property in which the outcomes of each step depend only on the last state of the model. Examples of probabilistic livestock herd models are dairy and swine models described by Jalvingh et al. (1992a, 1993).

Another type of stochastic herd models are Monte Carlo simulation models. A Monte Carlo simulation model typically uses the animal as the simulation unit. Discrete events are controlled by pseudorandom number generators linked to suitable probability distributions. "Whole animals" are maintained through a simulation by a set of status variables referring to the individual animal. It is necessary to make repeated runs from a specific data input in order to establish a mean herd value by a stochastic model. Examples of Monte Carlo simulation models are the dairy herd model SIMHERD (Sørensen et al., 1992) and the sow herd model Pig-Oracle (Marsh, 1986).

The distinction between empirical and mechanistic models relates to the conception of the world as a hierarchy of systems. In a hierarchy of systems any system is composed of subsystems and at the same time is part of a larger system as illustrated in Fig. 1. An empirical model output relates to input within the same hierarchy level. A mechanistic model describes a certain level indirectly from the behavior of one or more sublevels. The advantage of a mechanistic model is that parameters estimated at one level can predict results at the next higher level. Thus, the model becomes a medium for communication between levels of a system. It is important, however, to be aware that when a system is described by lower levels in the hierarchy, the degree of reduction is higher. A mechanistic model implies, in principle, a higher degree of reduction than an empirical model. Livestock herd simulation models are typically mechanistic in the sense that they simulate the production of the herd indirectly by simulation of subsystems (classes of animals or individual animals).

V. DEVELOPMENT OF A LIVESTOCK HERD SIMULATION MODEL

Development of a computer model is typically described in textbooks (e.g., Dent and Blackie, 1979) to include these steps: model building, model validation, sensitivity analysis, and application. The key question concerning the relevance of a certain model is the degree of agreement between the outcome from the computer model and the actual behavior of the system of concern. This has been called the problem of validation (Sørensen, 1990). The traditional validation method compares real-world data to simulated data, i.e., goodness-of-fit tests. For a livestock herd simulation model this would require several years of data from farms applying the same production strategy throughout the years. Since it is not realistic to expect farmers to practice uniform management across time, goodness-of-fit tests have not been an efficient method for validation of livestock herd simulation models (Bywater, 1990; Sørensen, 1990).

It is therefore necessary to apply other methods for gaining confidence in the model. The model should be constructed to make it possible for the user to analyze the results in sufficient detail to accept the final results as a consequence of accepted assumptions. The specification of output in the model should allow the user such possibilities, and the results should not be confined to the level needed for analyzing simulated experiments. This type of validation procedure makes it impossible to distinguish between validation and application. Model application is necessary to obtain an acceptable degree of validation.

VI. THE COMPUTER MODELS SIMHERD AND SIMCOW

When evaluating the application of livestock herd simulation models in research, it is important to bear in mind that models are typically developed for decision support for livestock producers or livestock production advisors. Recent reviews of livestock models for decision support are given by Bywater (1990) and Jalvingh (1992). Some models simulating the livestock herd for research purposes have been developed. To illustrate this type of model, the models SIMHERD, which simulates a diary herd, and SIMCOW, which simulates a dairy cow, are described in this section.

SIMHERD is a mechanistic, dynamic, and stochastic model simulating production of a dairy herd with its attendant young stock (Sørensen et al., 1992; Sørensen and Enevoldsen, 1994). The core of SIMHERD is a net energy model called SIMCOW that describes relations between feed intake and production at the individual cow level (Sørensen et al., 1994; Østergaard et al., 1994). State changes and production of the herd are simulated by state changes of the individual cows and heifers in time steps of 7 days. All discrete events at individual animal level such as individual inherent and lactational milk production potential, heat detection, conception, fetus death, sex and viability of the calf, and involuntary culling are triggered stochastically. As regards milk production potential at birth, each animal is assigned an individual value triggered from a normal distribution with herd-specific average and standard deviation. To form lactational variation, the active milk production potential is triggered from another normal distribution with the inherent individual value as average and a herd-specific standard deviation. The cow is described by a set of state variables that are shown in Table 1. The equations determining the relation between intake and production are shown in Table 2. Energy is measured in Scandinavian feed units (SFU), with 1 SFU being equivalent to the net energy for milk production in 1 kg of barley (Weisbjerg and Hvelplund, 1993).

The requirements for maintenance and pregnancy are always satisfied either by energy intake or by mobilization. Milk production is not allowed to be negative. Milk production and liveweight changes are calculated from a milk production potential, a growth potential, available energy from feed intake, and a rule for distribution between lactation and growth. The equations are shown in Table 2.

Growth potential is described by a Gompertz growth function [Eq. (2)]. The equation expresses that the cow is aiming to reach mature weight. The growth potential depends on the actual weight and the mature weight. If the actual weight is far less than the mature weight, then the growth potential is high.

The milk yield potential describes the maximum daily milk production possible given parity stage of lactation and inherent milk

Table 1 Individual Characteristics and State Variables of the Individual Animal in the Model

Variable	Description
Individual characteristics	
Identification	
Inherent milk production capacity	Expected FCM/day 1–24 wpp in 3rd lactation
State variables	
Age	Days
Lactation stage	Weeks postpartum
Lactation number	
Estrus status	Cycling or not and weeks to estrus
Pregnancy status	Pregnant or not and weeks to calving
Decision for culling	Decision made or not
Milk production potential	Maximum yield (FCM/day) in the present week
Milk production	Yield (FCM/day) in the present week
Liveweight	Kilograms

Table 2 Equations for Production and Feed Intake of the Cow in the Model

$$g_{pot} = -nW \ln(W/A) \tag{1}$$

$$y_{pot} = aT^b e^{-cT} \tag{2}$$

$$c_y(y - y_{pot})/[c_g(g - g_{pot})] = k_0 + k_1 T \tag{3}$$

$$c_y y + c_g g = E_{lg} \tag{4}$$

$$K = \begin{cases} 5.55 - 2.22e^{-0.28T} & \text{(1st lactation)} \\ 7.08 - 2.95e^{\,0.329T} - 0.0231T & \text{(older cows)} \end{cases} \tag{5}$$

$$E_{av} = -1.47 + 1.43E_{in} - 0.0261E_{in}^2 \tag{6}$$

g_{pot} = potential daily gain, kg/day; y_{pot} = potential milk yield, kg/day; A = mature weight; T = weeks postpartum; W = actual liveweight; y = actual milk yield, kg/day; g = actual growth rate, kg/day; c_y = net energy requirement per kg milk; c_g = net energy requirement per kg gain; E_{lg} = available net energy for lactation and growth, SFU/day; K = bulk capacity, units/day; E_{av} = available net energy, SFU/day; E_{in} = net energy intake, SFU/day; a, b, c, n, k_0, k_1 are constants.

production capacity. The relationship between stage of lactation and milk production potential is described by an incomplete gamma function (Wood, 1967) [Eq. (1)].

The milk production potential of a first lactation cow depends on the size at first calving (Foldager and Sejrsen, 1991). Equation (1) is consequently adjusted by size at first calving. For each kilogram increase in weight, the aggregate milk production potential 1–252 days postpartum is increased by 6.9 kg of 4% fat-corrected milk (FCM) (Østergaard et al., 1994).

Feed intake capacity is based on the Danish feed intake system (Kristensen, 1983; Kristensen and Ingvartsen, 1986; Kristensen and Kristensen, 1987). A parameter is attached to each feedstuff in the model describing a bulk content per unit net energy content. Each cow has a bulk capacity that is described in Eq. (5), using parameters given by Kristensen and Kristensen (1987). The equation is based on feeding systems with a high level of forage or a total mixed ration given ad libitum.

Feed intake is assumed to decrease during lactation. This decrease is assumed to be less for first lactation cows than for older cows because first lactation cows grow during lactation.

The feed intake capacity is assumed to increase with increasing weight at calving (Kristensen and Ingvartsen, 1986). However, it is found that fat cows at calving exhibit a relatively slow increase in feed intake in early lactation (Treacher et al., 1986; Holter et al., 1990). SIMCOW therefore chooses to increase feed intake capacity not by weight at calving but by the size of the cow at calving. Size of cow is calculated on the basis of weight at first calving, mature weight, and age. In addition to size, Kristensen and Ingvartsen (1986) found that feed intake capacity depends on milk production capacity. In SIMCOW, feed intake capacity is scaled according to inherent milk production capacity and a specified benchmark.

Available net energy is calculated from net energy intake using Eq. (6) given by Kristensen and Aaes (1989). The relation E_{av}/E_{in} expresses the efficiency rate for net energy. The efficiency rate expresses the relationship between energy input and output for a certain genotype. Cows with a high inherent milk production capacity are expected to have a higher efficiency rate than low production capacity cows at a fixed energy input. The efficiency rate is therefore scaled according to the inherent milk production capacity and a specified benchmark. The efficiency rate can further be controlled by a factor (b_{eff}), which can be specified for first and later lactations.

Equations (3) and (4) describe the distribution rules used to calculate milk production and growth if available energy is insufficient to meet the potentials. Similar distribution rules have been applied in other models (Doyle, 1985, Bruce et al., 1984; Williams et al., 1989). It is assumed that the distribution between growth and milk production changes during lactation. Growth increases in priority and milk production decreases in priority during lactation. Available net energy for lactation and growth, E_{lg} in Eq. (4), is calculated as available net energy minus energy requirements for maintenance and pregnancy. If available energy exceeds the requirements for all potentials, the residual energy is transformed to gain, however, with a net energy content, c_{gl}, other than that used for "normal growth" (c_g).

All the constants in Eqs. (2)–(4) can be defined by the model user. The default parameter values are presented in Table 3. The constant n in Eq. (1) was estimated by Taylor (1968) to be $(36A^{0.27})$ [1], where A is mature weight. The values for energy requirements for milk production and growth in Table 3 are according to AFRC (1993). The standard value for gain above the growth potential, c_{gl}, is higher than the standard value for c_g, following the value of energy concentration in fat (ARC, 1980).

The shape of the lactation curve [Eq. (2)] is determined by the constants a, b, and c, which are specified for first and later lactations, respectively. The default values for these constants are based on cal-

Table 3 Standard Values for Constants in Eqs. (1)–(4)

Constant	All cows	1st lactation	Older cows
n	0.0049		
a		0.9356	1.3110
b		0.0611	0.0995
c		1.22×10^{-2}	2.56×10^{-2}
k_0	0.90		
k	10.06		
b_{eff}		0.92	1.04
c_y	0.40		
c_g	2.65		
c_{g1}	5.00		

culations made by Østergaard et al. (1994). The value for the constant a in Eq. (2) is scaled proportionally to the inherent milk production capacity of the cow, so variation in milk yield between cows appears as a consequence of variation in the values of inherent milk production capacity. The effect of changing the feeding level from medium to high on milk production has been found to be approximately 1 kg of 4% fat-corrected milk (FCM) per SFU in the first part of lactation, when the cows were given silage ad libitum plus a constant level of concentrates (Kristensen, 1983). The constants in Eq. (3) express this effect of energy level on milk production in the first 24 weeks of lactation, given all other standard values, silage ad libitum, and flat rate feeding with concentrate. Milk production in the first 24 weeks of lactation is 25% lower for first lactation cows than for older cows, when energy intake and initial liveweight are 13% and 17% lower, respectively. These figures are in agreement with empirical herd data (Hindhede and Thysen, 1985).

The feed efficiency was assumed to be similar for first lactation and older cows given a typical difference in production and feed intake. This was ensured by the chosen values for efficiency factors b_{eff}. The value 0.92 for first lactation compared to 1.04 for later lactations makes a 4% unit decrease in the feed efficiency for first lactation cows (Table 3).

The specifications for feeding cows are shown in Table 4. It is possible to use four different feeds, two forages and two concentrates, for eight groups of cows. If the amount of feed offered to a specific cow exceeds the cow's feed intake capacity, the forage intake is scaled down.

Table 4 Example of a Winter Feeding Plan (SFU/day) for Cows in the Model

Feedstuff	Bulk capacity	1st lact. cows 1–24 wpp[a]	Older cows 1–24 wpp[a]	Yield threshold (kg 4% FCM/day)[b]					
				1st lact. cows		Older cows		Dry period cows	Cows to be culled
				21.5 SFU/day	16.5 SFU/day	24.0 SFU/day	19.0 SFU/day		
1. Silage	0.70	6.0	8.0	6.0	6.0	6.0	6.0		8.0
2. Straw	1.50							4.0	
3. Sugarbeets	0.22	5.0	5.0	5.0	4.0	5.0	4.0		5.0
4. Concentrate	0.18	8.0	8.0	3.5	2.5	3.5	2.5	2.0	8.0
		19.0	21.0	14.5	12.5	14.5	12.5	6.0	21.0

[a]Feeding level is maintained regardless of actual milk yield.
[b]After 24 wpp feeding plan is controlled by actual milk yield. The thresholds indicate when the cows qualify for the specified feeding.

Feedstuffs are defined by name, bulk per energy content, and feed type (forage or concentrate). The six groups of cows are first lactation and older cows 1–24 weeks postpartum (wpp); first lactation and older cows at two feed levels for feeding according to milk yield after 24 wpp; dry period cows; and cows pointed our for culling (sale for butchering). The milk yield thresholds for the feed levels in late lactation and the drying-off decision are determined in the feeding strategy for first lactation and older cows, respectively. In the feeding strategy it is also possible to define two seasons when the cows can be fed with two different feeding plans. The feeding plan for heifers specifies energy level for two groups of feedstuffs in five age groups. Grazing periods and average pasture intake for two groups of heifers can be specified.

With SIMCOW it is possible to simulate effects on lactational production of an average cow and to evaluate the result with graphs and tables. Evaluation of full herd strategies can be made by SIM-HERD. Further details are given by Sørensen et al. (1996a).

Systems behavior in SIMHERD is controlled by the state of the initial herd and assignment of 168 decision variables describing the production system and the management strategy concerning feeding, reproduction, culling, and disease. In the model, a cow can be a candidate for replacement if days open exceed a threshold that can be specified for high and low yielding cows, respectively, using their milk production in current lactation. A "candidate" cow is culled when a down-calving heifer is available for replacement and there is no free stable place.

VII. COMPUTER SIMULATION APPLIED IN LIVESTOCK PRODUCTION SCIENCE

A. Reproduction and Replacement

Reproduction and replacement in a livestock herd have many delayed and side effects. These complex problems can be assessed by livestock herd model simulations evaluating the technical and economic impact of different reproduction and culling strategies (e.g., Marsh et al., 1987; Dijkhuizen and Stelwagen, 1988; Jalvingh et al., 1992b). Clausen et al. (1995) evaluated nine culling strategies with the combination of different heat detection rates, initiation of insemination after calving, and heifer purchase policies using SIMHERD. The results show that the best culling strategy varies between herds depending on the total management strategy of the farm.

B. Feeding Management

Feeding management has typically been evaluated by feed ration evaluations at the animal level. In a herd the effect of a certain feeding strategy will be highly influenced by, for example, the management strategy in general. Sørensen et al. (1992) found that an increase in culling of open (non-pregnant) cows led to a relatively high increase in milk production per cow at a low forage quality compared to a high forage quality. Østergaard et al. (1996), who evaluated single versus multiple Total Mixed Ration (TMR) feeding groups under different management characteristics, found that the optimum energy content in the TMR was affected by the culling strategy applied on the farm.

C. Grassland Management

Grassland management is a highly complex interaction between animals and pasture growth. It is therefore a classical objective for computer simulation development and application (Spreen and Laughlin, 1986; Maxwell, 1990). Grass growth and quality follow a seasonal pattern. It is therefore often decided to use a certain calving pattern in dairy herds to synchronize feed requirement and grass growth. Sørensen et al. (1992) showed that the effect of seasonal calving and grass utilization at herd level very much depend on the production constraints of the herd. If the production is constrained by herd size, the positive effect of spring calving is very little compared to a constraint on milk production per year. The effect of different stocking rate and supplemental feeding levels on pasture was found to be highly dependent on the herd management system (Kristensen et al., 1997).

D. Herd Health Management

Health issues at herd level are very complex because the cause-and-effect relationships between production and health are very complex (Sørensen and Enevoldsen, 1992). The effect of a certain management change will therefore be almost unique for a certain herd. It therefore seems more relevant to develop simulation models that allow the advisor to describe the unique herd and evaluate certain herd-specific strategies than to generate general rules of thumb. SIMHERD and a set of other PC tools were therefore developed into a program package available for herd health management (Enevoldsen et al., 1995).

Health management is further complex due the contagious pattern of some diseases. Such problems are often dealt with by Markov chain models (Carpenter, 1988). A model based on SIMHERD describing the contagious cattle disease BVD has been developed (Sørensen et al., 1995), illustrating the relevance of combining interactions between herd management and contagious disease pattern.

E.   Combination of Case Studies and Simulation Experiments

Output from a herd during a time interval represents, theoretically, one of many possible outcomes given the initial circumstances. A stochastic herd simulation model allows the user to examine the distribution and possible outcomes for the applied production strategy and input variables specified for the herd. A herd simulation model with a detailed management strategy can be used to describe a certain herd. A group of such herds in a simulation framework allows the user to evaluate what is expected to happen if all farms make a certain change but leave all other parts of management as they were before. Such simulation experiments allow the user to examine the generality of a certain management pattern (Sørensen et al., 1996b).

F.   Examination of Animal Level Experiments at Herd Level

A mechanistic model (at herd or farm level) provides a medium of communication between research groups. Experimental research results at the animal level can be included in animal level relationships in a livestock herd model to improve the herd model. Experimental results can also be presented by a livestock herd model (Marsh et al., 1988; Sørensen et al., 1993). Kristensen et al. (1997) transformed data from 12 dairy herds into a SIMHERD input framework. Each management strategy characterized a certain herd. The effect of grazing intensity and supplemental feeding on feed intake and milk production at cow level estimated in an experimental station experiment was then imposed on each of the 12 management strategies, and the possible effect was estimated at herd level by simulation. Kristensen et al. (1996) found that the effect varied very much from herd to herd and the general recommendation based on the grazing experiment showed a positive economic effect in only 9 out of 12 herds.

REFERENCES

AFRC. 1993. Energy and protein requirements of ruminants. AFRC Technical Committee on Responses to Nutrients. Wallingford, UK: CAB International.

Agassi, J., and R. Cohen. (Eds.). 1982. *Scientific Philosophy Today. Essays in Honour of Mario Bunge.* Dordrecht, The Netherlands: Reidel.

ARC. 1980. *The Nutrient Requirement of Ruminant Livestock.* London: Commonwealth Agriculture Bureau.

Baldwin, R. L., and M. D. Hanigan. 1990. Biological and physiological systems: Animal sciences. In: *Systems Theory Applied to Agriculture and the Food Chain.* J. G. W. Jones and P. R. Street (Eds.). Elsevier Applied Science: London, pp 1–22.

Bertalanffy, L. von. 1951. General systems theory: A new approach to unity of science. *Hum. Biol.* 23:303–361.

Bertalanffy, L. von. 1973. *General System Theory: Foundations, Development, Application.* Rev. ed. New York: George Braziller.

Boulding, K. E. 1956. General systems theory—The skeleton of science. *Manage. Sci.* 2:197–208.

Bruce, J. M., P. J. Broadbent, and J. H. Topps. 1984. A model of the energy system of lactating and pregnant cows. *Anim. Prod.* 38:351–362.

Bywater, A. C. 1990. Exploitation of the systems approach in technical design of agricultural enterprises. In: *Systems Theory Applied to Agriculture and the Food Chain.* J. G. W. Jones and P. R. Street (Eds.). pp 61–88.

Carpenter T. E. 1988. Microcomputer programs for Markov and modified Markov chain disease models. *Prev. Vet. Med.* 5:169–179.

Clausen, S. M., J. T. Sørensen and A. R. Kristensen. 1995. Technical and economic effects of culling and reproduction strategies in dairy cattle herds estimated by stochastic simulation. *Acta Agric. Scand. Sect. A, Anim. Sci.* 45: 63–73.

Dent, J. B., and M. J. Blackie. 1979. *Systems Simulation in Agriculture.* London: Applied Science.

Dijkhuizen, A. A., and J. Stelwagen. 1988. An economic comparison of four insemination and culling policies in dairy herds by method of stochastic simulation. *Livest. Prod. Sci.* 18:239–252.

Doyle, C. J. 1983. Evaluating feeding strategies for dairy cows: A modelling approach. *Anim. Prod* 36:47–57.

Enevoldsen, C., J. T. Sørensen, I. Thysen, C. Guard, and Y. T. Gröhn. 1995. A diagnostic and prognostic tool for epidemiologic and economic analysis in dairy herd health management. *J. Dairy Sci.* 78:947–961.

Foldager, J., and K. Sejrsen. 1991. Rearing Intensity in Dairy Heifers and the Effect on Subsequent Milk Production. Report 693, National Institute of Animal Science, Denmark. (With English summary and subtitles.)

France, J., and J. H. M. Thornley. 1984. *Mathematical Models in Agriculture.* London: Butterworth.

Hindhede, J., and I. Thysen. 1985. Milk yield and body weight gain in different housing systems for the dairy cow. In: *The Effect of Diary Cow Housing Systems on Health, Reproduction, Production and Economy.* V. Østergaard (Ed.). Rep. 588. National Institute of Animal Science, Denmark, pp. 141–182. (With English summary and subtitles.)

Holter, J. B., M. J. Slotnick, H. H. Hayes, C. K. Bozak, W. E. Urban, and M. L. McGilliard. 1990. Effect of prepartum dietary energy on condition score, postpartum energy, nitrogen partitions, and lactation production responses. *J. Dairy Sci.* 73:3502–3511.

Huirne, R. B. M., T. H. B. Hendriks, A. A. Dijkhuizen, and W. J. Giesen. 1988. The economic optimization of sow replacement decisions by stochastic dynamic programming. *J. Agric. Econ.* 39:420–438.

Jalvingh, A. W. 1992. The possible role of existing models in on-farm decision support in dairy cattle and swine production. *Livest. Prod. Sci. 31:* 351–365.

Jalvingh, A. W., A. A. Dijkhuizen, and J. A. H. Arendonk. 1992a. Dynamic probabilistic modeling of reproduction and replacement management in sow herds. General aspects and model description. *Agric. Syst.* 39:133–152.

Jalvingh, A. W., A. A. Dijkhuizen, J. A. H. Arendonk, and E. W. Brascamp. 1992b. An economic comparison of management strategies on reproduction and replacement in sow herds using a dynamic probabilistic model. *Livest. Prod. Sci.* 32:331–:331–350.

Jalvingh, A. W., J. A. M. van Arendonk, and A. Dijkhuizen. 1993. Dynamic probabilistic simulation of dairy herd management practices. I. Model description and outcome of different living. *Livest. Prod. Sci.* 37:107–132.

Kristensen, A. R. 1987. Optimal replacement and ranking of dairy cows determined by a hierarchic Markov process. *Livest. Prod. Sci.*16:131–144.

Kristensen, A. R. 1993. Markov decision programming techniques applied to the animal replacement problem. Dissertation. Copenhagen: The Royal Veterinary and Agricultural University.

Kristensen, T., J. T. Sørensen, and S. M. Clausen. 1997. Simulated effect on daily cow and herd production of grazing intensity. *Agric. Syst.* (accepted).

Kristensen, V. F. 1983. The regulation of feed intake through the composition of the ration and the feeding principle. Rep. 551. National Institute of Animal Science, Denmark. pp. 7.1–35. (English summary and subtitles.)

Kristensen, V. F., and O. Aaes. 1989. The effect of feeding level on feed efficiency. Rep. 660. National Institute of Animal Science, Denmark, pp. 10–44. (With English summary and subtitles.)

Kristensen, V. F., and K. L. Ingvartsen. 1986. Prediction of feed intake. In: *New Developments and Future Perspectives in Research on Rumen Function. A.* Neimann-Sørensen (Ed.). CEC Seminar, June 25–27, 1985, Forskningscenter Foulom. Rapport EUR 10054 EN. Luxembourg, pp. 157–181.

Kristensen, V. F., and T. Kristensen. 1987. Systemer til forudsigelse af foderoptagelsen hos kvæg. Vurdering og justering of systemet til milkekøer. Paper presented at the annual meeting of the National Institute of Animal Science, Denmark, May 20, 1987. (In Danish.).

Le Moigne, J. L. 1989. Systems profile: First, joining. *Syst. Research* 6:331–343.

Marsh, W. E. 1986. Economic decision making on health and management in livestock herds: Examining complex problems through computer simulations. Ph.D. Thesis. University of Minnesota.

Marsh, W. E., A. A. Dijkhuizen, and R. S. Morris. 1987. An economic comparison for four culling decision rules for reproductive failure in United States dairy herds using Dairy ORACLE. *J. Dairy Sci.* 70:1274–1280.

Marsh, W. E., D. T. Galligan, and W. Chalupa. 1988. Economics of recombinant bovine somatotropin use in individual dairy herds. *J. Dairy Sci.* 71: 2944–2958.

Maxwell, T. J. 1990. Plant–animal interactions in northern temperature sown grasslands and seminatural vegetation. In: *Systems Theory Applied to Agriculture and the Food Chain.* J. G. W. Jones and P. R. Street (Eds.). Elsevier Applied Science: London, pp. 23–60.

Mertens, D. R. 1977. Principles of modeling and simulation in teaching and research. *J. Dairy Sci.* 60:1176–1186.

Østergaard, S., J. T. Sørensen, V. F. Kristensen, and T. Kristensen. 1994. Modelling of the production of a dairy cow in a new energy system. Presentation and documentation of the PC model SIMCOW. Rep. 24. National Institute of Animal Science, Denmark. (With English summary and subtitles.)

Østergaard, S., J. T. Sørensen, J. Hindhede, and A. R. Kristensen. 1996. Technical and economic effects of feeding dairy herds TMR in one group compared to multiple groups under different herd and management characteristics estimated by stochastic simulation. *Livest. Prod. Sci.* 45:23–33.

Rountree, J. H. 1977. Systems thinking—Some fundamental aspects. *Agric. Syst.* 2:247–254.

Sanders, J. O., and T. C. Cartwright. 1979. A general cattle production systems model. I. Structure of the model. *Agric. Syst.* 4:217–227.

Sørensen, J. T. 1990. Validation of livestock herd simulation models: A review. *Livest. Prod. Sci.* 26:79–90.

Sørensen, J. T., and C. Enevoldsen. 1992. Modelling the dynamics of the health-production complex in livestock herds: A review. *Prev. Vet. Med. 13:* 287–297.

Sørensen, J. T., and C. Enevoldsen. 1994. SIMHERD: A dynamic, mechanistic and stochastic simulation model for dairy herd health management decision support. *Kenya Vet.* 18:431–434.

Sørensen, J. T., and E. S. Kristensen. 1992. Systemic modelling: A research methodology in livestock farming. In: *Global Appraisal of Livestock Farming Systems and Study on Their Organizational Levels: Concepts, Methodology and Results.* A. Gibon and G. Matheron (Eds.). EUR14479. The Commission of the European Communities, pp. 45–57.

Sørensen, J. T., E. S. Kristensen, and I. Thysen. 1992. A stochastic model simulating the dairy herd on a PC. *Agric. Syst.* 39:177–200.

Sørensen, J. T., C. Enevoldsen, and T. Kristensen. 1993. Effects of different dry period lengths on production and economy in the dairy herd estimated by stochastic simulation. *Livest. Prod. Sci.* 33:77–90.

Sørensen, J. T., S. Østergaard, V. F. Kristensen, and T. Kristensen. 1994. A net energy model describing the effect of feed management on live weight changes along the lactational cycle in dairy cattle. Paper presented at the

45th Meeting of the European Association for Animal Production, Sept. 5–8, 1994.

Sørensen, J. T., C. Enevoldsen, and H. Houe. 1995. A stochastic model for simulation of the economic consequences of bovine virus diarrhoea virus infection in a dairy herd. *Prev. Vet. Med.* 23:215–227.

Sørensen, J. T., C. Enevoldsen, and S. Østergaard. 1996a. SIMHERD User Manual. Internal Report No. 70, Danish Institute of Animal Science.

Sørensen, J. T., S. M. Clausen, T. Kristensen, J. Hindhede, E. S. Kristensen, and C. Enevoldsen. 1996b. Dynamic stochastic simulation as an analytic tool in dairy cattle case studies. In: Livestock farming systems: research, development, socio-economics and the land manager. Proceedings of the Third International Symposium on Livestock Farming Systems. EAAP Publication No. 79, pp. 270–280.

Spedding, C. R. W. 1988. General aspects of modelling and its application in livestock production. In: *Modelling of Livestock Production Systems.* S. Korver and J. A. M. Van Arendonk (Eds.). Boston: Klüwer Academic, pp. 3–12.

Spreen, T. M., and D. M. Laughlin. (Eds.). 1986. *Simulation of Beef Cattle Production Systems and Its Use in Economic Analysis.* Boulder, CO: Westview Press.

Taylor, St. C. S. 1968. Time taken to mature in relation to mature for sexes, strains and species of domesticated mammals and birds. *Anim. Prod. 10:* 157–169.

Treacher, R. J., I. M. Reid, and C. J. Roberts. 1986. Effects of body condition at calving on the health and performance of dairy cows. *Anim. Prod. 43:* 1–6.

Van Arendonk, J. A. M. 1985. Studies on the replacement policies in dairy cattle. Ph.D. Thesis. Wageningen, The Netherlands: Wageningen Agricultural University.

Weinberg, G. N. 1975. *An Introduction to General Systems Thinking.* New York: Wiley.

Weisbjerg, M. R., and T. Hvelplund. 1993. Estimation of net energy content (FU) in feeds for cattle. Report 3 from the National Institute of Animal Science, Denmark. (With English abstract and subtitles.)

Williams, C. B., P. A. Oltenacu, and C. J. Sniffen. 1989. Application of neutral detergent fibre in modeling feed intake, lactation response, and body weight changes in dairy cattle. *J. Dairy Sci.* 72:652–663.

Wood, P. H. P. 1967. Algebraic model of the lactation curve in cattle. *Nature (Lond.)* 216:164–165.

13

The Plant/Animal Interface in Models of Grazing Systems

M. Herrero, J. B. Dent, and R. H. Fawcett
University of Edinburgh, Edinburgh, Scotland

I. INTRODUCTION: THE PLANT/ANIMAL INTERFACE IN A SYSTEMS CONCEPT

Grazing systems are one of the main types of agroecological systems for food production in the world. These systems comprise about half of the world's land area (Stuth and Stafford-Smith, 1993), and increased interest exists in improving their management and ensuring sustainability. There is concern to prevent degradation of the resource base and its consequent environmental, social, and economic effects.

A common approach in studying grazing systems has been by way of mathematical modeling (Stuth and Stafford-Smith, 1993). Unfortunately, many scientists have looked at specific and detailed phenomena within a part of the system without taking into account interactions and effects at the whole-system level (Demment et al., 1995). The plant/animal interface is one of many subsystems of a grazing system. It is a key element but is subject to a number of forward and backward interactions with the other subsystems. Consequently, it cannot be fully understood in isolation nor can it be

495

modeled without reference to all relevant components. Only recently have more integrated approaches, linking concepts from different disciplines across a variety of levels of knowledge, been implemented. The integration of various levels exposes enormous complexity within a grazing system. The degree of detail in these levels depends on the purpose for which the model is meant to be used, the users, the level of accuracy required, and the planning horizon (i.e., operational, strategic, or tactical).

This chapter reviews these approaches to obtain an adequate definition of the plant/animal interface in a systems context.

II. MODELING THE FORAGE RESOURCE

A. Modeling Primary Production

A wide range of approaches to modeling the forage resource are found in the literature. These vary widely in degree of complexity, number of species represented, variable and parameter definitions, and simulated output.

Some of the simplest representations are based on the functional form of plant growth curves [see Thornley and Johnson (1990) for a review]. For example, Brougham (1956), Morley (1968), Noy-Meir (1975, 1976, 1978), Christian et al. (1978), and Woodward et al. (1993, 1995) used logistic growth curves to represent pasture growth in their models:

$$\frac{dW}{dt} = mW \left(1 - \frac{W}{W_{max}} \right) \tag{1}$$

where m = the maximum relative growth rate, W = initial plant biomass, W_{max} = the asymptote plant biomass, and t = time. This description assumes that growth is proportional to plant biomass, rate of growth is proportional to amount of substrate, and substrate is finite (Thornley and Johnson, 1990).

Equation (1) is easy to parameterize when smooth experimental data of the dynamics of plant growth are available. The appropriateness of the parameters will depend on the quality of the experimental data and will reflect only the particular conditions in which the plants were growing for that particular data set. Nevertheless, the shape of the curve can be explained physiologically. Exponential growth occurs due to increased irradiance captured by increases in leaf area index (LAI) during early stages of development, while growth progressively decreases to a plateau as respiration losses due to senescence equal photosynthesis. It has the advantage that it is a simple curve with biologically meaningful parameters that can represent changes in the growing environment by modification of m and W_{max}. However, it is limited in that it does not represent the physiological mechanisms (e.g., photosynthesis, LAI development, nitrogen uptake) underlying sward growth and is therefore not flexible enough to represent effects of management interventions. Such models also fail to describe biomass in different botanical fractions (leaves, stems, dead material) or species compositions (grass–legume mixtures or rangelands) and their vertical distribution within the sward, which are important elements in predicting diet selection and/or species succession caused by disturbances (e.g., grazing or fire) in grasslands. The fact that only one sward component is represented [i.e., total herbage dry matter (DM)] implies that diet selection can be studied only by superimposing selectivity coefficients on total DM (Christian et al., 1978).

These limitations have led to the construction of several more detailed grassland models for single pasture species (Johnson and Thornley, 1983, 1985; Thornley and Veberne, 1989; Smith et al., 1985; Lopez-Tirado and Jones, 1991a, 1991b; Doyle et al., 1989; Sheehy et al., 1996; Guerrero et al., 1984; Charles-Edwards et al., 1987; Rodriguez et al., 1990; Murtagh, 1988; Veberne, 1992; van Keulen et al., 1981; Seligman et al., 1992; Herrero, 1995) or multiple species (Gilbert, 1975, Innis, 1978; Parsons et al., 1991; Hanson et al., 1988, 1994; Coughenour, 1984; Coughenour et al., 1984; Hunt et al., 1991; Hacker et al., 1991; Blackburn and Kothmann, 1989; Detling et al., 1979; Lauenroth et al., 1993; Richardson et al., 1991; Moore et al., 1997). The former come mostly from the agricultural sciences, while some of the latter also have a strong ecological background (e.g., Innis, 1978; Coughenour, 1984; Hanson et al., 1988, 1994).

The majority of these models represent plant growth as a function of one or more environmental, soil, and/or management variables. The simplest analyses use only one environmental driving variable (e.g., rainfall or irradiance) to determine sward growth rates. Charles-Edwards et al. (1987) used Monteith's (1972) factorial approach to determine growth rate of a sward, with ample supply of nutrients and water, based on daily irradiance intercepted by the pasture, the efficiency of light utilization by the plant to produce new material, and a partitioning coefficient for above-ground material. Shiyomi et al. (1986) used a similar approach to study energy flows in grasslands in Japan. Guerrero et al. (1984) and Hacker et al. (1991) determined plant growth as a function of rainfall. Due to the large effects on growth caused by severe water stress in the regions of their studies, they were able to use simple soil water balance budgets as primary predictors of herbage production. In even more complex models (e.g., Hanson et al., 1988, 1994), the basic components of water balance submodels include rainfall, evapotranspiration, transpiration, runoff, and infiltration, and these are modeled using well-recognized principles (van Keulen and Wolf, 1986; Thornley and Johnson, 1990). Moisture indices are derived from these variables to scale the growth rates of forage.

On the other hand, several models estimate biomass production as functions of a number of environmental variables. This usually results in models representing carbon (C) and nitrogen (N) fluxes in grassland ecosystems. The level of detail and empirical representations varies widely between models, although this is usually due to the original objectives of the model or their implicit site-specificity.

Inputs to the carbon cycle are usually represented by photosynthesis, and one of the most common methods is to integrate single-leaf photosynthesis over the canopy LAI using Beer's law (Monsi and Saeki, 1953) as the light attenuation factor through the depth of the canopy (Johnson and Thornley, 1983, 1985; Thornley and Veberne, 1989; Hanson et al., 1988, 1994; Sheehy et al., 1996). Single-leaf photosynthesis is commonly represented by rectangular (Innis, 1978; Johnson and Thornley, 1983; Doyle et al., 1989) or nonrectangular hyperbolas (Johnson and Thornley, 1985; Thornley and Veberne, 1989; Herrero, 1995). Other authors (Coughenour, 1984; Hunt et al., 1991) also include CO_2 concentrations and stomatal, internal, and leaf boundary layer resistances to account for water use and CO_2 effects on photosynthesis. Temperature and leaf N content (Thornley and Veberne, 1989; Hanson et al., 1994; Herrero, 1995) are used to scale the photosynthetic capacity of the sward. Outputs from the carbon pool are represented by fractions used for new growth, senescence, respiration, and grazing. Recycling of nutrients from senescent tissues also contributes to the carbon cycle.

The approaches to representing the nitrogen cycle of the grazing system are also diverse, but the basic factors are demonstrated in a simple model by Scholefield et al. (1991) (Fig. 1).

Although Scholefield et al.'s (1991) model is empirical, the same processes can receive a mechanistic treatment. However, the complexity of the model and its subsequent validation are increased severalfold (e.g., Veberne, 1992).

Thornley and Veberne (1989) argue that data to validate soil–plant mechanistic models are scarce or incomplete and that experi-

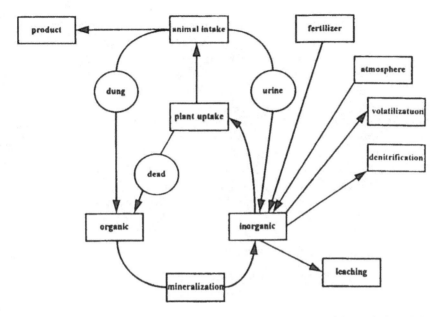

FIGURE 1 Basic nitrogen cycle in pasture ecosystems. [From Scholefield et al. (1991).]

ments are difficult to design, and therefore a subjective assessment on the behavior of the model is sometimes made. While the C cycle is relatively easy to validate due to the wide availability of field methods, there is still a considerable amount of progress to be made in designing soil submodels that are easy to parameterize at field level.

B. Sward Structure and Composition

It is well recognized that apart from herbage availability, sward structure plays an important role in determining the intake of grazing ruminants (Stobbs, 1973; Freer, 1981; Hodgson, 1985; Demment et al., 1995) and that ruminants select preferentially for the leaf component of the sward (Laredo and Minson, 1973; Hendricksen and Minson, 1980; Cowan et al., 1986; Penning et al., 1994). In rangelands (i.e., Coughenour, 1984; Hanson et al., 1994), species composition plays a major role in determining the diet selected by the animal (O'Reagain and Schwartz, 1995), and, depending on animal numbers and environmental conditions or disturbances, this selection modifies the subsequent species composition of the sward (Humphreys, 1991; Hacker and Richmond, 1994). Therefore, a basic requirement of pasture models, apart from their ability to predict total biomass, is that they should be capable of differentiating between plant parts and their density across the sward's vertical strata and/or species composition in the case of multispecies systems. Table 1 shows some of the main differences in the representation of sward morphology and composition between models. It can be observed from the table that several models do not differentiate between plant parts.

The most common fractionation of morphological composition is between leaf, stem, and dead material, and this is usually linked to age characteristics of the sward. This fractionation occurs partly because the models rely on photosynthesis and require an estimate

Table 1 Differences in the Representation of Sward Composition in Some Models

Model	No. of species[a]	Total biomass	Plant parts[b]	Phenology or age of plant parts	Speci comp
Noy-Meir (1975, 1976, 1978)	1	√	—	—	—
Johnson and Thornley (1983, 1985)	1	√	l/s/d	√	—
Herrero (1995)	1	√	l/s/d	√	—
Parsons et al. (1991)	2	√	l/s/d	√	√
Charles-Edwards et al. (1987)	1	√	—	—	—
Hunt et al. (1991)	1	√	—	—	—
Hanson et al. (1988, 1994)	3+	√	g/d/p	√	√
Christian et al. (1978)	1	√	g/d	√	—
Coughenour et al. (1984)	3+	√	b/sh/s/f	√	√
Lauenroth et al. (1993)	3+	√	l/s/d	√	√
Lopez-Tirado and Jones (1991a)	1	√	g/d	—	—
Guerrero et al. (1984)	1	√	—	—	—
Rodriguez et al. (1990)	1	√	l/s/d	√	—
Woodward et al. (1993, 1995)	1	√	—	—	—
Doyle et al. (1989)	2	√	l/s/d	√	—
Smith et al. (1985)	1	√	g/d	√	—
Hacker et al. (1991)	3+	√	—	—	√
Seligman et al. (1992)	1	√	l/s/d	√	—

[a]1 = single species, 2 = grass/legume, 3+ = rangelands.
[b]1 = leaves, s = stems, d = dead, p = propagules, g = green, sh = sheats, f = flowers.

of LAI for pasture growth calculations. This is a convenient attribute, since removal of LAI by grazing can link, in a physiological sense, the effect of grazing on total resource capture, pasture growth, and subsequent sward composition (Johnson and Parsons, 1985; Parsons et al., 1994).

The models of Johnson and Thornley (1983, 1985) and Thornley and Veberne (1989) did not represent the vertical distribution and bulk density (bd) within the sward, but they do have a convenient structure to model them. These models divide the leaf and stem structural mass into four distinct age categories, from new material down to senescent. The move from one category to the next is determined by the rates of appearance of these components, therefore making it possible to distribute them, separately, across the height of the sward according to the pasture species modeled. We developed a flexible method to estimate sward structure from these models using the following simple statements (Fig. 2):

1. A sward with a determined total height h and total herbage mass dm can be described as a series of i discrete horizons (h_i) of herbage mass dm_i, where dm_i is composed of variable proportions of leaf and/or stem (L/S) and dead material. The concept of sward horizons or layers in grassland modeling is well recognized and is useful for the representation of grazing processes (Ungar and Noy-Meir, 1988; Ungar et al., 1992; Demment et al., 1995).

2. As h decreases, the amount of dm_i in h_i increases (Fig. 2a), as is commonly observed (Stobbs, 1975; Illius and Gordon, 1987; Mayne et al., 1987).

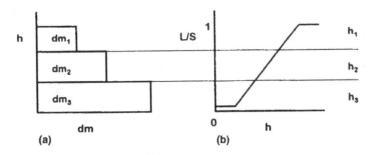

FIGURE 2   Representation of the vertical distribution of (a) herbage mass and (b) plant parts within a single-species pasture sward.

3. As h increases, the leaf/stem ratio increases (Fig. 2b). The proportion of dead material decreases, but for simplicity we consider only leaf and stem.

Following the nomenclature in the Thornley papers described above, we divided the dry matter (dm_i) of a sward into three horizons as follows:

$$dm_1 = WL_1 + f_1 Ws \qquad (2)$$

$$dm_2 = WL_2 + Wsh_1 + Wsh_2 + f_2 Ws \qquad (3)$$

$$dm_3 = WL_3 + WL_4 + Wsh_3 + Wsh_4 + f_3 Ws \qquad (4)$$

where WL_j and Wsh_j are the structural weights of leaves and stems of different ages, respectively, where $j = 1$ is new material and $j = 4$ is senescent tissues. In this three-horizon example, note that the first horizon (top of the sward) contains only leaf, which is commonly observed in many pasture species (Hodgson, 1985; Humphreys, 1991). Ws is the storage compartment, and f_i the fraction of Ws associated with the plant material in dm_i. Bulk density can be estimated as

$$bd_i = \frac{dm_i}{h_i} \qquad (5)$$

and

$$bd = \frac{dm}{h} \qquad (6)$$

where bd_i = bulk density in the ith horizon and bd = total sward bulk density.

The flexibility of the modeling structure presented in this chapter permits representation of a large number of sward structures, since the number of horizons and the type and quantity of plant parts comprising them can be changed without altering the functionality and robustness of the original plant models. At the same time, the structure has useful attributes in modeling intake and diet selection by different methods. The approach has been specifically developed for tropical and temperate monospecific swards, but it can be extended to grass–legume swards.

Separation between green and dead material in other models is mainly for the purpose of determining the sward's nutritive value. For example, to be able to represent diet quality, some models (Christian et al., 1978; Guerrero et al., 1984; Smith et al. 1985) subdivide the biomass empirically into three or four compartments representing new, mature, senescent, and/or dead material without describing them morphologically and assign quality characteristics (i.e., digestibility, cell wall) within these categories. However, the definition of physiological states without a morphological description presents three problems. First, they are difficult to handle in diet selection studies, where the different components of the diet selected by ruminants are usually identified by botanical fractions (Hendricksen and Minson, 1980; Humphreys, 1991), therefore reflecting morphological differences in the sward (Arnold, 1981; Hodgson, 1985). Physiological state of the sward does not represent its morphological structure. Second, it is difficult to accommodate different diet selection patterns of different animal species. For example, sheep are able to select more leaf than cattle (Arnold, 1981; Forbes and Hodgson, 1985; Penning et al., 1994), and these differences cannot be predicted if the sward is only divided into physiological state compartments. Third, even at a similar chemical composition, botanical fractions have different physical structures (Wilson, 1994), which affect the rates of breakdown from large to smaller particles of forage in the rumen (Kennedy and Murphy, 1988; Wilson and Kennedy, 1996) and therefore affect passage rates and pasture intake. This concept is difficult to model when sward physiological states are used because the botanical composition of each compartment (e.g., new material) is not known. From the diet selection viewpoint, in rangeland models the discrimination between species becomes more important (Baker et al., 1992) than within species, and most rangeland models only discriminate between, rather than within, species to represent sward biomass.

III. GRAZING PROCESSES AND DIET SELECTION

A. Intake and Grazing Processes

Intake prediction is one of the most important elements in grazing systems models because the prediction of animal responses to nutrients (Blaxter, 1989) [see Forbes and France (1993) for reviews] are largely dependent on it. In addition, pasture intake influences the regrowth of the sward (Brougham, 1956; Vickery, 1981; Parsons et al., 1988), the efficiency of fertilizer use (Humphreys, 1991; Herrero, 1995), supplementation strategies (Allden, 1981; Ørskov, 1994; Rook et al., 1994), nutrient cycling (Simpson and Stobbs, 1981; Scholefield et al., 1991), land use practices via the area required to maintain stock

(Olney and Kirk, 1989; Herrero et al., 1996a, 1996b), and the spatial distribution of pasture species in rangeland landscapes (Senft et al., 1987; Demment et al., 1995; O'Reagain and Schwartz, 1995).

A number of methods of simulating intake and grazing processes have been reported, but three distinct approaches can be observed:

1. Prediction of intake from systems of energy requirements
2. Establishment of relations between herbage mass and intake
3. Prediction of intake from grazing behavior measurements

The flexibility of studying different nutritional and management strategies and their effects on the whole system will largely depend on the method chosen to represent intake.

1. Intake as a Function of Energy Requirements

One method used to represent intake assumes that estimates of pasture intake can be derived from the energy requirements of the animal and the energy content of the pasture consumed. This last parameter has been usually derived from in vivo or in vitro digestibility estimates. The energy value of the forage as well as the animal's requirements have been expressed most commonly as digestible energy (DE), metabolizable energy (ME), or net energy (NE) (McDonald et al., 1995).

Two approaches are commonly used. The first is to estimate intake from the "inverse" of the nutrient requirements, and the second is to use regression equations, which are often included in requirements systems. These methods of intake estimation have been widely used in livestock models (Sanders and Cartwright, 1979; Konandreas and Anderson; 1982: Guerrero et al., 1984; Gartner and Hallan, 1984; Olney and Kirk, 1989). However, although it is accepted that nutrient requirements represent one of the most important driving forces of eating, these systems per se [apart from SCA (1990) and NRC (1996)] fail to take into account constraints on intake imposed by herbage availability and sward structure (Hodgson, 1985). This has already been discussed by Whelan et al. (1984). However, the following points should also be considered.

1. Classic work by Conrad et al. (1964) demonstrated that intake was proportional to energy requirements when the digestibility of the diet was higher than 67%. Below this threshold, intake was constrained by physical limitations of the reticulorumen. Therefore, for low digestibilities, when

the "reverse" calculation of intake from requirements is applied, intake is usually overestimated because animals are not physically able to eat sufficient quantities of forage. More recently, Forbes (1993) suggested that, for cows, this digestibility threshold may be higher depending on the level of production. In view of these problems, several models have incorporated static physical fill limitation constraints on intake (Forbes, 1977; Kahn and Spedding, 1984; Mertens, 1987; Finlayson et al., 1995).

2. Maintenance energy requirements scale with metabolic weight (Brody, 1945), but rumen size scales with body weight (Demment and van Soest, 1985; Illius and Gordon, 1991), thus partly explaining why digestibility, which is a crucial parameter in requirements systems, is not a good predictor of intake for low quality forages (Laredo and Minson, 1973; Poppi et al., 1981; Kibon and Ørskov, 1993). In the trial of Laredo and Minson (1973), sheep consumed more leaf than stem with both plant fractions having the same digestibility, suggesting that other factors, such as the physical structure of plant parts, which influence particle breakdown and passage rates (Poppi et al., 1981, Kennedy and Murphy, 1988; McLeod et al., 1990) play an important role in the control of feed intake and also that dynamic models of digestion that consider these factors may yield better estimates of potential intake.

2. Empirical Relations Between Herbage Mass and Intake

A variety of models have simulated the effect of herbage availability on intake using empirical relations (Freer et al., 1970; Noy-Meir, 1975, 1976; Arnold et al., 1977; Vera et al., 1977; Edelsten and Newton, 1975, 1977; Christian et al., 1978; Sibbald et al., 1979; White et al., 1983; McCall, 1984; Johnson and Parsons, 1985; Thornley and Veberne, 1989; Rodriguez et al., 1990; Blackburn and Kothmann, 1991; Richardson et al., 1991; Seman et al., 1991; Finlayson et al., 1995). These models use three basic steps for the calculation of intake at grazing.

Step 1 Estimate potential intake of the animal. Potential intake is usually defined as the intake of herbage without the constraints imposed by herbage availability, as a function of animal and plant characteristics. It is usually an input (Johnson and Parsons, 1985; Thornley and Veberne, 1989) or calculated in another submodel from the knowledge of body weight, the energy requirements of the animal, and the digestibility or metabolizability of the diet (Arnold et al., 1977; Christian et al., 1978; Richardson et al., 1991) and physical fill limitations (Kahn and Spedding, 1984; Doyle et al., 1989; Finlayson et al., 1995).

Step 2 Calculate the constraints on intake imposed by herbage availability. This is usually done by estimating "scaling factors" with empirical functions and leads to a term often called relative intake. Table 2 summarizes the functions used in different models to scale intake on the basis of different measures of herbage availability.

Common features of these scaling factors are their general shape, often expressed as Michaelis–Menten equations (Noy-Meir, 1975, 1976; Johnson and Parsons, 1985; Blackburn and Kothmann, 1991) and exponential or quadratic functions (see Table 2). However, large discrepancies occur between authors in the slopes of these functions (Fig. 3), which are caused by the animal and sward characteristics for which the equations were derived. Nevertheless, marked decreases in intake appear to occur if less than 1000–1500 kg/ha DM is available.

Herbage availability is described in different ways by different authors. The most common relationship between intake and herbage availability is derived from herbage mass per unit of area, while others derive functions on the basis of herbage available per animal (Zemmelink, 1980; Loewer et al., 1987) and yet others use both measures (McCall, 1984). An exception is the function of Johnson and Parsons (1985), which uses LAI to estimate relative intake. This is an interesting concept, since LAI provides an appropriate physiological interface between pasture removal (grazing) and regrowth (resource capture by photosynthesis). However, under most practical circumstances LAI is not measured, and some types of animals (e.g., cattle) remove not only LAI, which is associated with the leaf components of the sward only, but also the stem fraction. We have adapted this function and expressed it on the basis of leaf or total herbage mass while at the same time keeping the physiological relationship with LAI. This can be done with the knowledge of three easily measured parameters: the specific leaf area (SLA) of leaves, the leaf mass (LM), and the leaf-to-stem ratio of the sward being grazed (pL). The adaptation can be done in two simple steps. First,

$$LAI = LM * SLA \qquad\qquad (7)$$

Table 2 Functions to Estimate the Effect of Herbage Availability on Dry Matter Intake of Grazing Ruminants

Source	Function to estimate relative intake (RI)[a]
Freer et al. (1970); Arnold et al. (1977)	$RI = 1 - \exp(-0.001^*DM)$
Noy-Meir (1975, 1976); Blackburn and Kothmann (1991); Woodward et al. (1993, 1995)	$RI = [imax^*(DM/(DM + X))]/imax$
Vera et al. (1977)	$RI = 1 - \exp(-0.002503^*DM)$
Edelsten and Newton (1975, 1977)	$RI = 1 - \exp(-2.4^*10^{-7*}DM^2)$
Christian et al. (1978)	$RI = 1 - \exp(-0.000008^*DM^2)$
Sibbald et al. (1979)	$RI = DM/(DM + 250)$
White et al. (1983); Bowman et al. (1989)	$RI = 1 - \exp(-0.000002^*DM^2)$
Zemmelink (1980); Konandreas and Anderson (1982); Doyle et al. (1989)	$RI = [imax^*(1 - \exp(-DMH/imax)^{1.23})^{(1/1.23)}]/imax$
Johnson and Parsons (1985); Thornley and Veberne (1989); Parsons et al. (1991); Richardson et al. (1991)	$RI = [imax^*(LAI/K)^Q/(1 + (LAI/K)^Q)]/imax$, where $RI = imax/2$ for $LAI = K$; $K = 1$ and $Q = 3$ for sheep grazing ryegrass
Loewer et al. (1987); Rodriguez et al. (1990)	$RI = 2^*FA/B - FA^2/B^2$, where $B = 750$
Seman et al. (1991)	$RI = 1 - ((1 - 0.1)/(HI - LOW)^2)^*(HI - SH)^2$, where $HI = 20$ and $LOW = 5$
McCall (1984); Finlayson et al. (1995)	$RI = \theta^* \exp[-1.016^* \exp(-1.038^*A)]$, where $A = (DM/imax)^*(area/animals)$; $\theta = 1 - 1.42^* \exp(-0.00198^*DM)$

[a]imax = potential intake [kg/(animal·day)], DM = pasture dry matter (kg/ha); X = Michaelis constant for consumption (g/m²); LAI = leaf area index (m² leaf/m² soil); K = half-maximal response of LAI; Q = constant; DMH = available dry matter/animal [kg/(animal·day)]; FA = forage available/kg bodyweight (g DM/kg BW); B = threshold level of forage availability (g DM/kg BW); HI = height above which additional increases in sward height do not affect intake (cm); LOW = height below which forage is unavailable for grazing (cm); SH = total sward height (cm) area = grazing area (ha); animals = number of animals.

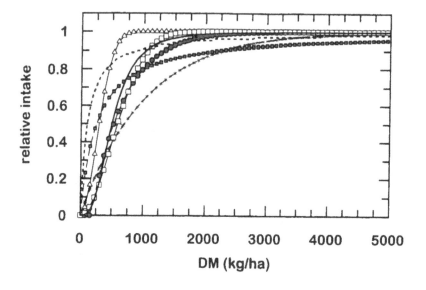

—✳— Freer et al. (1970), Arnold et al. (1977)

—△— Christian et al. (1978)

—●— McCall (1984), Finlayson et al. (1995)

—■— Sibbald et al. (1979)

—□— White et al. (1983), Bowman et al. (1989)

——— Johnson and Parsons (1985),Thornley and Veberne (1989), Richardson et al. (1991)

- - - - Noy-Meir (1975, 1976)

FIGURE 3   The relationship between herbage availability and intake at graz-ing in different models.

where LM = leaf mass in the sward (g/m²) and SLA = specific leaf area (m²/kg leaf). The total dry matter of the sward is then

$$DM = \frac{LM}{pL} \tag{8}$$

where pL = proportion of leaf in the sward.

The analysis provided here is more flexible than that offered by other authors (e.g., Johnson and Parsons, 1985). It can be linked to previously validated pasture growth models, and it can be extended for different horizons within the sward by obtaining the above-mentioned parameters on a horizon basis. The functional responses can be estimated on the basis of LAI, leaf mass, or total herbage mass, and the model is sensitive to changes in leaf-to-stem ratio, therefore representing the effect of morphological changes in the grazed sward (Fig. 4).

The functional response to pasture availability should also be modified by animal size, because animals of different sizes have different abilities to harvest forage under different sward conditions (Stephens and Krebs, 1986; Belovsky, 1987; Illius and Gordon, 1987; Ungar and Noy-Meir, 1988). For example, smaller ruminants can

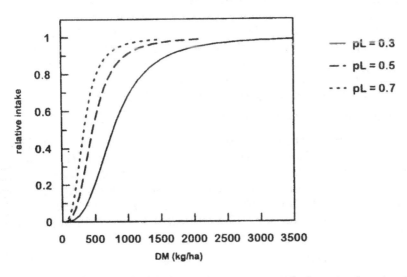

FIGURE 4 Effect of changes in the proportion of leaf on the functional response between herbage availability and intake at grazing.(——) pL = 0.3; (− − −) pL = 0.5; (—) pL = 0.7.

graze shorter swards more efficiently, and therefore swards can reach lower herbage masses before intake is reduced (Illius and Gordon, 1987). Although all models modify potential intake largely on the basis of body weight or a function of it, few models (Zemmelink, 1980; McCall, 1984) modify explicitly the functional response according to body size. Johnson and Parsons (1985) claim a value of $K = 1$ for ewes (i.e., 80 kg body weight) grazing ryegrass while a value of $K = 2$ provides a suitable relation for mature dairy cattle (i.e., 600 kg body weight) (Parsons, personal communication). However, they do not provide a specific relation of this parameter with body weight. We scaled parameter K to body weight using an allometric relationship derived from Illius and Gordon (1987), who claim that differences in the ability of animals of different sizes to graze are caused by differences in incisor breadth, hence mouth size. The relationship between parameter K and body weight then becomes

$$K = 0.229 \ BW^{0.36} \tag{9}$$

Figure 5 shows this relation for three body sizes.

Step 3   The third and final step in calculating intake in these models is to multiply potential intake by the relative intake factor and by the number of grazing animals. This approach is probably the most commonly used to represent the effect of herbage availability on intake because of its simplicity and the ease of obtaining appropriate data for validation. However, these systems fail to represent the mechanics of grazing and therefore fail to provide full understanding about the sward variables affecting intake. Therefore, for some purposes, more detailed models, usually based on grazing behavior measurements, are used to represent these relations.

3.   Prediction of Intake from Grazing Behavior Measurements

The prediction of intake from grazing behavior measurements [for recent reviews see Hodgson et al. (1994), Demment et al. (1995), and Laca and Demment (1996)] has been largely based on the early work of Allden (1962), Arnold and Dudzinski (1967a, 1967b), Allden and Whittaker (1970), Stobbs (1970, 1973, 1974), and Chacon and Stobbs (1976). Allden and Whittaker (1970) postulated that intake at grazing could be predicted as

$$Intake = IB \ RB \ GT \tag{10}$$

where IB = bite size, RB = biting rate, and GT = grazing time.

Intake per bite is the variable most sensitive to sward characteristics, while biting rate and grazing time are partly dependent on

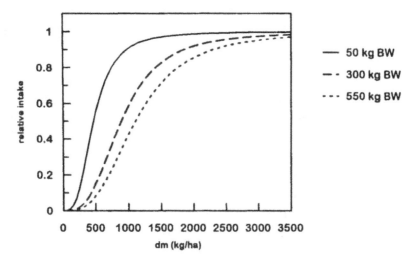

FIGURE 5 Effect of different body weights on the shape of the functional response between herbage mass and intake. (——) 50 kg BW; (– – –) 300 kg BW; (---) 550 kg BW.

bite size and act as compensatory mechanisms when bite size is too small to obtain the desired intake level (Hodgson, 1981). Chambers et al. (1981) and Newman et al. (1994a) suggested that biting rate declines at high bite sizes because of an increase in the ratio of manipulative to biting jaw movements and therefore it is also partly dependent on sward characteristics. This subject has been clearly depicted by Laca et al. (1994), who found that time per bite (TB) was linearly associated with the total number of jaw movements per bite (JM) (Fig. 6a):

$$TB = 0.43 + 0.682 \text{ JM}, \qquad r^2 = 0.96 \tag{11}$$

The proportion of total manipulative jaw movements that performed manipulation and mastication (MJM) increased asymptotically with bite size (Fig. 6b) according to the relation

$$MJM = \frac{1.028 \text{ IB} - 0.246}{0.234 + \text{IB}}, \qquad r^2 = 0.69 \tag{12}$$

As bite size decreases, biting rate and/or grazing time increase to compensate for this reduction. However, this compensation is sometimes partial (Allden and Whittaker, 1970; Stobbs, 1973; Jamieson and Hodgson, 1979; Hendricksen and Minson, 1980; Hodgson, 1981); thus

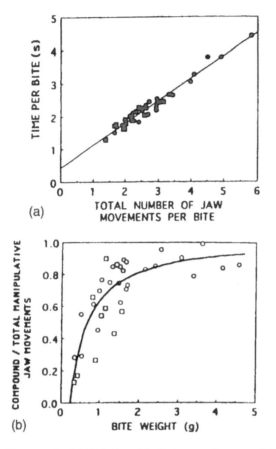

FIGURE 6   (a) Relationship between time per bite and total number of jaw movements per bite. (b) Relationship between bite size (IB) and the proportion of manipulative jaw movements relative to total jaw movements. [From Laca et al. (1994).]

potential intake cannot be attained. Hodgson (1986) claims that this is the reason variations in daily herbage intake frequently reflect closely the observed variations in bite size.

In most modeling studies, maximum values of biting rate and grazing time obtained from experimental studies are often used as behavioral limits of the grazing process, while most efforts are concentrated on modeling bite dimensions. Maximum biting rate is close to 36,000–40,000 bites/day (Stobbs, 1973; Chacon and Stobbs, 1976; Jamieson and Hodgson, 1979), while maximum grazing time is about

12–13 h/day (Fig. 7) (Arnold, 1981). These values are similar for cattle and sheep (Hodgson, 1982, 1985; Forbes, 1988; Demment et al., 1995).

Bite size is positively related to herbage mass or sward height (Black and Kenney, 1984; Hodgson, 1985; Forbes, 1988; Burlison et al., 1991). Burlison et al. (1991), working in swards ranging from 5 to 55 cm in height, explained 78% of the variation in IB of sheep with the relation

$$IB = 33 + 5.2H \tag{13}$$

where H = sward height.

The slope of this relationship was similar to those reported by Hodgson (1981) and Forbes (1982) when expressed on the basis of bite size per kilogram body weight. Burlison et al. (1991) also argued that due to the bias caused by changes in bulk density across grazed horizons, the responses often found were asymptotic, thus confirming the results of other authors (Penning, 1986; Ungar and Noy-Meir, 1988; Baker et al., 1992).

A better understanding of how changes in sward characteristics affect bite size can be achieved by describing this variable at a lower, more detailed, level of aggregation. Burlison et al. (1991) describe the components of bite size in Fig. 8.

Bite depth is generally proportional to sward height (Milne et al., 1982; Wade et al., 1989; Laca et al., 1992; Ungar et al., 1992; Demment et al., 1995), but it may decrease depending on the relative height of stem material in the grazed horizons (Barthram and Grant, 1984; Forbes, 1988; Flores et al., 1993). Burlison et al. (1991) found that the following relation for sheep explained 93% of the variation in bite depth:

$$\text{Bite depth} = -1.0 + 0.37H \tag{14}$$

Since sward height is a good predictor of LAI (Parsons et al., 1994), which in turn reflects leaf mass [see Eq. (7)], these results are in close agreement with the relation between leafiness and bite size found in several pastures (Stobbs, 1975; Chacon and Stobbs, 1976; Hendricksen, and Minson, 1980; Hodgson, 1986).

Bite area increases with sward height (Burlison et al., 1991; Laca et al., 1992). However, it also increases with decreasing bulk density (Burlison, 1987), especially on higher swards (Laca et al., 1992), which might be explained by the limitations posed by the shearing strength required to harvest a bite (Hodgson, 1985). Nevertheless, bite area is less sensitive to sward characteristics than bite depth (Hodgson, 1986)

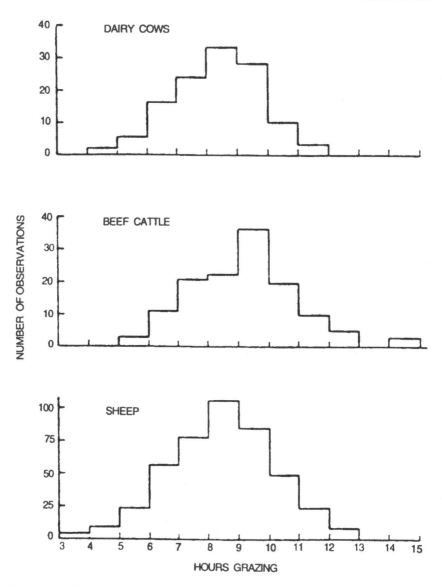

FIGURE 7 Distribution of grazing times of cattle and sheep. [From Arnold (1981.)]

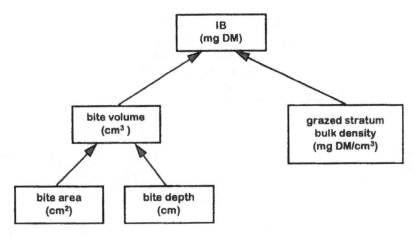

FIGURE 8   The components of bite size. [From Burlison et al. (1991.)]

and has been considered constant for a given body weight in some models (Parsons et al., 1994).

Bite volume is also positively related to sward height (Ungar and Noy-Meir, 1988). Burlison et al. (1991) obtained the following relationship for sheep, thus explaining 83% of the variation in bite volume:

$$\text{Bite volume} = -32 + 8.0H \qquad (15)$$

There are some factors unrelated to sward structure that affect grazing behavior, mainly grazing time. For example, Brumby (1959) and Journet and Demarquilly (1979) showed that cows increased their grazing time by 5 min/kg milk between yields of 5 and 25 kg of milk, and by 12 min/kg milk between 20 and 35 kg of milk. Similarly, Arnold and Duzinski (1967a) and Arnold (1975) found increases of 7–12% in grazing time during early lactation of sheep. Dougherty et al. (1988) found no difference in biting rate, bite size, and grazing time when cattle were supplemented with ground corn at levels up to 4.5 kg/animal. However, other authors (Holder, 1962; Marsh et al., 1971; Leaver, 1986; Mayne and Wright, 1988; Rook et al., 1994) suggest that grazing time is reduced with supplementation, with the level of reduction being dependent on the type and level of supplementation. For example, Marsh et al. (1971) found reductions in grazing time of 22 min/kg concentrate fed, while Mayne and Wright (1988) found reductions of 43 min/kg when silage was fed. The type of supplement and its interactive effects with the basal diet might be the reason why Dougherty et al. (1988) could not find differences in grazing time at various levels of supplementation. In temperate regions, the largest proportion (70–90%) of grazing time occurs during daylight (Penning et al., 1991; Rook et al., 1994); however, in the tropics night grazing is frequently observed due to high ambient temperatures during the day (Humphreys, 1991). The largest proportion of rumination time also occurs during daylight (Rook et al., 1994). Mastication and rumination increase with increased neutral detergent fiber (NDF) concentration in forages (Demment and Greenwood, 1988) or reduced digestibility (Arnold, 1981), in order to reduce particle sizes of the forage consumed for passage through the gastrointestinal tract. Fasting increases grazing time and reduces rumination time (Greenwood and Demment, 1988).

Comparison of predicted intakes from grazing behavior measurements with limited experimental data has shown close agreement (Gordon, 1995). However, these models still require considerable effort to be widely validated; thus their range of application has been limited mostly to research purposes. Nevertheless, they have provided a significant contribution to the understanding of intake from grazed grasslands in the past two decades and have given valuable insight to assist design of appropriate management strategies in some grazing systems.

A recognized criticism of this approach is that a large part of grazing behavior is caused by the animals' need for nutrient supply (Ungar and Noy-Meir, 1988). Most models do not integrate the two processes, and those that do have not integrated mechanistic models of digestion and metabolism at the same level of aggregation as they treat grazing behavior. Considerable research needs to be done to address these issues.

B. Diet Selection

Diet selection is one of the crucial elements in grazing systems models for appropriate prediction of animal performance (see above). The two basic distinctions that are made are (1) selection within pasture species and (2) selection between plant species. Following Thornley et al. (1994), it is possible to describe the approaches for modeling diet selection as (1) empirical (descriptive), (2) goal-oriented (teleonomic), and (3) mechanistic (reductionist). The application of a particular approach is dependent on the type of pasture and animal models used.

There is general agreement that ruminants prefer to eat leaf instead of stem or dead material and that the material eaten is usually of a higher nutritive value than the material on offer. Ruminants also tend to avoid plants with antinutritional compounds (e.g., tannins, alkaloids). In rangelands, abundance, nutritive value, and spatial distribution are interrelated. For example, plants of a higher nutritive value are less abundant than low quality ones (Belovsky, 1987). Management practices and the spatial distribution of plant species create grazing routes that animals follow and thus have an influence on the diet selected. Animals also tend to graze closer to the water source in arid and semiarid environments (Arnold, 1981). Even when some basic empirical rules, such as those previously mentioned, appear to exist, the mechanisms used by animals to select their diet have not been fully elucidated.

Empirical representations of diet selection are the most common in grazing models, and in general they use the basic principles described above. Examples of these can be found in Christian et al. (1978), Illius (1986), Blackburn and Kothmann (1991), Baker et al. (1992), and Freer et al. (1997). These models assign "selectivity coefficients" on the basis of digestibility or palatability of different morphological units (Illius, 1986; Blackburn and Kothmann, 1991) or physiological states (e.g., Freer et al., 1997). A problem that arises with assigning these types of coefficients on a plant species basis is that they are modified according to the species composition of the patch and therefore may modify diet selection. For example, in terms of acceptability for an animal, the selectivity coefficient of the species depends on the other species present. Arnold (1981) argues that little progress is going to be made in understanding diet selection as long as nutritive value is expressed with traditional analyses (e.g., digestibility, cell wall constituents, nitrogen), because these cannot be described at a molecular level and therefore the substances determining "palatability" cannot be fully determined.

Goal-seeking diet selection models are based on foraging theory (Stephens and Krebs, 1986). The general principle behind them is optimization of the diet selected using the predator–prey concept. The ruminant (predator) will try to maximize its *benefits* (e.g., energy retention in most cases) relative to the *costs* of obtaining them (e.g., energy expenditure due to searching, handling, and walking) by optimally selecting between plant species and/or plant parts (prey). These models are used mostly for ecological research (Belovsky, 1987; Thornley et al., 1994; Newman et al., 1994b).

Few mechanistic models of diet selection are available (Parsons et al., 1994), and it is recognized that a mechanistic representation is still far from complete due to the lack of knowledge to describe mechanistically some of the factors affecting diet selection. Certainly, this is an area that requires more research to improve understanding of the mechanisms involved and to make better predictions of the diets selected by grazing ruminants.

IV. MODELING ANIMAL PERFORMANCE

From the viewpoint of whole grazing systems, it is now clear that the plant/animal interface is not completely represented if the consequences of grazing and other nutritional management practices (e.g., production) on the animal are not modeled. A series of papers and books relate to the subject (e.g., Forbes and France, 1993; van Soest, 1994; Journet et al., 1995), but there appears to be no consensus on the best approaches to modeling these processes. Nevertheless, those discussed in the following subsections represent, broadly, the most common approaches.

A. Empirical Relations Between Stocking Rate and Animal Production

It is widely recognized that stocking rate (SR) is one of the major determinants of animal production from pastures and the sustainability of the grazing system. There have been a number of mathematical descriptions of the relationship between SR and animal performance (e.g., Mott, 1961; Petersen et al., 1965; Edye et al., 1978), but the one most commonly used was derived by Jones and Sandland (1974), who suggested that (1) the relation between animal performance per head (kg/hd) and SR could be described by a linear regression and (2) the relation between animal production per hectare (kg/ha) and SR was quadratic.

Apart from SR, other authors have used different statistical relations between animal performance and herbage availability [see Humphreys (1991)], level of N fertilization (Karnezos et al., 1988), rainfall (Bransby, 1984), pasture species (McCaskill and McIvor, 1993; McIvor and Monypenny, 1995), and others. These relationships will not be considered further, since we believe they are not appropriate for use in grazing systems models because they are statistical relationships of specific datasets and as such represent only the data from which they are derived (slopes and intercepts vary significantly between studies); they do not provide an understanding of the factors influencing animal performance, and they do not have the flexibility to represent changes in management practices within the system.

Therefore they are not suitable to test alternative strategies on the behavior of the system and its parts. However, for a full explanation of these types of relationships see Humphreys (1991).

B. Systems of Nutritional Requirements

The energy requirements of ruminants have been estimated with reasonable accuracy, and differences between the systems used in different countries (i.e., Jarrige, 1989; NRC, 1989, 1996; SCA, 1990; AFRC, 1993) seem to be small (McDonald et al., 1995). Traditional "requirements systems" were not designed to predict intake but to assess the nutritional and productive consequences of different feedstuffs for the animal once their intake was known. Therefore, a criticism that often arises is that the effective calculation of nutrient supply to the animal, and hence the quality of the predictions of animal performance, are largely dependent on the accuracy of the intake estimate used for the calculations. Hence the importance of the representation of intake prediction in grazing systems models.

Several models of grazing systems, whether designed for sheep or beef or dairy cattle, rely on one form or another of an energy requirements system to represent animal performance (Vera et al., 1977; Christian et al., 1978; Sibbald et al., 1979; Konandreas and Anderson, 1982; White et al., 1983; Doyle et al., 1989; Walker et al., 1989; Richardson et al., 1991; Seman et al., 1991; Hanson et al., 1994; Thornley et al., 1994; Freer et al., 1997). However, from the nutritional management viewpoint, these systems per se present some inadequacies that need to be addressed by other mechanisms to improve their flexibility.

1.   These systems are static, and digestibility estimates are central to the calculation of energy in feedstuffs in the appropriate units (e.g., DE, ME, NE). In requirements systems, these estimates are an input and are fixed for a particular feedstuff. However, effective digestibility is a consequence of degradation and passage through the gut, and therefore it is dependent on plant and animal characteristics (Demment and Greenwood, 1988; Illius and Gordon, 1991). Due to the inherent selection by grazing animals on the basis of chemical and physical characteristics of different plants and/or plant parts (see Section III), it is necessary to model degradation and passage before describing digestibility and consequent nutrient supply. This requires dynamic models.

2.   Even the most recent requirements systems do not take into account explicit protein–energy interactions (Oldham, 1984; Preston

and Leng, 1987). Lack of rumen-degradable protein reduces microbial growth and depresses the rate of structural carbohydrate digestion (Ørskov, 1992). Therefore, the effect of some supplementation strategies on animal performance cannot be predicted adequately (Preston and Leng, 1987).

3. Most requirements systems do not take into account interactions between different feeds [except limited interactions modeled by Sniffen et al. (1992) and NRC (1996)]. For example, the reduction in cell wall digestibility is a well-known consequence of reduced rumen pH caused by feeding large quantities of concentrates (Istasse et al., 1986; Argyle and Baldwin, 1988), and this, and the subsequent forage/concentrate substitution rates, cannot be predicted adequately by some requirements systems.

4. Requirements systems require knowledge of the current level of production to calculate requirements and are therefore not *predictors* of animal performance. Since they were designed mainly from observations of stall-fed animals, they were implemented to calculate the quantities and types of feeds to give to an animal at a known level of production. In other words, animal performance was not predicted, it was usually an input to the calculation (even when using the intake prediction equations in these systems). The rationale behind prediction of animal performance in grazing systems should be exactly the opposite: What level of production can be attained, relative to the potential production of an animal of a given size and in a given physiological state, by following a particular grazing and overall nutritional strategy? Potential production is a function of the animal's genetic characteristics (Oldham and Emmans, 1988), while actual production is dependent on the resources available to the animal, the way it can utilize them, and the overall management of the grazing system.

We believe that the place of these systems in grazing systems models lies in the estimation of only the potential requirements, which are dependent mainly on body weight, physiological state, and level of production. However, the estimation of the supply of nutrients to meet those requirements needs a different approach, namely, dynamic models of digestion.

C. Dynamic Models of Digestion

A wide range of dynamic models of digestion can be found in the literature (e.g., Waldo et al., 1972; Mertens and Ely, 1979; Forbes, 1980; Black et al., 1980; Bywater, 1984; Fisher et al., 1987; Hyer et al., 1991;

Illius and Gordon, 1991, 1992; Sniffen et al., 1992; Fisher and Baumont, 1994). These types of models have been recently reviewed by Illius and Allen (1994), and Baldwin (1995) reviewed research models representing metabolism and the formation of end products of fermentation in ruminants[*] (e.g., Baldwin et al., 1970, 1977, 1987; Gill et al., 1984; Murphy et al., 1986; Danfaer, 1990; Dijkstra et al., 1992, 1993; Poppi et al., 1994). A range of approaches to model digestive processes can also be found in Forbes and France (1993).

The basic objectives of dynamic models of digestion are to predict potential intake, digestibility, and animal performance as a function of the nutritional quality of plants on offer and a range of animal characteristics. There is evidence that such models are better at predicting nutrient supply and animal performance than requirements systems (Fox et al., 1992, 1995; Ainslie et al., 1993). However, there are certain basic aspects that need to be considered that define the accuracy and flexibility of the model. A discussion of these follows.

Description of Feed Fractions   The basic fractionation of feedstuffs is represented in Fig. 9. The separation of dry matter into its basic chemical entities is important because different feed fractions of different forages have different degradation and passage rates (Illius and Gordon, 1991; Russell et al., 1992) and therefore have different digestibilities. Consequently, they supply different amounts of nutrients to the animal (Murphy et al., 1982; Gill et al., 1990). These fractionations are also important into predicting effects of supplementation on the rate of cell wall digestion (Argyle and Baldwin, 1988), modeling protein–energy interactions, and using recent standards of protein requirements (e.g., Fox et al., 1992; O'Connor et al., 1993; AFRC, 1993). Nevertheless, other authors consider that the nutritional description of the potentially degradable fractions of feedstuffs requires yet further fractionations (Mertens and Ely, 1979; Sniffen et al., 1992), although it is questionable that they provide better predictions than simpler approaches (Illius and Allen, 1994). The fractionation presented here is robust, simple, and suitable for use in whole grazing system models.

Degradation Kinetics   The concentration and potential degradation kinetics of the cell wall of forages are among the important determinants of intake and digestibility (Mertens, 1987; Illius and

---

[*]Readers are referred to Baldwin (1995) for further information, since metabolic models are not covered in this chapter.

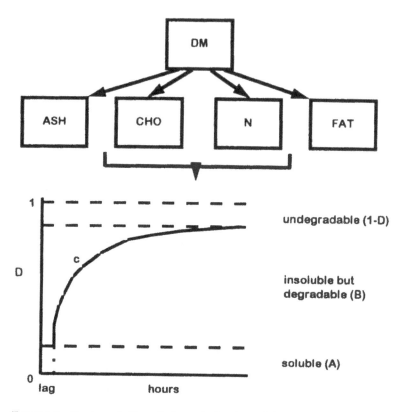

FIGURE 9   Basic nutritional characterization of forages.

Gordon, 1991). The degradation kinetics of the crude protein fraction reflect nitrogen supply to rumen microbes (Czerkawski, 1986; Ørskov, 1992; AFRC, 1993) and therefore have an effect on cell wall degradation rates. Potential degradation characteristics are a function of plant characteristics (Russell et al., 1992; Ørskov, 1994).

Parameters reflecting the potentially degradable fraction can be obtained from dacron bag studies (McDonald, 1981; Dhanoa, 1988) or in vitro gas production measurements (Herrero and Jessop, 1996, 1997; Herrero et al., 1996b; Jessop and Herrero, 1996; Jessop et al., 1996) by fitting the first-order model described by Waldo et al. (1972) and McDonald (1981), where the rate of degradation is proportional to the amount of substrate:

$$D = A + B(1 - \exp^{-c(t-lag)}) \tag{16}$$

where D = degradation, A = soluble fraction (usually determined as the washing loss in degradation studies), B = insoluble but potentially degradable fraction, degraded at a fractional rate c (h^{-1}), lag = lag phase before degradation begins (h), and t = time. See Jessop and Herrero (1996) and Herrero and Jessop (1996) for a description of the method when gas production measurements are used. Second-order models are also used to describe degradation kinetics, but the complexity of the analysis increases, since several microbial pools need to be represented to ensure proper biological behavior of the model (Illius and Allen, 1994). For alternative descriptions of degradation kinetics see Mertens (1993).

Passage Rates While there have been great efforts to describe degradation characteristics of forages, substantially less information is available on passage rates. This is surprising, since digestibility and intake are functions of the competition between these two processes, and therefore passage rates are equally important. Passage rates are closely related to the mechanics of breakdown, largely caused by rumination, from large to small particles in the rumen (Kennedy and Murphy, 1988; Wilson and Kennedy, 1996), and this is why several models (Mertens and Ely, 1979; Fisher et al., 1987; Illius and Gordon, 1991) use different compartments reflecting pools of large to small particles to describe the different carbohydrate fractions. However, the required number of compartments to describe adequately particle dynamics and whether fractionation is really necessary to improve intake predictions are not known (Illius and Allen, 1994). Obviously, with this approach the understanding of the processes controlling the flow of material through the gut is greater, and this (depending on modeling objectives) should be seen as an advantage. It is convenient to represent rates of breakdown and passage as a function of animal characteristics (Illius and Gordon, 1991) since this improves the accuracy of predictions of intake (Illius and Allen, 1994). Illius and Gordon (1991) found the following relations for breakdown of large to small particles (BR):

$$BR = 0.144 \text{ ICW}^{-0.144} \text{ BW}^{-0.27}, \qquad r^2 = 0.62 \tag{17}$$

where ICW = indigestible cell wall (g/kg) and BW = body weight (kg).

They also found the following relationships for passage through the whole gut (PWG, the inverse of mean retention time) and passage of small particles (SPR) from the rumen, respectively:

$$PWG = 0.071 \text{ BW}^{0.27}, \qquad r^2 = 0.76 \tag{18}$$

$$\text{SPR} = 0.75 \text{ PWG}, \quad \text{c.v. } 15.5\% \tag{19}$$

We analyzed the data of Shem et al. (1995) on 17 forages and applied a scaling rule similar to that of Illius and Gordon (1991). We found that particle passage rate (PR) from the rumen could be described by

$$\text{PR} = (0.0256 - 0.00007B$$

$$+ 0.00127cB)(3.96 \text{ BW}^{-0.27}), \quad r^2 = 0.82 \tag{20}$$

For nomenclature see Eqs. (16) and (17). These passage rates apply to the B fraction and the indigestible fraction of cell wall of forages and represent mostly small particles. Passage rates also depend on the feeding level of the animal (AFRC, 1993). To account for effects of feeding level (FL, in multiples of maintenance), these should be multiplied by 0.25 FL, and a scaling rule similar to that claimed by Sniffen et al. (1992) is obtained. Other relations can be found in Sauvant et al. (1995) but for total DM. Certainly more work is required on this subject to understand the factors affecting passage rates (e.g., buoyancy and its relation to particle fermentation and density). For concentrated feeds, Sniffen et al. (1992) describe the passage rate (PRC) as

$$\text{PRC} = -0.424 + (1.45 \text{ PR}) \tag{21}$$

Rumen Size   For intake predictions, most dynamic models require that a threshold value be set for the maximum capacity of the rumen or total gut. Accurate allometric relations for these parameters are found in the literature [see Peters (1983), Demment and van Soest (1985), Mertens (1987), Demment and Greenwood (1988), and Illius and Gordon (1991)], Illius and Gordon (1991) determined this allometric relation for 18 species and found that the relationship between the weight of dry matter in the rumen (DMR) and body weight could be described by

$$\text{DMR} = 0.021 \text{ BW}, \quad r^2 = 0.98 \tag{22}$$

Mertens (1987) found a very similar relation when expressing rumen content on the basis of neutral detergent fiber.

Dynamic models of digestion and their descriptions of feed and animals are useful for the integration of other processes within the grazing system. Since models of this nature monitor the flow of feed components through the gastrointestinal tract, they predict their excretion patterns and the composition of excreta, which are integral to linking the animal with the soil fertility subsystems and their consequent effects on pasture growth in grazing models.

V. OTHER RELATIONS BETWEEN PLANTS AND ANIMALS

There are other relations that need to be taken into consideration when modeling the plant/animal interface that are particularly difficult to model. For example, treading, poaching, and fouling can reduce herbage availability and the subsequent regrowth of the sward (Brockington, 1972; Christian, 1981; Wilkins and Garwood, 1986).

Forage is damaged due to trampling and poaching, especially in wet soils (Wilkins and Garwood, 1986) and/or in very high swards (Herrero, unpublished), but, as Christian (1981) states, it would be difficult to account for these effects in a grazing model.

Dung pats and urine affect herbage availability and modify diet selection patterns of ruminants (Brockington, 1972), thus leading to the spatial effect of patchy swards in some cases. The effects of excreta are greater than those of urine and are mediated by the number of dung pats, the area they cover, and the stocking rate. The most common way to model these effects is by empirical relations that are usually dependent on stocking rate (Brockington, 1972; Hanson et al., 1994) to scale the amount of herbage available to the animal. Dung pats can also affect herbage growth by excluding light from the patches for several months (Wilkins and Garwood, 1986), but this effect can be reduced by management practices, at least in intensive systems. In rangelands it is difficult to control but stocking rates are also lower and therefore the effect of dung on animal consumption is, perhaps, less important.

Nevertheless, quantification of these aspects is important, because significant amounts of dung can be deposited in pastures. Its contribution to nutrient cycles cannot be neglected due to its key role in the sustainability of the grazing system and the overall dynamics of the biology of grazing cycles. However, better approaches are required to quantify the fate of minerals from ruminant excretions to the soil, water, and atmosphere (Scholefield et al., 1991; Humphreys, 1994).

VI. FUTURE RESEARCH AND DEVELOPMENT NEEDS

Future research needs can be divided into two groups: research on aspects dealing with the knowledge acquisition and representation of the main biological processes and definition of an integrated approach for the selection of management interventions leading to sustainable grazing systems.

A. Biological Processes

In terms of biological processes, there are aspects in both the plant and animal sciences that need to be addressed. For example, although mechanistic models of single pasture species are available and have proven robust in their predictions, there is a need for better understanding of the processes controlling growth in grass–legume associations and rangelands. Understanding competition for resources (light and nutrients) by different species and finding suitable mathematical definitions that reflect the biological processes is not a trivial task. This is closely linked with the representation of grazing processes and diet selection, even in single-species pastures. Most representations of diet selection have been empirical, and until the mechanisms that control diet selection have been elucidated, little progress is going to be made in modeling diet selection. A key issue in solving this problem is the need to link the behavioral aspects of grazing to digestion and metabolism models, since the release and balance of nutrients and pattern of supply play an important part in controlling rates of intake and what the animal chooses to eat (Gill and Romney, 1994). In terms of intake prediction, it appears that the weakest information is related to the flow of material through the gut. More efforts should be directed toward research into the factors controlling the passage of feed particles through the gastrointestinal tract.

B. Decision Support Systems: An Integrated Approach

Models can be built solely to increase our understanding about systems under study and the nature of the functional relations between parts of the system. For example, many authors have investigated the effect of certain variables on the stability and steady states of grazing systems (Noy-Meir, 1975, 1976, 1978; Johnson and Parsons, 1985; Thornley and Veberne, 1989), and other scientists have looked at more fundamental relationships between the animal and plant communities in terms of body weight effects (allometry) (Belovsky, 1987; Demment and van Soest, 1985; Illius and Gordon, 1987; Taylor et al., 1987), grazing behavior (Ungar and Noy-Meir, 1988, Ungar et al., 1991; Laca et al., 1992; Parsons et al., 1994), diet selection (Belovsky, 1987; Parsons et al., 1994; Newman et al., 1994b), or digestive processes (Illius and Allen, 1994; Baldwin, 1995). Increased understanding of these processes has led to improved methods for modeling grazing systems

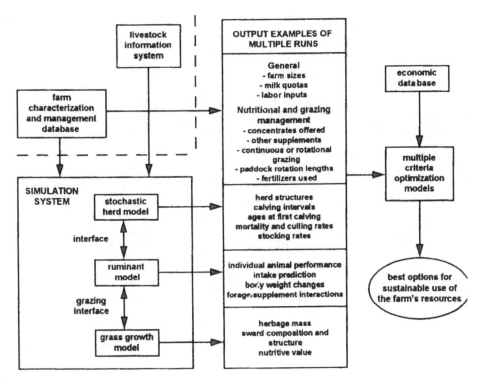

FIGURE 10 A decision support system for grazing systems. [From Herrero et al. (1996a).]

at various levels of detail. However, it is important to emphasize the trade-off between the objective of the model and its accuracy and level of detail if cost-effective models are to be built.

For some purposes, the use of grazing systems models is not complete if mechanisms to select between alternative grazing strategies are not available (Herrero et al., 1996a, 1997). This is specially valid if the models are to be used in farm management or in a regional planning context.

The classical approach to selection of management strategies has been to use linear programming (LP) models with the objective of optimizing economic performance (Dent et al., 1986; Conway and Killen, 1987; Kleyn and Gous, 1988; Olney and Kirk, 1989), but these have not used whole system simulation models to provide the inputs for the LP models. Two important things need to be considered. First, it is well recognized that economic optimization is only one and not necessarily the main objective of farmers (Gasson and Errington, 1993; Perkin and Rehman, 1994; Dent et al., 1994; Dent, 1996). Therefore, the selection mechanism needs to consider several simultaneous objectives and trade-offs between them. Herrero et al. (1996a, 1997) consider that the use of multiple-criteria decision-making models (MCDM), which are extensions of linear programming models, can be linked to a simulation system to create a DSS and provide options for the management of the grazing system (Fig. 10).

A similar approach was used by Veloso et al. (1992) with crop models. Since a range of multiple objectives can be represented, they have the flexibility of dealing with different types of farmers and their managing capacities. The simulation system provides the dynamics of the system under a variety of management scenarios, and the MCDM selects the best alternatives according to the farmer's objectives. Improved selection of strategies can be gained if the objectives of the farmers are better represented, and this requires a further substantial input from the social and behavioral sciences (Dent, 1996).

REFERENCES

AFRC. 1993. *Energy and Protein Requirements of Ruminants*. An advisory manual prepared by the AFRC Technical Committee on Responses to Nutrients. Wallingford, UK: CAB International.

Ainslie, S. J., D. G. Fox, T. C. Perrey, D. J. Ketchen, and M. C. Barry. 1993. Predicting amino acid adequacy of diets fed to Holstein steers. *J. Anim. Sci.* 71:1312–1319.

Allden, W. G. 1962. Rate of herbage intake and grazing time in relation to herbage availability. *Proc. Aust. Soc. Anim. Prod.* 4:163–166.

Allden, W. G. 1981. Energy and protein supplements for grazing livestock. In: *Grazing Animals*. F. H. W. Morley (Ed.). World Animal Science, B1, Amsterdam: Elsevier, pp. 289–307.

Allden, W. G., and I. A. M. Whittaker. 1970. The determinants of herbage intake by grazing sheep: The interrelationship of factors influencing herbage intake and availability. *Aust. J. Agric. Res.* 21:755–766.

Argyle, J. L., and R. L. Baldwin. 1988. Modeling of rumen water kinetics and effects of rumen pH changes. *J. Dairy Sci.* 71:1178–1188.

Arnold, G. W. 1975. Herbage intake and grazing behaviour in ewes of four breeds at different physiological states. *Aust. J. Agric. Res.* 26:1017–1024.

Arnold, G. W. 1981. Grazing behaviour. In: *Grazing Animals*. F. H. W. Morley (Ed.). World Animal Science, B1. Amsterdam: Elsevier, pp. 79–104.

Arnold, G. W., and M. L. Dudzinski. 1967a. Studies on the diet selected of the grazing animal. II. The effect of physiological status in ewes and pasture availability on herbage intake. *Aust. J. Agric. Res.* 18:349–359.

Arnold, G. W., and M. L. Dudzinski. 1967b. Studies on the diet selected of the grazing animal. III. The effect of pasture species and pasture structure on the herbage intake of sheep. *Aust. J. Agric. Res.* 18:657–666.

Arnold, G. W., N. A. Campbell, and K. A. Galbraith. 1977. Mathematical relationships and computer routines for a model of food intake, liveweight change and wool production in grazing sheep. *Agric. Syst.* 2:209–225.

Baker, B. B., R. M. Bourdon, and J. D. Hanson. 1992. FORAGE: A model of forage intake in beef cattle. *Ecol. Modelling* 60:257–279.

Baldwin, R. L. 1995. *Modelling Ruminant Digestion and Metabolism*. London: Chapman & Hall.

Baldwin, R. L., H. L. Lucas, and R. Cabrera. 1970. Energetic relationships in the formation and utilization of fermentation end-products. In: *Physiology of Digestion and Metabolism in the Ruminant*. A. T. Phillipson (Ed.). Newcastle Upon Tyne, UK: Oriel Press, pp. 319–334.

Baldwin, R. L., L. J. Koong, and J. Ulyatt. 1977. A dynamic model of ruminant digestion for evaluation of factors affecting nutritive value. *Agric. Syst.* 2: 255–288.

Baldwin, R. L., J. H. M. Thornley, and D. E. Beever. 1987. Metabolism of the lactating cow. II. Digestive elements of a mechanistic model. *J. Dairy Res.* 54:107–131.

Barthram, G. T., and S. A. Grant. 1984. Defoliation of ryegrass-dominated swards by sheep. *Grass Forage Sci.* 39:211–219.

Belovsky, G. E. 1987. Foraging and optimal body size: An overview, new data and a test of alternative models. J. Theor. Biol. 129:275–287.

Black, J. L., and P. A. Kenney. 1984. Factors affecting diet selection by sheep. II. Height and density of pasture. Aust. J. Agric. Res. 35:565–578.

Black, J. L., D. E. Beever, G. J. Faichney, B. R. Howarth, and N. M. Graham. 1980. Simulation of the effects of rumen function on the flow of nutrients from the stomach of sheep: Part 1. Description of a computer program. *Agric. Syst.* 6:195–219.

Blackburn, H. D., and M. M. Kothmann. 1989. Forage dynamics model for use in range or pasture environments. *Grass Forage Sci.* 44:283–294.

Blackburn, H. D., and M. M. Kothmann. 1991. Modelling diet selection and intake for grazing herbivores. *Ecol. Modelling* 57:145–163.

Blaxter, K. L. 1989. *Energy Metabolism in Animals and Man*. Cambridge, UK: Cambridge Univ. Press.

Bowman, P. J., D. A. Wysel, D. G. Fowler, and D. H. White. 1989. Evaluation of a new technology when applied to sheep production systems. *Agric. Syst.* 29:35–47.

Bransby, D. I. 1984. A model for predicting livemass gain from stocking rate and annual rainfall. *J. Grassland Soc. Southern Africa* 1:22–26.

Brockington, N. R. 1972. A mathematical model of pasture contamination by grazing cattle and the effects on herbage intake. *J. Agric. Sci. Camb.* 79: 249–257.

Brody, S. 1945. *Bioenergetics and Growth.* New York: Reinhold.

Brougham, R. W. 1956. The rate of growth of short-rotation ryegrass pastures in the late autumn, winter and early spring. *N. Z. J. Sci. Technol.* A38:78–87.

Brumby, P. J. 1959. The grazing behaviour of dairy cattle in relation to milk production, liveweight and pasture intake. *N. Z. J. Agric. Res.* 4:798–807.

Burlison, A. J. 1987. Sward canopy structure and ingestive behaviour in grazing animals. Ph.D. Thesis. University of Edinburgh, Scotland.

Burlison, A. J., J. Hodgson, and A. W. Illius. 1991. Sward canopy structure and the bite dimensions and bite weight of grazing sheep. *Grass Forage Sci.* 46:29–38.

Bywater, A. C. 1984. A generalised model of feed intake and digestion in lactating cow. *Agric. Syst.* 13:167–186.

Chacon, E., and T. H. Stobbs. 1976. Influence of progressive defoliation of a grass sward on the eating behaviour of cattle. *Aust. J. Agric. Res.* 27:709–727.

Chambers, A. R. M., J. Hodgson, and J. A. Milne. 1981. The development and use of equipment for automatic recording of ingestive behaviour in sheep and cattle. *Grass Forage Sci.* 36:97–105.

Charles-Edwards, D. A., P. Tow, and T. R. Evans. 1987. An analysis of the growth rates of pastures and animal production. *Agric. Syst.* 25:245–259.

Christian, K. R. 1981. Simulation of grazing systems. In: *Grazing Animals.* F. H. W. Morley (Ed.). World Animal Science, B1. Amsterdam: Elsevier, pp. 361–377.

Christian, K. R., M. Freer, J. R. Donnelly, J. L. Davidson, and J. S. Armstrong. 1978. *Simulation of Grazing Systems.* Wagenigen, The Netherlands: Pudoc.

Conrad, H. R., A. D. Pratt, and J. W. Hibbs. 1964. Regulation of feed intake in dairy cows. 1. Change in importance of physical and physiological factors with increasing digestibility. *J. Dairy Sci.* 47:54–62.

Conway, A. G., and L. Killen. 1987. A linear programming model of grassland management. *Agric. Syst.* 25:51–71.

Coughenour, M. B. 1984. A mechanistic simulation analysis of water use, leaf angles, and grazing in East African graminoids. *Ecol. Modelling* 26:203–230.

Coughenour, M. B., S. J. McMaughton, and L. L. Wallace. 1984. Modelling primary production of perennial graminoids—uniting physiological processes and morphometric traits. *Ecol. Modelling* 23:101–134.

Cowan, R. T., T. M. Davison, and R. K. Shephard. 1986. Observations on the diet selected by Friesian cows grazing tropical grass and grass-legume pastures. *Trop. Grasslands* 20:183–192.

Czerkawski, J. W. 1986. *An introduction to Rumen Studies.* Oxford, UK: Pergamon.

Danfaer, A. 1990. *A Dynamic Model of Nutrient Digestion and Metabolism in Lactating Dairy Cows.* No. 671. Beretning. Foulum, Denmark.

Demment, M. W., and G. B. Greenwood. 1988. Forage ingestion: Effects of sward characteristics and body size. *J. Anim. Sci.* 66:2380–2392.

Demment, M. W., and P. J. Van Soest. 1985. A nutritional explanation for body-size patterns of ruminant and nonruminant herbivores. *Am. Nat.* 125: 641–671.

Demment, M. W., J. L. Peyraud, and E. A. Laca. 1995. Herbage intake at grazing: A modelling approach. In: *Recent Developments in the Nutrition of Herbivores.* M. Journet et al. (Eds.). Paris, France: INRA Editions, pp.121–141.

Dent, J. B. 1996. Theory and practice in FSR/E: Consideration of the role of modelling. In: *Systems-Oriented Research in Agriculture and Rural Development.* M. Sebillotte (Ed.). Montpellier, France: CIRAD-SAR, pp. 100–110.

Dent, J. B., S. R. Harrison, and K. B. Woodford. 1986. *Farm Planning with Linear Programming: Concept and Practice.* London, UK: Butterworths.

Dent, J. B., M. J. McGregor, and G. Edwards-Jones. 1994. Integrating livestock and socio-economic systems into complex models. In: *The Study of Livestock Farming Systems in a Research and Development Framework.* A. Gibbon and J. C. Flamant (Eds.). EAAP Publication No. 63. Wageningen, The Netherlands: Wageningen Pers, pp. 25–36.

Detling, J. K., W. J. Parton, and H. W. Hunt. 1979. A simulation model of *Boutelona gracilis* biomass dynamics on the North American shortgrass prairie. *Oecologia* 38:167–191.

Dhanoa, M. S. 1988. On the analysis of dacron bag data for low degradability feeds. *Grass Forage Sci.* 43:441–444.

Dijkstra, J., H. D. St. C. Neal, J. France, and D. E. Beever. 1992. Simulation of nutrient digestion, absorption and outflow in the rumen. Model description. *J. Nutri.* 122:2239–2256.

Dijkstra, J., H. Boer, J. van Bruchem, M. Bruining, and S. Tamminga. 1993. Absorption of volatile fatty acids from the rumen of lactating dairy cows as influenced by volatile fatty acid concentration, pH and rumen liquid volume. *Br. J. Nutri.* 69:385–396.

Dougherty, C. T., T. D. A. Forbes, P. L. Cornelius, L. M. Lauriault, N. W. Bradley, and E. M. Smith. 1988. Effects of supplementation on the ingestive behaviour of grazing steers. *Grass Forage Sci.* 43:353–361.

Doyle, C. J., J. A. Baars, and A. C. Bywater. 1989. A simulation model of bull beef production under rotational grazing in the Waikato region of New Zealand. *Agric. Syst.* 31:247–278.

Edelsten, P. R., and J. E. Newton. 1975. A simulation model of intensive lamb production from grass. Technical Report No. 17. Grassland Research Institute. Hurley, UK.

Edelsten, P. R., and J. E. Newton. 1977. A simulation model of a lowland sheep system. *Agric. Syst.* 2:17–32.

Edye, L. A., W. T. Williams, and W. H. Winter. 1978. Seasonal relations between animal gain, pasture production and stocking rate on two tropical grass-legume pastures. *Aust. J. Agric. Res.* 29:103–113.

Finlayson, J. D., O. J. Cacho, and A. C. Bywater. 1995. A simulation model of grazing sheep. I. Animal growth and intake. *Agric. Syst.* 48:1–25.

Fisher, D. S., and R. Baumont. 1994. Modeling the rate and quantity of forage intake by ruminants during meals. *Agric. Syst.* 45:43–54.

Fisher, D. S., J. C. Burns, and K. R. Pond. 1987. Modeling ad libitum dry matter intake by ruminants as regulated by distension and chemostatic feedbacks. *J. Theor. Biol.* 126:407–418.

Flores, E. R., E. A. Laca, T. C. Griggs, and M. W. Demment. 1993. Sward height and vertical morphological differentiation determine cattle bite dimensions. *Agron. J.* 85:527–532.

Forbes, J. M. 1977. Interrelationships between physical and metabolic control of voluntary feed intake in fattening, pregnant and lactating sheep: A model. *Anim. Prod.* 24:90–101.

Forbes, J. M. 1980. A model of the short term control of feeding in the ruminant: Effect of changing animal or feed characteristics. *Appetite* 1:21–41.

Forbes, J. M. 1993. Voluntary feed intake. In: *Quantitative Aspects of Ruminant Digestion and Metabolism.* J. M. Forbes and J. France (Eds.). Wallingford, UK: CAB International, pp. 479–494.

Forbes, J. M., and J. France. 1993. Quantitative aspects of ruminant digestion and metabolism. Wallingford, UK: CAB International.

Forbes, T. D. A. 1982. Ingestive behaviour and diet selection in grazing cattle and sheep. Ph.D. Thesis. University of Edinburgh, Scotland.

Forbes, T. D. A. 1988. Researching the plant-animal interface: The investigation of ingestive behavior in grazing animals. *J. Anim. Sci.* 66:2369–2379.

Forbes, T. D. A., and J. Hodgson. 1985. Comparative studies of the influence of sward conditions on the ingestive behaviour of cows and sheep. *Grass Forage Sci.* 40:69–77.

Fox, D. G., C. J. Sniffen, J. D. O'Connor, J. B. Russell, and P. J. van Soest. 1992. A net carbohydrate and protein system for evaluating cattle diets. III. Cattle requirements and diet adequacy. *J. Anim. Sci.* 70:3578–3596.

Fox, D. G., M. C. Barry, R. E. Pitt, D. K. Roseler, and W. C. Stone. 1995. Application of the Cornell net carbohydrate and protein model for cattle consuming forages. *J. Anim. Sci.* 73:267–277.

Freer, M. 1981. The control of food intake by grazing animals. In: *Grazing Animals.* F. H. W. Morley (Ed.). World Animal Science, B1, Amsterdam: Elsevier, pp. 105–124.

Freer, M., J. L. Davidson, J. S. Armstrong, and J. R. Donnelly. 1970. Simulation of summer grazing. Proc. 11th. Int. Grassland Congress. Queensland: Surfer's Paradise.

Freer, M., A. D. Moore, and J. R. Donnelly. 1997. GRAZPLAN: Decision support systems for Australian grazing enterprises. II. The animal biology model for feed intake, production and reproduction and the GrazFeed DSS. *Agric. Syst.* (in press).

Gartner, J. A., and D. Hallam. 1984. A quantitative framework for livestock development planning. Part 3. Feed demand and supply. *Agric. Syst. 14*: 123–142.

Gasson, R., and A. J. Errington. 1993. *The Farm Family Business*. Wallingford, UK: CAB International.

Gilbert, B. J. 1975. RANGES: Grassland simulation model. Fort Collins, CO: Colorado State University.

Gill, M., and D. Romney. 1994. The relationship between the control of meal size and the control of daily intake in ruminants. *Livest. Prod. Sci. 39*:13–18.

Gill, M., J. H. M. Thornley, J. L. Black, J. D. Oldham, and D. E. Beever. 1984. Simulation of the metabolism of absorbed energy-yielding nutrients in young sheep. *Br. J. Nutr. 20*:13–23.

Gill, M., D. E. Beever, and J. France. 1990. Biochemical basis needed for the mathematical representation of whole animal metabolism. *Nutr. Res. Rev. 2*:181–200.

Gordon, I. J. 1995. Animal-based techniques for grazing ecology research. *Small Ruminant Res. 16*:203–214.

Greenwood, G. B., and M. W. Demment. 1988. The effect of fasting on short-term cattle grazing behaviour. *Grass Forage Sci. 43*:377–386.

Guerrero, J. N., H. Wu, E. C. Holt, and L. M. Schake. 1984. Kleingrass growth and utilization by growing steers. *Agric. Syst. 13*:227–243.

Hacker, R. B., and G. S. Richmond. 1994. Simulated evaluation of grazing management systems for arid chenopod rangelands in Western Australia. *Agric. Syst. 44*:397–418.

Hacker, R. B., K. M. Wang, G. S. Richmond, and R. K. Lindner. 1991. IMAGES: An integrated model of an arid grazing ecological system. *Agric. Syst. 37*:119–163.

Hanson, J. D., J. W. Skiles, and W. J. Parton. 1988. A multi-species model for rangeland plant communities. *Ecol. Modelling 44*:89–123.

Hanson, J. D., B. B. Baker, and R. M. Bourdon. 1994. SPUR2 Documentation and User Guide. GPSR Tech. Report No. 1. Fort Collins, CO: USDA-ARS, Great Plains Systems Research Unit.

Hendricksen, R., and D. J. Minson. 1980. The feed intake and grazing behaviour of cattle grazing a crop of *Lablab purpureus*. *J. Agric. Sci., Camb. 95*: 547–554.

Herrero, M. 1995. Grassland modelling: A decision-support tool. In: *Central America: Conservation and Sustainable Development*. Proceedings of a workshop on Sustainability of Livestock Production Systems, Aug. 8–11, 1995, San José, Costa Rica. *Cienc. Vet. 17*:72–79.

Herrero, M., and N. S. Jessop. 1996. Relationship between in vitro gas production and neutral detergent fibre disappearance in three tropical grasses. *Anim. Sci. 62*:682.

Herrero, M., and N. S. Jessop. 1997. In vitro gas production of tropical pasture legumes. Proceedings of the XVIII International Grassland Congress, Saskatchewan, Canada. (In press).

Herrero, M., R. H. Fawcett, and J. B. Dent. 1996a. Integrating simulation models to optimise nutrition and management for dairy farms: A methodology. In: *Livestock Farming Systems: Research, Socio-Economics and The Land Manager*. J. B. Dent et al. (Eds.). EAAP Publication No. 79. Wageningen, The Netherlands: Wageningen Pers, pp. 322–326.

Herrero, M., I. Murray, R. H. Fawcett, and J. B. Dent. 1996b. Prediction of the chemical composition and in vitro gas production of kikuyu grass by near-infrared reflectance spectroscopy. *Anim. Feed Sci. Technol.* 60:51–67.

Herrero, M., R. H. Fawcett, E. Perez, and J. B. Dent. 1997. The role of systems research in grazing management: applications to sustainable cattle production in Latin America. In: *Applications of Systems Approaches of the Farm and Regional Levels*. D. P. Teng et al. (Eds.). London, UK: Kluwer Academic, pp. 129–136.

Hodgson, J. 1981. Variations in the surface characteristics of the sward and the short-term rate of herbage intake by calves and lambs. *Grass Forage Sci.* 36:49–57.

Hodgson, J. 1982. Influence of sward characteristics on diet selection and herbage intake by the grazing animal. In: *Nutritional Limits to Animal Production from Pastures*. J. B. Hacker (Ed.). Wallingford, UK: CAB International, pp. 153–166.

Hodgson, J. 1985. The control of herbage intake in the grazing ruminant. *Proc. Nutr. Soc.* 44:339–346.

Hodgson, J. 1986. Grazing behaviour and herbage intake. In: *Grazing*. J. Frame (Ed.). British Grassland Society Occasional Symposium No. 19. pp. 51–64.

Hodgson, J., D. A. Clark, and R. J. Mitchell. 1994. Foraging behaviour in grazing animals and its impact on plant communities. In: *Forage Quality, Evaluation and Utilization*. G. C. Fahey et al. (Eds.). Madison, WI: American Society of Agronomy, pp. 796–827.

Holder, J. M. 1962. Supplementary feeding of grazing sheep—Its effects on pasture intake. *Proc. Aust. Soc. Anim. Prod.* 4:154–159.

Humphreys, L. R. 1991. *Tropical Pasture Utilisation*. Cambridge, UK: Cambridge Univ. Press.

Humphreys, L. R. 1994. *Tropical Forages: Their Role in Sustainable Agriculture*. Essex, UK: Longman Scientific and Technical.

Hunt, H. W., M. J. Trlica, E. F. Redente, J. C. Moore, J. K. Detling, T. G. F. Kittel, D. E. Walter, M. C. Fowler, D. A. Klein, and E. T. Elliott. 1991. Simulation model for the effects of climate change on temperature grassland ecosystems. *Ecol. Modelling* 53:205–246.

Hyer, J. C., J. W. Oltjen, and M. L. Galyean. 1991. Development of a model to predict forage intake by grazing cattle. *J. Anim. Sci.* 69:827–835.

Illius, A. W. 1986. Foraging behaviour and diet selection. In: *Grazing Research at Northern Latitudes*. O. Gudmundsson (Ed.). Proceedings of NATO Workshop, Reykjavik, Iceland: Nato Publishers, pp. 227–236.

Illius, A. W., and M. S. Allen. 1994. Assessing forage quality using integrated models of intake and digestion in ruminants. In: *Forage Quality, Evaluation and Utilization.* G. Fahey et al. (Eds.). Madison, WI: American Society of Agronomy, pp. 869–890.

Illius, A. W., and I. J. Gordon. 1987. The allometry of food intake in grazing ruminants. *J. Anim. Ecol. 56*:989–999.

Illius, A. W., and I. J. Gordon. 1991. Prediction of intake and digestion in ruminants by a model of rumen kinetics integrating animal size and plant characteristics. *J. Agric. Sci., Camb. 116*:145–157.

Illius, A. W., and I. J. Gordon. 1992. Modelling the nutritional ecology of ungulate herbivores: Evolution of body size and competitive interactions. *Oecologia 89*:428–434.

Innis, G. S. 1978. *Grassland Simulation Model.* Ecol. Stud. No. 26. New York: Springer-Verlag.

Istasse, L., R. I. Smart, and E. R. Ørskov. 1986. Comparison between two methods of feeding concentrate to sheep given a diet high or low in concentrate with or without buffering substances. *Anim. Feed Sci. Technol. 16*: 37–49.

Jamieson, W. S., and J. Hodgson. 1979. The effects of variation in sward characteristics upon the ingestive behavior and herbage intake of calves and lambs under a continuous stocking management. *Grass Forage Sci. 34*: 273–282.

Jarrige, R. (Ed.). 1989. *Ruminant Nutrition. Recommended Allowances and Feed Tables.* Paris, France: INRA Editions.

Jessop, N. S., and M. Herrero. 1996. Influence of soluble components on parameter estimation using the in vitro gas production technique. *Anim. Sci. 62*:626–627.

Jessop, N. S., S. Nagadi, and M. Herrero. 1996. Use of in vitro gas production method to study ruminal protein–energy interactions. *J. Dairy Sci. 79*(Suppl. 1):182.

Johnson, I. R., and A. J. Parsons. 1985. A theoretical analysis of grass growth under grazing. *J. Theor. Biol. 112*:346–368.

Johnson, I. R., and J. H. M. Thornley. 1983. Vegetative crop growth model incorporating leaf area expansion and senescence, and applied to grass. *Plant, Cell and Environ. 6*:721–729.

Johnson, I. R., and J. H. M. Thornley. 1985. Dynamic model of the response of a vegetative grass crop to light, temperature and nitrogen. *Plant, Cell Environ. 8*:485–499.

Jones, R. J., and R. L. Sandland. 1974. The relation between animal gain and stocking rate. Derivation of the relation from the results of grazing trials. *J. Agric. Sci. 83*:335–342.

Journet, M., and C. Demarquilly. 1979. Grazing. In: *Feeding Strategies for the High Yield Dairy Cow.* W. H. Broster and H. Swan (Eds.). London, UK: Granada Publishers, pp. 295–321.

Journet, M., E. Grenet, M.-H. Farce, M. Theriez, and C. Demarquilly. 1995. *Recent Developments in the Nutrition of Herbivores.* Paris, France: INRA Editions.

Kahn, H. E., and C. R. W. Spedding. 1984. A dynamic model for the simulation of cattle herd production systems. 2. An investigation of various factors influencing the voluntary intake of dry matter and the use of the model in their validation. *Agric. Syst.* 13:63–82.

Karnezos, T. P., N. M. Tainton, and D. I. Bransby. 1988. A mathematical model used to describe animal performance on kikuyu and coastcross II pastures. *J. Grassland Soc. S. Africa* 5:38–41.

Kennedy, P. M., and M. R. Murphy. 1988. The nutritional implications of differential passage of particles through the ruminant alimentary tract. *Nutr. Res. Rev.* 1:189–208.

Kibon, A., and E. R. Ørskov. 1993. The use of degradation characteristics of browse plants to predict intake and digestibility by goats. *Anim. Prod.* 57: 247–251.

Kleyn, F. J., and R. M. Gous. 1988. Mathematical programming model for the optimisation of nutritional strategies for a dairy cow. *S. African J. Anim. Sci.* 18:156–160.

Konandreas, P. A., and F. M. Anderson. 1982. Cattle herd dynamics: An integer and stochastic model for evaluating production alternatives. ILCA Res. Rep. No. 2. Addis Ababa, Ethiopia: International Livestock Centre for Africa.

Laca, E., and M. W. Demment. 1996. Foraging strategies of grazing animals. In: *The Ecology and Management of Grazing Systems.* J. Hodgson and A. W. Illius (Eds.). Wallingford, UK: CHB International, pp. 137–158.

Laca, E. A., E. D. Ungar, N. Seligman, and M. W. Demment. 1992. Effects of sward height and bulk density on bite dimensions of cattle grazing homogeneous swards. *Grass Forage Sci.* 47:91–102.

Laca, E. A., E. D. Ungar, and M. W. Demment. 1994. Mechanisms of handling time and intake rate of a large mammalian herbivore. *Appl. Anim. Behav. Sci.* 39:3–19.

Laredo, M. A., and D. J. Minson. 1973. The voluntary intake, digestibility, and retention time by sheep of leaf and stem fractions of five grasses. *Aust. J. Agric. Res.* 24:875–888.

Lauenroth, W. K., D. L. Urban, D. P. Coffin, W. J. Parton, H. H. Shugart, T. B. Kirchner, and T. M. Smith. 1993. Modeling vegetation structure–ecosystem process interactions across sites and ecosystems. *Ecol. Modelling* 67:49–80.

Leaver, J. D. 1986. Effects of supplements on herbage intake and performance. In: *Grazing.* J. Frame (Ed.). British Grassland Soc. Occasional Symp. No. 19, pp. 79–88.

Loewer, O. J., K. L. Taul, L. W. Turner, N. Gay, and R. Muntifering. 1987. GRAZE: A model of selective grazing by beef animals. *Agric. Syst.* 25: 297–309.

Lopez-Tirado, Q., and J. G. W. Jones. 1991a. A simulation model to assess primary production and use of *Bouteloua gracilis* grasslands. Part I. Model structure and validation. *Agric. Syst.* 35:189–208.

Lopez-Tirado, Q., and J. G. W. Jones. 1991b. A simulation model to assess primary production and use of *Bouteloua gracilis* grasslands. Part II. Experimentation. *Agric. Syst.* 35:209–227.

McCall, D. G. 1984. A systems approach to research planning for North Island hill country. Ph.D. Thesis. Massey University, New Zealand.

McCaskill, M. R., and J. G. McIvor. 1993. Herbage and animal production from native pastures and pastures oversown with *Stylosanthes hamata*. 2. Modelling studies. *Aust. J. Exp. Agric.* 33:571–579.

McDonald, I. 1981. A revised model for the estimation of protein degradability in the rumen. *J. Agric. Sci. Camb.* 96:251–252.

McDonald, P., R. A. Edwards, J. F. D. Greenhalgh, and C. Morgan. 1995. *Animal Nutrition*. 5th ed. London, UK: Longman.

McIvor, J. G., and R. Monypenny. 1995. Evaluation of pasture management systems for beef production in the semi-arid tropics: Model development. *Agric. Syst.* 49:45–67.

McLeod, M. N., P. M. Kennedy, and D. J. Minson. 1990. Resistance of leaf and stem fractions of tropical forage to chewing and passage in cattle. *Br. J. Nutr.* 63:105–119.

Marsh, R., R. C. Campling, and W. Holmes. 1971. A further study of a rigid grazing management system for dairy cows. *Anim. Prod.* 13:441–448.

Mayne, C. S., and I. A. Wright. 1988. Herbage intake and utilization by the dairy cow. In: *Nutrition and Lactation in the Dairy Cow*. P. C. Garnsworthy (Ed.). London: Butterworths, pp. 280–293.

Mayne, C. S., R. D. Newberry, S. C. F. Woodcock, and R. J. Wilkins. 1987. Effect of grazing severity on grass utilization and milk production of rotationally grazed dairy cows. *Grass Forage Sci.* 42:59–72.

Mertens, D. R. 1987. Predicting intake and digestibility using mathematical models for ruminal function. *J. Anim. Sci.* 64:1548–1558.

Mertens, D. R. 1993. Rate and extent of digestion. In: *Quantitative Aspects of Ruminant Digestion and Metabolism*. J. M. Forbes and J. France (Eds.). Wallingford, UK: CAB International, pp. 13–52.

Mertens, D. R., and L. O. Ely. 1979. A dynamic model of fiber digestion and passage in the ruminant for evaluating forage quality. *J. Anim. Sci.* 49:1085–1095.

Milne, J. A., J. Hodgson, R. Thompson, W. G. Souter, and G. T. Barthram. 1982. The diet ingested by sheep grazing swards differing in white clover and perennial ryegrass content. *Grass Forage Sci.* 37:209–218.

Monsi, M., and T. Saeki. 1953. Uberden Lichtfaktor in den Pflanzengesellschaften und seine Bedentung für die Stoff produktion. *Jap. J. Botany* 14:22–52.

Monteith, J. L. 1972. Solar radiation and productivity in tropical ecosystems. *J. Appl. Ecol.* 9:747–766.

Moore, A. D., J. R. Donnelly, and M. Freer. 1997. GRAZPLAN: Decision support for Australian grazing enterprises. III. Pasture growth and soil moisture submodels and the GrassGro DSS. *Agric. Syst.* (In press).

Morley, F. H. W. 1968. Pasture growth curves and grazing management. *Aust. J. Exp. Agric. Anim. Husbandry* 8:40–45.

Mott, G. O. 1961. Grazing pressure and the measurement of pasture production. Proceedings of the VIII International Grassland Congress, Hurley, UK, pp. 606–611.

Murphy, M. R., R. L. Baldwin, and L. J. Koong. 1982. Estimation of stoichiometric parameters for rumen fermentation of roughage and concentrate diets. *J. Anim. Sci.* 55:411–421.

Murphy, M. R., R. L. Baldwin, and M. J. Ulyatt. 1986. An update of a dynamic model of ruminant digestion. *J. Anim. Sci.* 62:1412–1422.

Murtagh, G. J. 1988. Factors affecting the growth of kikuyu. II. Water supply. *Aust. J. Agric. Res.* 39:43–51.

Newman, J. A., A. J. Parsons, and P. D. Penning. 1994a. A note on the behavioral strategies used by grazing animals to alter their intake rates. *Grass Forage Sci.* 49:502–505.

Newman, J. A., A. J. Parsons, J. H. M. Thornley, P. D. Penning, and J. R. Krebs. 1994b. Optimal diet selection by a generalist grazing herbivore. *Funct. Ecol.* 9:255–268.

Noy-Meir, I. 1975. Stability of grazing systems: An application of predator–prey graphs. *J. Ecol.* 63:459–481.

Noy-Meir, I. 1976. Rotational grazing in a continuously growing pasture: A simple model. *Agric. Syst.* 1:87–112.

Noy-Meir, I. 1978. Grazing and production in seasonal pastures: Analysis of a simple model. *J. Appl. Ecol.* 15:809–835.

NRC. 1989. *Nutrient Requirements of Dairy Cattle*. 6th rev. ed. Washington, DC: National Research Council. National Academy Press.

NRC. 1996. *Nutrient Requirements of Beef Cattle*. 7th rev. ed. Washington, DC: National Research Council. National Academy Press.

O'Connor, J. D., C. J. Sniffen, D. G. Fox, and W. Chalupa. 1993. A net carbohydrate and protein system for evaluating cattle diets. IV. Predicting amino acid adequacy. *J. Anim. Sci.* 71:1298–1311.

Oldham, J. D. 1984. Protein–energy interrelationships in dairy cows. *J. Dairy Sci.* 67:1090–1114.

Oldham, J. D., and G. C. Emmans. 1988. Prediction of responses to protein and energy yielding nutrients. In: *Nutrition and Lactation in the Dairy Cow*. P. C. Garnsworthy (Ed.). London, UK: Butterworths, pp. 76–96.

Olney, G. R., and G. J. Kirk. 1989. A management model that helps increase profit on Western Australian dairy farms. *Agric. Syst.* 31:367–380.

O'Reagain, P. J., and J. Schwartz. 1995. Dietary selection and foraging strategies of animals on rangeland. Coping with spatial and temporal variability. In: *Recent Developments in the Nutrition of Herbivores*. M. Journet et al. (Eds.). Paris, France: INRA Editions, pp. 407–423.

Ørskov, E. R. 1992. *Protein Nutrition in Ruminants*. 2nd ed. London: Academic Press.

Ørskov, E. R. 1994. Plant factors limiting roughage intake in ruminants. In: *Livestock Production in the 21st Century—Priorities and Research Needs*. P. A.

Thacker (Ed.). Saskatoon, Saskatchewan, Canada: University of Saskatchewan, pp. 1–10.

Parsons, A. J., I. A. Johnson, and A. Harvey. 1988. Use of a model to optimize the interaction between frequency and severity of intermittent defoliation and to provide a fundamental comparison of the continuous and intermittent defoliation of grass. *Grass Forage Sci.* 43:49–59.

Parsons, A. J., A. Harvey, and I. R. Johnson. 1991. Plant–animal interactions in a continuously grazed mixture. II. The role of differences in the physiology of plant growth and of selective grazing on the performance and stability of species in a mixture. *J. Appl. Ecol.* 28:635–658.

Parsons, A., J. H. M. Thornley, J. Newman, and P. D. Penning. 1994. A mechanistic model of some physical determinants of intake rate and diet selection in a two-species temperate grassland sward. *Funct. Ecol.* 8:187–204.

Penning, P. D. 1986. Some effects of sward conditions on grazing behaviour and intake by sheep. In: *Grazing Research at Northern Latitudes.* O. Gudmundsson (Ed.). Proceedings of NATO Workshop. Reykjavik, Iceland: Nato Publishers, pp. 219–226.

Penning, P. D., A. J. Parsons, R. J. Orr, and T. T. Treacher. 1991. Intake and behaviour responses by sheep to changes in sward characteristics under continuous stocking. *Grass Forage Sci.* 46:15–28.

Penning, P. D., A. J. Parsons, R. J. Orr, and G. E. Hooper. 1994. Intake and behaviour responses by sheep to changes in sward characteristics under rotational grazing. *Grass Forage Sci.* 49:476–486.

Perkin, P., and T. Rehman. 1994. Farmers objectives and their interactions with business and life styles: Evidence from Berkshire, England. In: *Rural and Farming Systems Analysis. European Perspectives.* J. B. Dent and M. J. McGregor (Eds.). Wallingford, UK: CAB International, pp. 193–212.

Peters, A. 1983. *The Ecological Implications of Body Size.* Cambridge, UK: Cambridge Univ. Press.

Petersen, R. G., H. L. Lucas, and G. O. Mott. 1965. Relationship between rate of stocking and per animal and per acre performance on pasture. *Agron. J.* 57:27–30.

Poppi, D. P., D. J. Minson, and J. H. Ternouth. 1981. Studies of cattle and sheep eating leaf and stem fractions of grasses. 1. The voluntary intake, digestibility and retention time in the reticulo-rumen. *Aust. J. Agric. Res.* 32:99–108.

Poppi, D. P., M. Gill, and J. France. 1994. Integration of theories of intake regulation in growing ruminants. *J. Theor. Biol.* 167:129–145.

Preston, T. R., and R. A. Leng. 1987. *Matching Ruminant Production Systems with Available Resources in the Tropics and Sub-tropics.* Armidale, Australia: Penambul Books.

Richardson, F. D., B. D. Hahn, and P. I. Wilke. 1991. A model for the evaluation of different production strategies for animal production from rangeland in developing areas: An overview. *J. Grassland Soc. S. Africa* 8:153–159.

Rodriguez, A., J. N. Trapp, O. L. Walker, and D. J. Bernardo. 1990. A wheat grazing systems model for the US southern plains: Part I. Model description and performance. *Agric. Syst.* 33:41–59.

Rook, A. J., C. A. Huckle, and P. D. Penning. 1994. Effects of sward height and concentrate supplementation on the ingestive behaviour of spring-calving dairy cows grazing grass-clover swards. *Appl. Anim. Behav. Sci. 40:* 101–112.

Russell, J. B., J. D. O'Connor, D. G. Fox, P. J. van Soest, and C. J. Sniffen. 1992. A net carbohydrate and protein system for evaluating cattle diets. I. Ruminal fermentation. *J. Anim. Sci.* 70:3551–3561.

Sanders, J. O., T. C. Cartwright. 1979. A general cattle production systems model. Part 2. Procedures used for simulating animal performance. *Agric. Syst.* 4:289–309.

Sauvant, D., J. Dijkstra, and D. Mertens. 1995. Optimisation of ruminal digestion: A modelling approach. In: *Recent Developments in the Nutrition of Herbivores*. M. Journet et al. (Eds.). Paris, France: INRA Editions, pp. 143–165.

SCA. 1990. Feeding systems for Australian livestock: Ruminants. Standing Subcommittee on Agriculture, Commonwealth Scientific and Industrial Research Organisation (CSIRO). Melbourne, Australia: CSIRO Publications.

Scholefield, D., D. R. Lockyer, D. C. Whitehead, and K. C. Tyson. 1991. A model to predict transformations and losses of nitrogen in UK pastures grazed by beef cattle. *Plant Soil* 132:165–177.

Seligman, N. G., J. B. Cavagnaro, and M. E. Horno. 1992. Simulation of defoliation effects on primary production of a warm-season, semiarid perennial-species grassland. *Ecol. Modelling* 60:45–61.

Seman, D. H., M. H. Frere, J. A. Stuedemann, and S. R. Wilkinson. 1991. Simulating the influence of stocking rate, sward height and density on steer productivity and grazing behaviour. *Agric. Syst.* 37:165–181.

Senft, R. L., M. B. Coughenour, D. W. Bailey, L. R. Rittenhouse, O. E. Sala, and D. M. Swift. 1987. Large herbivore foraging and ecological hierarchies. *Bioscience* 37:789–795.

Sheehy, J. E., F. Gastal, P. L. Mitchell, J. L. Durand, G. Lemaire, and F. I. Woodward. 1996. A nitrogen-led model of grass growth. *Ann. Bot. 77:* 165–177.

Shem, M. N., E. R. Ørskov, and A. E. Kimambo. 1995. Prediction of voluntary dry matter intake, digestible dry matter intake and growth rate of cattle from the degradation characteristics of tropical foods. *Anim. Sci.* 60:65–74.

Shiyomi, M., T. Akiyama, and S. Takahashi. 1986. Modelling of energy flows and conversion efficiencies in a grassland ecosystem. *Ecol. Modelling* 32: 119–135.

Sibbald, A. R., T. J. Maxwell, and J. Eadie. 1979. A conceptual approach to the modelling of herbage intake by hill sheep. *Agric. Syst.* 4:119–134.

Simpson, J. R., and T. H. Stobbs. 1981. Nitrogen supply and animal production from pastures. In: *Grazing Animals*. F. H. W. Morely (Ed.). World Animal Science, B1. Amsterdam: Elsevier, pp. 261–287.

Smith, E. M., L. M. Tharel, M. A. Brown, C. T. Dougherty, and K. Limbach. 1985. A simulation model for managing perennial grass pastures. Part I. Structure of the model. *Agric. Syst.* 17:155–180.

Sniffen, C. J., J. D. O'Connor, P. J. van Soest, D. G. Fox, and J. B. Russel. 1992. A net carbohydrate and protein system for evaluating cattle diets. II. Carbohydrate and protein availability. *J. Anim. Sci.* 70:3562–3577.

Stephens, D. W., and J. R. Krebs. 1986. *Foraging Theory*. Monographs in Behavior and Ecology. Princeton, NJ: Princeton Univ. Press.

Stobbs, T. H. 1970. Automatic measurement of grazing time by dairy cows on tropical grass and legume pasture. *Trop. Grasslands* 4:237–244.

Stobbs, T. H. 1973. The effect of plant structure on the intake of tropical pastures I. Variation in the bite size of grazing cattle. *Aust. J. Agric. Res.* 24:809–819.

Stobbs, T. H. 1974. Components of grazing behaviour of dairy cows on some tropical and temperate pastures. *Proc. Aust. Soc. Anim. Prod.* 10:299–302.

Stobbs, T. H. 1975. Factors limiting the nutritional value of grazed tropical pastures for beef and milk production. *Trop. Grasslands* 9:141–150.

Stuth, J. W., and M. Stafford-Smith. 1993. Decision support for grazing lands: An overview. In: *Decision Support for the Management of Grazing Lands. Emerging Issues*. J. W. Stuth and B. G. Lyons (Eds.). Man Biosphere Ser. No. 11. Paris, France: UNESCO, pp. 1–35.

Taylor, St. C. S., J. I. Murray, and A. W. Illius. 1987. Relative growth of incisor arcade and eating rate in cattle and sheep. *Anim. Prod.* 45:453–458.

Thornley, J. H. M., and I. R. Johnson. 1990. *Plant and Crop Modelling. A Mathematical Approach to Plant and Crop Physiology*. Oxford, UK: Oxford Univ. Press.

Thornley, J. H. M., and E. L. J. Veberne. 1989. A model of nitrogen flows in grassland. *Plant, Cell Environ.* 12:863–886.

Thornley, J. H. M., A. J. Parsons, J. Newman, and P. D. Penning. 1994. A cost–benefit model of grazing intake and diet selection in a two-species temperate grassland sward. *Funct. Ecol.* 8:5–16.

Ungar, E. D., and I. Noy-Meir. 1988. Herbage intake in relation to availability and sward structure: Grazing processes and optimal foraging. *J. Appl. Ecol.* 25:1054–1062.

Ungar, E. D., N. G. Seligman, and M. W. Demment. 1992. Graphical analysis of sward depletion by grazing. *J. Appl. Ecol.* 29:427–435.

van Keulen, H., and J. Wolf. 1986. *Modelling of Agricultural Production, Weather, Soils and Crops*. Wageningen, The Netherlands: Pudoc.

van Keulen, H., N. G. Seligman, R. W. Benjamin. 1981. Simulation of water use and herbage growth in arid regions. A re-evaluation and further development of the model "Arid Crop." *Agric. Syst.* 6:159–193.

van Soest, P. J. 1994. *Nutritional Ecology of the Ruminant*. 2nd ed. Cornell, NY: Cornell Univ. Press.

Veloso, R. F., M. J. McGregor, J. B. Dent, and P. K. Thornton. 1992. Farm planning for the Brazilian "Cerrado" region: Application of crop simulation models for yield forecasting. In: *Computers in Agricultural Extension Programs*. Proceedings of the 4th International Conference, Florida, pp. 13–18.

Vera, R. R., J. G. Morris, and L. J. Koong. 1977. A quantitative model of energy intake and partition in grazing sheep in various physiological states. *Anim. Prod.* 25:133–153.

Verberne, E. L. J. 1992. Simulation of the nitrogen and water balance in a system of grassland and soil. Nota 258. DLO-Institut voor Bodemurucht baaried. Haaren, The Netherlands, p. 56.

Vickery, P. J. 1981. Pasture growth under grazing. In: *Grazing Animals*. F. H. W. Morley (Ed.). World Animal Science, B1. Amsterdam: Elsevier, pp. 55–77.

Wade, M. H., J. L. Peyraud, G. Lemaire, and E. A. Comeron. 1989. The dynamics of daily area and depth of grazing and herbage intake of cows in a five day paddock system. Proceedings of the XVI Grassland Congress, Nice, France, pp. 1111–1112.

Waldo, D. R., L. W. Smith, and E. L. Cox. 1972. Model of cellulose disappearance from the rumen. *J. Dairy Sci.* 55:125–129.

Walker, J. W., J. W. Stuth, and R. K. Heitschmidt. 1989. A simulation approach for evaluating field data from grazing trials. *Agric. Syst.* 30:301–316.

Whelan, M. B., E. J. A. Spath, and F. W. H. Morley. 1984. A critique of the Texas A&M model when used to simulate beef cattle grazing pasture. *Agric. Syst.* 14:81–84.

White, D. H., P. J. Bowman, F. H. W. Morley, W. R. McManus, and S. J. Filan. 1983. A simulation model of a breeding ewe flock. *Agric. Syst.* 10:149–189.

Wilkins, R. J., and E. A. Garwood. 1986. Effects of treading, poaching and fouling on grassland production and utilisation. In: *Grazing*. J. Frame (Ed.). British Grassland Society Occasional Symposium No. 19, pp. 19–31.

Wilson, J. R. 1994. Cell wall characteristics in relation to forage digestion by ruminants. *J. Agric. Sci. Camb.* 122:173–182.

Wilson, J. R., and P. M. Kennedy. 1996. Plant and animal constraints to voluntary feed intake associated with fibre characteristics and particle breakdown and passage in ruminants. *Aust. J. Agric. Res.* 47:199–225.

Woodward, S. J. R., G. C. Wake, A. B. Pleasants, and D. G. McCall. 1993. A simple model for optimizing rotational grazing. *Agric. Syst.* 41:123–155.

Woodward, S. J. R., G. C. Wake, and D. G. McCall. 1995. Optimal grazing of a multipaddock system using discrete time model. *Agric. Syst.* 48:119–139.

Zemmelink, G. 1980. Effect of selective consumption on voluntary intake and digestibility of tropical forages. Agric. Res. Rep. No. 896. Wageningen, The Netherlands: Pudoc.

14

Field Machinery Selection Using Simulation and Optimization

John C. Siemens
University of Illinois at Urbana–Champaign, Urbana, Illinois

I. INTRODUCTION

The technology is available to determine the optimum machinery set for a farm (Bowers, 1987; ASAE, 1994); however, it is seldom used in making farm machinery purchases. The determination, because of the great number of variables involved, is very tedious without a computer and appropriate software. Several machinery selection software programs have been written.

Whole-farm profit-maximizing linear programs have been developed (Doster, 1981; Black and Harsch, 1976). Linear programming for machinery selection, however, has limitations. Because of the requirement of an integer number of machines, a conditional optimization approach must be used in which the user provides input for a specific machinery complement, examines the consequences, and provides input for a revised complement until a suitable complement is determined.

Farm machinery selection models have been developed that select machinery on the basis of time constraints of various operations on the farm. Hughes and Holtman (1976) developed such a model that selects and matches machinery implements with power sources for a multicrop farm. Their time constraint model was further developed by Singh (1978) and Wolak (1981). The model has worked well in determining the "best" size for a machinery complement given the time constraints, suitable workdays available, and operation requirements. Rotz et al. (1983) extended the work of Wolak so that the model selected a "best" machinery complement based on both time constraints for proper matching of implements and power units and total cost including cost of timeliness to determine the least-cost complement. The model was used to select a machinery complement for different tillage systems on a range of soils for a variety of crop rotations. Rotz et al. (1983) also used the model to study the influence of changes in probability level for suitable weather, farm size, and economic parameters. This analysis was done by changing one group of parameters at a time and noting the effects on the machinery set selected. Economic parameters that varied included the inflation of all costs to future value and discounted to present value based on given inflation and discount rates. Annual tax deductions on fixed and operating costs are subtracted from the total cost. Both inflation and income tax deductions reduce the cost of owning large, more expensive equipment.

Several programs on farm machinery have been developed using expert systems. Schueller et al. (1986) developed an expert system for troubleshooting grain combine performance. McKinion and Lemmon (1985) developed a cotton management expert system that uses a knowledge base of extension service expertise coupled with information generated from a dynamic cotton crop simulation model. Kline et al. (1989) presented a knowledge-based expert system for farm machinery sizing and selection. They expanded the knowledge-based expert system of Bender et al. (1985) to help in structuring, modifying, running, and interpreting a farm management linear programming model.

The purposes of this chapter are to present the major factors involved in selecting an optimum machinery set and to review the farm machinery selection and management program previously presented by Siemens et al. (1990). The program is written in the C programming language and especially for midwest U.S. farms. For most midwest farms the major crops are corn and soybeans. However, the program is written in such a way that it can also be used for several other crops and locations.

II. METHODOLOGY

Optimum machinery size is defined as that machinery set that results in the minimum total cost, as shown in Fig. 1. In this chapter "optimum" and "least-cost" are used synonymously.

For corn or soybean farms and many other farms, machinery size can be varied by changing tractor sizes and matched implements and/or combine sizes and matched attachments. This means that Fig. 1 can be visualized as three-dimensional, the dimensions being tractor size, combine size, and total cost. Total cost is the sum of the costs for machinery, labor, and timeliness. The least-cost machinery combination is determined by the low point on the three-dimensional surface.

A. Machinery Costs

If a specific set of machinery is used to perform a set number of field operations on a farm of a given size, the annual use of each machine

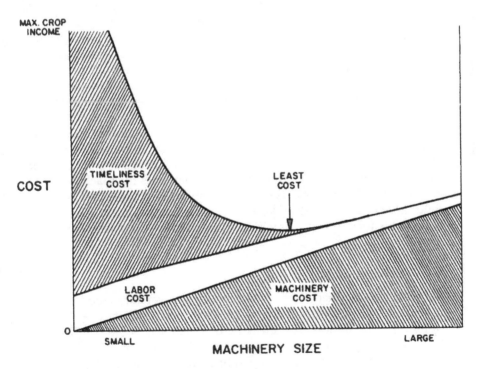

FIGURE 1   Costs related to machinery size for specific farm size.

and the fixed and variable costs can be estimated. Several sizes of farm machinery sets could be used on this farm. As the size of the machinery set increases, both productivity and initial price increase. The annual machinery costs, fixed and variable, would increase with increasing machinery size, as estimated by the lower line in Fig. 1.

B. Labor Costs

It is necessary to know the relation between productivity of machinery and labor cost. For the owner-operator of a farm producing field crops, the time used for operating machinery need not be considered in selecting the optimum machinery size unless other enterprises compete for the time of the owner-operator and a value can be placed on his or her time. Other enterprises might include a hog or beef operation or a part-time job. Then the owner-operator's time is important, and this cost should be considered.

If hired labor is available and used, the cost may be on an hourly or annual basis. On an hourly basis, total labor cost is directly proportional to machine operating time and inversely proportional to machine productivity. When labor is hired on an annual basis, total labor cost is independent of machine operating time and productivity. The question in either case is how to apportion labor costs between field operations and other farm tasks. If the only reason for hiring labor on a particular farm is to operate machinery, then the entire annual cost of hired labor should be assessed against the field operations.

Figure 1 assumes that one man is paid on an hourly basis only when he is operating field machinery. Labor cost is constant until the machinery is large enough to reduce total field time required below the level of full-time employment for all the hours the machinery can be operated. When this occurs, labor costs will decrease because total field hours decrease. Field operations are completed when they should be (decrease in timeliness cost) rather than when labor is available.

C. Timeliness Costs

Timeliness is defined here as the ability of available labor, using a given set of machinery, to complete each field operation within an optimum time period. The measure of timeliness is the cost accrued because the operation was not completed in the optimum time period. Some operations, such as planting, may have a penalty assessed directly to them because of untimely completion. Other operations, such as primary tillage, may have only an indirect influence on timeliness cost as they may affect the completion of subsequent operations.

It has been well established in the literature that yield decreases if corn is not planted before the middle of May in the U.S. Corn Belt. The value (or cost) of the yield decrease is the timeliness cost. The date at which this cost begins has also been reasonably well established in the literature of the various state agricultural experiment stations. There is little if any influence of factors other than the latitude of the location on this date. And there is general agreement that the rate of yield loss is approximately 1 bushel per acre per day.

No such penalty has been established for soybean planting. Indeed, there seems to be general disagreement as to whether one exists as long as time is available for the soybeans to reach maturity. A realistic timeliness penalty for soybean planting and for planting of other crops is needed.

It is logical that harvesting also has a timeliness cost associated with it. Timeliness cost for harvest of both corn and soybeans depends on the variety, harvest losses, drying costs, and many other factors. This means that the many causes and their interrelationships make the determination of a timeliness cost for harvesting complex. For this reason, there is no general agreement on how to set a penalty for late harvesting. For purposes of the program, a timeliness cost for late harvesting is an option and is commonly used.

III. A MACHINERY SELECTION PROGRAM

A. Stored Files

A primary goal in writing the program was to require as little input as possible from the program user. Therefore, several files of data contain default values that can be changed by the program user.

1. Equipment File

The equipment file is a stored file that contains a list of tractors with matched implements and combines with matched attachments that are available within the program. The program has space for nine tractor sizes, commonly varied in increments of 20 hp from 60 to 220 hp, and four combine sizes of 125, 140, 180, and 240 hp. For each tractor and combine and appropriate attachments, the file contains the size and name, estimated productivity for each tractor–implement and combine–attachment combination, and a list price.

Table 1 Portion of Stored File of Combines with Matched Corn
Heads and Grain Platforms and Tractors with Matched Implements
and Corresponding Productivity Values and Assumed List Prices

Machine	Productivity (acres/h)	List price ($)
180 hp combine		80,300
8 row corn head	4.50	23,700
22 ft grain platform	4.96	11,400
140 hp tractor		55,500
5 × 18 moldboard plow	3.48	8,700
11 ft chisel plow	6.08	6,955
22 ft disk	11.70	13,900
22 ft field cultivator	11.46	7,600
14 ft shredder	6.13	6,500
21 ft rotary hoe	20.36	4,300
Nine-knife applicator	10.22	0
40 ft fertilizer applicator	15.60	0
42 ft sprayer	18.32	3,500
21 ft drill	14.61	19,300
Subsoiler	3.33	6,900
Paraplow	4.12	8,500
Ro-till	4.12	10,500
19 ft combination tool	10.31	14,300
7 ft mower	3.60	4,300
11 ft wheel rake	4.80	2,900
9.25 ft mower-conditioner	4.04	10,000
Baler (pto)[a] wire	5.43	15,200
160 hp tractor		63,000
6 × 18 moldboard plow	3.94	9,100
13 ft chisel plow	7.08	8,700
25 ft disk	13.30	15,100
25 ft field cultivator	13.34	8,800
14 ft shredder	6.13	6,500
21 ft rotary hoe	20.36	4,300
11-knife applicator	12.27	0
40 ft fertilizer applicator	15.60	0
42 ft sprayer	18.32	3,500
24 ft drill	16.26	21,200
Subsoiler	3.75	6,900
Paraplow	4.38	8,500
Ro-till	4.38	10,500
24 ft combination tool	13.16	17,200
7 ft mower	3.60	4,300
11 ft wheel rake	4.80	2,900
9.25 ft mower-conditioner	4.04	10,000
Baler (pto)[a] wire	5.43	15,200

[a]pto = Powered through power take-off of tractor.

The essential methodology used to match the size of the implements to tractor power and to calculate productivity is given in ASAE (1994) or based on user experience. Table 1 lists an example of the file contents for the 180 hp combine with attachments and the 140 and 160 hp tractors with matched implements. The files are similar for other combines and tractors. The information in the file (Table 1) may be changed by the program user and stored.

A list price of zero means that the implement is not purchased by the farmer but is otherwise available. For example, fertilizer spreaders are often made available by fertilizer dealers and the cost is included in the cost of the fertilizer. If a tractor does not have sufficient power to pull an implement, the productivity is set to 0.0 in the file. For example, for the 60 hp tractor (not shown here), the productivity for the chisel plow is 0.0, which means that the program will not allow the 60 hp tractor to be used for chisel plowing.

2. Planter and Row Cultivator File

There are two parts to the planter and row cultivator file. The first part includes matched pairs of planters and cultivators along with estimated productivity and the list price of each (Table 2). The second part of the file is an array used to match the size of the tractors, combine corn heads, and planters (Table 3). The array provides a choice of (1) the same number of rows on the planter as on the corn head or (2) twice as many rows on the planter as on the corn head.

Table 2 Stored File of Planters and Row Cultivators with Productivity Values and Assumed Purchase Prices

Machine	Productivity (acres/h)	List price ($)
4-Row planter	4.24	9,400
4-Row cultivator	3.70	2,500
6-Row planter	7.00	13,050
6-Row cultivator	5.56	3,600
8-Row planter	9.33	16,700
8-Row cultivator	7.40	4,750
12-Row planter	14.00	27,000
12-Row cultivator	11.10	9,300
16-Row planter	21.00	36,000
16-Row cultivator	14.00	12,500
24-Row planter	28.00	60,000
12-Row cultivator	11.10	9,300

Table 3 Stored Array Used to Match Size of Tractors, Combine Corn Heads, and Planters

	Corn head (rows)							
	4	6	8	12	4	6	8	12
	Planter sizes (rows)*							
Tractor (hp)	4	6	8	12	8	12	16	24
60	4	—	—	—	—	—	—	—
80	4	6	8	—	8	—	—	—
100	4	6	8	12	8	12	16	—
120	4	6	8	12	8	12	16	24
140	4	6	8	12	8	12	16	24
150	4	6	8	12	8	12	16	24
160	4	6	8	12	8	12	16	24
180	4	6	8	12	8	12	16	24
200	4	6	8	12	8	12	16	24
220	4	6	8	12	8	12	16	24

*Planters may have either the same number of rows as the combine corn head or twice as many rows.

A constraint imposed is that the tractor, planter, and row cultivator sizes must be compatible in regard to available or required power.

3. Machinery Cost Factors File

The machinery cost factors file contains the constants for the equations used to estimate the fixed and variable machinery costs (Table 4). The constants and formulas were obtained from ASAE (1994). The formulas for estimating remaining farm value are as follows:

Tractors	$RFV = LP \times 0.68 \times (0.920)^Y$
Combines	$RFV = LP \times 0.64 \times (0.885)^Y$
Balers, forage harvesters, and self-propelled sprayers	$RFV = LP \times 0.56 \times (0.885)^Y$
All other machines	$RFV = LP \times 0.60 \times (0.885)^Y$

where

RFV = remaining farm value at end of year y, $

LP = list price of machine, $

Y = age of machine, years

Table 4 Values of Repair Cost Constants RC1 and RC2 and Life

Machine	RC1	RC2	Life (h)
Tractor	0.007	2.00	12,000
Moldboard plow	0.290	1.80	2,000
Chisel plow	0.280	1.40	2,000
Disk stalks	0.180	1.70	2,000
Field cultivator	0.270	1.40	2,000
Stalk chopper	0.440	2.00	1,200
Plant corn	0.320	2.10	1,500
Row cultivator	0.170	2.20	2,000
Rotary hoe	0.230	1.40	2,000
Anhydrous applicator	0.630	1.30	1,200
Fertilizer applicator	0.630	1.30	1,200
Sprayer	0.410	1.30	1,500
Drill	0.320	2.10	1,500
Subsoiler	0.280	1.40	2,000
Paraplow	0.280	1.40	2,000
Forage chopper	0.150	1.60	2,500
Combination tool	0.270	1.40	2,000
Mower	0.460	1.70	2,000
Hay rake	0.170	1.40	2,500
Mower and conditioner	0.180	1.60	2,500
Baler	0.230	1.80	2,000
Self-prop. combine	0.040	2.10	3,000
Corn head	0.040	2.10	3,000
Grain platform	0.040	2.10	3,000

Source: ASAE (1994).

Depreciation is obtained by subtracting the remaining farm value from the purchase price.

Housing, interest, and insurance costs are estimated by multiplying the percentage entered by the program user for these costs times the remaining farm value at the beginning of the year.

Repair costs are estimated using the formula

$$TAR = LP \times RC1 \times (h/1000)^{RC2}$$

where

TAR = total accumulated repair cost, $
LP = list price of machine, $
h = total machine use, h
RC1, RC2 = constants (Table 4)

The life of each machine is used as a maximum limit for the time of ownership.

Fuel consumption for tractors and combines is estimated by using the equation for fuel efficiency in ASAE (1994):

$$\text{Diesel fuel, gal}/(\text{hp}\cdot\text{h}) = 0.52X + 0.77 - 0.04(738X + 173)^{0.5}$$

where X = the ratio of the equivalent pto power required to the maximum available from the pto. X is set in the program at 0.85.

Fuel and lubrication costs are estimated from the fuel consumption value, the price of fuel, and the assumption that lubrication cost is 10% of the fuel cost.

4. Workday Probability File

The workday probability file lists the probabilities of a day being suitable for field work based on historical data for weeks of the year. Data are included for northern, central, and southern Illinois (Table 5). File space is provided for other locations. The program user must provide and store the workday probability data for his or her location and when running the program must select the location to be used. The probabilities for northern, central, and southern Illinois are from Schwart (1981) and represent the fraction of time during the weeks listed in which field work has been feasible at least 5 out of 6 years (83.3% of the time). The workday probability data may be changed by the user.

5. User Input Files

User input files contain data that are likely to vary between users or between farm situations. After the data are entered, these files may be stored for a specific farm.

6. Economic Data File

In addition to the list price of each machine and the field capacity of each power unit–attachment combination, economic data common to each machine are needed to estimate the fixed and variable costs of each machine. These data are entered in the economic data file (Table 6). Data include the availability and cost of machinery operators; price of diesel fuel; a housing, interest, and insurance charge as a percent of remaining farm value; machinery purchase price as a percent of list price; and percent inflation. The maximum number of machinery operators the program will utilize is six, and their cost may be entered on an hourly basis. The program does not optimize the number of machinery operators.

Table 5   Stored Workday Probability Data for Illinois

Week	Probability of workday		
	Northern	Central	Southern
1–13	0.00	0.00	0.00
14–15	0.12	0.14	0.07
16–17	0.36	0.26	0.17
18–19	0.44	0.36	0.30
20–21	0.39	0.28	0.30
22–23	0.59	0.46	0.43
24–25	0.46	0.46	0.50
26–27	0.61	0.51	0.57
27–28	0.61	0.51	0.57
29–31	0.64	0.60	0.62
32–35	0.64	0.70	0.62
36–37	0.66	0.70	0.68
38–39	0.47	0.51	0.59
40	0.47	0.56	0.54
41	0.48	0.56	0.54
42–43	0.63	0.71	0.58
44–45	0.46	0.51	0.54
46	0.56	0.58	0.40
47	0.56	0.50	0.40
48–49	0.42	0.26	0.11
50–52	0.00	0.00	0.00

Source: Schwart (1981).

Also entered in the economic factors file (Table 6) are the crop prices, crop yields, and penalty dates for planting and harvesting. These data are used to compute timeliness costs when the field operations are scheduled. These data, including the crops grown, may be changed by the program user. For each crop the yield is assumed to equal the yield entered for a specific crop if the planting and harvesting operations are completed on or before the penalty date entered. When a planting or harvesting operation occurs after the penalty date, a yield decrease is assumed and a timeliness penalty is calculated. The timeliness penalty is calculated by multiplying the crop yield, appropriate timeliness factor, acres delayed, and days of delay. The timeliness cost is equal to the timeliness penalty times the crop price. An example of the timeliness factors is given below.

	Timeliness factor (% yield loss per day of delay)	
Crop	Planting	Harvesting
Corn	1.0	0.5
Soybeans	1.0	1.0
Wheat	0.25	1.0
Oats	1.0	1.0
Sorghum	0.25	1.0

7. Desired Field Operations File

The most critical information input by the user for a specific farm is the list of desired field operations to be performed and the data related to these operations. The main objective of the program is to select the set of machinery required to complete the desired operations in a timely fashion at least cost. A list of desired field operations for an example farm is presented in Table 7.

Code numbers are used to enter the field operations. The list of available code numbers for the 30 different field operations included in the program can be viewed on the screen. Other data to be entered

Table 6   Economic Data File with Default Values

Operators: Two @ $7.50/h
Economic factors:
   Purchase price, % of list   90.0
   Housing, interest, and insurance   12.0
   Percent inflation   3.5
   Fuel price ($/gal)   1.00
Crop prices, yields, and penalty dates:

	Price ($/bu)	Yield (bu/acre)	Penalty dates (M/D)	
			Planting	Harvesting
Corn	2.50	150	5/15	11/15
Soybean	7.50	40	5/31	10/15
Wheat	3	60	10/30	7/20
Oats	7.50	40	5/31	10/15
Sorghum	4.50	100	5/31	10/15

Table 7 User Input of Field Operations for Example Farm

Code No.	Field operations	Start date (M/D)	Finish date (M/D)	Area (acres)	Labor (h/day)	Land area No.
2	Combine soybeans	9/15	0/0	500	8.0	1
1	Combine corn	10/15	0/0	500	10.0	2
7	Chisel plow	10/15	0/0	500	10.0	2
8	Disk harrow	4/1	0/0	500	10.0	1
9	Field cultivate	4/25	0/0	500	10.0	1
21	Plant corn	4/25	0/0	500	10.0	1
8	Disk harrow	4/1	0/0	500	10.0	2
9	Field cultivate	5/1	0/0	500	10.0	2
22	Plant soybeans	5/1	0/0	500	10.0	2
18	Row cultivate	6/1	0/0	500	10.0	1
18	Row cultivate	6/10	0/0	500	10.0	2

for each operation include the earliest start date; latest finish date; acres to be covered; available labor hours per day or the hours per day the operation can be performed, whichever is least; and a land area number. Entry of a latest finish date is optional and is needed only if an operation must be completed by the date entered. The land area numbers are used to help ensure that the operations are scheduled correctly; they are explained in more detail below.

The first operation listed in Table 7 is "Combine soybeans" with an earliest start date of 9/15. This is interpreted to mean that September 15 is the earliest date soybeans are commonly ready for harvest. A latest finish date is not specified for soybean harvest. The next entries are the acres to be covered for the operation and the number of hours per day the operation can be performed. The latter figure is assumed to be the number of hours spent in the field operating machinery. No allowance is made for time used to adjust and service equipment or for traveling to fields.

The second operation listed for the example farm is "Combine corn," with an earliest start date of October 15. The third operation, "Chisel plow," is to be performed on the same land from which corn is harvested. Thus, the land area number is the same for both combine corn and chisel plow operations. Land area numbers are used in scheduling the field operations. For the example farm, the chisel plow operation will not get ahead of the combine soybean operation when the operations are scheduled.

The desired spring operations for corn are "Disk harrow," "Field cultivate," and "Plant corn." All of these operations are to be performed in sequence on the land that was in soybeans the previous year, and therefore the land area number is 1 for each of these operations. For soybeans, the spring operations are disk harrow, field cultivate, and plant soybeans, and the land area number is 2 for these operations. Both crops are to be cultivated once.

A primary objective of the program is to determine the optimum set of machinery and the machinery-related costs for the list of desired field operations (Table 7).

B. Matching of Implements to Tractors

As explained later, three options are provided to the user in running the program. These options may require that the user set the number and size of the tractors and combines as shown in Table 8.

For farms that can justify two or more tractors, if the tractors are different in size a method is needed for determining which field operations the different size tractors will perform. For large farms with several tractors, the number of alternatives is tremendous. Of course, a complete farm machinery selection program would include the solution to this problem. The method used by Freesmeyer (1985) is one possibility. For the program described herein the user must assign the implements needed to perform the field operations to ei-

Table 8 Specifying Number and Size of Tractors and Combines

	Quantity	Size[a]		Quantity	Size[a]
Tractors (max. 6):			Combines (max. 3)	1	2
Large	1	5	Corn heads	1	
Small	1	1	Grain platforms	1	
Available tractor sizes			Available combine sizes		
0 = 60 hp 5 = 160 hp			0 = 140 hp (4-row head)		
1 = 80 hp 6 = 180 hp			1 = 180 hp (6-row head)		
2 = 100 hp 7 = 200 hp			2 = 215 hp (8-row head)		
3 = 120 hp 8 = 220 hp			3 = 260 hp (12-row head)		
4 = 140 hp					

[a]Needed to run without optimization.

Table 9 Match of Implements to Tractors with Values for Example Farm

Planter size (T or S)? S			Large tractors (L)		1
T = twice the combine size			Small tractors (S)		1
S = same as combine size			Total		2

Implement	L	S	Implement	L	S
Moldboard plow	0	0	Spray pesticide	0	0
Chisel plow	1	0	Drill	0	0
Disk	1	0	Subsoiler	0	0
Field cultivate	1	0	Paraplow	0	0
Chop stalks	0	0	Chop forage	0	0
Plant	0	1	Combination tool	0	0
Row cultivate	0	1	Mow hay	0	0
Rotary hoe	0	0	Rake hay	0	0
Apply anhydrous	0	0	Mow and condition	0	0
Apply fertilizer	0	0	Bale hay	0	0

ther large or small tractors as shown for an example farm in Table 9. The user must also specify whether the number of rows on the planter is to be the same as or twice the number of rows on the combine corn head.

IV. THE SEARCH FOR THE OPTIMAL MACHINERY SET

Three options are available when running the program. The first option is an attempt to provide the user with an estimate of the machinery costs, work schedule, etc., for a specified number and specified sizes of tractors and combines. The machinery set specified might be the current set of machinery on a farm or a set in which the user is interested. Most farmers have a set of machinery currently being used and may be interested in an estimate of the total cost including potential timeliness penalties and the work schedule. Or they may want to evaluate the effects of changing only one or two machines, or they would like to know the effects of using the present machinery on a larger farm. The first option of the program may be used for these purposes.

The second and third options are programmed to determine the optimum set of machinery for the desired field operations. For the second option the user specifies the number of tractors, combines, and laborers, and the program optimizes the sizes of the tractors and combines. The third option is an attempt to determine the optimum set of machinery including number and sizes of tractors and combines and number of laborers. The search for the optimal machinery set is explained here using the second option.

Assume that the program user has entered the list of field operations as shown in Table 7, specified the number of tractors and combines as shown in Table 8, and matched tractors to implements as shown in Table 9. Note that in Table 8 for the example farm, the number of tractors has been set to one large tractor and one small tractor and the number of combines to one. Now the goal of the second option is to find the sizes of the tractors and combines that result in the least cost.

The program first sets the size of the large tractor to 220 hp, the largest available in the program; the small tractor to one tractor size increment less than the largest tractor size; and the size of the combine to 225 hp, the largest in the program. With this set of machinery, the input data, the stored data files, and the field operations are scheduled, and the costs for machinery, labor, and timeliness are calculated. These costs are summed to get the total average annual machinery cost.

Then both the small tractor and combine sizes are decreased, and again the work schedule and costs are computed. The size of the small tractor and combine continue to be decreased until the total annual cost increases due to increased timeliness costs. Then combinations of small tractor and combine sizes are tried until the least-cost machinery set has been found with the size of the large tractor fixed at 220 hp. Table 10 duplicates the screen that shows the costs in dollars per acre for the different machinery sets evaluated for an example farm with the large tractor fixed at 220 hp. The xxxxxx's on the screen indicate a machinery set that is not able to complete the operations within a calendar year, a tractor of insufficient size to pull an implement, or an operation that could not be completed by the latest finish date listed in the input data.

Next the size of the large tractor is reduced by one size increment, in this case to 200 hp, and the above procedure is repeated. The size of the large tractor is reduced until the least cost increases.

During this process, no trials are made with both tractors the same size, which in some situations would result in the least cost. Therefore, the final set of trials is made with both tractors the same size. Tractor and combine sizes are varied until the least-cost combination is found.

Thus, the total cost is determined for all possible size combinations of two tractor sizes and combine sizes that could result in the

Table 10 Initial Search for Optimal Machinery Set
for Example Farm

1 Large tractor, 220 hp
Small tractor
hp

hp	140	180	215	260
60	xxxxxx	xxxxxx	xxxxxx	
80	66.32	60.23	xxxxxx	
100	68.79	62.54	64.88	
120		63.88		
140		64.66		
160	72.09	65.50		
180		66.46		
200				69.51
220				

Combines, hp

least cost. In this manner the optimum or least-cost machinery set is
determined.

The third program run option is an attempt to determine the
optimum machinery set including the number of operators and number and sizes of tractors and combines. For this option the program
starts with one machine operator, one tractor, and one combine, and
the least-cost sizes of the tractor and combine are determined. Then
the number of operators and number of tractors, and eventually the
number of combines, are increased, and for each case the least-cost
machinery set is determined. After many trials, the optimum machinery set is determined.

V. LOW COST MACHINERY SETS

In searching for the optimal machinery set for a farm it is common
for the program to determine the costs for over 40 combinations of
tractor and combine sizes. For several of these combinations the total
annual cost varies by only a small amount. For this and other reasons
the user may be interested in a machinery set other than the set that
results in the least cost. Therefore, when the optimal search is completed the program presents the eight lowest cost machinery sets
found during the search (Table 11). The first set presented is the least-
cost machinery set found during the search for the optimal machinery

Table 11 The Eight Low-Cost Machinery Sets Found for Example Farm During Search for Optimal Set

	Size (hp)			Cost ($/acre)			
Set[a]	Large tractor	Small tractor	Combine	Machinery	Labor	Timeliness	Total
1	160	80	215	49.95	6.10	1.19	57.24
2	180	80	215	51.61	5.93	0.00	57.53
3	200	80	215	52.55	5.79	0.00	58.34
4	160	100	215	52.26	6.10	1.19	59.55
5	180	100	215	53.92	5.93	0.00	59.85
6	140	80	215	49.16	6.37	4.64	60.17
7	220	80	215	54.84	5.40	0.00	60.24
8	200	100	215	54.86	5.79	0.00	60.66

[a]Each set consists of one large tractor, one small tractor, and one combine each with appropriate attachments.

set. The presentation includes for each set only the sizes of the large and small tractors and the size of the combines; costs for machinery, labor, and timeliness; and total cost.

VI. OUTPUT INFORMATION

The program user may obtain a presentation of the output information for any of the eight lowest cost machinery sets found during the search. In this section we discuss the output information for the least-cost machinery set for the example farm.

A. Machinery Set

The first output screen is a list of the machinery set including the assumed purchase price and annual use of each machine (Table 12). The annual number of hours each operator spends operating machinery is also given.

B. Work Schedule

The computed work schedule is the most critical portion of the program output (Table 13). Considerable effort has been devoted to making the computed work schedule realistic in terms of utilizing operators and machinery.

Table 12 Optimum Machinery Set and Labor Use Determined for Example Farm

Machine	Purchase price ($)	Annual use (h)	Annual use (acres)
215 hp combine	106,020	173	
8-Row corn head	25,200	91	500
22 ft grain platform	13,320	82	500
160 hp tractor	67,500	231	
11 ft chisel plow	8,190	76	500
25 ft disk harrow	15,930	89	1,000
29 ft field cultivator	11,970	66	1,000
80 hp tractor	35,100	236	
8-Row planter	20,430	107	1,000
8-Row cultivator	6,390	129	1,000

Number of machine operators = 2: Operator 1 would work 553 h; operator 2 would work 260 h

Table 13 Work Schedule Using Optimum Machinery Set for Example Farm

Field operation	Start date	Finish date	Calendar days	Work days	Acres
Combine soybeans	9/15	10/8	24	10.3	500
Combine corn	10/15	10/31	17	9.1	500
Chisel plow	11/1	11/16	16	7.6	500
Disk harrow	4/1	4/24	24	4.5	500
Field cultivate	4/25	5/6	12	3.3	500
Plant corn	4/25	5/14	20	5.4	500
Disk harrow	4/24	5/22	29	4.5	500
Field cultivate	5/22	5/30	9	3.3	500
Plant soybeans	5/22	6/5[a]	15	5.4	500
Row cultivate	6/5	6/21	17	6.4	500
Row cultivate	6/21	7/5	15	6.4	500

[a]Timeliness penalty applies.

The program schedules the field operations on a day-to-day basis, giving priority to the order in which the operations are listed in the input data. For the desired field operations listed in Table 7, the first operations scheduled are the harvesting operations. When only one machinery operator is available for harvesting it is assumed that the operator spends 50% of the time hauling and processing the grain. Thus, the number of hours per day one operator actually spends operating the combine is reduced. When two operators are available, the combine operates the full number of hours per day listed in the input data and the operation requires the use of both operators. It is realized that large variations exist in the time spent in unloading, hauling, and processing grain during harvest operations. Revisions of the computer program may be needed to account for these variations.

Each field operation starts on the date listed in the input data unless an operator and machine needed to carry out the operation are not available. Other operations could be under way utilizing all operators or machines required. In that case the start date for the operation in question is the first date that both an operator and required machines are available. The acres completed on a given date equal the productivity (acres per hour) for the machinery being used times the hours per day labor is available times the probability of that date being suitable for field work.

If an operation is not completed by the latest finish date listed in the input data or if all operations are not completed within one calendar year, the machinery set is regarded as being unacceptable.

For the example field operations listed in Table 7, the operations are scheduled beginning with the fall operations and then the spring operations are scheduled. No timeliness penalties occur in conjunction with combine soybean, combine corn, or plant corn operations as these are completed before their respective penalty dates. Soybean planting is not completed until after the specified penalty date (Table 6), and therefore a timeliness penalty is calculated.

C. Cost for Each Field Operation

As the field operations are scheduled, the annual hours of use are accumulated for each machine. The fixed and variable costs for each machine are calculated as previously explained using the annual use input data and data from the stored files. The fixed and variable costs are summed, and the result is used to calculate the cost per acre for each field operation (Table 14).

Table 14 Estimated Cost for Each Field Operation

Machine	Years of use	Annual hours	Power and machine cost ($/acre)	Machine cost only ($/acre)
215 hp combine	9	173		
8-Row corn head	10	91	25.60	6.75
22 ft grain platform	10	82	20.70	3.55
160 hp tractor	10	231		
11 ft chisel plow	10	76	9.60	2.50
25 ft disk harrow	10	89	6.50	2.30
29 ft field cultivator	10	66	4.90	1.75
80 hp tractor	10	236		
8-Row planter	7	107	6.40	3.90
8-Row cultivator	8	129	4.20	1.10

D. Detailed Costs for Each Machine

If desired, the detailed costs for each machine are provided for each year up to a maximum of 10 years of ownership. For the example farm the costs of the combine and corn head are presented in Tables

Table 15 Detailed Costs for 215 hp Combine[a]

Year	Total use (h)	Percent of life	Accumulated costs ($)			Average cost ($/acre)
			Depreciation	HII[b]	Repairs	
1	173	6	39,298	12,722	118	312
2	346	12	46,971	19,928	522	167
4	692	23	59,771	31,950	2,250	147
5	866	29	65,090	36,944	3,597	133
6	1,039	35	69,797	41,365	5,278	123
7	1,212	40	73,963	45,277	7,297	115
8	1,385	46	77,649	48,739	9,660	109
9	1,558	52	80,912	51,803	12,371	104

[a]List price = $117,800; purchase price = $106,020; fuel price = $1/gal; fuel use = 10.88 gal/hr; machine use = 173 h/yr.
[b]HII = Housing + interest + insurance.

Table 16   Detailed Costs for Eight-Row Corn Head[a]

Year	Total use (h)	Percent of life	Accumulated costs ($)			Average cost ($/acre)
			Depreciation	HII[b]	Repairs	
1	91	3	9,341	3,024	7	24.75
2	181	6	11,165	4,737	32	15.90
3	272	9	12,779	6,253	75	12.75
4	363	12	14,207	7,594	138	12.00
5	454	15	15,471	8,781	220	9.80
6	544	18	16,590	9,832	323	8.90
7	635	21	17,580	10,762	447	8.20
8	726	24	18,457	11,585	591	7.65
9	817	27	19,232	12,313	757	7.20
10	907	30	19,918	12,958	945	6.75

[a]List price = $28,000; purchase price = $25,200; machine covers 500 acres/yr; fuel cost = $1.97/acre; productivity = 5.51 acres/h; annual use = 90.7 h/yr.
[b]HII = Housing + interest + insurance.

15 and 16. The program presents the costs of other machines in a similar manner.

E.   Total Machinery-Related Costs for the Example Farm

Finally, the total machinery-related cost for the machinery set is given for the farm (Table 17). The machinery fixed cost includes estimated depreciation, housing, interest, and insurance. The estimated fuel and repair costs are added to the fixed cost to get the total machinery cost. Labor cost is figured using the labor rates entered. Timeliness

Table 17   Average Annual Machinery-Related Costs

Cost item	$/yr	$/acre
Machinery fixed cost	41,874	41.87
Fuel cost	4,708	4.71
Repair cost	3,368	3.37
Total machinery cost	49,950	49.95
Labor cost	6,098	6.10
Timeliness cost	1,190	1.19
Total cost	57,238	57.24

cost is calculated for the acres of any crop planted or harvested after the penalty dates specified. From the work schedule explained previously for the example farm, the "Plant soybeans" operation was completed late. Thus, a timeliness cost is calculated for the operation. The costs for machinery, labor, and timeliness are summed to provide an estimate of the total annual machinery-related costs for the machinery set.

REFERENCES

ASAE: 1994. *Agricultural Machinery Management Data*. St Joseph, MI: ASAE.

Bender, D. A., B. A. McCarl, J. K. Schueller, D. E. Kline, and S. H. Simon. 1985. Expert system interpreter for a farm management linear program. Paper No. 85-5518. St Joseph, MI: American Society of Agricultural Engineers.

Black, J. R., and S. B. Harsh. 1976. *Corn-Soybean-Wheat Farm Planning Guide*. Telplan 26. East Lansing, MI: Cooperative Extension Service, Michigan State University.

Bowers, W. 1987. *Machinery Management*. Moline, IL: John Deere.

Chen, L. H. 1987. A Machinery Selection, Crop Budgeting, and Scheduling Model. Tech. Bull. 145. Mississippi Station, MS: Mississippi Agricultural and Forestry Experimental Station, Mississippi State University.

Doster, H. 1981. Top Crop Farmer Planning Model. West Lafayette, IN: Cooperative Extension Service, Purdue University.

Freesmeyer, S. R. 1985. A time constrained optimal farm machinery selection program for microcomputers. M.S. Thesis. Urbana, IL: Univ. of Illinois.

Hughes, H. A., and J. B. Holtman. 1976. Machinery complement selection based on time constraints. *Trans. ASAE* 19(5):812–814.

Hunt, D. R. 1967. A Fortran program for selecting farm equipment. *Agric. Eng.* 48(6):332–335.

Kline, D. E., D. A. Bender, and B. A. McCarl. 1989. FINDS: Farm-Level Intelligent Decision Support System. *Appl. Eng. Agric.* 5(2):273–280.

McKinion, J. M., and H. E. Lemmon. 1985. Symboli computers and AI tools for a cotton expert system. Paper No. 85-5520. St. Joseph, MI: American Society of Agricultural Engineers.

Rotz, C. A., H. A. Muthar, and J. R. Black. 1983. A multiple crop machinery selection algorithm. *Trans. ASAE* 26(6):1644–1649.

Schueller, J. K., R. M. Slusher, and S. M. Morgan. 1986. An expert system with speech synthesis for troubleshooting grain combine performance. *Trans. ASAE* 29:342–344, 350.

Schwart, R. B. 1981. *Farm Machinery Decisions*. Worksheet No. 1. *What Size Machinery for Your Farm?* Circular 1065. Urbana, IL: Illinois Cooperative Extension Service, University of Illinois.

Siemens, J. C., K. Hamburg, and T. Tyrrell. 1990. A farm machinery selection and management program. *J. Prod. Agric.* 3:212–219.

Singh, D. 1978. Field machinery system modeling and requirements for selected Michigan cash crop production systems. Ph.D. Thesis. East Lansing, MI: Michigan State University.

Wolak, F. J. 1981. Development of a field machinery selection model. Ph.D. Thesis. East Lansing, MI: Michigan State University.

15

Whole-Farm Simulation of Field Operations: An Object-Oriented Approach

Harbans Lal

Pacer Infotec Inc., Portland, Oregon

I. INTRODUCTION

Over the past 30 years scientists have developed several computer simulation models that can be employed for analyzing complex farm situations. Using these models, managers make planning and/or management decisions including the selection and optimum utilization of machinery, labor, and other farm resources to increase their profits. These models can be broadly classified into (1) biological plant scale, (2) field scale, and (3) whole-farm models. The biological plant scale models simulate the behavior of an individual crop based on its physiological development and growth process and predict its yield under different climatic and soil conditions (Jones et al., 1989). The field-scale models capture the behavior of an individual field in terms of its adaptability to different cropping systems (a combination of crops over time) to suit its environment and soil characteristics. The field-scale models, generally, do not consider the operational constraints (machinery and labor requirements) to implement the candidate or selected cropping systems (Tsai et al., 1987).

The whole-farm models capture the operational constraints and behavior of a farm. Such models predict the performance of the farm in response to different management strategies under different climatic and soil conditions. The stochastic nature of weather and the complexity of other factors, both endogenous and exogenous, make modeling a whole-farm system much more complex and challenging than other industrial production systems. The magnitude of this complexity further increases in multicrop production systems where different crops compete for the limited farm resources (machinery and labor) during the cropping season.

II. SIMULATION MODELING APPROACHES

Simulation models capture the dynamic behavior—the changes over time—of the system. These changes could constitute discrete events and/or continuous processes.

An operational farm system involves both discrete events and continuous processes. The decision to start an operation in a cropping sequence is an example of a discrete event. On the other hand, once an operation has started it becomes a continuous process and its execution is controlled by the rate process, which depends on the machinery capacity and other crop- and soil-related factors. Most field operations such as planting, plant protection, and harvesting need to be completed within a short span of time to be effective for the crop production. These spans, defined by earliest start time and latest finish time, are referred to as agronomic windows. Scheduling of field operations and selection and allocation of machinery and labor to finish them within their respective agronomic windows are critical decisions that farmers face on a daily basis.

Several researchers have developed whole-farm simulation models that can facilitate decision making under such complex and conflicting multichoice situations (Smith, 1985; Tsai et al., 1987; Chen and McClendon, 1985; Glunz, 1985; Thai and Wilson, 1988; Miles and Tsai, 1987; Labiadh and Frisby, 1987). These models range from simulating a single operation, usually harvesting (Glunz, 1985; Thai and Wilson, 1988; Miles and Tsai, 1987; Labiadh and Frisby, 1987), to simulating the complete growing season with one or more crops (Tsai et al., 1987; Chen and McClendon, 1985; Smith, 1985). The ability of these models to duplicate the real-world situation depends on their formulation and the incorporation of factors responsible for the system's behavior in a manner as realistic as possible (van Elderen, 1981, 1987).

The approaches used for whole-farm simulation models can be classified into two broad categories: (1) algorithmic or procedural and (2) object-oriented. The procedural approach emphasizes understanding and modeling the system based on the processes involved in the system. In this approach, an algorithm or a procedure is first defined to solve the problem at hand. A computer program is then written using one of the procedural languages such as Fortran, Basic, or Pascal to implement the algorithm. A major limitation to this approach to simulation is the difficulty of modeling human decision patterns responsible for the system's behavior. In the case of a whole-farm system, it is difficult to capture the farmer's priorities and preferences for allocating the available machinery and labor to different operations on a daily basis. Instead, simple approximations and averages are used (O'Keefe and Roach, 1987) that do not represent the real-world situations.

Most conventional models use a relatively large simulation time step (even up to a week) and allocate total amount of work done to different fields rather than estimating day-by-day and field-by-field work based on the available farm resources (operators, power units, and implements) for the specific operation. Furthermore, they don't permit changing the allocation of operators to a tractor, allocation of tractors to an implement, and allocation of implements to an operation on the basis of their availability and the farmer's preferences. In the real-world setting, farmers with multiple units of tractors and implements use some sort of heuristic for assigning tractors to different implements and switch units to get the most work done. Thus the actual capacity of an operation such as plowing depends on the components of the machinery system (implement, power source, and operator) selected and used for the operation rather than the average capacity of all the plowing units available on the farm. The availability of a machinery system will vary according to the priorities for other operations requiring the same machinery or operator.

The object-oriented approach to modeling can capture many of these operational requirements of a farm system, thus resulting in a more powerful model. Within this approach, the system is described as a collection of objects. These objects are enumerated by assigning values to the attributes of an object-class. Objects within an object-oriented environment communicate with each other by passing and receiving events (Rumbaugh et al., 1991). This process of communication is a one-way transmission of information that could change their status or result in the generation of new objects. It is not like a subroutine call that returns a value. In the real world, all objects occur concurrently. An object sending an event to another may expect a reply, but the reply is a separate event under the control of a second object, which may or may not choose to send it. The object-oriented approach to data modeling is fairly recent. There are only limited applications of this approach in the field of agriculture. However, some pioneering work using these techniques for agricultural decision aids has been carried out by Freeman et al. (1989), Stone et al. (1986), Helms et al. (1987), Whittaker et al. (1987), and Nute et al. (1996).

III. AN OBJECT-ORIENTED MODEL OF THE WHOLE-FARM SYSTEM

Objects in the "object-oriented" paradigm are discrete entities that incorporate both data structure and behavior. Objects can be concrete, such as a file in a file system, or conceptual, such as scheduling policy in a multiple processing system (Rumbaugh et al., 1991). A whole-farm system consists of several objects such as Field_1 or Tractor_1. They can be grouped into several object-classes such as fields, tractors, and implements. Each of these object-classes is defined with a set of attributes. An instance of an object-class or an object can be generated by assigning specific values to these attributes. For example, an "implement" object-class is defined with Number, Type, Name, Tractors/OperatorList, Width, Speed, and Availability as its attributes. A specific object (DiskHarrow) of this object-class is generated by assigning Implement_2 to Number, LandPrep to Type, DiskHarrow to Name, Tractor_1 and Tractor_2 to Tractors/OperatorList, 4.74 to Width, 7 to Speed, and Avail to Availability. This particular object (DiskHarrow) of the "implement" object-class is a land preparation implement that has a working width of 4.74 m. It is currently available and can be operated by Tractor_1 or Tractor_2 at an average speed of 7 km/h. The example below depicts the representation of the implement object-class and that of object Disk-Harrow in PROLOG data structure (the language used for this project and discussed later in the chapter). Quotes are used to indicate string data type in PROLOG.

PROLOG predicate (object-class):

implement(Number, Type, Name, Tractors/

OperatorList, Width, Speed, Availability)

PROLOG clause (a specific item):

implement("Implement_2","LandPrep","DiskHarrow",

["Tractor_1", "Tractor_2"],4.74,7,"Avail")

The other object-classes for modeling a whole-farm system could include farm, labor, tractor, irrigation equipment, crop, field, operation, scheduled work hours, and daily weather. A schematic of an object data model for the whole-farm system consisting of these object-classes is depicted in Fig. 1. The attribute set for each of these classes is presented in Table 1.

According to this model the farm consists of crops, fields, operators, tractors, and implements. A crop can be grown on one or more fields. Each crop has a list of operations that need to be carried out in proper sequence for it to grow and yield. Each operation requires an implement for its execution and would need to meet certain agronomic requirements to be effective for the crop production. Implements can be self-propelled or require a separate power source. A

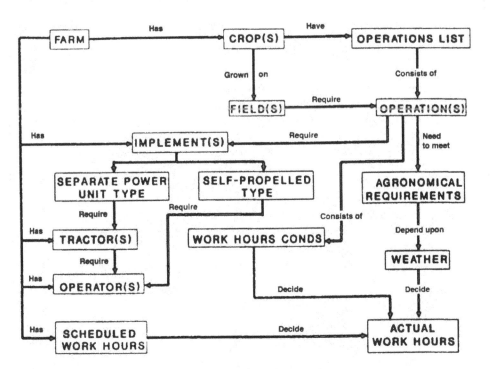

FIGURE 1   Semantic representation of farm knowledge.

Table 1   Farm Object-Classes and Their Attributes

Object-class	Attributes
Farm	Name, location, total area, cultivated area, principal activity, farm code
Labor	Number, name, sex, functions, availability
Tractor	Model, horsepower, operators list, availability
Implement	Number, type, name, tractors/operators list, width, speed, availability
Irrigation equipment	Number, specification, capacity, operators list, availability
Crop	Number, name, operations list
Field	Number, identification, area, soil type, crop list, variety list, maturity days list, fertilizer type list, irrigation type, production cost list, sale price list, availability
Operations	Operation type, operation name, crop, earliest start time, latest finish time, work hours conditions list, implement list, efficiency
Scheduled work hours	Month, scheduled daily working hours

self-propelled implement becomes functional with just an operator, whereas an implement requiring a separate power source would need an external power unit such as a tractor in addition to an operator for it to be functional. Therefore, to be complete and functional for an operation, a machinery system could consist of an implement, external power source, and an operator or just an implement and an operator, depending on its type. The model farm also has a work schedule and weather conditions under which it operates. These factors along with other work rules (WORK HOURS CONDS, Fig. 1) associated with the operation influence its actual work hours. The actual field work in hectares is in turn calculated based on the actual work hours for the day and the width, speed, and operational efficiency of the machinery system selected for the operation.

IV.   SIMULATING A WHOLE-FARM SYSTEM

The object-oriented model just described is a static representation of a farm system. It depicts the structure of its objects and their relationships at a single moment in time. Attribute values and links held by objects are called their state. Over time, objects change their states in response to external or internal stimuli. Simulation models examine these changes to the objects and their relationships over time. In developing such a model, we need to define and understand the dynamic aspects of the system and how they influence different objects. These aspects, referred to as controls by Rumbaugh et al. (1991), describe the sequences of operations that occur in the system in response to stimuli. An individual stimulus from one object to another is an *event*. The response of an event depends on the state of the object receiving it and can include a change of state or the sending of another event to an original sender or to a third object.

Controls and operations for a whole-farm system with multicrop production activities are summarized below. A simulation model should capture these controls in as realistic a manner as possible. These controls are performed on each field every day in our model.

1. Search through the farm knowledge base, find pending operations, and select the operation that can be performed on the field.

2. Check the conditions that need to be satisfied to make the operation effective for crop production, for example, the simulation day must be within the agronomic window of the operation, that is, the time window when this operation is appropriate for this crop.

3. Identify the machinery system (implement and an operator for self-propelled machinery systems; implement, power source, and an operator for the machinery systems that require a separate power source) for the operation, and check its availability.

4. Estimate the actual work hours for the day based on the operation type, scheduled work hours, climate, and soil- and crop-related factors.

5. Estimate the operational capacity for the day based on the selected machinery system—width, speed, and efficiency—and the actual work hours.

6. Calculate the actual work done, and update the farm knowledge base to reflect the changes in field conditions—completed and pending operations and availability status of the machinery system (implement, power source, and operator) for the day.

7. Prepare and store the simulation reports, including the operations completed and the machinery system used.

V. PROLOG

A. Introduction

PROLOG (PROgramming in LOGic) is one of the two artificial intelligence (AI) languages (LISP is the other) and provides an environment for developing computer models with an object-oriented data structure. It is defined as a declarative language, in contrast to procedural languages such as Basic, Fortran, Pascal, and C. Of course, many other software development programs are available, including the various "Visual" packages for developing Windows-based programs. PROLOG contains a basic scheme for representing knowledge and also has a built-in inference engine. A program in PROLOG is a collection of facts and rules (knowledge base) about the system. These facts and rules are assembled in two principal components of the program code, namely predicates and clauses. In a broader sense, predicates are analogous to subroutine definitions of procedural languages, and clauses are similar to calls for a subroutine with specific argument values. From the collection of facts and rules, PROLOG derives solutions to the questions posed interactively by the user or listed as "goals" within the program code. The search for solutions to the question or the "goal" is made through all the clauses. This permits achieving multiple solutions to a single query. An expert system written in PROLOG will check all the rules for possible solutions, whereas some expert system development programs stop and present only the first solution found. The capabilities such as symbol processing, list manipulation, and recursion facilitate handling heuristic, in addition to quantitative and procedural, computations necessary for simulating dynamic behavior of a system.

PROLOG has been employed mostly for writing expert systems. There have been some efforts to develop simulation programs, simulation languages, and object-oriented programming languages using PROLOG (Adelsberger, 1984; Futo and Gergely, 1986, 1987; Kornecki, 1986; Yokoi, 1986; Fan and Sackett, 1988). However, there is no reference in the literature to the use of this language for agricultural simulation modeling.

Turbo PROLOG (Borland, 1986) was used in this project. This version of PROLOG provided a convenient programming environment. It permits writing the code in natural-language-like sentences, which facilitates better understanding of program logic by nonprogrammers. The Turbo PROLOG code can be compiled into an executable program and can be distributed freely to different users.

B. Simulation with PROLOG

The whole-farm simulation model was structured in two distinct components: (1) factual farm and regional knowledge and (2) procedural knowledge for simulation. The factual farm and regional knowledge consists of object-classes and their individual instances (objects), weather data, and the regional expert knowledge as depicted in the object data model (Fig. 1). They are stored as PROLOG databases and can be developed for each farm using the user interface (known as Info Manager) of the simulator.

The procedural knowledge for carrying out the simulation is a PROLOG program that consists of a group of predicates and clauses. This program is designed to search through the farm knowledge base and apply the controls and operations discussed previously for emulating the dynamic behavior of the farm on a daily basis. One of the most critical tasks of this program was to use PROLOG (mainly a searching language) to perform time increments for the simulation process. This was achieved by employing the backtracking (repeat—fail combination) property of PROLOG discussed by Flowers (1988). This property allows PROLOG to keep searching through the clauses of a predicate or database(s) until all possible options have been evaluated. PROLOG repeats all the tasks defined within the predicate if the value of any variable is changed during backtracking. Some specific examples of this property used for the simulation model are discussed later in this section.

A simplified version of the PROLOG code for the simulator is presented in Fig. 2. It consists of three principal hierarchical modules (predicates): doSimulation, simulateSeason, and simulateADay, which work in a hierarchical manner.

The predicate doSimulation creates the environment for the simulation, carries out the simulation, and prepares reports. The farm and other knowledge is loaded into computer memory using predicates getFarmKnowledge and getOtherKnowledge. The season's simulation is carried out by predicate simulateSeason, which has two arguments: start day (SDay) and finish day (FDay). These two days represent the time span of the simulation based on the earliest start time of the first operation and the latest finish time of the last operation within the cropping system.

Within the simulateSeason predicate, "asserta" assigns the start day "SDay" to the database "currentday." "SDay" is also assigned to variable "SimDay." The predicate simulateADay is then called five times with the current day (SimDay) and different operation

```
    doSimulation:-
      getFarmKnowledge(labor, field, equipment,
        crop,irrigate, operation, tractor,
        implement),
      getOtherKnowledge(expertfile, weatherfile,
        workhrsfile),
      findSimulationDuration(SDay, FDay),
      simulateSeason(SDay,FDay),
      writeResultFiles(result1,result2,result3,result4).

    simulateSeason(SDay,FDay):-
      asserta(currentday(Sday)),
      repeat,
      currentday(SimDay),
      makeAvailAll,
      simulateADay(Simday,"Irrigation"),
      simulateADay(Simday,"HarvestingOps"),
      simulateADay(Simday,"PlantProtectionOps"),
      simulateADay(Simday,"PlantFertiOps"),
      simulateADay(Simday,"LandPrepOps"),
      Sday1=Simday+1,
      changeDay(Sday1),
      Sday1>Fday.

    simulateADay(SDay,OpsType):-
      fieldc(FieldN,Crop),
      chkCondAndWork(FieldN,Crop,Sday,OpsType),
      fail.

    simulateADay(_,_):-!.

    chkCondAndWork(FieldN,Crop,_,_):-
      fieldo(FieldN,_,_,_,Crop,OpsList,_,
        _,_,_,_,_),
      OpsList=[],!.
```

FIGURE 2   Simplified code of operations simulation in PROLOG.

types, namely Irrigation, HarvestingOps, PlantProtectionOps, PlantFertiOps, and LandPrepOps. The priority for an operation is determined by the order in which it is called by this predicate. Under the current format, irrigation gets the highest priority, followed by harvesting, plant protection, planting and fertilizer application, and land preparation operations.

On the successful completion of the day's simulation, the simulation day is updated by one (Sday1 = Simday + 1). The content of the "currentday" database is also changed to "Sday1" by the use of predicate changeDay. As a last step within the simulateSeason predicate, the updated day is compared to the last day for the simulation using the condition (Sday1 > Fday). This condition is satisfied when the simulation for the entire cropping season is completed. Else the condition fails, thus forcing PROLOG to backtrack. The backtracking continues through the predicate until "repeat." The "repeat" is a nondeterministic predicate that always reverses the backtracking. On the reversal, PROLOG finds the content of the currentday database increased by 1 day. As explained earlier, this forces PROLOG to reexecute all the successive clauses (makeAvailable and simulateADay) with the new day. This process is repeated until the condition Sday1 > Fday becomes true, which results in successful completion of predicate simulateSeason.

The simulateADay predicate carries the day's simulation. It has two arguments: the current day (Sday) and the operation type (OpsType) and uses two predicates, fieldc and chkCondAndWork. "fieldc" provides values to variables FieldN and Crop based on the first record of the fieldc database. This database consists of field names and the crops being grown on those fields for the entire farm. The entries in the database are made in order of priority for the fields to be attended to during the simulation. For example, if Field5 has higher priority than Field4, then Field5 should be listed prior to Field4 in the fieldc database. During the execution of this predicate, these values are assigned to "FieldN" and "CropN" in the predicate chkCondAndWork, which checks for different conditions to be met prior to starting the operation and also prepares simulation reports.

Figure 3 lists the conditions that are checked by the predicate chkCondAndWork for scheduling different field operations for the day. This predicate calculates the work based on the selected machinery system and actual work hours for the day. It prepares and writes reports and keeps track of the status of different fields, implements, labor and power sources (tractors, etc.) available on the farm.

On the successful completion of tasks of chkCondAndWork for the current field, crop and, operation type, the predicate fail causes the predicate simulateADay to fail and forces it to backtrack. In this case, the backtracking occurs up to the predicate fieldc, which now points to the next entry in the database. PROLOG assigns new values to the variables FieldN and CropN from the next record of the fieldc database and repeats all the steps performed by simulateADay in an attempt to succeed. However, the fail predicate causes it to fail again. This process is repeated until all entries of the fieldc database

1) Calculate soil moisture of the day based upon yesterday's ET, rain and irrigation, if any.

2) Check if the field has an irrigation facility and if so find out the irrigation equipment used on the field.

3) Check the availabilities of irrigation equipment and any operators needed to operate the equipment.

4) Check number and amount of pending irrigations

5) Irrigate the field, if soil moisture is below threshold, number or amount of pending irrigations is not zero and irrigation equipment and operator are available.

6) Make the field, the equipment and the operator associated with the irrigation non-available for the day, else do not change their status. Update the soil moisture status of the field irrespective of decision about the irrigation.

7) If the field is not being irrigated, get the list of operations associated with the crop being grown on the field.

8) If the operations list is not empty, pick up first operation from the list.

FIGURE 3 Tasks performed by the Operations Simulator for scheduling irrigation and other field operations for a field on a daily basis.

have been processed. This ensures that all the fields listed in the fieldc database are checked daily for every type of operation. Having processed all entries of this database, the clause fails even prior to calling the predicate chkCondAndWork. At this stage PROLOG searches for an alternative simulateADay clause or rule and finds "simulateADay(_,_):-!.", which always succeeds for any value of SDay and OpsType. It thus successfully completes the task of checking all fields on the farm for carrying out the operation type instantiated (the assigned value of OpsType) in the simulateADay predicate. The same process is repeated for other operation types by calling the simulateADay predicate with different values of "OpsType" for the current day.

```
9) Check if the operation is the type of operation
being tried now.

10) Get details for the operation such as
agronomic window, implement list, etc.

11) Check suitability of the "current day" for the
operation with respect to its agronomic window.
The "current day" should be within the work
window of the operation.

12) Check the availability of the required machinery
set for the operation; the implement and the
operator in case of self-propelled implements, and
the implement, the tractor (power unit) and the
operator in the case of other types of implements.

13) If all above conditions are satisfied carry
out the operation based upon the implement
characteristics; working width, speed of operation
and actual work-hours and operation efficiency.

14) Record the work and no work as applicable and
remove the operation from the list, if it has been
completed or its latest finish time has passed.

15) Make the machinery set (Implement, Tractor and
Operator) and the field not available for the day.
```

FIGURE 3   Continued

chkCondAndWork is the most demanding predicate. It checks several conditions to perform its task. However, its clauses are structured to perform efficiently. They avoid deep level searches for the operations that have been already completed and also for the operations for which the field is not ready. The time required for the day's simulation decreases as crops on different fields are harvested and the operations lists associated with the crop become empty.

The simulator produces three reports: work report, no-work report, and summary report. The work report contains daily information about the work performed during the simulation, and the no-work report contains daily information about the operations that were attempted but could not be done because of factors such as nonavailability of the machinery or excessive rain during the day. The summary report summarizes the work and no-work reports for each operation for the entire cropping season. Table 2 presents the contents of these reports.

Table 2 Simulation Reports and Their Contents

Report description	Contents
Work report	Julian day, month, field, crop, operation, total area, accumulated done area, day's work, implement, tractor, operator, and number of hours worked
No-work report	Julian day, month, field, crop and operation, total area, done area, reason for no-work, implement list and its availability report, operator list and its availability report, tractor list and its availability report
Summary report	Field, crop, operation, scheduled start and finish times, actual start and finish times, number of days and hours worked, total done area, number of nonworking days

VI. SIMULATOR: A COMPONENT OF FARMSYS

The simulator, discussed in this section, is one of the four components of an integrated decision support system called FARMSYS for the whole-farm multicrop production system (Lal, 1989). The other components of FARMSYS are Expert Analysis System, Information Management System, and Yield Estimation System. The Expert Analysis System (Lal et al., 1991a) analyzes the simulation reports in the context of the available farm resources. It makes recommendations for improving the timeliness of field operations and/or the overall utilization efficiency of farm labor and machinery. The Information Management System (Lal et al., 1990) provides an intelligent interface between the user and the simulator. It allows the user to enter and update farm information in a user-friendly manner. The Yield Estimation System estimates crop yields and profits from different fields, crops, and the whole farm (Lal et al., 1992).

FARMSYS is operated using a pulldown menu developed with the tools of Turbo PROLOG Toolbox (Borland, 1987). The schematic representation of the screen layout and the set of menus along with their final actions are depicted in Fig. 4. The time required for a simulation run depends on farm size and type of computer used. The other components of FARMSYS responded quickly and interactively without noticeable time delays.

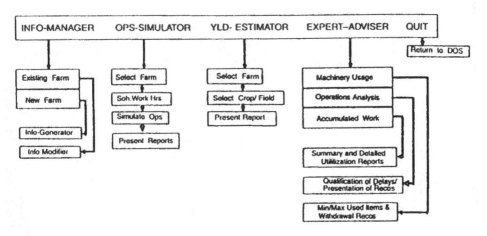

FIGURE 4 Schematic of FARMSYS screen layout and its principal menus.

VII. TESTING AND EVALUATION

The overall goal of FARMSYS was to provide a planning and/or management tool for researchers, educators, and farmers to test different combinations of resources such as equipment, crop mixes, and labor for different management strategies (no-till, minimum till, etc.) over a variety of weather-years. It can evaluate timeliness of different operations, the utilization efficiency of farm resources (labor and machinery), and their effects on crop yields. Performance of different farms for a particular weather-year or that of the same farm over different weather-years can be evaluated using FARMSYS. The flow diagram for such an evaluation is shown in Fig. 5.

The Operations Simulator, along with other components of FARMSYS, was subjected to two types of testing and evaluation: professional qualification and operational verification.

Professional qualification involved letting a team of authorities, mainly agricultural engineering professors interested in knowledge-based systems, evaluate and critique the logic, functioning, and user interface of the system. This session was primarily aimed at identifying the strengths and weaknesses of the Information Management and Expert Analysis systems of FARMSYS. The term "qualification" is used to indicate that this phase involved evaluating the system

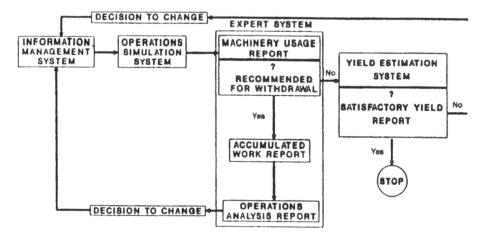

FIGURE 5   Using FARMSYS as a decision aid tool.

more on the basis of subjective judgments of the authorities rather than as a quantitative test.

The detailed procedure and results for the professional qualification are discussed by Lal (1989) and Lal et al. (1990, 1991a, 1991b). The experience with FARMSYS indicated that the process of seeking subjective evaluation from qualified professionals offers a good and practical alternative for assessing the quality of knowledge-based systems.

For the operational verification of FARMSYS, we used the knowledge base (data) of an operational farm in northern Florida. The term "verification" is used here in its usual sense, that of testing the quantitative results of the simulations for correct logic and for calculations performed as designed in the program.

The test farm consists of about 400 cultivated hectares divided into 23 fields. It has four operators, five tractors, and 25 implements. It grows four crops (cotton, peanuts, soybeans, and wheat) within 10 cropping systems of up to 2 years duration (Table 3). Each of these crops requires a varying number of operations ranging from six for wheat to 16 for cotton. Each operation has specific agronomic window and implement requirements.

There is no limitation on the number of fields FARMSYS can simulate. However, for the sake of simplicity for running the test simulations and reporting the results, we grouped the fields with the same cropping system (such as Cotton-Peanut) into a single field.

Table 3 Cultivated Area and Cropping Systems of Different Fields on the Test Farm

Field	Area (ha)	Cropping system
Field_1	102.2	Cotton-Cotton
Field_2	76.8	Cotton-Wheat
Field_3	6.0	Cotton-Soybeans
Field_4	41.6	Cotton-Peanuts
Field_5	79.6	Soybeans-Cotton
Field_6	16.4	Peanut-Wheat
Field_7	10.0	Peanut-Cotton
Field_8	11.8	Soybeans-NoCrop
Field_8	8.0	NoCrop-Soybeans
Field_9	26.0	NoCrop-Cotton
Field_10	26.4	NoCrop-NoCrop
Total	404.8	

This led to a total of 10 fields (Table 3) with the total cultivated area of the farm unchanged.

Wheat covered approximately 25% of the cultivated area as a second crop on the farm. It is planted in the second half (October–November) of a year and harvested in the following year. Therefore, it became necessary to simulate for two consecutive years to cover the entire cropping season. The second year crops were suffixed with the character "2," and their agronomic windows for different operations for the second year were adjusted by adding 365 to their values of the first year.

Weather (especially daily rainfall) and scheduled work hours are required inputs for the simulation. The actual weather of Tallahassee, Florida, for the years 1954 (dry), 1960 (normal), and 1964 (wet) were used for the test runs. The runs were made by selecting "Dry" followed by another "Dry" year.

The first two simulations were made with all the machinery and labor available on the farm for 6 and 8 h of daily scheduled work for the entire simulation period. These simulations resulted in zero "done area" for a number of operations on several fields. This indicated that a fixed daily work schedule of 6 or 8 h for all calendar months during the cropping season would not be sufficient to attend to the operational requirements with the available machinery and labor on the farm. A flexible schedule with more daily work hours during agronomic windows of uncompleted operations would be required. The Expert Analysis System of FARMSYS was used to identify those periods in which it recommended increasing scheduled work hours for the months of April, May, June, and September.

The third simulation was made with 12 h of daily work schedule for the months of April, May, June, and September and 8 h in the remaining months. In this simulation, all operations were completed within their agronomic window on all fields. These initial simulations showed and emphasized the need for a flexible work schedule for different months of the cropping season of the farm, just as farmers do in the "real world."

Figure 6 presents an example of the report produced for every day during the third simulation. It indicates that on Julian day 100, the simulator tried to carry out five operations. It successfully completed three but failed to carry out the other two. The completed operations were

1. Plowing 14.2 ha on Field_1 with a total cultivated area of 102.2 ha for Cotton using BottomPlow, Tractor_1, and Jerry by working 12 h
2. Harrowing 30.2 ha on Field_2 for Cotton using DiskHarrow, Tractor_1, and Keith by working 12 h
3. Fertilizing 27.9 ha on Field_4 for Cotton using Ferti-SpreaderLP, Tractor_4, and Joe by working 12 h

The operations that were tried but could not be done on the same day were fertilization on Field_5 and Field_8 for the Soybeans crop. The implement FertiSpreaderLP needed for the operation was not available (MachinesNotAvail). This implement was being used for applying fertilizer to cotton on Field_4 as indicated in the work report. This field (Field_4) had higher priority than the soybean fields (Field_5 and Field_8).

In addition to a work schedule that allows timely completion of operations within their agronomic windows, farmers are also interested in identifying machinery and labor that are underutilized. FARMSYS can be used to identify such machines. It would help farmers reduce their capital outlay and also decrease their operating costs. Specific rules for identifying underutilized machinery and labor can vary from one farm to another. However, for the present testing, FARMSYS was coded with the following rules, which can be easily adapted for specific farm situations.

1. Recommend removal of the least utilized item of an object-class (implement, tractor, or labor) if it worked less than 20%

Work Report
General format
workreport(julianDay, Month, Field, Crop,
   Operation, CultArea, Day DoneArea, TotDoneArea,
   Implement, Operator, Tractor, ActWorkHr)

For Julian Day 100
workreport(100, "April", "Field_1", "Cotton",
   "Plowing", 102.2, 14.2, 14.2, "BottomPlow",
   "Jerry", "Tractor_1", 12)

workreport(100, "April", "Field_2", "Cotton",
   "Harrowing", 76.8, 30.2, 30.2, "DiskHarrow",
   "Keith", "Tractor_2", 12)

workreport(100, "April", "Field_4", "Cotton",
   "Fertilize", 41.6, 27.9, 27.9, "FertiSpreaderLP",
   "Joe", "Tractor_4", 12)

No Work Report
General format
noworkreport(julianDay, Month, Field, DoneArea,
   Crop, Operation, DoneArea, ReasonOfNoWork,
   ImplementList, AvailStatus, WhereAboutImple-
   mentList, TractorList, AvailStatus, WhereAbout-
   TractorList, OperatorList, AvailStatus,
   whereAboutOperatorList)

For the Julian Day 100
noworkreport(100, "April", "Field_5",
   79.6, "Soybeans", "Fertilize", 0,
   "MachinesNotAvail",
   ["FertiSpreaderLP"], "NotAvail",
   [["FertiSpreaderLP", "field4", "Cotton",
   "Fertilize"]],
   ["Tractor_4", "Tractor_5"], "NotChecked", [],
   ["Keith", "Ray", "Joe", "Jerry"], "NotChecked", [])

noworkreport(100, "April", "Field_8",
   11.8, "Soybeans", "Fertilize", 0,
   "MachinesNotAvail",
   ["FertiSpreaderLP"], "NotAvail",
   [["FertiSpreaderLP", "field4", "Cotton",
   "Fertilize"]],
   ["Tractor_4", "Tractor_5"], "NotChecked", [],
   ["Keith", "Ray", "Joe", "Jerry"], "NotChecked", [])

FIGURE 6   Examples of work and no-work reports of the FARMSYS Operations Simulator.

Table 4 Actual Start and Finish of the Operations Affected by Withdrawal of Ray and Tractor_5 from the Test Farm

Operation[a]	S. window		4th run		5th run		6th run	
	Sst	Sfn	ASt	Afn	ASt	Afn	ASt	AFn
Field_1 (102.2 ha), first year (cotton)								
FertiSprding	167	223	173	181	173	181	179	186
Spraying4	167	223	182	183	182	183	187	189
Spraying5	172	243	184	184	184	184	190	191
Spraying6	183	244	186	186	186	186	192	192
Field_2 (76.8 ha), first year (cotton)								
FertiSprding	167	212	182	187	182	189	190	197
Spraying4	167	223	188	189	190	192	198	200
Spraying5	172	243	190	191	193	193	202	202
Spraying6	183	244	192	192	194	194	203	203
Field_3 (6.0 ha), first year (cotton)								
FertiSprding	167	212	188	188	190	190	202	202
Spraying4	167	223	190	190	193	193	203	203
Spraying5	172	243	193	193	195	195	204	204
Spraying6	183	244	194	194	196	196	206	206
Field_4 (41.6 ha), first year (cotton)								
FertiSprding	167	212	189	192	191	193	204	209
Spraying4	167	223	193	193	194	194	210	210
Spraying5	172	243	195	195	197	197	211	211
Spraying6	183	244	196	196	198	198	212	212
Field_5 (79.6 ha), first year (soybeans)								
Spraying1	197	228	197	197	200	200	197	197
Field_5 (79.6 ha), second year (cotton)								
FertiSprding	533	580	547	552	547	553	548	555
Spraying4	533	578	553	553	554	554	556	556
Spraying5	538	609	554	554	555	556	557	557
Spraying6	548	610	555	556	557	557	558	558
Field_6 (16.4 ha), first year (peanuts)								
Spraying4	183	223	184	184	184	184	183	183
Spraying5	192	243	192	192	195	195	192	192
Fertilize	289	350	289	289	290	290	290	290

Table 4 Continued

Operation[a]	S. window		4th run		5th run		6th run	
	Sst	Sfn	ASt	Afn	ASt	Afn	ASt	AFn
Field_7 (10.0 ha), first year (peanuts)								
Cultivation2	151	197	151	151	151	151	152	152
Spraying4	183	223	186	186	186	186	184	184
Spraying5	192	243	194	194	196	196	193	193
Field_7 (10.0 ha), second year (cotton)								
FertiSprding	533	580	553	553	554	554	556	556
Spraying4	533	578	557	557	558	558	559	559
Spraying5	538	609	558	558	559	559	560	560
Spraying6	548	610	559	559	560	560	561	561
Field_8 (11.8 ha), first year (soybeans)								
Fertilize	092	136	106	106	109	109	117	117
Harrowing	092	136	117	117	117	117	118	118
Bedding	106	136	118	118	118	118	119	119
Spraying1	197	228	198	198	202	202	208	208
Field_10 (26.4 ha), second year (cotton)								
Cultivation3	533	578	547	548	548	552	547	548
FertiSprding	533	580	554	556	555	557	557	558
Spraying4	533	578	560	560	561	561	567	567
Spraying5	538	609	561	561	567	567	568	568
Spraying6	548	610	567	567	568	568	569	569

SSt = Earliest start time, SFn = latest finish time, ASt = actual start time, AFn = actual finish time. [a]The timings of other operations on different fields remain unchanged. [b]4th run: With Ray and Tractor_5 available for work. [c]5th run: With Ray withdrawn from the farm. [d]6th run: With Ray and Tractor_5 withdrawn from the farm.

as much as the most used item of the same class and there were one or more complementary units to perform the same operations.

2. Recommend increasing the machinery capacity to improve the timeliness for the delayed operations (1) by increasing scheduled daily work hours, (2) by increasing the speed of operation for the machinery, or (3) by acquiring a bigger machine if neither of the first two options is possible.

Lal et al. (1990) present these rules in detail and describe how they were used for searching through the simulation reports and farm knowledge base for developing recommendations by the Expert Analysis System. The subsequent simulation scenarios were created by incorporating recommendations of the Expert Analysis System for the previous run.

The fourth simulation was made after removing ChiselPlow, NoTillPlanter, PDSprayer2, GeneralPlanter, and PDSprayer. These implements were not used at all during the third simulation and were recommended for removal by the Expert Analysis System. This simulation produced exactly the same reports as were produced during the previous run.

The fifth simulation was made after removing "Ray" (the least utilized operator during the fourth simulation and recommended for removal) in addition to other removals. All operations were completed within their agronomic windows during this simulation also. However, certain operations such as Spraying1 on Field_5 previously performed by Ray were performed by Keith and got shifted by 3 days (from day 197 to day 200) in their start and completion (Table 4). This happened because the substituted operator, Keith, during the period of these operations was busy elsewhere and was not available to start the operations on the days carried out during earlier simulations.

The removal of Ray during the simulation also affected the performance of Tractor_5. This tractor was mostly operated by Ray during earlier simulations. There was a slight change in the operations performed by Tractor_5. Its overall utilization decreased from 120 h in 13 days to 108 h in 11 days. This reduction can be attributed mainly to the nonavailability of an operator for Tractor_5 when it became a possible choice for the work.

The sixth simulation was made after removing Tractor_5 in addition to Ray and other removals. This was done based on the recommendation of Expert Analysis System after the fifth simulation. All operations were completed within their agronomic windows during the sixth simulation also. There were some additional shifts in the start and/or completion days of certain operations such as first year Cultivation2 for the peanut crop and second year Ferti-Spreading for the cotton crop on Field_7. It was further interesting to note that certain other operations such as Spraying1 for the first year soybean crop on Field_5 were expedited with the combined withdrawal of Ray and Tractor_5 (Table 4). At first glance, it seemed to be an erroneous simulation. However, a critical review of the simulation code revealed that it was doing what it was designed to do. The complex matrix of rules and priorities for different fields, operation types, and assignments of implements to an operation, tractors to an implement, and operators to a tractor in a cropping season can easily cause the farm system to behave in this particular manner. During the sixth simulation, for example, the spraying (Spraying1) on Field_5 was expedited because it got precedence over the cotton fields (Field_1, Field_2, Field_3, and Field_4). These cotton fields, though with higher priority than the soybean field, were not ready for spraying due to delayed fertilizer application without Ray and Tractor_5. This made the equipment available for spraying on the soybean field (Field_5). In the previous simulations, Field_5 had to wait for this equipment until the completion of spraying on all cotton fields. This shows how the simulator is flexible to reschedule operations on different fields, as a farm manager would do in a real situation, based on the availability of machinery and labor. During this simulation, Tractor_2 became the least utilized tractor on the farm, but it was not recommended for removal because it was used more than 20% as much as the most utilized tractor (Tractor_4) on the farm. However, some plant protection and cultivation equipment such as PCultivator were still underutilized and were recommended for removal.

The seventh simulation was made after removing PCultivator in addition to the other removals. All operations were successfully completed within their agronomic windows with some additional shifts in start and/or completion of certain operations. At this time MCultivator was identified as the least utilized plant protection implement and was recommended for removal.

The eighth simulation was made after removing MCultivator in addition to other removals. All operations were completed within their agronomic windows. At this stage, no other machinery or labor was recommended for removal. They were either unique items on the farm and did not have any complementary unit to perform their tasks or they were used more than 20% as much time as the most utilized item of their class. Table 5 presents the usage of the most and least utilized items of different object-classes and the recommendations of the Expert Analysis System after each simulation. Machinery and labor available on the test farm and items recommended for removal by FARMSYS are presented in Table 6.

Table 5 Most and Least Utilized Objects During Simulations 1–5 and Recommendations of FARMSYS Expert Analysis System

Simulation	Object specification	Hours worked	Remarks and recommendations
1. With Ray, Tractor_5, PCultivator, and MCultivator	Jeff	2820	Most utilized object
	Ray	142	Least utilized object and recommended for removal
	Tractor_3	1432	Most utilized object
	Tractor_5	120	Least utilized object and recommended for removal
	Sprayer	552	Most utilized object
	Pcultivator	24	Least utilized object and recommended for removal
2. After removing Ray, but with Tractor_5, PCultivator, and Mcultivator	Jeff	2844	Most utilized object
	Keith	734	Least utilized object but not recommended for removal
	Tractor_3	1428	Most utilized object
	Tractor_5	108	Least utilized object and recommended for removal
	Sprayer	560	Most utilized object
	Pcultivator	24	Least utilized object and recommended for removal
3. After removing Ray and Tractor_5, but with PCultivator, MCultivator	Jeff	2924	Most utilized object
	Keith	654	Least utilized object but not recommended for removal

Tractor_4	1492	Most utilized object
Tractor_2	516	Least utilized object but not recommended for removal
Sprayer	560	Most utilized object
Pcultivator	24	Least utilized object and recommended for removal

4. After removing Ray, Tractor_5, and PCultivator but with MCultivator

Jeff	2912	Most utilized object
Keith	654	Least utilized object but not recommended for removal
Tractor_4	1480	Most utilized object
Tractor_2	516	Least utilized object but not recommended for removal
Sprayer	560	Most utilized object
Mcultivator	72	Least utilized object and recommended for withdrawal

5. After removing Ray, Tractor_5, PCultivator, and MCultivator

Jeff	2996	Most utilized object
Keith	630	Least utilized object but not recommended for removal
Tractor_3	1468	Most utilized object
Tractor_2	516	Least utilized object but not recommended for removal
Sprayer	560	Most utilized object
SprayCoupe	289	Least utilized object but not recommended for removal

Table 6 Machinery and Labor Available on the Test Farm and
Items Recommended for Removal by FARMSYS

Implements available		
DiskHarrow	PDSprayer[a]	Peanut Combine
ChiselPlow[a]	PDSprayer2[a]	Cotton Picker
SCultivator	SprayCoupe	BottomPlow
MCultivator1	General PLanter[a]	FertiSpreaderLP
MCultivator2[a]	NoTilPlanter[a]	SubSoiler
LBCultivator	GrainDril	Planter_77
PCultivator[a]	Bedder	BMower
Sprayer	General Grain Combine	FertiSpreaderPF

Operators available
Laborer_3
Laborer_4
Laborer_1[a]
Laborer_2

Tractors (HP) available
JD4450 (148)
JD4430 (125)
JD2950 (90)
JD3020 (70)
JD2640 (70)[a]

[a]Identified for withdrawal.

VIII. CONCLUDING REMARKS

The object-oriented approach to simulation of field operations
showed considerable advantages over the conventional approach to
simulation. The new approach helped in organizing, representing,
and manipulating farm knowledge in different object-classes. Farmers' preferences and priorities for different fields, operation types, and
assigning implements to an operation, tractors to an implement, and
operators to a tractor were captured and used in the simulation process. The approach resulted in a flexible simulation model that re-
scheduled operations on different fields, as a farm manager would
do in a real situation, based on the availability of machinery and
labor. The model can also be used for a variety of farming situations
ranging from a highly mechanized farm (such as the test farm) to a
labor-intensive farm with little or no mechanization. It is feasible be-

cause of the structure of the model in two distinct components: (1) factual farm and regional knowledge and (2) procedural knowledge about timing, priorities, rules, and methods for simulating the system.

Traditionally, different computer languages are used for simulation models, expert systems, and database management systems. This complicates the process of combining these powerful tools into an integrated decision support system. The object-oriented approach using PROLOG helped in developing all these components within a single environment, leading to a seamless decision support system. The modular nature of programming in PROLOG also permits its easy modification and upgrading.

The reports generated by the simulator showed scheduling of operations on different fields. They also identified situations when an operation was attempted but could not be completed because of non-availability of one or more components of the machinery system (implement, tractor, and operator) needed for the operation.

In general, the simulator is much more versatile and robust than most traditional whole-farm models. However, the following additions would further enhance its capabilities. The simulator presently assumes that all farm resources are available exclusively for crop production. Most farms, especially small-scale enterprises, have animal production as an integral component of their farming systems. Therefore, a criterion that partitions the available resources between the animal and crop production would result in a more versatile model.

Actual work hours for an operation on a given day are estimated on the basis of soil and crop conditions, operation type, scheduled work hours, and time remaining within the agronomic window. Farmers also consider past and expected future weather in deciding actual working hours for the day. Therefore, rules that incorporate these factors should be developed and included in the system for deciding the actual work hours for different operations.

REFERENCES

Adelsberger, H. H. 1984. PROLOG as a simulation language. In: *Proceedings of the 1984 Winter Simulation Conference*. S. Sheppard, V. Pooch, and D. Pegden (Eds.). San Diego, CA: SCS, pp. 501–504.

Borland. 1986. *Turbo PROLOG: The Natural Language for Artificial Intelligence*. Scotts Valley, CA: Borland International Inc.

Borland. 1987. *Turbo PROLOG Toolbox—User's Guide and Reference Manual*. Scotts Valley, CA: Borland International Inc.

Chen, L. H., and R. W. McClendon. 1985. Soybean and wheat double cropping simulation model. *Trans. ASAE* 28(1):65–69.

Fan, I. S., and P. J. Sackett. 1988. A PROLOG simulator for interactive flexible manufacturing systems control. *Simulation* 50(6):239–247.

Flowers, E. B. 1988. Failing with grace. *Turbo Technix (Borland Language J.)* 1(5):76–85.

Freeman, S., A. D. Whittaker, and J. M. McGrann. 1989. Knowledge-based machinery management aid. ASAE Paper 89-7576. St. Joseph, MI: ASAE.

Futo, I., and T. Gergely. 1986. TS-PROLOG: A logic simulation language, *Trans. SCS* 3(4):112–119.

Futo, I., and T. Gergely. 1987. Logic programming in simulation. *Trans. Soc. Comput. Simu.* 3(3):195–216.

Glunz, D. J. 1985. Comparative analysis of Florida citrus harvest systems. MS Thesis. Gainesville, FL: University of Florida.

Helms, G. L., J. W. Richardson, M. E. Rister, N. D. Stone, and D. K. Loh. 1987. COTFLEX: A farm-level expert system to aid farmers in making farm policy decisions. Annual Meeting of Am. Agric. Econ. Assoc. East Lansing, MI.

Kornecki, A. 1988. Simulation and artificial intelligence as tools in aviation education. In: *AI Papers*. R. J. Uttamsingh (Ed.). Simulation Series Vol. 20, No. 1. San Diego, CA: Society for Computer Simulation, pp. 121–126.

Labiadh, S., and J. C. Frisby. 1987. A simulation model to select alfalfa harvesting machines. ASAE Paper 87-1047. St. Joseph, MI: ASAE.

Lal, H. 1989. Engineering farm knowledge for a seamless decision-support system. Ph.D. Dissertation. Gainesville, FL: University of Florida.

Lal, H., R. M. Peart, J. W. Jones, and W. D. Shoup. 1990. An intelligent information manager for knowledge-based systems. *Appl. Eng. Agric.* 6(4): 525–531.

Lal, H., R. M. Peart, W. D. Shoup, and J. W. Jones. 1991a. Expert result analyzer for a field operations simulator. *Comput. Electron. Agric.* 6:123–141.

Lal, H., R. M. Peart, J. W. Jones, and W. D. Shoup. 1991b. An object-oriented field operations simulator in PROLOG. *Trans. ASAE* 34(3):1031–1039.

Lal, H., J. W. Jones, R. M. Peart, and W. D. Shoup. 1992. FARMSYS: A whole-farm machinery management decision support system. *Agric. Syst.* 38: 257–273.

Miles, G. E., and Y. J. Tsai. 1987. Combine systems engineering by simulation. *Trans. ASAE* 30(5):1277–1281.

Nute, D., G. Zhu, and M. Rauscher. 1996. DSSTOOLS: A toolkit for development of decision support systems in PROLOG. Proc. 6th Int. Conf. on Computers in Agriculture, Cancun, Mexico. St Joseph, MI: ASAE.

O'Keefe, R. M., and J. W. Roach. 1987. Artificial intelligence approaches to simulation. *J. Opt. Res. Soc.* 38(8):713–722.

Rumbaugh, J., M. Blaha, W. Premerlanim, F. Eddy, and W. Lorensen. 1991. *Object-Oriented Modeling and Design*. Englewood Cliffs, NJ: Prentice-Hall.

Smith, R. D. 1985. Biomass storage, decision support systems and expert systems in crop production and processing. Ph.D. Thesis. West Lafayette, IN: Purdue University.

Stone, N. D., R. E. Frisbie, J. W. Richardson, and R. N. Coulson. 1986. Integrated expert system applications for agriculture. Proc. Int. Conf. on Computers in Agric. Ext. Programs, February 1986. Orlando, FL, pp. 836–841.

Thai, C. N., and C. P. Wilson. 1988. Simulation of peach post harvest operations. ASAE Paper 88-6058. St. Joseph, MI: ASAE.

Tsai, Y. J., J. W. Jones, and J. W. Mishoe. 1987. Optimizing multiple cropping systems: A systems approach. *Trans. ASAE* 30(6):1554–1561.

van Elderen, E. 1981. Scheduling of field operations: Research report. *IMAG* 81(3):865–871.

van Elderen, E. 1987. *Scheduling Farm Operations: A Simulation Model.* Wageningen, Netherlands: Pudoc.

Whittaker, D. A., E. J. Monke, and F. R. Foster. 1987. ADAM: An ADaptive Assembler of Models. ASAE Paper 87-5536. St. Joseph, MI: ASAE.

Yokoi, S. 1986. A PROLOG based object-oriented language and its compiler. In: *Lecture Notes in Computer Science.* G. Goos and J. Hartmanis (Eds.). Logic Programming '86 Proc. 5th Conf., Tokyo, Japan.

16

Fundamentals of Neural Networks

William D. Batchelor
Iowa State University, Ames, Iowa

I. INTRODUCTION

Many agricultural systems are composed of highly complex biological components. The previous chapters introduce methods of using traditional simulation techniques to describe agricultural systems, which include growth of crops, animals, and diseases, and agricultural operations including machinery selection and management. Mechanistic approaches used to describe biological processes are based on mathematically describing the biological processes of the system at a level of detail that is sufficient to satisfy the objectives of the model. In some instances, the underlying mechanisms of biological processes as well as the response of processes to a wide range of environmental conditions may be understood very well. In this case, traditional mathematical models can successfully describe these processes. In other cases, the underlying mechanisms governing biological processes of interest or the response of processes to environmental changes may not be well understood. In both cases, traditional modeling techniques are limited, and empirical approaches are often used to represent biological processes. Often, empirical approaches are bound by restricted ranges for which the equations were derived.

Furthermore, many empirical equations cannot adequately describe the response of the system to a range of conditions.

The foundation of traditional modeling is to develop the mathematical relationships describing the biological processes of the system. One limitation to traditional mathematical modeling is that the mathematical relationships describing each process of the system must be known. Limitations in our understanding of the system introduce error into the model. New mathematical techniques such as neural networks (NN) have been developed by artificial intelligence researchers to overcome this problem. Neural networks are highly sophisticated pattern recognition systems that are capable of learning relationships in patterns of information. They mathematically mimic the biological human learning process and are capable of learning the hidden relationships between system inputs and responses. This learning is accomplished by repeatedly presenting the NN with many patterns containing the system inputs and responses. The network then learns the relationships between the patterns defining the inputs to the system and the system response through an iterative optimization technique that attempts to minimize the error between the measured system response and the response computed by the NN. Once a network has been trained to recognize patterns, it can then be used to classify new patterns according to its knowledge of existing patterns. This mathematical modeling technique can be used as a "blackbox" approach to predict the response of a biological system to system inputs. A significant advantage of neural networks is that the mathematical relationships describing the processes do not have to be known because the network derives these relationships from the patterns of inputs and outputs through an iterative learning procedure.

Neural networks can be used in a stand-alone mode to predict biological responses, or they may be integrated as subcomponents of larger biological models. Neural networks can be used within simulation models to represent complex biological processes where the response of the process to system inputs are known but the details of the process cannot easily be quantified. Thus, they can be used to supply missing relationships where physical principles are not well understood. In addition to this, they are currently being used to gain an understanding of relationships by looking at data in a very different way. Neural networks are similar to nonlinear regression, but they are much more robust (Thai and Shewfelt, 1990) and can expose hidden relationships in large bodies of information by using pattern recognition theory. They have successfully been used in biological applications to predict processes such as soybean phenology (Elizondo et al., 1994), aflotoxin concentrations in peanuts (Parmar et al., 1994), optimum temperatures for greenhouses (Seginer and Sher, 1992), insect pest treatment thresholds (McClendon and Batchelor, 1995), and recognition of patterns from digital images (Deck et al., 1991).

There are many commercial neural network packages that are available on both PC and workstation platforms. The most basic and inexpensive software packages are simply collections of neural network algorithms that can be incorporated into other programs (see Rummelhart and McClelland, 1986). The user must supply code to input data into the algorithms and to output data in useful forms. The advantage of this approach is that the algorithms can easily be incorporated as components of larger models. These algorithms can also be rewritten in many programming languages and can be implemented on different computer platforms. The most sophisticated PC-based software packages are highly sophisticated Windows-based programs. These programs typically require little time to learn, and users can very quickly begin to develop a neural network. The only thing that the user must do is provide input data, select an architecture, and set up the number of hidden nodes and learning rates. These packages typically allow data to be input directly, or data can be imported from spreadsheets, ASCII datafiles, or various databases. Often, these packages allow the user to select from among several different architectures. Outputs vary from tabular to graphical, and outputs can often be exported into spreadsheet, ASCII, or database files. Many of these packages can compile a neural network into an executable code that can be called from simulation models.

II. FUNDAMENTALS OF NEURAL NETWORKS

A. Biological Neural Networks

The mathematics of neural networks are derived from understanding how biological neural networks in the human brain memorize patterns and learn information. Individual neurons in the human brain (there are over 100 billion) are connected to many other neurons to form a complex network that has an amazing capacity to recognize patterns and learn information. The basic computational element of a biological neural network is the neuron, which consists of synapses, dendrites, soma, and axons (Fig. 1). The synapse is an area of electrochemical contact between neurons that transfers electrochemical

Synapse

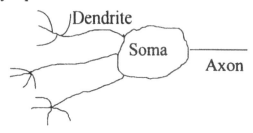

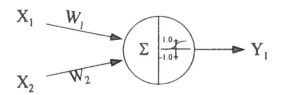

FIGURE 1   Mathematical representation of a biological neuron.

voltages from axons of nearby neurons to the dendrite, which serves as an input channel to the neuron body (soma). Potassium ions are the primary source for the electrochemical voltage. The dendrite can change the membrane permeability to potassium ions, effectively adding a resistance, or weighting factor, that modifies the signal transmitted to the soma. The soma evaluates the inputs from the dendrites and determines whether to give an output. If the sum of the signals from the input dendrites exceeds 70 mV, the soma outputs a signal of approximately 100 mV; otherwise it does not fire (0 mV). Thus, the cell body acts as an on/off switch that transfers multiple voltage inputs into a single voltage output by comparing the sum of the inputs to a threshold value.

Learning in a biological neural network is accomplished through an iterative training process in which the pattern is repeatedly presented to the biological network as it forms interconnections with other neurons and derives the appropriate weights for the interconnected dendrites required to memorize the information. When patterns of information are presented to the biological neural network, information is converted to an electrochemical representation. This representation is repeatedly presented to the network until the necessary interconnections with other neurons are formed and the weighting along each of the interconnected dendrites is found. At this point, the neural network acts as a permanent storage device for the information it has learned. The memory is simply a result of the structure of the network and the weights (permeability to potassium ions) along each dendrite. When this new information is presented to the network at a later date, the network can rapidly recall the memorized information.

B. Artificial Neural Networks

An artificial neuron that mimics a biological neuron is the basic computational element in an artificial neural network (Fig. 1). Each artificial neuron contains input channels, an activation function, and an output that mimics the synapse, dendrite, soma, and axon of a biological neuron. Input signals representing information are transmitted to the neuron by multiplying the input signal (X_i) and the weight of the input connection (W_i). The sum of the input signals (X_iW_i) is then transformed to an output signal (Y_i) by a transfer function that is typically a sigmoidal function ranging from 0 to 1. This transformation function mimics the firing mechanism of the biological neuron body.

The most common method of connecting artificial neurons together is the feedforward architecture, referred to as a feedforward neural network. Neural networks of this class map patterns of inputs to associated output patterns using decision region theory. Feedforward NN are able to generalize relationships between inputs and outputs very well. In addition to this, they can effectively extract nonlinear relationships to extrapolate to other similar patterns. Neural networks are often referred to as heteroassociative memory devices. An example of a feedforward NN is shown in Fig. 2. Information is processed from left to right through the layers of nodes until it reaches the final output layer. It is assumed that all of the nodes in a layer fire simultaneously. Thus, processing is said to occur in parallel for each layer. These neural networks are often referred to as parallel processors of information.

Figure 2 shows a three-layer NN with i inputs, j hidden nodes, and k output nodes. The number of layers refers to the number of

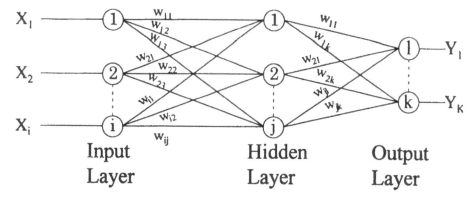

FIGURE 2 Three-layer feedforward neural network with i input nodes, j hidden nodes, and k outputs.

levels of neurons that are connected together. The number of layers also determines the dimension of the decision region that can be described. In a three-layer NN, three layers of neurons—an input layer, a middle (or hidden) layer, and an output layer—are connected together. Each node in the input layer is connected to each node in the hidden layer. Similarly, each node in the hidden layer is connected to each node in the output layer. In a typical NN, input information is normalized into a range of [0,1], and the input values are transformed to an output value for each node by using a transfer function. The input to the network can be represented as a vector X, where

$$X = [x_1 \quad x_2 \quad \cdots \quad x_i] \tag{1}$$

for i inputs. Input neurons typically pass the value of the input directly to its output using a linear transfer function with a range of [0,1]. Each interconnection between the input layer and the hidden layer has a weight associated with it. The weights from input node i to hidden node j follow the convention W_{ij}. The matrix describing this information is

$$W = \begin{bmatrix} w_{11} & w_{12} & \cdots & w_{1j} \\ w_{21} & w_{22} & \cdots & w_{2j} \\ \vdots & & & \\ w_{i1} & w_{i2} & \cdots & w_{ij} \end{bmatrix} \tag{2}$$

The input signal for each node in the hidden layer is computed by summing the products of the output from each input layer and the weight of the interconnections:

$$I = \sum_{m=1}^{i} \sum_{n=1}^{j} X_m W_{mn} \tag{3}$$

The sum of the input signals is then transformed to an output value using a sigmoidal transfer function that transforms the output to a range from [0,1]. The output signals from the hidden nodes are then multiplied by the weights of the interconnections between the hidden layer and the output layer. The nodes in the output layer then compute the sum of the input signals and transform the output using a sigmoidal function to compute the output of the neuron.

The output can be described by the vector

$$Y = [y_1 \quad y_2 \quad \cdots \quad y_k] \tag{4}$$

Thus, the resulting matrix representation of a feedforward NN is

$$XW = Y \tag{5}$$

The relationships between inputs and outputs, or memory, are stored in the weights of the interconnections. Increasing the number of hidden nodes and interconnections increases the memory capacity of the neural network. This system could be modified to include more or fewer hidden nodes; however, at least two are required. By including more hidden nodes, the network has more interconnections available to map inputs to outputs.

The number of layers in an NN affects the dimension of the decision region. A two-layer NN can solve problems with a two-dimensional decision region. A three-layer NN can solve problems that have a three-dimensional decision region. It can be shown that three layers are adequate for solving any n-dimensional problem because of the properties of matrix addition.

Neural networks must be trained to recognize patterns. The goal of training an NN is to determine the values of the interconnection weights that minimize the error between predicted and measured outputs over many patterns of inputs X_i and outputs B_i. The threshold function employed by each neuron adds nonlinearity to the solution; thus the weight matrix cannot be found by linear algebra techniques. An alternative technique based on numerical solutions must be used to solve for the optimum weight matrix. The back-propagation algorithm, which uses the method of steepest descent (Rumelhart and McClelland, 1986), is the most common method of solving for the optimum values of the weights in W that minimize errors between measured output patterns and those computed by the neural network. Training a neural network involves presenting the network with sample patterns, called training patterns, consisting of known inputs and outputs. The network then solves for the best weight matrix that describes the relationship between the input and output for all the training patterns. Using this process, the network actually learns the relationships between the inputs and outputs of the training patterns without being told what the relationships are.

The back-propagation algorithm is the most widely used method of computing the best set of weights and activation thresholds for the training patterns presented to the network. Some forms of the algorithm compute both weights and activation thresholds, while other methods set the neuron activation threshold at 1.0 and compute only the optimum weights of the interconnections. The process begins by randomly initializing the weights along each interconnection. Each input pattern is presented to the network, and the output is computer based on these initial weights. The sum square error is then computed by squaring the difference between the desired and computed output over all output nodes according to

$$E = \sum_{i=1}^{k} (y_k - y_k')^2 \tag{6}$$

where y_k and y_k' are the actual and computed output values and k is the number of output nodes. If the sum square of error is greater than some threshold value, the weights are adjusted and the process is repeated.

The adjustment of each weight, θ, is computed by the back-propagation algorithm according to

$$\theta = \delta E + \sigma \theta_{-1} \tag{7}$$

where δ is the learning rate, σ is the momentum of the change, θ_{-1} is the previous weight change for the connection, and E is the error. The new weight for each connection is computed from right to left back through the network—hence the term back-propagation.

The learning rate, which is a dimensionless factor (0–1.0), has the greatest impact on the adjustment. This is the rate of change in relation to the total error in the network. The larger the learning rate, the faster the network will converge. If the error surface follows a relatively smooth path to the global minimum error and does not contain many local minima, a large learning rate (1.0) may converge quickly to the global minimum. In some cases, a large learning rate should be used initially, but as the global minimum is approached the learning rate should be reduced so that the algorithm can converge to the global minimum rather than oscillating around it. If the error surface is not smooth but has many steep slopes leading to local minima, large learning rates may cause the system to converge to a local minimum. In this case, the network should be trained several times so that the starting point will occur at different points on the error surface. The training exercise that produces the lowest minimum sum square error should be accepted. The momentum term influences the change of weights based on the previous weight change. This increases the time required for the network to converge and smooths out the path to the global minimum.

III. APPLICATIONS OF NEURAL NETWORKS IN BIOLOGICAL SYSTEMS

Neural networks are excellent tools to solve problems where large bodies of data exist but the relationships between system inputs and responses are not well understood. In agricultural problems, NN have been used to describe biological processes such as disease levels and to recognize information from digital images. Neural networks can also be used to solve many types of problems that have been previously solved by simulation, multivariate analysis, and expert systems (Table 1). In general, a neural network should be used if (1) the problem requires qualitative or complex quantitative reasoning, (2) the solution is derived from highly interdependent parameters that have no precise quantification, (3) data are readily available but are multivariate and intrinsically noisy or error-prone, and (4) project development time is short but sufficient neural network training time is available.

Thai et al. (1990) developed an NN to model consumer preferences for honeydew melons based on Brix value, firmness, and a sensory score derived from trained taste panelists and consumer panelists. Their goal was to use an NN to determine if evaluation by consumer preference (score) could be related, or predicted, by scores from a trained panelist. Ten melons were harvested during five different maturity stages and were evaluated by 10 trained panelists and 60 consumer panelists to determine the sensory score. They also measured the Brix value, and firmness was measured with an instron materials testing device. Firmness was characterized by the peak load and the area under the force–time curve derived from the compression test. The inputs for the NN were physical characteristics of Brix value, peak load, and work, and the output was the trained panelist score. The NN was trained using half of the data (five melons) and

Table 1 Potential Applications of Neural Networks in Agriculture

Problem	Applications
Pattern recognition	Extraction of patterns from large databases
	Identification of trends in data
	Capture of human decision making
	Identification of patterns in images
	Analysis of satellite imagery
Process monitoring	Electrical signal analysis
	Wear detection
	Quality control
	Food inspection
Trend analysis	Stock, commodity, futures markets
	Weather prediction
	Weather patterns
	Chemical reactions rates
	Pest population levels
	Electrical loads
Robotic control	Object identification
	Learning motion
Optimization	Harvest date
	Pest treatment thresholds
	Greenhouse environments

validated using the remaining patterns. The number of hidden nodes was varied to find the best architecture for the NN. The panelists evaluated each NN based on the slope of a linear regression line fitting predicted and measured expert score. Using this approach, the NN architecture that gave a slope closest to 1.0 was considered the best architecture. They were successfully able to use the slope of the predicted vs. measured output for each NN to make judgments about the correlation between the inputs and outputs for each network.

In another attempt to predict physical processes using neural networks, Elizondo et al. (1994) developed an NN to predict daily solar radiation based on other readily available weather data variables. Inputs to the NN included daily precipitation, maximum and minimum temperature, clear sky radiation, day length, and day of year. The output was daily solar radiation. Twenty-three years of measured weather data were divided into 11 years for training and 12 years for validating the NN. Thus, there were approximately 4015 training patterns and approximately 4380 validation patterns consisting of daily measured weather data. Several NN architectures were tested, and a sensitivity analysis was conducted on the learning rate and momentum coefficients. The panel compared each NN based on the coefficient of multiple determination (R^2) computed for predicted and measured daily solar radiation for the validation patterns. They were able to obtain an R^2 value of 0.635 for the best network.

Priest et al. (1994) developed a neural network to predict the temperature inside a naturally ventilated broiler house based on environmental and biological factors. Their NN consisted of inputs of wind speed, outside temperature, bird age, inlet cycle time, and inlet configuration. The output was inside temperature. They used a four-layer NN architecture and conducted a sensitivity analysis on the number of hidden nodes in each of the hidden layers. They compared NN architectures on the basis of percent error. Their best NN predicted the correct temperature 94.7% of the time.

Parmar et al. (1994) developed an NN to predict aflatoxin contamination levels in preharvest peanuts in the southeastern United States. Inputs included soil temperature, number of drought days, crop age, and accumulated heat units, and the output was aflatoxin level (ppb). They used data collected over 8 years to train and validate the NN. The best architecture gave R^2 values of 0.95 and 0.94 for the training and validation scenarios, respectively.

In a similar effort, Batchelor and Yang (1995) developed an NN to predict the severity level of soybean rust. They had 577 measurements of weather conditions and soybean and disease development that formed the scenarios available for training and validation. The scenarios were randomly divided into training (70%) and validation (30%) scenarios. Inputs for the NN were planting date, days to soybean maturity, first day that disease was observed, crop age, degree-days for rust and soybean development, and number of days that relative humidity exceeded 90%. The output was percent disease severity. Overall, the NN gave excellent predictions of soybean rust severity, with R^2 values for the validation scenarios of over 0.90 for many NN architectures.

Bolte (1989) discussed the development of a feedback network for selecting an alfalfa cultivar. Each cultivar was defined by name, resistance level to several pests, and type of cultivar. Several cultivars were used to train the network. The user then input the characteristics of the cultivar desired, and the network recommended the cultivar that most closely matched the desired characteristics. Bolte (1989) also discussed the development of a feedback neural network to predict future grain prices. The inputs to the network were average monthly grain price, yearly loan rate, target price, white wheat production, white wheat exports, white wheat carryover/use ratio, world carry-over/use, a GNP deflation index, and the prime rate. The three-layer network was trained using 120 historical datasets. The network pre-dicted trends in the market reasonably well.

There have been many successful attempts to use neural net-works to analyze digital images from a machine vision system. Rig-ney and Kranzler (1989) developed a five-layer NN to determine the grade of pine tree seedlings on the basis of parameters measured from a machine vision system. Four inputs consisted of shoot diameter, seedling height, root area, and shoot area. The output was a numer-ical classification that corresponded to a standard classification scheme. The network was trained using images of 164 plants. Once trained, the network was validated using images from over 1600 ad-ditional plants. The grade of each plant was determined manually. The absolute error was used to evaluate the NN. They also conducted a sensitivity analysis on the number of hidden layers (one to three).

Deck et al. (1991) developed an NN to explore the possibility of using neural networks for discrimination of tasks in machine vision inspection. Normalized hue histogram data of 20 green potatoes and 20 good potatoes were used for training. The network output a value ranging from 0.1 to 0.9, where 0.1 represented a green potato and 0.9 represented a good potato. The greatest discrimination accuracy for the NN was 95%. These results were slightly better than previous methods of linear discrimination, which gave an accuracy of approx-imately 90%.

Zhuang et al. (1992) developed an NN to identify the location of muskmelons from a digitized color video image. One image was used to train a three-layer NN. Inputs consisted of binary color input, and three output nodes corresponded to classifications of leaves, melon, and soil. One image was used to train the NN, and two ad-ditional images were used to validate the NN. Input and output pat-terns consisted of individual digital pixels consisting of a binary coded color. The network was able identify melons in the validation images. However, some leaf pixels were misclassified as melons. This was corrected by filtering the pixels and setting lower limits on the number of contiguous pixels required to define a melon.

Das and Evans (1992) developed a neural network to recognize shape differences in gray level histograms of images of fertile and infertile eggs at early incubation stages. Gray levels (0–255) were divided into eight classifications. The frequencies of the individual gray levels served as eight different inputs to the three-layer NN. A single output node was defined, with output values of 0.9 representing a fertile egg and 0.1 an infertile egg. A threshold of 0.5 was applied to the network output to determine if an egg was infertile (0.1) or fertile (0.9).

Murase et al. (1994) developed an NN to relate textural features to the growth stage of lettuce grown in environmentally controlled growth chambers for a space station. Their goal was to use an NN to determine lettuce growth stage as a biofeedback mechanism for the environmental control system. The three-layer NN used textural features from digital images as inputs including contrast, homogeneity, and local homogeneity. The output was leaf size. They trained the NN and found that it gave reasonably good estimates of leaf size compared to measured values.

Uhrig et al. (1992) developed an NN to predict corn yields in the midwestern United States. The inputs were weekly maximum and minimum temperature, soil moisture from 0–1 m and 1–2 m depths, cumulative growing degree-days, and yield trend. The output was corn yield (bushels per acre). The NN was trained using weekly inputs. Twenty-nine years (1960–1989) of data from an Indiana crop reporting district were used for training and validation. The authors used 1990 data for validation. Their results were encouraging.

Ruan et al. (1994) developed an NN to predict rheological properties of dough from the torque developed during mixing. The properties considered were farinograph, extensibility, and extensional flow force. They trained and validated the NN using 62 different batches of dough with variable flour/water ratio and percent protein. The best NN gave relative errors less then 2.3%, 2.5%, and 7.8% for farinograph, extensibility, and extensional flow force, respectively.

Thai and Shewfelt (1990) used a neural network to quantify the color of tomatoes and peaches. Hue, chroma, and value of lightness were the inputs to the network. The network was trained using judgments from experts on the color quality of tomatoes and peaches. The network was then tested against experts and statistical evaluation method. The neural network gave good predictions of color quality in both peaches and tomatoes.

IV. DEVELOPMENT OF NEURAL NETWORKS

Development of a neural network requires extensive patterns or pairs of inputs and outputs. In addition to this, the appropriate inputs must be selected, and the inputs must be converted into a meaningful form for input into a network. Selection of the correct inputs is important.

If inputs that highly influence the process being predicted are left out, the NN has little chance of adequately predicting the process. For instance, crop yield is highly influenced by weather information. If a NN was developed to predict crop yield and weather information was not an input to the NN, it would likely not perform well. However, NN can discriminate between inputs and essentially ignore inputs that have minimal influence on the process being predicted. Neural network architecture must be determined, and development usually results in a sensitivity analysis of error in prediction over a range of architectural configurations. The steps involved in development are

1. Define the problem.
2. Determine inputs and outputs.
3. Collect data.
4. Select the form of representing inputs and outputs.
5. Select hidden nodes.
6. Select learning rate and momentum.
7. Develop training and validation scenarios.
8. Train the network with training scenarios.
9. Test the network with independent validation scenarios.
10. Conduct a sensitivity analysis of the network.

These steps are discussed through the following case study taken from McClendon and Batchelor (1995).

A. Case Study 1: Soybean Insect Pest Treatment Thresholds

McClendon and Batchelor (1995) developed a neural network to determine the economic treatment threshold when up to four different insects are present in soybean fields. The insects considered were velvetbean caterpillar (VBC), soybean looper (SL), corn earworm (CEW), and southern green stinkbug (SGSB). Economic treatment thresholds based on population levels and crop growth stage exist for individual insect populations; however, these thresholds cannot be used when insect complexes are present in the field. Entomologists who have extensive field experience in relating damage from pest complexes to yield reduction often must be called to make such complex decisions. They use intuition about population numbers of insects present and soybean development stage, age of insects present, and historic damage to make a treatment decision. McClendon and Batchelor proposed to use an NN to capture the expert's intuitive decision process in making these decisions. They proposed over 200 scenarios consisting of variable levels of the four insects at different soybean growth stages with different levels of historic damage. An expert entomologist then made one of three recommendations for each scenario: (1) do not treat, (2) wait 3 days and check the populations again, or (3) treat the populations. Next, they divided the scenarios into training and validation datasets and developed and tested the NN. The following discussion shows the steps involved in development of this NN.

1. Define the Problem

The first step is to clearly define the problem to be solved. For this problem, we want to develop an NN that can capture the knowledge and decisions of the expert entomologist in making pesticide treatment recommendations for four primary insects. Static thresholds exist for individual insects; however, when a combination of these pests are present, an expert has to determine when a treatment should be given. Thus, if only a single pest is present, the solution is easy to determine, but under realistic field conditions the standard guidelines cannot be followed and an expert must be called to determine the recommendation. The goal of this NN is to use the same information that the expert considers in the decision process and map this information to a treatment decision.

2. Determine the Inputs and Outputs

An expert must consider many factors before making a recommendation. Interviews were conducted with the expert to determine what information is typically used to make treatment recommendations. The most important factor is the population levels of each pest. If the populations are at a sufficient level, a treatment is always recommended, but if each pest is present at a low level, the combination of estimated damage from each pest must be evaluated. Timing of the infestation is also critical. If the pests are feeding on foliage, the treatment threshold is higher than if they are feeding on pods and seeds. Thus, plant age is a critical decision factor. Any historic damage must also be considered in the decision. Thus, input nodes should define the plant age, stage of plant development, pest levels, and historic damage. The following inputs were considered:

Input node 1:	Plant age
Input node 2:	Vegetative stage
Input node 3:	Reproductive stage before seed expansion
Input node 4:	Reproductive stage after seed expansion
Input node 5:	VBC population level
Input node 6:	SL population level
Input node 7:	CEW population level
Input node 8:	SGSB population level
Input node 9:	Historic defoliation

These input nodes reflect the inputs that are important to the expert. However, not all of the inputs may be important in predicting the decisions of the expert. After the NN is developed, a sensitivity analysis on the inputs can be performed by eliminating one or more inputs and retraining and validating the NN. If elimination of an input does not change the NN results, then the input is not considered important in duplicating the expert's recommendations using the NN.

The expert may make one of three recommendations: (1) treat immediately, (2) wait 3 days and check again, or (3) do not treat. Thus, we could have three output nodes that reflect these three decisions.

Output node 1:	Do not treat.
Output node 2:	Wait 3 days and check the populations again.
Output node 3:	Treat immediately.

Alternatively, we may have a single output node whose value [0, 1.0] represents the decision to be made.

3. Collect Data

Neural networks require patterns of input and output information for training and validation. For this example, we proposed 200 scenarios consisting of different insect population levels, crop growth stages, and historic damage. The inputs for each scenario are listed above. The expert then made a recommendation to either treat, wait 3 days and check the populations again, or not treat. These scenarios formed the database the NN needed to determine how the expert makes treatment recommendations.

4. Select the Form of the Input and Output Values

Early neural network software required that the inputs and outputs be binary; however, newer software allows analog input and output values. The software scales the input and output values from 0 to 1 and uses this transformed input for computations. The computed output is then transformed back to the scale set by the user. Most NN software also requires the user to establish a range for the values used for input and output that is used to scale the inputs and outputs. Inputs or outputs that can be described by a number, such as populations and days after planting, are already in a form that can be input and scaled by the NN. Inputs or output that require a nonnumeric value such as yes or no must be transformed into numeric values such as 0 = no and 1 = yes. For this case study, a range for each input and output node was established just outside of the range of the data in the training and validation scenarios. The range should be larger than the range in the data, but it should not be too large or the NN will not be able to distinguish between values for specific scenarios. The final definitions for the input and output nodes are shown in Table 2.

5. Select Hidden Nodes

For this case study, we focus on a three-layer feedforward architecture. One uncertainty related in this architecture is how to determine the optimum number of hidden nodes required for a given set of patterns. The optimum number of hidden nodes determines the number of interconnections in the network. The number of hidden nodes is a function of the relationships to be learned. Too few hidden nodes will not provide enough interconnections for the NN to learn the relationships, while too many hidden nodes will often cause the NN to memorize the training patterns and it will not be able to generalize well for other scenarios. Since the memory of a network is stored in the interconnections, it may seem to be an advantage to have as many hidden nodes as possible. However, it has been shown that too many hidden nodes results in excellent recall of training scenarios but gives poor predictions of validation scenarios because the NN has a sufficient number of interconnections to memorize specific scenarios in the training patterns. In contrast, too few hidden nodes can also decrease the accuracy of the network because there is not enough memory to completely define all the relationships between the input and output patterns. Often, a sensitivity analysis on error in predicting training and validation scenarios versus number of hidden nodes is performed to determine the optimum number of hidden nodes for a

Table 2 Inputs and Outputs for the Soybean Insect Pest Management Neural Network

Node		Low	High	Comments
Input 1	Crop age	0	150	Days after planting
Input 2	Veg. stage	0	25	Number of nodes on mainstem
Input 3	Rep. < R5 stage	0	1	0 = no; 1 = yes
Input 4	Rep. > R5 stage	0	1	0 = no; 1 = yes
Input 5	VBC	0	20	Actual population/m^2
Input 6	SL	0	20	Actual population/m^2
Input 7	CEW	0	20	Actual population/m^2
Input 8	SGSB	0	20	Actual population/m^2
Input 9	Historic defoliation	0	50	Percent defoliation
Output 1	Do not treat.	0	1	0 = no; 1 = yes
Output 2	Wait 3 days.	0	1	0 = no; 1 = yes
Output 3	Treat.	0	1	0 = no; 1 = yes

given problem. Most commercial NN software packages provide a default number of hidden nodes using an algorithm based on the number of input and output nodes. However, the user can always override this value.

6. Select Learning Rate and Momentum

Selecting the proper learning rate and momentum is also a trial-and-error procedure, and these values can vary for different problems, architectures, and training scenarios. There is no mathematical way to determine the best values. Recall that both of these are dependent on the shape of the error surface of the problem. For smooth error surfaces, convergence is rapid and there is little error of converging to a local minimum. Typically, it is recommended to first train the network with a learning rate and momentum of 0.9 (fast). If the network appears to oscillate rather than converge, then both factors should be reduced to 0.5 or lower.

7. Select Training and Validation Scenarios

Selecting the training and validation scenarios is crucial to the final accuracy of the network. You should begin by obtaining a reasonable number of scenarios that are representative of the overall range of input and output values. In this network, we obtained treatment recommendations for 200 scenarios that spanned the entire range of input and output variables. The scenarios were predesigned to cover the expected ranges of all input variables and treatment recommendations. They were randomly divided into training and validation patterns, based on percentages. This technique is often used; however, there is no standard method for determining the ratio of training to validation scenarios. It is important to note that validation scenarios should not be used during the training process because they would bias the training process. The idea of validation scenarios is to determine the accuracy of the network in predicting scenarios that it was not trained to predict. Of the total number of scenarios available for training and validation, 50–75% of them are often used for training and the remainder for validation.

8. Train the Network

Once the network has been defined and the training and validation scenarios have been selected, the network should be trained to learn the relationships between inputs and outputs of the training patterns. Training time depends on the number of nodes, number of training scenarios, learning rate and momentum, and the speed of the computer. Convergence criteria can often be specified in terms of the number of epochs or a minimum error between predicted and actual output values. Training time can range from a few seconds to several days, depending on the number of training patterns; input, hidden, and output nodes; learning rate; and complexity of the relationships being learned. Note that training a network is very time-intensive; however, obtaining an output from a trained network requires only a minimal amount of time.

9. Validate the Network

A criterion must be established to determine if the computed output matches the actual output. There are several statistics that are commonly used to determine the error between predicted and measured output values over all scenarios. The coefficient of multiple determination (R^2) is a statistical indicator usually applied to multiple regression analysis. It compares the accuracy of the model to the accuracy of using the mean of all samples. A perfect fit would result in an R^2 value of 1, and an R^2 value near 0 would indicate the NN gives no better results than using the mean of all samples. The mean square error, which is the average of the square of the difference between predicted and actual output, is often the primary statistic used to evaluate the performance of an NN. Although this statistic is useful, it should be used along with other statistics to determine overall NN performance. A similar statistic is the root mean square of error, which is the square root of the mean square error statistic. Three additional measures are minimum, maximum, and mean absolute error between predicted and actual output. These statistics give an overview of the range of error in terms of absolute values over all scenarios and can be used to determine whether the worst scenarios are within an acceptable range of error. Finally, some researchers use the linear correlation coefficient to determine the strength of the linear relationships between predicted and actual outputs. Values near 1 or −1 indicate strong positive or negative linear correlation, while values near 0 indicate that no linear relationship exists.

For this example, we must convert numeric outputs back to a nonnumeric recommendation. We established the criterion that the output node with the highest value is the recommendation of the network. Once the validation scenarios have been presented to the NN, this criterion will be applied to the output nodes to determine the overall recommendation of the NN. Statistics are then computed for the predicted and actual output values. Each validation scenario that was incorrectly predicted should be analyzed to determine why it was not predicted correctly. In some cases, the scenario may be very close to a threshold, and the NN may give the wrong recommendation because the boundary of the decision region near that threshold is unclear from the training patterns. More training scenarios should be developed around threshold values to better define the decision region. In other cases, the NN may give the wrong recommendation because the expert gave inconsistent recommendations for very similar patterns. If this is the case, conflicting scenarios should be presented to the expert again to determine the appropriate recommendations. Another problem could be that the training cases do not represent the incorrectly predicted test case. In this case, more training scenarios should be developed and the network should be retrained.

10. Perform Sensitivity Analysis of the Network

A sensitivity analysis should be performed on the number of hidden nodes to determine the optimum architecture for the problem. The NN should be retrained using different numbers of hidden nodes over a range of values. A plot showing the error in validation scenarios versus the number of hidden nodes will be useful to show the number of hidden nodes that minimizes the error in the validation scenarios. A second sensitivity analysis should be performed on the initial values of the weights at the beginning of training. This will allow the NN to begin training from several points on the error surface. It is possible that for some starting points the NN converges to a local minimum rather than a global minimum. If the error in the validation scenarios is different for different initial weight values, the error surface is "bumpy," and the initial weights that give the lowest error should be considered the optimum network.

B. Case Study 2: Predicting Soybean Rust Severity

Soybean rust (*Phakopsora pachyrhizi* Syd.) occurs in both the eastern and western hemispheres and is a major disease of tropical and subtropical areas (Bromfield, 1984; Sinclair, 1989; Tschantz, 1984). The disease can cause severe premature defoliation. It causes considerable yield loss in many Asian countries, with losses as high as 40% reported in Japan (Bromfield, 1984) and up to 80% reported in Taiwan (Yang et al., 1991b). The importance of the disease has increased owing to increased soybean production and expansion of the crop to new regions, and the disease was recently found in Hawaii (Killgore et al., 1994). Soybean rust has been considered a potential exotic threat to soybean production of the United States (Kuchler et al., 1984; Sinclair, 1994). Epidemiology of the disease has been studied extensively over the past two decades (Yang et al., 1991a). Regression or simulation models have been developed (Yang et al., 1991a, 1991b) to describe and predict the severity of soybean rust based on many sources of data describing the environmental effects on population dynamics and disease severity (Casey, 1979; Marchetti et al., 1976; Melching et al., 1979; Tschantz, 1984). Yang et al. found difficulties in predicting disease development using traditional modeling techniques for some datasets for the purpose of risk assessment. In this case study, we discuss the development of an NN to predict the disease progress for soybean rust (Batchelor and Yang, 1995).

1. Experimental Data

Data for developing the NN were taken from experiments conducted in 1980 and 1981 at the Asian Vegetable Research and Development Center (AVRDC), Shanhua, Taiwan, where soybeans are grown throughout the year and rust occurs in all seasons with urediniospores present as initial inoculum (Tschantz, 1984). The determinant soybean cultivar TK 5 was selected for this study. In order to create different environmental and plant physiological windows, a sequential planting experiment was performed. Each planting date was considered as an experiment. In 1980, planting began at day 7 of the year and continued through day 343 at weekly intervals with a total of 46 planting dates. In the 1981 experiment, planting started on December 21, 1980 and continued through day 264, for a total of 27 planting dates.

For each planting date, disease rating began as soon as rust was first observed. Each plot was divided into four equal sections. Disease severity, defined as percentage of diseased leaf area to total leaf area, was recorded at weekly intervals, eight times per planting date experiment. A weather station was established in the field. Rainfall, daily maximum and minimum temperature, relative humidity, and average temperature were recorded at 2 h intervals. For detailed information, see Yang et al. (1990) and Tschanz (1984). This gave a total of 577 observations of plant and disease conditions. For the purpose of NN modeling, each observation serves as a pattern that can be used for training, testing, or validation.

2. Define Neural Network Input and Output

The seven inputs shown in Table 3 were used for each of the neural networks (Fig. 3). The output of each network was percent soybean rust severity. The first five inputs were determined from the individual experiments and weather conditions. Degree-days of pathogen (CDDR) and crop (CDDS) development were computed based on

$$D_i = \begin{cases} 0 & \text{if} \quad T_i < T_{min} \\ (T_{opt} - T_i)/(T_{opt} - T_{min}) & \text{if} \quad T_i < T_{opt} \\ (T_i - T_{opt})/(T_{max} - T_{opt}) & \text{if} \quad T_i > T_{opt} \\ 0 & \text{if} \quad T_i > T_{max} \end{cases} \tag{8}$$

where T_{max}, T_{opt}, and T_{min} are maximum, optimum, and minimum temperatures for the growth of the organisms, respectively, and D_i is the number of degree-days occurring on day i. In our study, the three parameters used were obtained from the literature values as T_{max} =

Table 3   Definition of Inputs Used for the Neural Network[a]

Input variable	Description
1. PLD	Planting date
2. DM	Days to soybean maturity
3. OSD	First day that disease was observed
4. AGE	Crop age on sampling date
5. CDDR	Cumulative degree days for rust development
6. CDDS	Cumulative degree days for soybean development
7. CRH	Cumulative days that relative humidity exceeded 90%

[a]The output node was percent soybean rust severity.

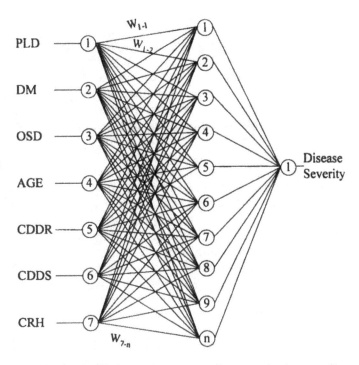

FIGURE 3 Architecture for a neural network that predicts soybean rust severity.

45°C, T_{opt} = 30°C, and T_{min} = 7°C for soybeans (Wilkerson et al., 1985) and T_{opt} = 22°C, T_{max} = 30°C, and T_{min} = 4°C for *P. pachyrhizi* (Casey, 1979; Marchetti et al., 1976). Cumulative physiological days for soybean development (CDDS) was computed for each of the 577 observation dates by summing the degree-days from the planting date to the observation date. Cumulative physiological days for rust (CDDR) were computed for each observation by summing the degree-days from 17 DAP to each observation date. For the rust pathogen, 17 days after planting was arbitrarily taken as the initial day because the natural inoculum caused infection as soon as the first leaf appeared.

3. Validate the Neural Network

A three-layer feedforward NN was developed using NeuroShell 2 neural network development software (Ward System Group, Frederick, MD). Two different validations were performed based on assumptions about the independence of the validation scenarios. In the first validation study, it was assumed that each observation of disease severity was independent of other observations for the same epidemic. Thus, the 577 scenarios were randomly divided into training (327), testing (125), and validation (125) scenarios, without consideration of the epidemic or planting date that each scenario was associated with. The 327 training scenarios were then used to train the NN, and the error between predicted and observed disease severity was used to adjust the weights in the network. The testing scenarios, a concept developed by Ward Systems Group (1993), were used to determine when the network training was completed. Periodically during the training process, the testing scenarios were presented to the network and the errors between predicted and observed soybean rust were computed for the testing scenarios. If the overall error was lower than the previous error in the testing scenarios, the weights were saved as the optimum values of the weights. This iterative process of presenting training and testing scenarios to the network continued until the network could no longer find new weights that minimized error in the testing scenarios. At this point it was assumed that the network had converged and that the optimum weights that minimized errors in the testing scenarios were found. At no time were the testing scenarios used to adjust the weights; rather, they were used to determine how well the weights computed during the training process predicted disease severity on an independent set of scenarios. This approach for training an NN is unique to the NeuroShell software and is an excellent technique to avoid overtraining the network; also it gives a network that can generalize relationships better than using training scenarios alone. The testing scenarios ensure that the network is trained to give the best predictions on scenarios independent of the scenarios used for weight adjustments.

Once the network was trained, the validation scenarios were presented to the network and the error in prediction (R^2) was computed for each network. The number of hidden nodes was altered over a range of 10–48 to determine the best architecture for the network that minimized the error in the validation scenarios.

In the second validation effort, it was assumed that scenarios within each of the 73 epidemics, or planting date studies, were not independent but that scenarios in different planting date studies were independent with respect to other planting date studies. The 73 disease epidemics or plantings were randomly divided so that 51 (70%) were used for training, 11 (15%) were used for testing, and the remaining 11 (15%) were used to validate the NN. The scenarios within each epidemic were then classified as training, testing, or validation scenarios based on the classification of the individual epidemics or plantings. The NN was trained for using different numbers of hidden nodes. The neural networks were then compared based on R^2, mean square error, mean absolute error, and the maximum absolute error between predicted and measured disease severity in the validation scenarios.

4. Validate Neural Network Performance

In the first validation of the NN, the 577 scenarios were randomly divided into 327 training, 125 testing, and 125 validation scenarios. Errors in predicting soybean rust severity are shown in Table 4. The best network had 14 hidden nodes and gave an R^2 value of 0.925 for the validation scenarios. The average error in predicting disease severity was 6.9% for each of the validation scenarios, and the largest

Table 4 Sensitivity Analysis of the Number of Hidden Nodes in Neural Network Predictions for a Three-Layer Feedforward NN with 320 Training, 125 Testing, and 125 Validation Scenarios

No. of hidden nodes	Training scenarios, R^2	Testing scenarios, R^2	Validation scenarios, R^2	Mean square error	Mean absolute error (%)	Maximum absolute error (%)
10	0.971	0.859	0.884	152	8.4	43.4
12	0.971	0.917	0.921	97	7.9	37.0
14	0.982	0.912	0.925	98	6.9	33.8
16	0.966	0.919	0.913	113	7.5	31.2
18	0.967	0.913	0.915	111	7.4	34.3
20	0.968	0.908	0.917	108	7.3	37.0
22	0.975	0.916	0.912	115	7.4	34.5
24	0.960	0.903	0.912	116	7.4	34.9
26	0.978	0.911	0.905	124	7.4	38.2
28	0.974	0.923	0.922	102	7.1	37.2
30	0.970	0.897	0.915	111	7.5	32.6
32	0.973	0.910	0.911	117	7.5	34.0
34	0.980	0.920	0.896	136	7.7	38.0
36	0.976	0.921	0.912	110	7.3	37.4
38	0.966	0.898	0.920	104	7.0	36.6
40	0.980	0.915	0.910	139	7.8	49.6
42	0.975	0.902	0.913	114	7.3	37.7
44	0.977	0.912	0.909	119	7.5	36.0
46	0.975	0.916	0.911	116	7.3	41.2
48	0.976	0.923	0.895	137	7.7	52.3

error was 33.8%. The network was sensitive to the number of hidden nodes, but there was no distinct pattern between R^2 and the number of hidden nodes for this example. Predicted and measured disease severity levels for the validation scenarios are shown in Fig. 4.

This NN gave such good results primarily because of the way we randomly selected our validation scenarios out of the planting date experiments. The data used in this study consisted of eight observations of plant and disease status during each planting date study. Thus, the observed disease severity can be viewed as a curve that progressed from 0 at planting to a high value (near 100%) at soybean maturity. The NN performed well in learning how the various inputs affected disease severity. When the scenarios were divided into training, testing, and validation scenarios, individual observations for each planting date study were randomly selected for training, testing, and validation. The training and testing scenarios were used by the NN to memorize how the inputs affected disease severity. The validation scenarios were not truly independent scenarios, since they were really a subset of the patterns that were used for training the NN. Even though the validation scenarios were not used for training, they were a subset of the training patterns in the biolog-

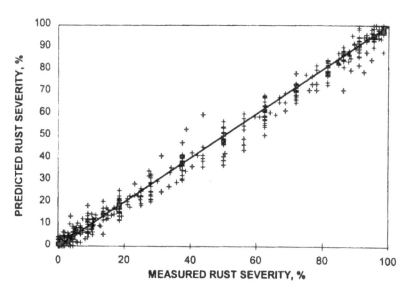

FIGURE 4   Error between predicted and measured soybean rust severity for validation scenarios using 32 hidden nodes. The 577 scenarios were divided into 70% training, 15% testing, and 15% validation scenarios.

ical system. Thus, the NN was able to predict disease severity for the training patterns very well because it was trained using other observations within the same set of planting date studies.

In the second validation effort, it was assumed that scenarios in each planting date study were not independent but scenarios in different planting date studies were independent. The planting date studies were divided into 70% training, 15% testing, and 15% validation scenarios. The three-layer NN was trained using different numbers of hidden nodes, and the results are shown in Table 5. The overall R^2 for each NN decreased compared to the previous validation effort, with R^2 values ranging from 0.625 to 0.754. The best architecture had 30 hidden nodes, with $R^2 = 0.754$. This NN gave a maximum absolute error of 71.2%, but the mean absolute error was 11.5% over all the validation scenarios.

In the second validation effort, it was assumed that scenarios within individual planting date studies were not independent of each other but that scenarios in different planting date studies were independent. This assumption decreased overall NN accuracy in predicting disease severity in the validation scenarios. This assumption is likely more realistic than assuming that all scenarios are independent. In view of disease forecasting, this validation is equivalent to projecting disease progress curves without referring to previous disease occurrence within a planting.

To use this neural network to predict severity of rust on a soybean crop, the user would have to supply the seven inputs listed in Table 3. Let's assume that the prediction is to be made when the crop

Table 5 Sensitivity Analysis of the Number of Hidden Nodes when Disease Epidemics (Total 75 Plantings) Were Divided into 70% Training, 15% Testing, and 15% Validation Plantings

Number of hidden nodes	R^2	Mean square error	Mean absolute error (%)	Max. absolute error (%)
26	0.668	421	14.3	67.3
28	0.689	420	14.5	72.1
30	0.754	332	11.5	71.2
32	0.698	408	13.2	76.8
34	0.689	421	14.1	76.8
36	0.625	507	15.4	68.5

is 50 days old. The user must first make some assumptions about future weather data, namely temperature and number of days that relative humidity exceeds 90%. This information can be generated for the remainder of the season using many different techniques, for example by using weather generators or long-term weather forecasts. The user must then compute the cumulative degree-days for soybean (CDDS) and rust (CDDR) development at a desired future date as well as the number of days that relative humidity exceeds 90%. The first date that disease was observed should also be known. This information can then be input into the NN, and the predicted soybean rust severity will be computed on the date that corresponds to the crop age entered by the user.

V. SUMMARY

Neural networks mimic the learning processes of the human brain and are capable of learning relationships between patterns. Once trained, they can deal with incomplete information very well because they try to match input patterns to output patterns. If one or more inputs are not known, the network can still map the input pattern to the appropriate output pattern.

Neural networks can require large amounts of data for training. The role of the developer is to obtain the inputs and outputs for the network. This can require some time for the expert to produce information that is to be learned. However, the learning algorithm takes care of determining the relationship between the input and output. The developer is required to select the type of network and network configuration and test the network. The developer must also be equipped to determine when the best network configuration has been found. It takes a minimum amount of time (in some cases, less than 1 day) to obtain knowledge and prepare the network for training. Training time may require from several hours up to a month of computation time depending on the complexity of the knowledge to be learned.

It is very easy to retrain neural networks if the knowledge changes. For instance, if a neural network is trained to recommend a particular insecticide based on population levels, the scenarios can be easily changed and the network can be retrained with minimum effort. Once the input and output parameters are identified by the expert and developer, the scenarios can be updated with minimal effort. Likewise, it is easy to add new information in the form of input or output nodes.

Neural networks are by design very flexible in representing knowledge. The developer can establish input nodes for all information required for the problem solution. Numeric inputs can easily be input to a node, and nonnumeric information can be accommodated by converting it to some type of numeric representation. For instance, if an input can be either red, blue, or green, the node for that input can be defined such that 1 = red, 2 = blue, and 3 = green. The neural network does not have to be told how this information relates to the output other than by example.

Neural networks were not designed to provide the user with an understanding of the relationships between the input and output. They cannot reason sequentially nor can they explain why a particular solution resulted. Expert systems use sequential reasoning and are subsequently able to explain why a particular conclusion was reached. This is a primary benefit in using expert systems to train new experts or students.

Once trained, neural networks are able to compute the output with just a few computations. All of the information required must be input before execution begins. The neural network can also be converted to hardware using logic chips in order to speed up the computations. This is beneficial if a network is being used for real-time operation in which speed is a factor. Hardware networks have the advantage of speed and compactness, but they are vulnerable to voltage fluctuations and require physical maintenance. They are also expensive and tedious to develop. Hardware networks can be developed using operational amplifiers that can be configured to learn by adjusting the resistance in the feedback loop. In this manner, the hardware neural network can be trained and used. Although software networks are slower, they are easier to develop and train, and they require no maintenance. However, software networks require a microprocessor-based computer system.

Many neural network software packages can automatically write a C or Fortran subroutine after the network has been trained. The inputs and outputs of the subroutine are the same as those of the network. This subroutine can then be incorporated into other programs and called when necessary. There are never any runtime distribution charges for these types of programs. Expert systems are often-stand alone systems that, with difficulty, can be made to communicate with other programs. Often there is a problem with RAM memory limitations. There is usually a question of runtime distribution fees.

Neural networks have several advantages over multivariate analysis techniques. They can handle noisy data with a higher degree of accuracy, and they can handle incomplete data, whereas multivariate analysis can blow up in either of these cases. Expert systems were not designed to handle trend data that can be represented with neural networks and multivariate analysis.

REFERENCES

Batchelor, W. D., and X. B. Tang. 1995. Predicting soybean rust severity using a neural network. ASAE Paper No. 95-3793. St. Joseph, MI: ASAE.

Bolte, J. P. 1989. Applications of neural networks in agriculture. ASAE Paper No. 89-7591. St. Joseph, MI: American Society of Agricultural Engineers.

Bromfield, K. R. 1984. *Soybean Rust.* Monograph No. 11. St. Paul, MN: American Phytopathological Society.

Casey, P. S. 1979. The epidemiology of soybean rust—*Phakopsora pachyrhizi* Syd. Ph.D. Thesis. University of Sydney.

Das, K. C., and M. D. Evans. 1992. Detecting fertility of hatching eggs using machine vision, II. Neural network classifiers. *Trans. ASAE* 35(6):2035–2041.

Deck, S., C. T. Morrow, P. H. Heinemann, and H. J. Sommer III. 1991. Neural networks versus traditional classifiers for machine vision inspection. ASAE Paper No. 91-3502. St. Joseph, MI: ASAE.

Elizondo, D., G. Hoogenboom, and R. W. McClendon. 1994. Development of a neural network model to predict daily solar radiation. *Agric. Forest Meteorol.* 71:115–132.

Killgore, E., R. Heu, and D. E. Gardner. 1994. First report of soybean rust in Hawaii. *Plant Dis.* 78:1216.

Kuchler, F., M. Duffy, R. D. Shrum, and W. M. Dower. 1984. Potential economic consequences of the entry of an exotic fungal pest: The case of soybean rust. *Phytopathology* 74:916–920.

McClelland, J. L., and D. E. Rumelhart. 1986. *Parallel Distributed Processing: Explorations in the Microstructure of Cognition.* Vol. 2. *Psychological and Biological Models.* Cambridge, MA: MIT Press.

McClendon, R. W., and W. D. Batchelor. 1995. An insect pest management neural network. ASAE Paper No. 95-3560. St. Joseph MI: ASAE.

Marchetti, M. A., J. S. Melching, and K. R. Bromfield. 1976. The effects of temperature and dew period on germination and infection by uredospores of *Phakopsora pachyrhizi*. *Phytopathology* 66:461–463.

Melching, J. S., K. R. Bromfield, and C. H. Kingsolver. 1979. Infection, colonization, and uredospore production on Wayne soybean by four cultivars of *Phakopsora pachyrhizi*, the cause of soybean rust. *Phytopathology 69*: 1262–1265.

Murase, H., Y. Nishiura, and N. Honami. 1994. Textural features/neural network for plant growth monitoring. ASAE Paper No. 94-4416. St. Joseph, MI: ASAE.

Parmar, R. S., R. W. McClendon, G. Hoogenboom, P. D. Blankenship, R. J. Cole, and J. W. Dorner. 1994. Prediction of aflatoxin contamination in preharvest peanuts. ASAE Paper No. 94-3562. St. Joseph, MI: ASAE.

Priest, J. B., J. McLaughlin, and R. R. Price. 1994. Neural network model of naturally ventilated broiler house. ASAE Paper No. 94-4014. St. Joseph, MI: ASAE.

Rigney, M. P., and G. A. Kranzler. 1989. Seedling classification performance of a neural network. ASAE Paper No. 89-7523. St. Joseph, MI: ASAE.

Ruan, R., S. Almaer, and J. Zhang. 1994. Prediction of dough rheological properties using neural networks. ASAE Paper No. 94-6523. St. Joseph, MI: ASAE.

Rumelhart, D. E., and J. L. McClelland. 1986. *Parallel Distributed Processing: Explorations in the Microstructure of Cognition*. Vol. 1. *Foundations*. Cambridge, MA: MIT Press.

Seginer, I., and A. Sher. 1992. Neural-nets for greenhouse climate control. ASAE Paper No. 927013. St. Joseph, MI: ASAE.

Sinclair, J. B. 1989. Threats to soybean production in the tropics: Red leaf blotch and leaf rust. *Plant Dis. 73*:604–606.

Sinclair, J. B. 1994.

Thai, C. N., and R. L. Shewfelt. 1990. Modeling sensory color quality: Neural networks vs. statistical regression. ASAE Paper No. 90-6038. St. Joseph, MI: ASAE.

Thai, C. N., A. V. A. Resurreccion, G. G. Dull, and D. A. Smittle. 1990. Modeling consumer preference with neural networks. ASAE Paper No. 90-7550. St. Joseph, MI: ASAE.

Tschanz, A. T. 1984. Soybean rust epidemiology: Final Report. Shanhua, Taiwan: Asian Vegetable Research and Development Center.

Uhrig, J. W., B. A. Engel, and W. L. Baker. 1992. An application of neural networks: Predicting corn yields. Proceedings of the 4th International Conference on Computers in Agricultural Extension Programs, Lake Buena Vista, FL.

Ward Systems Group. 1993. *Neuroshell 2*. Frederick, MD: Ward Systems Group, Inc.

Wilkerson, G. G., J. W. Jones, K. J. Boote, and J. W. Mishoe. 1985. SOYGRO V5.0: Soybean Crop Growth and Yield Model. University of Florida.

Yang, X. B., M. H. Royer, A. T. Tschanz, and B. Y. Tsia. 1990. Analysis and quantification of soybean rust epidemics from seventy-three sequential planting experiments. *Phytopathology 80*:1421–1427.

Yang, X. B., W. M. Dowler, and M. H. Royer. 1991a. Assessing the risk and potential impact of an exotic plant disease. *Plant Dis.* 75:976–982.

Yang, X. B., W. M. Dowler, and A. T. Tschanz. 1991b. A simulation model for assessing soybean rust epidemics. *J. Phytopathol.* 133:187–200.

Zhuang, X., B. A. Engel, and M. Benady. 1992. Locating melons using artificial neural networks. ASAE Paper No. 92-7014. St. Joseph, MI: ASAE.

17

Object-Oriented Programming for Decision Systems

John Bolte
Oregon State University, Corvallis, Oregon

I. THE EVOLUTION OF REPRESENTATION IN SIMULATION

Mathematical systems models and computer simulations are used by scientists, agriculturalists, and resource managers to allocate resources, enhance production management, and understand biological processes. These simulators are constructed in a series of steps as follows.

1. Identify the most important features of the system of interest.
2. Develop an abstract conceptual framework describing the system in terms of fundamental components and interaction between these components.
3. Formalize the conceptual framework by developing concrete representations of each of the system components, usually in the form of mathematical representations of the dynamics of each component with respect to other components of the system, resulting in the formulation of a simulation model.

4. Implement the simulation model in a computer representa-
    tion to execute and test the model's predictive capabilities.

Simulation models, then, become a vehicle for representing in abstract
terms a slice of reality consistent with our specific interests in enhancing our understanding of our world. Representation is the key
word here: Simulation models must provide us with an abstract representation of reality. How we represent the conceptual model we
derive from observation of reality therefore should be of considerable
interest to the modeler. It is here that various programming languages
and simulation tools have a central role to play.

Computer methods for the solution of simulation models had
their genesis with the introduction of the Fortran programming language in the 1950s. At that time, Fortran provided a powerful new
tool to allow scientists and engineers to solve complex mathematical
relationships. Both hardware and software tools for conducting simulations have continued to evolve since that time. Eventually, computing languages were designed specifically for simulation. As the
needs of the simulation community grew more complex, new simulation languages were developed for different types of models. These
languages provided direct support for higher level representation of
fundamental systems concepts and typically provided automated
methods for the solution of complex equations. The General Purpose
Systems Simulator (GPSS) was one of the first examples of such a
specialized language. Designed by G. Grudden for IBM, GPSS is a
very general language used primarily for queuing problems and inventory control. GPSS/PC and GPSS/H are two modern derivatives
of GPSS (Gordon, 1969; IBM, 1970; Schriber, 1972). Simscript, developed by H. M. Markowitz at the RAND Corporation and then at
CACI (Kiviat et al., 1983; Caci 1988) was the next major simulation
language. Simscript is a general-purpose simulation language used
primarily to program discrete-event simulations but also allowing for
a process-oriented approach.

Several contributions have been made to the advancement of
simulation by A. Alan B. Pritsker. The first simulation language written by Pritsker was General All-Purpose Simulation Program (GASP).
GASP had its origins at U.S. Steel, and Pritsker improved it in 1969.
Pritsker and P. J. Kiviat documented GASP II, and then Pritsker produced the latest version, GASP IV (Pritsker, 1974). It is largely based
on Fortran and requires the user to write a series of subroutines to
describe the simulation. Pritsker's next project was the development
of Simulation Language for Alternative Modeling (SLAM), which was

carried out with D. Pegden (Pritsker and Pegden, 1979). SLAM is a very powerful simulation language offering network orientation and capabilities for discrete-event and/or continuous simulation. Pegden, a student of Pritsker, developed SIMAN (SIMulation ANalysis) to model discrete-event, continuous, or combined systems. SIMAN uses block diagrams to describe the flow of entities through a system.

Although these special-purpose simulation languages are an improvement over programming models using general-purpose computer language such as Fortran, they are limited in the flexibility that they provide. Modelers had to drop down to some base language if they needed to do something beyond that which was provided by the simulation language. This made many simulation projects frustrating and tended to limit the variety of simulation projects that were undertaken. As simulations have become more complex, reflecting the higher level of complexity of the systems under consideration, the use of specialized simulation languages has become limited. The need for flexibility outstripped the improvements in the representation capabilities these languages provided.

General-purpose programming languages have continued to evolve. As computer scientists became more familiar with the constraints imposed by early programming languages, new paradigms were introduced to improve the efficiency of software development, the representational capabilities of computer languages, and the maintainability and reusability of software. An early successful paradigm was that of structured programming. Structured programming emphasizes developing algorithms as modular code components, and structured languages were developed that more strongly supported this paradigm by introducing rigid type checking, data abstraction structures, and programmer-defined abstract data types. These capabilities, not available in early Fortran-like languages, provided programmers with much more robust representational constructs for assembling programs. Programs became a collection of algorithms in the form of functions that operated on data in the form of abstract data types and structures. Well-known examples of these structured languages include C and Pascal and, more recently, Fortran-90.

The next major step in the development of programming paradigms was object-oriented programming (OOP). Object-oriented, or object-based, programming continues the trend established by structured programming in providing language-level support for high-level data abstraction and representation. It goes well beyond structured programming by establishing the concept of an *object*, a self-contained collection of data and algorithms that directly correspond to entities in the real world. Because these objects mirror reality at both the conceptual and implementation levels, object-based software tends to be easier to design and maintain than conventional software. This high level support for sophisticated representation, coupled with additional supporting language features, has moved OOP into the forefront of current programming paradigms.

Simula, the first object-oriented programming language, was developed in 1967 explicitly to provide simulation facilities for modeling real-world objects (Dahl and Nygaard, 1966, 1967a, 1967b; Kirkeud, 1989). Simula was not only the first simulation language, it was also the first object-oriented language. Simula introduced the concept of a class as the fundamental representational paradigm. Although it was never a commercial success, Simula developed a strong academic following in Europe and had great influence on programming language design. In addition to basic language features, Simula provides simulation for support through two key classes, SimSet and Simulation (Franta, 1977). The SimSet class provides list processing by implementing a double-linked list. The Simulation class provides scheduling, synchronization, and a simulated time flow mechanism.

There have since been many implementations of object-oriented simulation languages. DEMOS (Birtwistle, 1979) consists of approximately 30,000 lines of Simula code designed to model queuing networks. DISCO (Helsgaun, 1980) is a further extension of DEMOS that provides classes to assist in combined simulation and demonstrated the ease with which OOP languages could be extended. The Smalltalk "blue book" (Goldberg and Robson, 1983) explains the simulation constructs that were implemented in Smalltalk-80, a widely used object-based language. The design of Smalltalk was strongly influenced by Simula. MODSIM II, based on Modula-2, is an object-oriented general-purpose programming language that provides direct support for OOP and discrete-event simulation (Belanger et al., 1990). C with classes was introduced in 1980 (Stroustrup, 1983), and C++ followed shortly thereafter. C++ was originally developed by Stroustrup to provide support for event-driven simulations similar to those written in Simula but with the efficiency of code generation in C. A similar language, Objective-C, was developed but has been less successful in the commercial marketplace. Object-Pascal, an object-based successor to the Pascal programming language, has been widely used for a number of general software development efforts.

Of all these languages, C++ has quickly assumed a leading role and is currently the most widely used object-oriented language. Its popularity has been attributed to the fact that it builds on C, provides support for all the major characteristics defining the object-based paradigm, generates efficient executables, is widely available, and is portable across platforms.

II. OBJECT-ORIENTED PROGRAMMING CONCEPTS

Object-oriented programming can be considered a new programming paradigm (Budd, 1991). OOP is not simply a few new features added to programming languages; rather, it provides a new way of thinking about the process of decomposing problems and developing programming solutions (Meyer, 1988). OOP differs from previous programming paradigms in that it introduced a new program entity, the object. An object combines both data and functionality into a cohesive unit. An object is an instance, or specific realization, of a class. The class is a useful design and implementation abstraction that defines the data members and methods of objects (Ege, 1992). Data members store an object's state, and methods define an object's behavior and functionality. An object-oriented application is a set of objects that work together to achieve an overall goal.

Object-oriented programming embodies four key concepts that result in making software systems more understandable, modifiable, and reusable. These concepts are encapsulation, message passing, dynamic binding, and inheritance (Budd, 1991; Meyer, 1988; Nerson, 1992).

Encapsulation is the mechanism for information hiding and is one of the most important properties of objects (Meyer, 1988; Budd, 1991). Each object hides its own information, and whenever that information is needed it can only be obtained by means of the methods of the object. In this way, object data can be protected, and unwanted modification due to software errors is prevented. Encapsulation reduces the possibility of corrupting an object's data. The use of encapsulation therefore improves the modularity, reliability, and maintainability of system code (Wilde et al., 1993).

Message passing is a necessary result of encapsulation. Communication between objects is possible only through messages. Every action in an object-oriented program is initiated by objects sending messages to one another. The method executed by an object in response to a message is determined by the class of the receiver. All objects of a given class use the same method in response to a given message (Rumbaugh et al., 1991), operating on each specific instance's data. A message states "what" should be done by an object, whereas a method expresses "how" it will be done. A message is

"bound" to a method either statically (at compile time) or dynamically (at run time) (Ege, 1992).

Dynamic binding is the ability to determine what method to call
at run time rather than resolving the call at compile time. Frequently,
the specific receiver class of the message will not be known until run
time, so which method will be invoked cannot be determined until
then (Ege, 1992). Thus we say there is a *dynamic* or *late binding* between the message and the method used to respond to the message,
in contrast to *early binding* in traditional procedural languages such
as Fortran or C. This feature allows references to generic objects during compilation, with the run-time behavior of the system determined
by the actual (subclassed) objects that are active in the system at the
time. Dynamic binding is an extremely powerful tool for defining at
a generic level the specific types of behavior a class of objects must
provide. Dynamic binding provides an excellent mechanism for defining system-level interfaces for simulation objects, with specific behaviors appropriate to each subclass invoked during program execution.

Polymorphism is the ability to have different kinds of objects respond to the same message in different ways. It is based on dynamic
binding (Budd, 1991). For example, animal and plant simulation objects both might have the method Grow( ), but each one responds to
the Grow( ) message differently. Polymorphism allows this shared
code to be tailored to fit the specific circumstances of each individual
data type (Ege, 1992).

Inheritance denotes the ability of an object to derive its data and
behavior automatically from another object (Budd, 1991). When a
new class is defined as a subclass of an existing one, data members
and methods are automatically inherited from the parent class (superclass). This is based on creating new classes of objects as descen-
dants of existing ones, extending, reducing, or otherwise modifying
their functionality. Classes become organized as a hierarchy, with subclasses inheriting attributes from a superclass higher in the tree. Inheritance is a powerful concept that can greatly enhance and simplify
the design and implementation of a system by capturing common-
alities between similar components (Rumbaugh et al., 1991; Rubin
and Goldberg, 1992; Budd, 1991). It is an extremely valuable tool for
constructing and reusing complex simulation components, since the
inheritance hierarchy defines various levels of information and behavioral abstraction appropriate to the system being represented.
Coupled with dynamic binding, inheritance allows the development
of abstract frameworks for model construction defined at a high level

in the hierarchy, with specific implementations of system components occurring via subclassing these generic objects to provide specific component-level behavior.

Object-oriented software needs maintenance, just like conventional software. It has been estimated that maintenance costs typically represent up to 80% of the cost of software development (Shooman, 1983). Any successful software system will eventually enter a prolonged and costly maintenance phase (Rumbaugh et al., 1991; Ege, 1992). Most maintenance tasks involve enhancements requested by users or adaptations to a changing environment rather than error correction. One of the goals of object orientation is to make code maintenance and modification easier. Encapsulation provides an effective means for breaking up programs into modules that can be modified independently (Budd, 1991). Additionally, object-oriented programs are typically easier to understand because of the direct relationship between real-world entities and objects in the program (Rumbaugh et al., 1991).

III. OBJECT-ORIENTED SIMULATION

Simulation is one of the natural applications of object-oriented programming. From an object-oriented viewpoint, simulation models are readily envisioned as an assemblage of interacting objects representing specific entities in the real world (Zeigler, 1990). Although the mechanisms of OOP can be implemented with a traditional programming language supporting data abstraction, better results can be achieved with an object-oriented programming language (Bischak, 1991) that directly incorporates object orientation into the language specification. The object-based approach provides powerful tools for the representation of complex domains and results in very efficient programming environments (Rumbaugh et al., 1991). OOP's features and benefits are well documented (Booch, 1991).

The limitations of procedural approaches in simulation are well established. One of the most significant problems of traditional simulation modeling has been the lack of reusability of the simulation models and/or components (Rumbaugh et al., 1991). This problem stems from the fact that most simulation models are developed for a specific simulation experiment objective. When a new objective is encountered, a new model is developed from scratch even though it may include elements contained in earlier models (Rumbaugh et al., 1991; Sierra and Pulido, 1991). Inheritance, encapsulation, and dynamic binding provide mechanisms for enhancing reusability and maintainability of various models and model components.

These benefits have led researchers to develop object-based models in a variety of domains. Several object-oriented models have been reported in ecological and agricultural simulation (Sequeira et al., 1991; Whittaker et al., 1991; Van Evert and Campell, 1994; Folse et al., 1990). All indicated OOPs provide significant improvements in model conceptualization, program design, and component reuse.

Sequeira et al. (1991) used OOP in modeling cotton growth and development. Cotton was used as a case study in applying object-oriented techniques because a large body of knowledge focused on explicit descriptions of age and size for individual organs. Sequeira et al. proposed a hierarchical class structure, rooted in an "organism" class, that could in theory be extended to encompass multiorganism interactions with a cottonfield ecosystem. Their implementation focused on describing the cotton plant as a collection of organs (roots, stems, leaves, fruits, and storage classes). Each of these organs was represented as a class, either corresponding to a single entity (leaf, fruit) or as an assemblage (roots, stems, and stem storage). Each class contained associated data such as age and size as well as functional descriptions of processes controlling growth, generation, and interaction with other objects. A cotton plant was created by instantiating specific instances of the required classes. Populations of plants were created by instantiating multiple instances of the cotton plant class. Simulation of cotton plant growth was accomplished by starting initially with single root and stem objects and a simple branching structure. As the plant grew, its structural composition was augmented by the creation and association of new leaf and fruiting body instances, in a manner closely matching the formation of new plant structures in the actual plant.

This work demonstrated some of the key differences between procedural and object-based approaches. Of foremost interest was the representation of reality within the model, where the object conceptualization mirrored the real world one-for-one. Similarly, because objects are self-contained and are dynamically created and destroyed, growth can be represented in a very natural manner. It was also noted that while the two approaches offer similar mathematical developments at the individual process level, the procedural approach emphasizes differential equations and formula translations while the object-based approach emphasizes "direct expressions of individual behavior to stimulus outside itself" and resulted in model formulations that are clearer and more direct.

Van Evert and Campell (1994) reported the implementation of CropSyst, a crop systems model. The focus of this effort was to provide an abstract representation of a cropping system defined in terms of objects with minimal and well-defined interfaces. An object-based paradigm was chosen because of the capabilities for objects to provide these interfaces and because of the usefulness of the paradigm in providing easily understood representations of the cropping system components. The primary components developed in CropSyst include Time, Weather, Crop Planting, Crop, Crop Rotation, Soil, Tillage System, Soil Erosion, Aphid Population, Aphid Immigration, Pesticide Application, and Output, with subclasses being developed from these in some cases. Some of these objects correspond directly to physical components of the target system, while others represent functional operations, a difference from most object-based approaches. CropSyst provides for simulation of single-season crop growth cycles, consideration of crop rotation, and aphid dynamics. The authors concluded that object-oriented methods were superior to conventional approaches in developing a modular program structure and enhancing code reusability and maintenance via inheritance mechanisms.

Canpolat and Bolte (1993) developed an object-oriented version of the CERES-Wheat model, termed CropSim. They developed class descriptions for Plant, Leaf, Stem, and other related objects representing fundamental plant components using the original CERES algorithms. They reported that an object-based approach provided several advantages for crop modeling, including (1) improved conceptualization and implementation of the model due to the high degree of correspondence between software modeling components and the crop production environment they represent, (2) improved model modularity, (3) enhanced capabilities for modification and maintenance, and (4) expanded capability for customization and integration in broader decision support systems.

Other studies (Carruthers et al., 1988; Larkin et al., 1988) attempted to apply OOP in ecological domains. In addition, Crosby and Clapham (1990) used OOP in modeling nitrogen dynamics in the soil. Olson and Wagner (1992) reported an object-oriented knowledge-based system for cotton pest management. Whittaker et al. (1991) applied object-oriented concepts in modeling hydrologic processes. In all cases, similar results were reported. Object-based methods tended to promote a high degree of modularity and code reuse and provided superior representation capabilities. The benefits of OOP became more apparent as the complexity of the modeled system increased. This has been consistently noted in domains outside the realm of agriculture as well.

IV. OBJECT-ORIENTED DECISION SUPPORT SYSTEMS

Decision support systems are software systems designed to assist managers in assembling, integrating, and using knowledge from diverse sources to make decisions about complex processes. Decision support systems have received considerable attention in the agricultural community because of the complexities in managing these systems, which are composites of biological, physical, environmental, and economic interactions. Object-based methods are particularly appropriate for the construction of decision support systems because of the typically high degree of complexity of such software, the requirement for representation, and the ability to facilitate code reuse.

Construction of decision systems can be greatly enhanced through the use of high level object frameworks (Bolte, 1996). An object-based paradigm provides substantial capabilities for supporting the requirements of these frameworks by providing for virtualized interface descriptions, dynamic binding to these interfaces, and an extremely modular design approach. These fundamental object constructs can provide the building blocks for building integrative frameworks for decision support. A framework should additionally provide a number of high level services, including (1) standardized public interfaces defining object access and action initiation, (2) mechanisms for synchronizing the flow of execution among system components, (3) high level communications capabilities for components of the system to communicate with other components in a nonspecific manner, (4) support for standard representational constructs, including simulation constructs, inferencing, and other representational paradigms that are domain-independent, and (5) standardized methodologies for collecting and transferring information between components of the system. The object paradigm provides useful capabilities in all of these areas. The standardization of public interfaces is a central OOP concept and is readily implemented through the definition of generic objects with virtualized public interfaces. Virtualizing the interface allows direct communication to specific subclass objects without any knowledge of the type of that subclass, a critical requirement for the development of high level, domain-independent decision support frameworks. Similarly, high level communications between objects can be accomplished through a virtual interface in a manner that precludes requiring the caller to have specific knowledge about objects retrieving the communications. Additionally, interobject communication can be provided by a central "blackboard" maintained by the system. Synchronization and data collection can also be handled at a high level, with subclasses inheriting much of the behavior necessary to handle parallel execution between objects, data flow, and communication.

An integrative framework architecture should provide these components in a robust and flexible manner. An example of such a framework (Bolte et al., 1993) is described below.

A. Interface Standardization

In an object-oriented representation, objects representing real-world entities interact with other objects, frequently through time. These objects represent the fundamental simulation unit and descend from a common parent class. Simulation objects may come in different shapes and sizes, but they all conform to minimum interface requirements and have a common base level of functionality by virtue of the fact that they all derive from a single generic high level simulation object class that provides for and dictates certain behaviors. A simulation object typically contains one or more member variables describing the object's state, which is modified in response to interactions with other objects or time.

By defining a virtual simulation object class SimObj, the framework provides an abstract description and response to maintaining data, generating plots, and tracing simulation results for all derived child simulation objects. SimObj provides data and behavior that are common to all simulation objects and provides most of the required functionality for interacting with the simulation environment.

User-defined simulation object classes derive from SimObj. A simulation object has data and methods just like any object. Its data and methods are specifically designed to aid in the simulation of the object itself. These simulation objects provide the data and behavior specific to their uniqueness, while the parent class SimObj provides data and functionality that are common to all simulation objects. Simulation objects are designed to package data and behavior that are representative of the object being simulated.

The SimObj class provides certain data and behavior to a derived simulation object. For example, every simulation object has a *name*, identifying the object for reporting, debugging, and tracing. An object's *timestep* is the time interval between object updates and can be set on an object-by-object basis. Deterministic simulation objects are typically updated at a constant interval, while stochastic simulation objects may be updated at intervals established by their statistical properties. The priority of a simulation object establishes its ordering sequence in the simulation event list when two objects are being updated at the same time.

It is sometimes useful to have simulation objects store data. Because generalized data storage is a requirement of many systems, the framework defines a DataStore class that provides data storage capabilities. A DataStore is a repository for data. By generalizing the requirements for data storage into a specific class, any object can use that class to handle its data storage requirements. Thus, the data storage object definition becomes a standard way of sharing and manipulating data. A SimObj can acquire data storage capabilities by subclassing from a DataStore. Further, by extending the definition of a DataStore, any object that derives from DataStore automatically gets access to its extended capabilities. These might include statistics generation, file I/O, or visualization.

B.   Execution Synchronization

Execution synchronization in the context of object-based decision support typically requires the ability to handle messages, events, and parallel execution of simulation objects. The framework defines a simulation environment class (SimEnv) to handle the clock, event list, and communication threads. For differential equation-based continuous systems commonly used to represent biological systems, the SimEnv class also provides robust numerical integration capabilities. Individual objects participating in a simulation subclass form the SimObj class. The SimObj class defines the interface between the simulation environment and individual simulation objects representing various systems components. SimEnv maintains SimObj and provides overall coordination and data handling capabilities of the framework. It also provides a time flow mechanism (simulation clock) and basic features common to simulation engines for computer simulation (Fig. 1). Time flow mechanisms are critical for both continuous and event-driven simulations. Because different objects in the system may have widely varying time constants, each continuous object may define its own unique time step. Functionally, the simulation clock is initialized to the starting time for the simulation, and registered events are placed on the event list. The event list is a list of future events that is sorted from top to bottom by ascending time and descending priority. When the simulation begins, the clock removes the first event

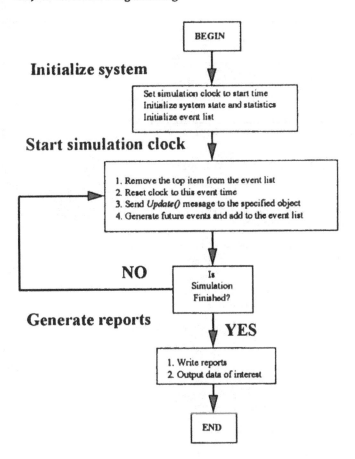

FIGURE 1   Flow diagram of simulation process.

on the event list, updates the current time to the event time, and then updates the event. When the system is updated in the event routine, new events can be scheduled and added to the event list. The process repeats by removing the next object at the top of the event list, resetting the clock, and then updating the object. The procedure of advancing the simulation clock continues until no more items are on the event list or a simulation stopping condition ends the simulation. Variable time increment simulations therefore jump through time from one event to the next, and the state of the system is updated for each event. This approach facilitates the coexistence of fixed-step and

```
// constants
const float a=0.3, b=0.2, rH=0.1, rP=0.3;

// parent Population class - derived from SimObj
class Population : public SimObj {
   protected:
      double pop;              // population size (state variable)

      // constructor - sets initial population to 0
      Population() : SimObj(), pop(0) { }

 public:
      // public "set" method for population
      void SetPopulation( double p ) { pop = p; }
};

// Prey class - derived from population
class Prey : public Population {
   protected:
      virtual void Update( SimEnv*, int, void* )
               { pop += ( rH*pop - a*pop*prey->pop ) * timeStep; }
};

// Predator class - also derived from Population
class Predator : public Population {
   protected:
      virtual void Update( SimEnv*, int, void* )
               { pop += ( -rP*pop - b*pop*pred->pop ) * timeStep; }
};


// declare global variables
Predator *pred;         // pointer to an instance of a Predator
Prey     *prey;         // pointer to an instance of a Prey

// main simulation program
int main() {
   pred = new Predator;        // create an instance of a Predator
   prey = new Prey;            // create an instance of a Prey
   pred->SetPopulation( 10 );  // Set initial conditions
   prey->SetPopulation( 50 );

   SimEnv simEnv;                     // create a simulation environment
   simEnv.RegisterSimObj( pred );  // tell it what to simulate
   simEnv.RegisterSimObj( prey );

   simEnv.SetStopTime( 100 );        // tell it how long to simulate

   simEnv.Run();                     // run the simulation
};                                   // all done!..
```

FIGURE 2   Simple Lotka–Volterra two-population competition model.

variable-step simulation objects, allowing the simulator to use which-ever approach is most appropriate for each object.

SimEnv and associated SimObj-derived classes provide the basis requirements for conducting a simulation. A simplified example based on a two-population Lotka–Volterra predator–prey simulation is implemented in Fig. 2. This simulation shows the creation of two objects (Predator and Prey) and a simulation environment (SimEnv). The virtual Update( ) method, implemented for each simulation object and called as necessary by the simulation environment, steps each object through a single time step. This basic simulation system can be readily extended by defining additional classes and methods to support more sophisticated simulation capabilities, as described next.

C. Interobject Communication

Communication among objects is a required component of any object-based system. Individual simulation objects send messages to other objects to implement some action or get some information. In a robust simulation environment, message passing should be implemented at several levels to provide either direct or indirect communication ca-pabilities. In Fig. 3, object 1 sends a message to object 2. Object 2 takes some action in response to the message being sent and replies to the sender of the message. This is a form of *object-dependent* com-munication because object 1 must have some reference to (direct knowledge of) object 2 to send it a message.

During the development of the framework, it was found that object-independent communication (i.e., a protocol available so that a simulation object can send messages to other objects in the system without requiring a direct reference to them) was needed. There are situations where the object sender may not know who the object re-ceivers are. In Fig. 4, the "ice cream man" simulation object broad-casts to anyone within hearing range that it has ice cream to sell, with no direct knowledge of who might be listening or how they may respond. Anyone interested will receive the message and respond accordingly. What is needed in this situation is an *object-independent* communication mechanism.

The implementation of such a scheme is relatively straightfor-ward. First, individual objects tell the simulation environment what messages they are interested in. An object accomplishes this by reg-istering a *notification filter* with the simulation environment, indicating that it wants to be notified whenever a particular message is broad-cast from one or more simulation objects. Second, objects must broad-

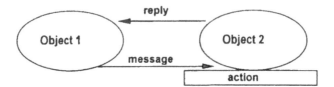

FIGURE 3   Dependent object communication.

cast messages to the simulation environment when it wants other objects in the system to receive those messages. To transmit a message, an object posts a *broadcast* message to the simulation environment, which will then send *notify* messages to all objects that have previously registered interest in that particular message.

The second step in object-independent communication is for individual simulation objects to broadcast their messages at appropriate times. When an object broadcasts a message to the system, the system takes the message and routes it to objects that have previously expressed interest in this message from this object. Figure 5 shows the interobject communication process.

An additional communication method involves a *blackboard*. A blackboard is a system object that can be used by any simulation object as a semipermanent repository of messages. Simulation objects post messages to the blackboard, and other simulation objects read these messages. The process of passing and reading messages is an additional form of interobject communication provided by the simulation environment. The difference between blackboard communication and the broadcast message passing scheme is in the speed of

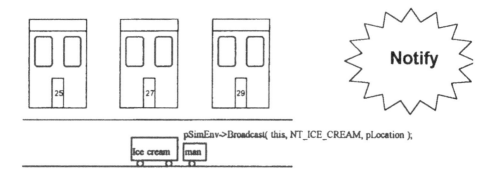

FIGURE 4   Ice cream man broadcasts a message.

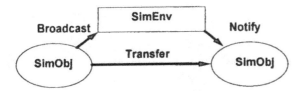

FIGURE 5   Broadcast/notify interobject communication.

delivery. Broadcast messages are sent "now" (in simulated time) and are then discarded from the system, whereas blackboard messages remain on the blackboard indefinitely until explicitly cleared. The amount of time between when a message is posted to the blackboard and when it is read is unspecified.

Simulation objects communicate with other simulation objects directly, via Broadcast/Notify and/or blackboard methodologies. Broadcasting and posting to the blackboard are two powerful message-passing schemes that allow for global or object-specific communication without introducing object dependencies.

D. Representation

Representation is a central concept for both decision and simulation systems, and modelers are generally well aware of issues of representation in simulation. The object paradigm provides an intrinsically compatible approach to representing conventional simulation objects. Equally important, however, is the ability to support additional representation paradigms, particularly in the area of knowledge-based systems. Because these heuristic systems stress a separation between generalized inferencing and representation capabilities and the domain in which these capabilities are applied, they are well suited for encapsulation using an object-oriented approach.

In this context, a potentially useful conceptual component of a decision system is an *agent*, an intelligent entity that might provide the system with diagnostic, control, or planning capabilities. In a resource management context, agents in the system might embody knowledge about the resources in question, using information derived from models and datasets to make suggestions regarding appropriate management tactics under different circumstances. Agents may be useful for guiding the trajectory of complex systems and may know how to answer questions and solve problems for other simulation objects. They may assist other simulation objects during a simulation run.

Agent capabilities were added to the framework by defining an Agent class, descending from SimObj, that can store and execute rules via an inference engine. Since agents are just another simulation object, albeit with somewhat different capabilities, they are added to the decision framework just like any other simulation object. How agents participate in a simulation is dependent on how quickly they are required to solve some problem. For example, consider the case of a simulation of a salmon hatchery facility. A salmon hatchery simulation is running, and one of the simulation objects (e.g., fish lot) has reached a minimum oxygen threshold value for sustaining growth. We might like an agent to interact with the simulation to monitor such problems and take corrective action. Several approaches are possible, depending on the requirements of the simulation. If the agents should immediately respond to any problems (rarely the case in the real world), the fish lot would either talk directly to the agent about the problem (object-dependent communication) or broadcast a message that the agent had previously been asked to listen for (object-independent communication). Alternatively, the intent of the simulation may be to explore the effects of scheduling and time management on the agent's availability for fixing problems. In this case, the agent would be assigned a time step proportional to its availability for action. Upon receiving an update message from the simulation environment, the agent would either directly observe each object it was monitoring (object-dependent communication) or look for problems previously posted on the blackboard that have not been resolved (object-independent communication). In either case, it would invoke its knowledge base for the particular problem, presenting options to the user or taking steps to correct the problem.

E. An Application: Aquacultural Facility Management

As an example of a decision system based on the preceding integrative framework, we have developed a decision system for aquaculture production facilities management, where a facility consists of multiple ponds, fish lots, and various supporting resources. At the lowest level, the system relies on simulation models to make predictions about the state of the facility through time. Individual components of the system are dynamically created (instantiated) from object templates reflecting the item in question and include fish ponds, fish lots, pumps, water conditioning units, and other entities important in the operation of actual facilities.

A high level, domain-independent framework provides most of the services described above, including synchronization of simulation objects, interobject messaging, a central bulletinboard for additional time-independent, nondirected communication, and centralized data sharing and statistics. The framework, written in C++, takes full advantage of the object paradigm. Within the system, agents are defined to provide expertise in areas of water quality analysis, assessment of fish performance, distribution and stocking of fish lots, pond diagnostics, and other areas roughly paralleling human experts in the field. Analyses of a particular facility are accomplished by running simulations of the system. In additional to "normal" simulation objects, a series of agents may be run in parallel with the simulation to monitor the system state at specific times, obtaining information about specific system components through the standardized interobject communication facilities provided by the system. The agents can take prescriptive action when problems are identified or present alternatives for the user to pursue. The agent can also make decisions on the allocation of resources to determine near-optimal pathways through dynamic solution space in these complex environments. Upon conclusion of a simulation, the agents can be queried for suggestions on improving the management strategy employed. An additional interesting benefit is that the effects of management "intensity," i.e., how intently the agents monitor the system, can be examined by varying the frequency with which an agent observes the system and takes action. This system has successfully demonstrated the usefulness of an integrative framework in allowing a mixture of representational paradigms to coexist in a synergistic manner.

V. CONCLUSIONS

The need for tools to synthesize and integrate knowledge sources for agriculture and resource management continues to increase. Fortunately, technologies for dealing with the problems encountered are becoming increasingly available. Many aspects of representation and analysis in these management systems can be abstracted to system-level services and provided through integrative frameworks. Similarly, interfaces between the components of these systems can be defined on an abstract basis using object-oriented tools. These tools are extremely valuable for constructing and reusing complex decision support components, providing mechanisms to define various levels of information and behavioral abstraction appropriate to the system being represented. They allow the development of abstract frameworks for model construction defined at a high level in the hierarchy, with specific implementations of the system components occurring via subclasses of these generic objects to provide specific component-level behavior. This aspect of OOP is critical to the design of complex software systems. Properly used, object-oriented techniques allow the development of decision support software that is more understandable, maintainable, and reusable than that developed using more traditional techniques.

REFERENCES

Belanger, R., B. Donovan, K. Morse, and D. Rockower. 1990. *MODSIM II Reference Manual*. Rev. 6. La Jolla, CA: CACI Products Co.

Birtwistle, G. 1979. *Discrete Event Modeling with SIMULA*. London: MacMillan.

Bischak, D. P. 1991. *Object-Oriented Simulation*. B. L. Nelson, W. D. Kelton, and G. M. Clark (Eds.). Proceedings of the 1991 Winter Simulation Conference. pp. 294–303.

Bolte, J. P. 1996. Integrative frameworks for decisionmaking in resource management. Proceedings, Ninth Florida Artificial Intelligence Research Symposium. (J. Stewman, Ed.) May 20–22, 1996. Key West: FL. pp. 395–399.

Bolte, J. P., J. A. Fisher, and D. H. Ernst. 1993. An object-oriented, message-based environment for integrating continuous, event-driven and knowledge-based simulation. Proceedings: Application of Advanced Information Technologies: Effective Management of Natural Resources. June 18–19, Spokane, WA. St. Joseph, MI: ASAE.

Booch, G. 1991. *Object-Oriented Design with Applications*. Redwood City, CA: Benjamin/Cummings.

Budd, T. 1991. *An Introduction to Object-Oriented Programming*. Reading, MA: Addison-Wesley.

CACI. 1988. *Simscript II.5 User Manual*. La Jolla, CA: CACI Products Co.

Canpolat, N., and J. P. Bolte. 1993. Object-oriented implementation of the CERES-Wheat model. ASAE Paper 934052. St. Joseph, MI: ASAE.

Carruthers, R., T. S. Larkin, and R. Soper. 1988. Simulation of insect disease dynamics: An application of SERB to a rangeland ecosystem. *Simulation* 51(3):101–109.

Crosby, J. C., and W. M. Clapham. 1990. A simulation modeling tool for nitrogen dynamics using object-oriented programming. *AI Appl.* 4(2):94–100.

Dahl, O. J., and K. Nygaard. 1966. SIMULA—An Algol-based simulation language. *CACM* 9(9):671–687.

Dahl, O. J., and K. Nygaard. 1967a. *SIMULA: Introduction and User's Manual*. Oslo: Norwegian Computing Center.

Dahl, O. J., and K. Nygaard. 1967b. *Simula: A Language for Programming and Description of Discrete Event Systems*. 5th ed. Oslo: Norwegian Computing Center.

Ege, R. K. 1992. *Programming in an Object-Oriented Environment.* San Diego, CA: Academic.

Folse, L. J., H. E. Meuller, and A. D. Whittaker. 1990. Object-oriented simulation and geographic information systems. *AI Appl.* 4(2):41–47.

Franta, W. R. 1977. *The Process View of Simulation.* Amsterdam: North-Holland.

Goldberg, A., and D. Robson. 1983. *Smalltalk-80: The Language and Its Implementation.* Reading, MA: Addison-Wesley Publishing Co.

Gordon, G. 1969. *System Simulation.* Englewood Cliffs, NJ: Prentice-Hall.

Helsgaun, K. 1980. DISCO—A SIMULA-based language for continuous, combined, and discrete simulation. *Simulation,* July, pp. 1–12.

IBM. 1970. *General Purpose Simulation System/360 User's Manual.* GH 20-0326. White Plains, NY: IBM.

Kirkeud, B. 1989. *Object-Oriented Programming with SIMULA.* Reading, MA: Addison-Wesley.

Kiviat, P. J., R. Villaneuva, and H. M. Markowitz. 1983. *SIMSCRIPT II.5 Programming Language.* La Jolla, CA: CACI Products Co.

Larkin, T. S., R. Carruthers, and R. Soper. 1988. Simulation and object oriented programming: The development of SERB. *Simulation* 51(3):93–100.

Meyer, B. 1988. *Object-Oriented Software Construction.* Englewood Cliffs, NJ: Prentice-Hall.

Nerson, J. M. 1992. Applying object-oriented analysis and design. *Commun. ATM* 35(9):63–74.

Olson, R. L., and T. L. Wagner. 1992. WHIMS: A knowledge based system for cotton pest management. *AI Appl.* 6(1):41–58.

Pritsker, A. A. B. 1974. *The GASP IV Simulation Language.* New York: Wiley-Interscience.

Pritsker, A. A. B., and D. Pegden. 1979. *Introduction to Simulation and SLAM.* Halsted.

Rubin, K. S., and A. Goldberg. 1992. Object behavior analysis. *Commun. ACM* 35(9):48–62.

Rumbaugh, J., M. Blaha, W. Premerlani, F. Eddy, and W. Lorensen. 1991. *Object Oriented Modeling and Design.* Englewood Cliffs, NJ: Prentice-Hall.

Schriber, T. 1972. *A GPSS Primer.* Ulrich's Books.

Sequeira, R. A., P. J. H. Sharpe, N. D. Stone, K. M. El-Zik, and M. E. Makela. 1991. Object-oriented simulation: plant growth and discrete organ to organ interactions. *Ecological Modeling* 58:55–59.

Shooman, M. L. 1983. *Software Engineering.* New York: McGraw-Hill.

Sierra, J. M., and J. A. Pulido. 1991. Doing object oriented simulations: Advantages, new development tools. 24th Annual Simulation Symposium Proceedings. New Orleans, LA. April 1–5, 1991. pp. 177–184.

Stroustrup, B. 1983. Adding classes to C: An exercise in language evolution. *Software Pract. Exper.* 13:139–161.

Van Evert, F. K., and G. S. Campell. 1994. CropSyst: A collection of object-oriented simulation models of agricultural systems. *Agron. J.* 86:325–331.

Whittaker, A. D., M. L. Wolfe, R. Godbole, and G. J. V. Alem. 1991. Object-oriented modeling of hydrologic process. *AI Appl.* 5(4):49–58.

Wilde, N., P. Matthews, and R. H. Hutt. 1993. Maintaining object-oriented software. *IEEE Software*, January, pp. 75–79.

Zeigler, B. P. 1990. *Object-Oriented Simulations with Hierarchical, Modular Models*. San Diego, CA: Academic.

18

Simulation of Crop Growth: CROPGRO Model

Kenneth J. Boote and James W. Jones
University of Florida, Gainesville, Florida
Gerrit Hoogenboom
University of Georgia, Griffin, Georgia

I. INTRODUCTION

Mathematical simulation of crop growth and yield processes was initiated about 30 years ago (de Wit, 1965; de Wit et al., 1978; Duncan et al., 1967; Keulen, 1975). These pioneering modelers provided a systems framework for modeling carbon balance, water balance, and energy balance in a complex crop production system. This is basically the approach that has been followed by most crop modelers since that time. Early crop modeling efforts were initially limited by the availability and capacity of computers, by inadequate understanding of physiological growth processes, and by lack of input/output standards and access to common file access and graphics tools. Early crop models also focused primarily on potential production, often not considering water, nutrient, and pest limitations. As a result, there was not much realism in the simulated outcomes. In the intervening years, computer technologies have advanced manyfold and considerable physiological information has been obtained to better describe crop growth mechanisms, to parameterize them, and to test crop models. There are now many examples of successful crop model use in research, and there is increased optimism that crop simulation models will have good on-farm applications to address the many interactions between the crop and its physical and biological environment, to predict the consequences on seed yield or total dry matter production (Boote et al., 1996).

In this chapter we describe approaches for mathematical simulation of plant growth and yield processes, following the example of the CROPGRO model (Boote et al., 1997a; Hoogenboom et al., 1992, 1993, 1994b). The goal of most crop growth simulation models is to predict the physical processes of plant carbon balance, soil–plant–water balance, soil–plant–nitrogen balance, and, in many cases, energy balance, in response to soil and aerial environment, crop management, and crop/genetic characteristics. The time step of crop simulation models is usually 1 day (corresponding to daily recording of weather information) and sometimes 1 h for some intermediate processes. The subsequent outcome is the prediction of the final seed yield or biomass or other desired outputs at maturity or other times. This chapter focuses primarily on crop development, carbon balance, and plant internal nitrogen balance aspects but illustrates how crop C and N balances are coupled to soil N and soil water balances.

II. BACKGROUND OF THE CROPGRO MODEL

Crop model development work leading to CROPGRO was initiated in 1980 at the University of Florida with the release of the original SOYGRO V4.2 in 1983 (Wilkerson et al., 1983). In 1982, the International Benchmark Sites Network for Agrotechnology Transfer (IBSNAT) project was initiated. As part of this effort, the Ritchie (1985) soil water balance and a preliminary phenology model were added and SOYGRO V5.0 was released (Wilkerson et al., 1985). Also during this time, a peanut crop growth model was developed based on modification of SOYGRO (Boote et al., 1986). As part of the IBSNAT project, standard input and output formats for climate, soil, and crop data files were developed. Subsequently, SOYGRO V5.4 and PNUTGRO V1.0 were released, with code improvements and compatibility with the IBSNAT standard input/output structure (Boote et al., 1987; Jones et al., 1987). SOYGRO V5.42 and PNUTGRO V1.02 releases were developed to fit within Version 2.1 of the Decision Support System for Agrotechnology Transfer (DSSAT) (Boote et al., 1989b; IBSNAT, 1989; Jones et al., 1989).

A model for common beans (BEANGRO V1.00, V1.01) was developed and released (Hoogenboom et al., 1990, 1994c). Between 1990 and 1994, soil N balance and N uptake features as well as N_2 fixation were added to these grain legume models. The decision was made to use one set of Fortran code for all three legume models in order to eliminate the need to make parallel changes in code of all three models. Species parameters and cultivar traits are entered as external input files. Early versions of CROPGRO that simulated all three legumes with added N balance were described by Hoogenboom et al. (1991, 1992, 1993). Recently, the model was adapted to simulate tomatoes, a nonlegume, also without changing the Fortran code (Scholberg et al., 1995, 1997). Version 3.0 of CROPGRO was released in August 1994. This chapter describes CROPGRO Version 3.1 released in December 1996.

III. OVERVIEW OF PROCESSES SIMULATED BY CROPGRO

CROPGRO is process-oriented and considers crop development, crop carbon balance, crop and soil N balance, and soil water balance. In this approach, state variables are the amounts, masses, and numbers of tissues whereas rate variables are the rates of inputs, transformations, and losses from state variable pools (Jones and Boote, 1987). For example, the crop carbon balance includes daily inputs from photosynthesis, conversion and condensation of C into crop tissues, C losses due to abscised parts, and C losses due to growth and maintenance respiration. The carbon balance processes also include leaf area expansion, growth of vegetative tissues, pod addition, seed addition, shell growth rate, seed growth rate, nodule growth rate, senescence, and carbohydrate mobilization. Addition of pods and seeds and their growth rates actually determine partitioning during the seed-filling phase.

Crop development includes important processes such as rate of vegetative stage development and rate of reproductive development, which together determine life-cycle duration, duration of root and leaf growth, and onset and duration of reproductive organs such as pods and seeds. These development processes greatly influence partitioning to plant organs over time.

The crop N balance processes include daily soil uptake, N_2 fixation, mobilization from vegetative tissues, rate of N use for new tissue growth, and rate of N loss in abscised parts. Soil water balance processes include infiltration of rainfall and irrigation, soil evaporation, distribution of soil water, root water uptake, drainage of water through the root zone, and crop transpiration.

The main time step in CROPGRO is 1 day (corresponding to daily recording of weather information), but the model has hourly time steps for some processes such as the photosynthetic calculations using the leaf-level hedgerow photosynthesis model. Model state variables are predicted and output on a daily basis for all seasonal crop, soil water, and soil N balance processes. Final seed yield, biomass, and other information are output at maturity. Examples of state variable equations are shown later for the crop carbon balance.

A. Crop Development

For many crop plants, the partitioning of dry matter to growth in the different plant parts depends on the state of development. To accurately predict the growth and yield of these crops, one must be able to accurately predict the timing and duration of the various crop growth phases and the total life-cycle duration.

Crop development in CROPGRO during various growth phases is differentially sensitive to temperature, photoperiod, water deficit, and N stresses. There is a vegetative development rate that affects the rate of leaf appearance. In addition there are 13 possible life-cycle phases from sowing to maturity, each having its own unique developmental accumulator starting at a unique prior endpoint growth stage and ending when the accumulator reaches a defined threshold when an event such as first flower or onset of seed growth occurs (Fig. 1). The physiological development rate, expressed as physiological days per calendar day (PD/day), during any phase is typically a function of temperature (T), photoperiod (P), and water deficit:

$$PD/day = f(T)f(P)f(\text{water deficit})$$

If conditions are optimum, one physiological day is accumulated per calendar day. The number of physiological days required for a phase to be completed is equal to calendar days if temperature, photoperiod, and water status are optimum, thus allowing the plant to develop at the maximum possible rate. If plants are grown in an optimum environment, one can actually measure the physiological day requirement for a given phase, say from emergence to first flower.

The crop development subroutine allows the use of different equations as well as different base and optimal temperatures. This generic subroutine works for different species and cultivars. For example, the species file for each crop defines those equation shapes

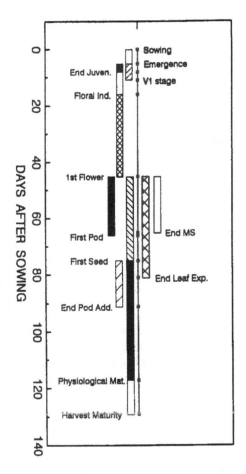

FIGURE 1 Ontogeny of vegetative and reproductive stages of grain legume crops. Timing of occurrence are shown for soybeans, MG 7 determinate cultivar, grown in Florida.

and cardinal temperatures [base temperature, first optimum, second (highest) optimum, and maximum temperature, as illustrated for four-point functions in Fig. 2]. The species file also defines the 13 phases, lists the starting point for accumulators for each phase, and indicates whether a given phase is sensitive to temperature, photoperiod, and water deficit. The threshold accumulation values for the phases (in PD) and the critical photoperiod parameters are given in

Boote et al.

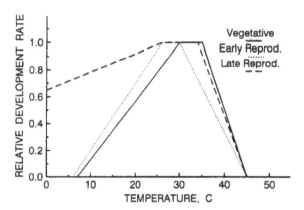

FIGURE 2 Effect of temperature on rate of vegetative node appearance, reproductive development rate before beginning seed stage, and reproductive development rate after beginning seed for soybeans.

the cultivar files. Figure 1 illustrates the time line for the 13 phases used for the three grain legumes—dry beans, peanuts, and soybeans. Note that several phases have a common starting point, such as time from first flower to first pod, first flower to first seed, and first flower to end of leaf expansion.

CROPGRO's flexibility to use different temperature functions before and after flowering was found to be very helpful for correctly predicting the reproductive phenology of soybeans. Using simplex optimization techniques with extensive data on time to flower, beginning seed, and physiological maturity, we discovered that soybean cardinal temperatures for reproductive growth differ for preflowering versus postflowering phases (particularly after beginning seed) and that both of these processes have cardinal temperatures different from those for rate of vegetative node (V stage) development (Fig. 2) (Grimm et al., 1993, 1994). V stage development has a higher base and higher optimum temperature than does rate of progress toward flowering. Of particular significance, the rate of development after beginning seed had a much lower derived base temperature (−48°C). Of course, soybeans do not survive freezing temperature, but their rate of progress from beginning seed (R5) to physiological maturity (R7) is scarcely slowed by cool temperature. This change (use of different temperature functions before and after seedset) greatly improved our ability to predict maturity dates of soybean cultivars grown farther north in the United States or in cool conditions. The flexibility of using different temperature and photoperiod sensitivities during various life-cycle phases has been helpful in modeling a number of species with the same basic model.

B. Carbon Balance and State Variable Equations

Total dry matter growth of the crop is based on a carbon balance that includes photosynthesis, respiration, partitioning, remobilization of protein and carbohydrate from vegetative tissues, and abscission of plant parts. Carbohydrate (glucose = CH_2O) is the basic "molecular unit" used in CROPGRO calculations of photosynthesis, maintenance respiration, and glucose conversion via growth respiration to plant tissue dry weight. Resulting plant part weights are expressed as dry weight.

Nonlinear first-order differential equations are used to describe the rates of change of component dry weights resulting from changes in crop carbon balance [see more detailed original explanations by Wilkerson et al. (1983)]. These equations are used in CROPGRO, with the inclusion of net carbohydrate pool dynamics (C_T) and pest removal of dry weight (P_T). Rate of change in total crop dry matter, W, in grams per square meter, is described as

$$\frac{dW}{dt} = \dot{W}^+ - \dot{S}_L - \dot{S}_S - \dot{S}_R - \dot{S}_{SH} - \dot{C}_T - \dot{P}_T$$

where $\dot{S}_L$, $\dot{S}_S$, $\dot{S}_R$, and $\dot{S}_{SH}$ are abscised parts of leaf (L), stem (S), root (R), and shell (SH), respectively, and

$$\dot{W}^+ = E(P_g - R_m)$$

where new growth, $\dot{W}^+$, is a function of gross photosynthesis (P_g), maintenance respiration (R_m), and the efficiency of conversion (E) of grams CH_2O to grams dry mass.

To represent dry matter change in the component plant parts, the following equations are used:

$$\frac{dW_L}{dt} = X_L \dot{W}^+ - \dot{S}_L - \dot{M}_L - \dot{C}_L - \dot{P}_L$$

$$\frac{dW_S}{dt} = X_S \dot{W}^+ - \dot{S}_S - \dot{M}_S - \dot{C}_S - \dot{P}_S$$

$$\frac{dW_R}{dt} = X_R \dot{W}^+ - \dot{S}_R - \dot{M}_R - \dot{C}_R - \dot{P}_R$$

$$\frac{dW_{SH}}{dt} = X_{SH} \dot{W}^+ - \dot{S}_{SH} - \dot{M}_{SH} - \dot{C}_{SH} - \dot{P}_{SH}$$

$$\frac{dW_{SD}}{dt} = X_{SD}W^+ + a(\dot{M}_L + \dot{M}_S + \dot{M}_R + \dot{M}_{SH}) - \dot{P}_{SD}$$

The terms $\dot{S}_L$, $\dot{S}_S$, $\dot{S}_R$, and $\dot{S}_{SH}$ describe daily abscission rates of leaf, petiole, root, and shell mass. The terms $\dot{M}_L$, $\dot{M}_S$, $\dot{M}_R$, and $\dot{M}_{SH}$ describe daily mass of protein mobilized from leaf, stem, root, and shell, respectively. The terms $\dot{C}_L$, $\dot{C}_S$, $\dot{C}_R$, and $\dot{C}_{SH}$ describe daily net mass of carbohydrate mobilized from leaf, stem, root, and shell, respectively. The terms $\dot{P}_L$, $\dot{P}_S$, $\dot{P}_R$, $\dot{P}_{SH}$, and $\dot{P}_{SD}$ describe daily pest losses of leaf, stem, root, shell, and seed mass, respectively. The partitioning coefficients X_L, X_S, X_R, X_{SH}, and X_{SD} depend on the phenological development of the crop and change on a daily basis. Partitioning is described and computed on a dry matter basis. Partitioning to shell and seed (SD), X_{SH} and X_{SD}, are more complex than these simple state variable equations imply and actually depend on addition and growth of shells and seeds as described later.

The CROPGRO model uses various more complex equations for predicting photosynthesis, maintenance respiration, and conversion efficiency; nevertheless, the coupling points to these state variable equations and the daily integration loop remain unchanged. Likewise, there are more complex equations and rules for prediction of partitioning, senescence, and protein mobilization, but they remain linked to these state variable equations and the daily integration loop. Indeed, the state variable equations and integration are the easy part. The difficult task is to describe the relationships of photosynthesis, respiration, senescence, N uptake, N_2 fixation, N mobilization, and carbohydrate mobilization processes to temperature, light, N supply, water deficit, and crop age. Because weather changes daily, these processes are dynamic and change from day to day. Therefore, closed-form solutions are impossible, and the dynamic processes are more suited to integration by computer simulation.

We next give a more detailed look at photosynthetic input, respiration, growth conversion efficiency, assimilation partitioning, and interactions of C balance with N uptake, N_2 fixation, and N balance.

C. Crop Photosynthesis

The CROPGRO model has two options for photosynthesis calculations: (1) daily canopy photosynthesis or (2) hourly leaf-level photosynthesis with hedgerow light interception. The daily canopy photosynthesis option is similar to, but modified from, the method used in SOYGRO V5.42. The leaf-level photosynthesis option, with hedgerow light interception, allows more mechanistic responses to temperature, CO_2, and irradiance and computes hourly values of canopy photosynthesis. The latter method gives more realistic light interception and growth response to row spacing and plant density.

Daily Canopy Photosynthesis. In the DAILY CANOPY photosynthesis option, daily photosynthesis is calculated as a function of daily irradiance for an optimum canopy, then multiplied by factors of 0 to 1 for light interception, temperature, leaf N status, and water deficit, similar to SOYGRO V5.42. In addition, there are adjustments for CO_2 concentration, specific leaf weight (SLW), row spacing, and cultivar variation in leaf photosynthesis rate. The daily canopy approach is fairly similar to radiation use efficiency (RUE) models, except that the RUE (slope of gross photosynthesis versus irradiance) is not constant and respiration is separately predicted.

Hourly Leaf Level, Canopy Photosynthesis. The HOURLY LEAF-level hedgerow photosynthesis light interception approach is described in more detail by Boote and Pickering (1994) and Pickering et al. (1995). Hourly distribution of solar radiation and photosynthetic flux density (PFD) are computed from daily inputs, using a function from Spitters (1986). Hourly temperatures are calculated from daily maximum and minimum air temperatures using a combined sine-exponential curve versus time of day (Parton and Logan, 1981; Kimball and Bellamy, 1986). Hourly total and photosynthetic irradiance are split into direct and diffuse components using a fraction diffuse versus atmospheric transmission algorithm (Erbs et al., 1982; Spitters et al., 1986). Each hour, light interception and absorption of direct and diffuse irradiance by hedgerow canopies is computed as a function of canopy height, canopy width, leaf area index (LAI), leaf angle, row direction, latitude, day of year, and time of day (Boote and Pickering, 1994). Crop height and width are predicted as a function of rate of vegetative node formation and internode length, which is further dependent on temperature, irradiance, water deficit, and vegetative stage. Individual plants are assumed to be shaped like ellipsoids, and light absorption accounts for effects of row spacing and spacing of ellipsoidal plants within the row. Absorption of PFD by sunlit and shaded classes of leaves is computed. Hourly canopy photosynthesis on a land area basis is computed from the sum of sunlit and shaded leaf contributions by multiplying the sunlit and shaded leaf photosynthetic rates by their respective LAIs. Hourly canopy assimilation values are summed to give the total daily gross photosynthesis.

Leaf photosynthesis rate of sunlit and shaded leaves is computed with the asymptotic exponential light response equation, where quantum efficiency (Q_E) and light-saturated photosynthesis rate (P_{max}) variables are dependent on CO_2, oxygen, and temperature. The Farquhar and von Caemmerer (1982) equations for the RuBP regeneration limited region are used to model the basic kinetics of the rubisco enzyme and to compute the efficiency of electron use for CO_2 fixation. This includes temperature effects on specificity factor of rubisco and on CO_2 compensation point in the absence of mitochondrial respiration [see Boote and Pickering (1994) and Pickering et al. (1995) for a full description of simplified algorithms for these computations]. Single-leaf P_{max} is modeled as a linear function of specific leaf weight (SLW) and as a quadratic function of leaf N concentration. The SLW and leaf N concentration (and thus P_{max}) are modeled to vary with LAI depth in the canopy. Leaf Q_E is modeled to have a mild dependence on leaf N concentration. Leaf carbohydrates are excluded from calculations of SLW and N concentration used to influence photosynthesis.

Daily gross canopy photosynthesis response to solar irradiance, temperature, leaf N concentration, and LAI are illustrated in Figs. 3–6, using the leaf-level assimilation compared to the daily canopy assimilation option. Responses are shown for soybean parameters for a closed canopy at 65 days after planting. Response to daily solar irradiance shows a gradual saturation of response beginning at 20 $MJ/(m^2 \cdot day)$, particularly so for the hourly version (Fig. 3). The greater degree of saturation of the hourly version is associated with the algorithm for computing fraction diffuse, the approach used for diffuse light absorption, and the fact that the solar constant remained the same for the day of simulation even though the "input" irradiance was varied (this does not happen in reality). Irradiance during midsummer months is typically between 14 and 24 $MJ/(m^2 \cdot day)$.

Simulated daily canopy photosynthesis for soybeans has a broad temperature optimum for both the hourly and daily options (Fig. 4). With the hourly version, the predicted canopy rate was within 5% of maximum for daytime mean temperature of 22–36°C (T_{max} 26–40°C). The leaf-level hourly version gives a broad temperature optimum because leaf photosynthesis depends on two temperature functions operating in opposite directions, Q_E decreases with increasing temperature, and P_{max} increases with increasing temperature up to 40°C, and

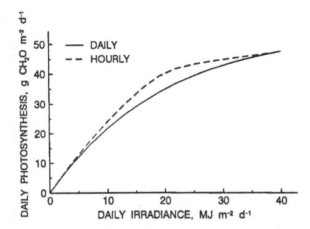

FIGURE 3 Daily canopy assimilation response to daily irradiance for the DAILY option and the HOURLY-LEAF option. Standard simulations for Figs. 3–6 used soybean species parameters, 350 ppm CO_2, 32°C maximum and 20°C minimum temperature, solar irradiance of 22 MJ/(m$^2 \cdot$ day), LAI = 5.85, simulated at 65 days after planting, except where the given factor was the variable of interest.

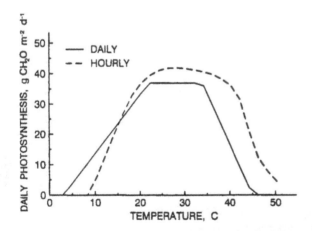

FIGURE 4 Daily assimilation response to temperature for the DAILY option and the HOURLY-LEAF option, using soybean parameters and conditions described in Fig. 3. Simulated with 12°C spread between maximum and minimum temperature and plotted versus mean daytime temperature.

also because many leaves in the canopy operate at low light flux, which increases the importance of the temperature effect on Q_E. The daily version temperature parameters were initially taken from SOY-GRO V5.42, based on relative growth rate data of Hofstra and Hesketh (1975), but we decreased the base temperature from 5 to 3°C and decreased the first (lowest) optimum temperature from 24 to 22°C, based on experience with SOYGRO V5.42 in the northern United States and in France. The hourly photosynthesis version has higher rates at high air temperature than does the daily version; however, this is supported by soybean measurements of Pan et al. (1994) taken in controlled environment studies. For leaf photosynthesis, the base temperature of 8°C and linear response up to 40°C for rate of electron transport were derived from Harley et al. (1985).

Canopy assimilation response to leaf N concentration shows initially rapid response from 18 to 30 mg N/g, with a gradually decreasing response at higher N concentration, reaching a saturating response above 52 mg N/g (Fig. 5). Canopy assimilation response to leaf N is expected to saturate more readily than single-leaf light-saturated photosynthesis because quantum efficiency is less sensitive to leaf N concentration. The leaf-level option has relatively higher rates than the daily option, especially at low leaf N concentration, a phenomenon that we believe may be caused by the modeled vertical distribution of leaf N (and SLW) about the canopy mean.

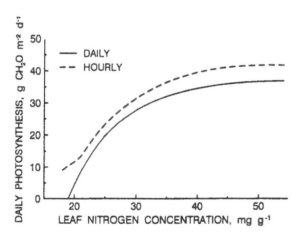

FIGURE 5 Daily assimilation response to mean leaf N concentration for the DAILY option and the HOURLY-LEAF option at a closed canopy stage, using soybean parameters and conditions described in Fig. 3.

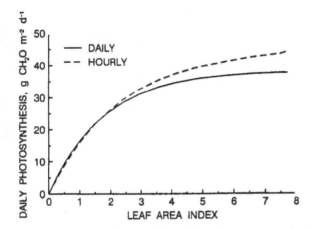

FIGURE 6 Daily assimilation response to leaf area index for the DAILY option and the HOURLY-LEAF option at a closed canopy stage, using soybean parameters and conditions described in Fig. 3.

Response to LAI (Fig. 6) is similar with the two assimilation options, although the daily version has a slightly steeper initial response and flattens sooner. The hourly option is closer to reported responses of canopy assimilation to LAI (Boote et al., 1984; Jeffers and Shibles, 1969). The response to LAI in Fig. 6 was computed for a closed canopy at 65 days after planting. Response to increasing LAI for an incompletely closed canopy would be less steep initially for the hedgerow option than for the daily option. The calculated light extinction coefficient for the daily version is 0.676 for 0.76 m row spacing.

D. Growth Conversion Costs, Growth, and Maintenance Respiration

Growth and maintenance respiration are handled similarly to SOY-GRO V5.42 (Jones et al., 1989; Wilkerson et al., 1983) and PNUTGRO V1.01 (Boote et al., 1986). Maintenance respiration depends on temperature, gross photosynthesis rate, and current crop biomass (minus oil and protein stored in the seed). The relative sensitivity of maintenance respiration is similar to the Q_{10} response (1.85) reported by McCree (1974).

Growth respiration and efficiency of conversion of glucose to plant tissue are computed using the approach of Penning de Vries and van Laar (1982, pp. 123–125) and Penning de Vries et al. (1974). This approach requires estimates of tissue composition in six types of compounds: protein, lipid, lignin, carbohydrate-cellulose, organic acids, and minerals [summarized for peanuts by Boote et al. (1986) and for soybeans by Wilkerson et al. (1983) and Jones et al. (1989)].

The glucose equivalent needed for biosynthesis depends on chemical composition of tissue and biochemical pathways of synthesis. The approach accounts for

1. Glucose respired to provide ATP, NADH, and NADPH for biosynthesis
2. Condensation and reduction and increased energy content of tissue

Table 1 shows the glucose cost of energy for synthesis versus the glucose equivalent that remains condensed in the product. Note that lipid synthesis is very costly (3.11 g glu/g product), in terms of both energy required and energy stored in the lipid. Proteins are also costly, particularly if N is assimilated via NO_3 or N_2 fixation. Even though the cost of protein from NH_4 appears to be less, we compute all costs of inorganic N assimilation as if originating from NO_3 because nitrification rapidly converts most of the NH_4 to NO_3 in most

Table 1　Glucose-Equivalent Cost for Growth Respiration and Synthesis of Six Classes of Plant Compounds

Plant compound	Glucose cost for growth and synthesis (g glu/g prod)			Conversion efficiency, E (g prod/g glu)
	Cost energy	C in product	Total cost	
Protein				
From NH_4	0.36	1.34	1.70	0.56
From NO_3	1.22	1.34	2.56	0.39
Cellulose (starch)	0.11	1.13	1.24	0.81
Lipid	1.17	1.94	3.11	0.32
Lignin	0.62	1.55	2.17	0.46
Org. acid	0.00	0.929	0.929	1.08
Mineral	0.05	0.00	0.050	—

Source: Adapted from Penning de Vries and van Laar (1982).

soils and because NH_4-fed plants have an uncertain added cost associated with cation balance. Proteins constructed with N from N_2 fixation cost 2.83 g glu/g protein, based on theoretical calculations for a hydrogenase-negative N_2-fixing soybean (K. J. Boote, unpublished calculations).

With these glucose-equivalent costs to synthesize six classes of compounds and knowledge of the approximate composition of the tissue, one can compute the total glucose-equivalent cost to produce 1 g of tissue (Penning de Vries and van Laar, 1982, p. 123):

$$E = 1/[2.56\,FP + 1.24\,FC + 3.11\,FF + 2.17\,FL + 0.93\,FO + 0.05\,FM]$$

where E = gram product per gram glucose, and FP, FC, FF, FL, FO, and FM are fraction protein, cellulose-starch, fat, lignin, organic acid, and mineral, respectively, and assuming NO_3 as the source of nitrogen.

The growth conversion efficiency (E) is computed daily in CROPGRO for all new tissues because the N balance method allows N (protein) concentration of newly produced tissues to vary with N supply. Seed lipid composition additionally varies with temperature. "Normal adequate N" composition values are entered in the species file for each plant component, except that protein concentration of new tissue depends on N supply relative to C supply. Any change in protein concentration is offset by an opposite change in the carbohydrate-cellulose fraction, to keep the total fraction composition at 1.00.

E. Assimilate Partitioning Relative to C Balance

The partitioning algorithms in CROPGRO are similar to those of SOY-GRO V5.42, PNUTGRO V1.02, and BEANGRO 1.0, except that N deficit can limit vegetative and reproductive growth and alter partitioning between root and shoot (described in Section III.F). New dry weight growth of each plant component (W_i) depends on the computed partitioning factor to each tissue type (X_i), conversion efficiency (E_i), gross photosynthesis, and maintenance respiration. The partitioning factor varies with crop developmental stage and addition of reproductive sites.

$$\dot{W}_i^+ = X_i E_i (P_g - R_m)$$

1. Partitioning to Vegetative Growth

During vegetative growth (prior to pod formation), all assimilate is allocated to vegetative tissues, with partitioning among leaf, stem,

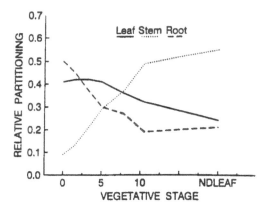

FIGURE 7 Partitioning of dry matter among leaf, stem, and root components as a function of vegetative growth stage of soybeans. Taken from soybean species file. When beginning bloom stage occurs (whether at low or high V stage), partitioning begins a gradual transition to the values at NDLEAF (date when the last branch or main axis leaves complete their expansion).

and root being dependent on V stage progression (Fig. 7), and further modified by water deficit and N deficit. As reproductive development progresses, new sinks (podwalls, seeds) are formed, and assimilate is increasingly partitioned to reproductive rather than vegetative tissue growth. Thereafter, the vegetative tissues share in the fraction of growth remaining after supplying reproductive tissues.

2. Partitioning to Reproductive Growth

At the beginning pod stage, CROPGRO starts adding new classes or cohorts of reproductive sinks on a daily basis. Fruits of each cohort increase in physiological age and pass through a shell growth phase. Partway through the shell growth phase, seeds in each fruit "set" and begin their rapid growth phase. Thus, reproductive "sink" demand comprises many individual reproductive tissues, all of different ages, each having a genetic potential assimilate demand dependent on temperature. The priority order for assimilate use is seeds, podwalls, addition of new pods, and vegetative tissues. Nitrogen fixation and nodule growth are actually given higher priority than all other tissues because CROPGRO computes N demand (based on tissue growth and "normal" N concentrations) and takes the required carbohydrate off the top to supply any required N_2 fixation if N uptake is inadequate. (Giving seed higher demand than nodules for assimilate creates model instability.) Once a full complement of reproductive sites have been added and are growing rapidly, there will be little assimilate remaining for vegetative growth. A genetic limit of fraction partitioning to pods and seeds (XFRUIT) is defined to account for the fact that some species are truly indeterminate and continue to grow vegetatively even during rapid seed filling. With this approach, a fraction $(1 - XFRUIT)$ of the assimilate can be "reserved" for vegetative growth only.

CROPGRO explicitly defines sink strength only for the reproductive tissues: podwalls and seeds. Leaves, stems (including petioles), roots, and nodules can each be viewed as having one aggregated or lumped class with no age or positional structure. The approach to computing sink strength of seeds and podwalls requires estimates of potential seed size, potential seed-filling duration, and threshing percentage under optimum conditions (cultivar traits). From these parameters, the potential growth rate per seed and per shell (podwall) under optimum conditions is internally computed. Actual growth rate per seed or shell subsequently is modified by temperature, C supply, and N supply. The approach to computing pod addition rate depends on the actual daily assimilate supply divided by actual growth requirement per seed (or shell) and a "normal" pod addition period. This method automatically adjusts for large- or small-seeded cultivars and for long or short seed-fill duration types. The pod addition period can be estimated experimentally as the physiological time elapsed from addition of the first pod (seed) to the point of maximum pod (seed) number. (Note, for example, the large species difference in seed addition period between peanuts and soybeans in Fig. 11a, shown in Section VI.)

CROPGRO is a source-driven model except under three circumstances: (1) during early V stage development (up to the VSSINK stage), when potential leaf area expansion and sink strength can be limiting, thus reducing growth and photosynthesis; (2) during severe N deficit, which limits growth of vegetative or seed components, causing carbohydrates to accumulate in vegetative tissue; and (3) if there are insufficient pods and seeds to use all the available assimilate. This rarely happens but is possible if conditions are adverse during the entire seed addition period and conditions then improve dramatically after the crop has completed its seed addition phase.

F. Nitrogen Uptake, Balance, and Fixation Relative to Carbon Balance

Soil nitrogen balance and root nitrogen uptake processes were included in CROPGRO using the code from CERES-Wheat (Godwin et al., 1989). The soil N balance processes—N uptake, mineralization, immobilization, nitrification, denitrification, leaching, etc.—are the same as described for the CERES models (Godwin and Jones, 1991). Uptake of NO_3^- and NH_4^+ are Michaelis–Menten functions of NO_3^- and NH_4^+ concentration, soil water availability, and the root length density in each layer. If N supply exceeds N demand, the actual daily crop N uptake is constrained to be equal to or less than crop N demand. Total crop N demand is computed from the dry matter increase in each organ type multiplied by the maximum N concentration allowed for each tissue type. The upper N concentrations for each tissue type are specified in the SPECIES file. This upper limit (for vegetative tissues) is subject to downregulation as the crop progresses through its reproductive cycle from beginning seed to maturity.

An N_2 fixation component has been incorporated into CROP-GRO. There is a thermal time requirement for initiating first-nodule mass. When N uptake is sufficient, nodule growth is slow, receiving a minimum fraction (bypass flow) of the total assimilate that is allocated to roots. When N uptake is deficient (less than N demand) for growth of new tissues, carbohydrates are used for N_2 fixation to the extent of the nodule mass and the species-defined nodule specific activity. If nodule mass is insufficient, then assimilates are used for nodule growth at a rate dependent on soil temperature and species-defined nodule relative growth rate. The N_2 fixation rate is further influenced by soil temperature, soil water deficit, soil aeration (flooding), and plant reproductive age.

G. Consequences of N Deficiency on Growth, Photosynthesis, Partitioning, and Composition

When the sum of N uptake and N_2 fixation is less than N demand, vegetative tissue is initially grown at full rate but at lower N concentration. As leaf N concentration declines, leaf and canopy photosynthesis are decreased, following a quadratic function from maximum to minimum leaf N concentration (Fig. 5). There is a minimum N concentration of new tissue (specified for tissue types in the SPECIES files) below which tissue growth rate is progressively decreased to hold a constant minimum N concentration. When tissue concentration approaches this lower N limit, then vegetative growth is reduced and carbohydrates begin to accumulate in vegetative tissues. In this situation, excess carbohydrates are exported to enhance nodule function. Accumulated carbohydrates are mobilized at a defined rate and can be used for growth as N becomes available from N uptake or N_2 fixation. If N deficiency is severe (when the ratio of N supply to N demand is less than 0.5), then assimilate partitioning to roots is increased. The most recent CROPGRO version allows seed protein and lipid composition to change with C and N supply. When N supply is more deficient than C supply, seed N concentration (protein) is allowed to decline, although single-seed growth rate is decreased only slightly. Alternatively, when C supply is more deficient than N, seed N concentration is allowed to increase.

H. Mobilization of N and Carbohydrates, Senescence, and Maturity

Carbohydrates accumulate in leaf and stem tissues under several conditions: (1) when N deficiency limits growth, (2) when there is insufficient sink during early sink-limited growth or after a full seed load is set, and (3) during "programmed" accumulation from flowering until rapid seed growth begins. During this programmed accumulation, up to 30% of the daily vegetative growth increment can be allocated to carbohydrate storage in stems (primarily) and leaves (secondarily), ending when a full seed load occurs and assimilate is no longer used for vegetative growth. Carbohydrate mobilization occurs continuously but is allowed to accelerate during seed growth.

Mobilization of protein (C and N) from old to new vegetative tissue occurs slowly during vegetative growth, but mobilization from vegetative tissue becomes up to twofold faster when seeds begin to grow (after 75 days in Figs. 8a and 8b). The maximum rate of protein mobilization depends on the rate of reproductive development. An advantage of using mobilized protein is that more seed mass can be produced per day than when all reduced N must come from current N assimilation or N_2 fixation. A disadvantage of protein mobilization is that as leaf N declines (Fig. 8a), leaf photosynthesis is decreased and some leaf area is abscised (Fig. 8b). Loss of nonprotein leaf and stem mass also occurs when protein mass is mobilized; thus leaf area, leaf mass, leaf protein, stem mass, and stem protein all decline.

Senescence of leaves and petioles is dependent on protein mobilization and is additionally enhanced by drought stress. All three legumes have gradual senescence of foliage and loss of leaf area and

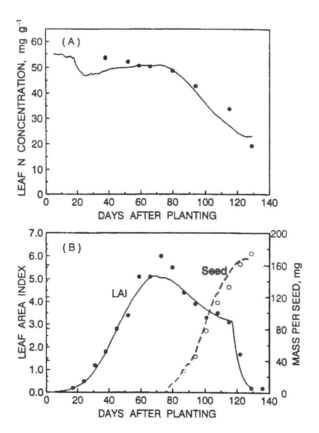

FIGURE 8 (a) Leaf N concentration and (b) LAI and average mass per seed for 'Bragg' soybeans (MG 7) planted June 12, 1984 at Gainesville, Florida. Lines are CROPGRO simulations, and symbols are field-measured values.

leaf protein as seed fill progresses, thus reducing seed growth rate; however, soybeans and common beans additionally have a grand senescence phase that starts at physiological maturity (R7 stage) and causes almost all remaining leaves to abscise (Fig. 8b). The grand senescence of soybean and dry bean foliage causes all seed growth to cease even if seeds have not filled the pods. Peanut plants, by contrast, are indeterminate, and their foliage normally remains green to maturity even though the growth of individual seeds will cease when they reach the limits of their individual pod cavities (maximum shelling percentage).

I. Soil Water Balance, Root Growth, and Evapotranspiration

The soil water balance in CROPGRO is the same as that in the CERES growth models and is described in detail by Ritchie (1985). The soil is divided into a number of layers, up to 10 or more. Water content in each layer varies between a lower limit [LL(J)] and a saturated upper limit [SAT(J)]. If water content of a given layer is above a drained upper limit [DUL(J)], then water is drained to the next layer with the "tipping bucket" approach, using a drainage coefficient specified in the soil file. Infiltration and runoff of rainfall and applied irrigation water depend on the Soil Conservation Service (SCS) runoff curve number. Vertical drainage may be limited by the saturated hydraulic conductivity (K_{sat}) for each layer. This feature allows the soil to retain water above layers that have been compacted or have natural impedance to water flow. In such cases, soil layers may become saturated for a period of time, causing root death, reduced root water uptake, and decreased N_2 fixation.

Root mass and root length in each soil layer are state variables that can change daily. New root length growth each day depends on the daily assimilate partitioned to roots and a length-to-weight parameter. The distribution of the new root length density into respective soil layers depends on (1) progress of downward root depth front, (2) a soil-rooting preference function [WR(L)] describing the probability distribution of roots growing in each layer of soil (definable late in the season at the end of the root growth period), and (3) soil water content in each zone. If the fraction available soil water content in a given zone is less than 0.25 or within 2% of the saturated value, then root growth in that zone is decreased. Thus, roots tend to grow into moist soil layers rather than drier layers or saturated layers. Root senescence in a given zone is accelerated when available soil water content is below a critical fraction (0.25) or near saturation.

1. Evapotranspiration Options and Root Water Uptake

CROPGRO allows several optional methods for computing climatic potential evapotranspiration (E_0): (1) The Priestley–Taylor method (Priestley and Taylor, 1972), also described by Ritchie (1985); (2) the FAO version of the Penman ET equation as described by Jensen et al. (1990); and (3) an hourly energy balance that also solves for foliage and soil temperature (Pickering et al., 1990, 1993, 1995). Only temperature and solar radiation are required to compute the Priestley–Taylor equilibrium evapotranspiration. The FAO–Penman method additionally requires windspeed and humidity data. The climatic potential evapotranspiration (E_0), from either option 1 or option 2, is then multiplied by an exponential function of LAI (K = KCAN + 0.15) to give climatic potential plant transpiration. (K is the extinction coefficient for total energy capture for transpiration, whereas KCAN is the light extinction coefficient for the daily photosynthesis option.) The water-supplying capacity of the soil–root system is calculated from root length and soil water content in each soil layer and then compared to climatic potential plant transpiration. Actual plant transpiration and water extraction by roots is the minimum of the two rates. Drought stresses on photosynthesis and growth processes occur when root water uptake capability cannot supply the climatic potential plant transpiration.

2. Water Deficit Effects on Processes

Processes sensitive to water deficit include photosynthesis, transpiration, N_2 fixation, leaf area increase (via decreased specific leaf area, SLA), V stage progress, internode elongation (height and width increase), and partitioning to roots. When root water uptake is unable to meet the potential transpiration of the foliage, total crop photosynthesis and transpiration are reduced equally in proportion to the decrease in water uptake. Decreases in V stage development, SLA, and internode elongation and increase in partitioning to roots begin when the ratio of root water uptake to potential evapotranspiration first falls below 1.5. These processes are considered more sensitive to water supply than photosynthesis. Rate of root depth progression is accelerated when the ratio drops below 1.0. The N_2 fixation process is decreased as the fraction available of soil water in the nodule zone (between 5 and 40 cm depth) falls below a species-dependent fraction available of water (0.35–0.40). Leaf senescence is accelerated (with a delay feature) following days when plant transpirational demand cannot be met.

J. Coupling of Pest Damage

Under field conditions, biotic pests frequently decrease crop yield below the possible yield predicted considering the limitations of temperature, solar radiation, water, nitrogen fertilization, and crop management conditions. A generic approach was developed in CROP-GRO to allow scouting data on observed pest damage to be input via pest-coupling modules to predict yield reduction from pests (Batchelor et al., 1993; Boote et al., 1993). With these generic pest-coupling modules, pest effects are allowed to modify crop state variables (mass or numbers of various tissues), rate variables (photosynthesis, water flow, senescence), or inputs (water, light, nutrients). Pest damage estimates can be input as "damage files." Examples of this approach are illustrated by Boote et al. (1993) for simulating the effects of defoliating insects, seed-feeding insects, and rootknot nematode on soybean growth and for simulating effects of leafspot disease on peanut growth and yield. Although specific damage files need to be developed for many more pests, the procedure has been developed, and most of the coupling points to crop carbon balance are already coded into CROPGRO. Standard input-output files for the pest damage aspects have been developed in the DSSAT software (Hoogenboom et al., 1994b).

IV. SPECIES AND CULTIVAR DIFFERENCES IN PROCESS RESPONSE TO CLIMATE

With the CROPGRO model, there is one common set of Fortran code, and all species attributes associated with dry beans, peanuts, and soybeans, are input from species files, as well as ecotype and cultivar files. There are no hardcoded, crop-specific subroutines in CROPGRO; rather, all species or cultivar differences are handled externally through input parameters and relationships described in these three input files. The species files describe the basic tissue compositions; photosynthetic, respiratory, N assimilation, partitioning, senescence, phenological, and growth processes; and the sensitivity of those processes to environmental factors (e.g., temperature, solar irradiance, plant N status, plant water deficit, or soil water deficit). Cultivar and ecotype aspects are found in another set of files. For example, the cultivar file describes, in one line, a minimum set of 15 important genetic attributes that primarily determine life-cycle attributes but also include traits such as leaf photosynthesis, seed size, and seed-fill duration.

The species files of CROPGRO are used to create soybean, peanut, or dry bean attributes and create species that differ in sensitivity to temperature, solar irradiance, water deficit, etc. Basic differences in dry matter accumulation ability of species are described by the curvilinear responses of daily assimilation versus daily solar irradiance (described by two parameters) when all other factors such as LAI, water, temperature, and leaf N are optimum. For the hourly version of canopy assimilation, the dry matter accumulation ability depends primarily on single-leaf photosynthesis rate (A_{max}), which is both a species attribute and a cultivar attribute. Species files have been calibrated/coupled so that dry matter accumulation ability is nearly the same whether the daily or hourly assimilation option is selected.

Vegetative processes that are sensitive to temperature include rate of germination and emergence, rate of vegetative node formation, duration of vegetative growth, photosynthesis, maintenance respiration, nodule growth rate, specific nodule activity, specific leaf area, and specific internode length. Such temperature sensitivities are described in the species files for each process by table look-up functions or by four-point temperature functions describing minimum cardinal temperature, with either linear or quadratic interpolation between these four points. Figure 2, for example, shows temperature response for rate of vegetative node appearance, reproductive development rate prior to beginning seed, and reproductive development rate after beginning seed. Similar functions are used to describe temperature sensitivity of other reproductive processes: rate of pod addition, limits on partitioning to reproductive, and rate of single-seed growth. This is how species differences in response to temperature were created.

Because of the importance of temperature and unusually cool or hot weather stresses on seed-yielding ability, we give several specific examples of literature sources and derivation of temperature effects on reproductive growth processes of these grain legumes. Temperature effects on soybean life-cycle duration have been described and shown in Fig. 2. Figure 9 illustrates the relative shape of single-seed growth rate versus temperature for soybeans, which have their optimum at 23.5°C according to the data of Egli and Wardlaw (1980), although we extended the lower optimum to 21°C. With this function, we can mimic the reported decline in final mass per seed and individual seed growth rate as temperature exceeds 23°C (Egli and Wardlaw, 1980; Baker et al., 1989; Pan et al., 1994). The same seed growth rate function works well for peanut plants, in which final seed size is also reported to decline as temperature increases (Cox, 1979). There are no data on single-seed growth rate versus temperature for dry beans although their seeds grow at cooler temperatures and are thought to be more sensitive to elevated temperature than soybeans.

The relative shape of the pod addition rate versus temperature curve (Fig. 9) for soybeans is presumed similar to the single-seed growth rate function except that the upper optimum is extended to 26.5°C and pod addition is reduced at temperatures below 21.5°C to a cutoff of zero podset below 14°C. Several studies indicate that soybeans do not form pods when the nighttime temperature is 14°C or

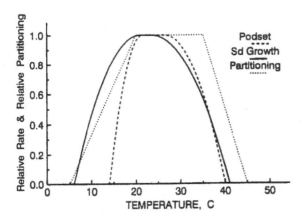

FIGURE 9   Effect of temperature on single-seed growth rate, relative pod addition rate, and relative partitioning limit for soybeans.

lower (Thomas and Raper, 1977, 1978) or even 18°C and lower (Hesketh et al., 1973; Hume and Jackson, 1981), although the response is cultivar-dependent (Lawn and Hume, 1985). The pod addition rate function for peanuts is also similar except that the pod addition rate declines to zero as the temperature drops from 21 to 17°C according to data of Campbell (1980). The pod addition function for dry beans is assumed to have its entire response curve shifted to cooler temperatures.

At higher than optimum temperature, these three grain legumes exhibit poor reproductive fruit formation, extended vegetative growth, and poor partitioning. Specific causes may be associated with increased production of vegetative primordia and the expression of more vegetative sites, high-temperature-induced delay in onset of reproductive sites, and failure of successful fertilization of reproductive sites. Thus, we have included a "partitioning limit" function (Fig. 9) that decreases the maximum fraction partitioning to pods as temperature exceeds 35, 33, and 28°C in soybeans, peanuts, and dry beans, respectively. We have since confirmed this general phenomenon with experimental data on soybeans (Boote et al., 1994; Pan et al., 1994). Harvest index declines progressively as temperature increases above the optimum for soybeans (Baker et al., 1989; Pan et al., 1994) and peanuts (Hammer et al., 1995).

For additional information on other traits and relationships in the individual species files, readers are referred to each species file, to definitions of the relationships in the species files, and to DSSAT publications (Hoogenboom et al., 1994b).

V. MODEL INPUTS AND OUTPUTS

A. Required Model Inputs

The CROPGRO model uses the standard input/output files specified by the IBSNAT project, used in common with other crop models within IBSNAT's Decision Support System for Agrotechnology Transfer (DSSAT) (Hoogenboom et al., 1994b). Required weather data include daily solar radiation, maximum air temperature, minimum air temperature, and rainfall. To run the Penman–Monteith option, daily wind run and dewpoint temperature (or minimum daily relative humidity) are also required.

Soils information can typically be obtained or extracted from U.S. Soil Conservation Service (SCS) publications. Using this SCS data, the soils file creation software in the DSSAT can be used to derive most of the soils traits needed to run the soil water balance. Caution is recommended, however, in relying on the LL, DUL, and SAT values obtained with this program, particularly for high clay content soils. Direct field measurements of LL, DUL, and SAT values for soil layers are better.

Crop cultural management practices such as planting date, row spacing, plant population, cultivar, fertilization, irrigation, crop residue, and other treatment information are entered into a File X used for running the crop model. The File X is used to specify cultural management conditions for actual site-specific fields and experiments. It can also be used to specify hypothetical "what-if" simulations. This file has cross-references that link a specific field or treatment to a given soil, to a given weather-year file, and to a given cultivar in the cultivar file. Cultivar characteristics, 15 in all, are described in the "read-in" cultivar file. Procedures are available for developing some of the more critical required cultivar traits from variety trial and growth analysis information. For more information, readers are referred to Hoogenboom et al. (1994b).

B. Typical Model Outputs

There are standard output files from CROPGRO for end-of-season output (OVERVIEW.OUT, NBAL.OUT, SUMMARY.OUT) and for in-season processes of growth (GROWTH.OUT), water balance (WATER.OUT), nitrogen balance (NITROGEN.OUT), and carbon balance (CARBON.OUT). These files are automatically read by the graphics program and provide time-series graphs of growth, water, nitrogen, and carbon balance processes for different treatments. The graphics also allow a comparison of simulated versus observed "end-of-season" traits.

In addition, there are standard DSSAT files for entering observed data that can be graphically compared to the simulated outputs. For example, observed plant growth analyses, leaf or canopy photosynthesis, light interception, N concentration, carbohydrate concentration, soil water by layer, and soil N by layer measured during the season are entered into a File T (actual prefix is determined by experimental identification). The graphics program reads this file and allows you to compare simulated model outputs to observed data. Likewise, the final end-of-season yield and component information is entered into a File A, which the model reads and places on the computer screen for you to compare to yield and other aspects predicted at the end of the season.

VI. SIMULATED RESPONSES OF THE CROPGRO MODEL

A. Simulating Species Differences in Growth Patterns

CROPGRO's species files allowed it to simulate the different growth patterns for peanuts and soybeans grown in the same experiment as shown in Figs. 10 and 11. 'Florunner' peanuts are nonsenescing indeterminate plants that retain their LAI to maturity; by contrast, 'Bragg' soybeans are senescing determinate plants that have faster decline in LAI during seed growth and rapid leaf abscission at maturity (Fig. 10a). Note in Fig. 10b that the peanuts have a steeper slope of dry matter accumulation than the soybeans (in both real data and simulations; this difference was also observed in a 1976 experiment). The greater crop growth rate of the peanuts is associated with greater leaf and canopy assimilation, presumed lower maintenance respiration, and lower N concentration of vegetative tissue. The indeterminate peanuts have an earlier onset of seed addition but a very long seed addition phase, which contrasts with the rapid increase in seed number of the determinate soybeans (Fig. 11a). 'Florunner' peanuts produced a greater total seed yield than 'Bragg' soybeans, primarily because of their longer seed growth period. The longer seed-filling period of peanuts is evident in the longer period of total seed mass

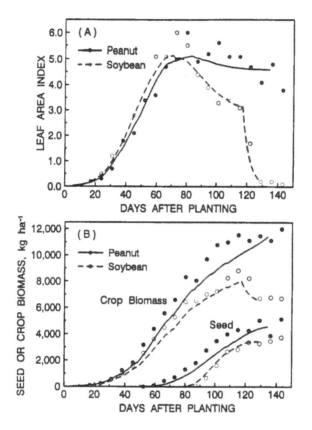

FIGURE 10   (a) Leaf area index and (b) total crop mass and seed mass of 'Bragg' soybeans and 'Florunner' peanuts planted June 12, 1984 at Gainesville, Florida under irrigated conditions. Lines are simulations, and symbols indicate field-observed values.

accumulation (Fig. 10b) and the longer duration of increase in average mass per seed (Fig. 11b).

B.   Simulating Cultivar (Maturity Group) Differences Within a Species

Cultivar differences within a species are typically traits that influence life-cycle duration and degree of determinancy. Soybean cultivars are broadly categorized into maturity groups (MG) from 000 to 12, based primarily on their sensitivity to day length, which influences their life-cycle duration. Soybean cultivars in MG 000 that have shorter life

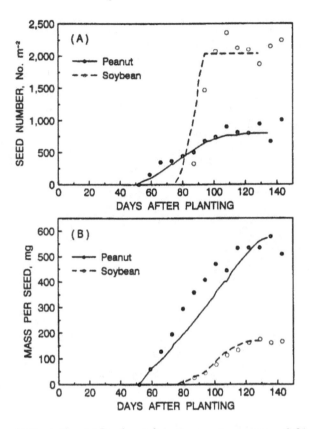

FIGURE 11 (a) Seed number per square meter and (b) average mass per seed of 'Bragg' soybeans and 'Florunner' peanuts planted June 12, 1984 at Gainesville, Florida under irrigated conditions. Lines are simulations, and symbols indicate field-observed values.

cycles and are less day length sensitive can be grown at northerly latitudes in the United States, Canada, and Europe, whereas cultivars grown in the southern United States and in the tropics have longer life cycles and greater day length sensitivity (approaching MG 12 near the equator). Table 2 shows critical short day length values (CSDL), photoperiod sensitivity slopes (PPSEN), and physiological day requirements to complete their life cycles (under short days and warm temperatures) for MG 00–9 soybean cultivars as solved by Grimm et al. (1993, 1994) from experimental data. Table 2 also shows simulated days to maturity and yield performance for MG 00–9 cultivars

Table 2 Effect of Soybean Maturity Group (MG) on Days to Maturity and Yield for Crops Grown in Gainesville, Florida and Ames, Iowa[a]

| | | | | | | Florida simulation | | Iowa simulation | |
| | CSDL | PPSEN | EM-FL | FL-SD | SD-PM | Maturity | Yield | Maturity | Yield |
MG	(h)	(days/h)	(days)	(days)	(days)	(days)	(kg/ha)	(days)	(kg/ha)
00	14.35	0.148	16.0	12.0	30.0	70.5	842	94.0	1514
0	14.10	0.171	16.8	13.0	31.0	73.7	916	101.4	1838
1	13.84	0.203	17.0	13.0	32.0	77.2	984	109.5	2086
2	13.59	0.249	17.4	13.5	33.0	84.3	1108	122.1	2399
3	13.40	0.285	19.0	14.0	34.0	94.2	1398	135.3	2463
4	13.09	0.294	19.4	15.0	34.5	107.9	1984	152.2[b]	2280
5	12.83	0.303	19.8	15.5	35.0	120.0	2261	166.8[c]	1730
6	12.58	0.311	20.2	16.0	35.5	132.7	2572	—[d]	
7	12.33	0.320	20.8	16.0	36.0	145.0	2655	—[d]	
8	12.07	0.330	21.5	16.0	36.0	159.3	2785	—[d]	
9	11.88	0.340	23.0	16.0	36.5	170.2	2660	—[d]	

[a]Crops were planted on day 123 at 30 plants/m^2 and grown under rain-fed conditions, using 10 years of historical weather at each site (1978–1987 in Florida and 10 of 1984–1995 in Iowa). Critical short day length (CSDL), photoperiod sensitivity (PPSEN), and physiological days from emergence to flowering (EM-FL), flowering to first seed (FL-SD), and first seed to physiological maturity (SD-PM) are given for each MG, as used by the CROPGRO soybean model. Physiological days requirement to emergence is 3.6 days.
[b]Freeze damage in 2 of 10 years.
[c]Freeze damage in 5 of 10 years.
[d]Freeze damage in 10 of 10 years and maturity not reached.

grown at Gainesville, Florida, and at Ames, Iowa that were planted on May 1. The low number MGs mature much too early in Florida and yield poorly. The optimum MGs in Florida for yield and most effective use of the season are MGs 6–9 (both in simulations and in actual production practice). In Iowa, the very low number MGs are also early-maturing and low-yielding, but MGs 2 and 3 are the highest yielding and more optimally fit the season at Ames, Iowa. MG 4 was only somewhat lower in yield, but it suffered freeze damage in two out of 10 seasons even for the early May planting date. MG 5 suffered freeze damage prior to maturity in five out of 10 seasons and would not be grown in Iowa. The shorter critical day length and greater photoperiod sensitivity of higher number MG cultivars prolongs their life cycles when they are grown in longer day length environments; thus MGs 5–12 will not mature prior to frost in the midwestern states. CROPGRO simulations of dry bean and peanut cultivars employ similar traits that influence life cycle and other genotype-specific processes. See Boote and Tollenaar (1994) for examples of other simulated cultivar traits that influence yield.

C. Response to Planting Date

Soybean yield response to planting date was simulated with CROP-GRO for three sites (Florida, North Carolina, and Iowa) using soybean cultivar characteristics described above (Fig. 12). In Florida, optimum planting date for highest yield of MG 7 under rain-fed conditions was April and May (days 93–123), with good yield levels for planting in early June. Yields were lower with earlier plantings because plants were shorter and flowered earlier, and plantings on days 63 and 78 suffered killing frost damage in three out of 10 and one out of 10 years, respectively. Later than optimum plantings had progressively lower yields because of shorter life cycles and smaller plants. The response to planting date for MG 6 in North Carolina was

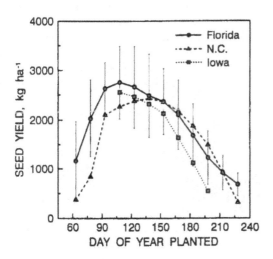

FIGURE 12 Soybean yield response to planting date simulated under rain-fed conditions for 10 years of historical weather (ten from 1984 to 1995) at Ames, Iowa; 5 years of weather (1984–1988) at Clayton, North Carolina; and 10 years (1978–1987) at Gainesville, Florida. Maturity groups 3, 6, and 7 were used in Iowa, North Carolina, and Florida, respectively, grown at 30 plants/ m^2 in 76 cm rows in Iowa and 91 cm rows at the other two sites. Vertical bars represent plus or minus one standard deviation, shown for Florida simulations.

similar to that in Florida, with a broad optimum between mid-April and early June. Plantings on days 63 and 78 incurred killing frosts in four out of 10 and two out of 10 years, respectively. Late planting dates were lower in yield for the same reasons as in Florida. In addition, plantings on days 213 and 228 in North Carolina suffered freeze damage prior to maturity in two out of 10 and four out of 10 years, respectively. By contrast, yield of MG 3 in Iowa was highest with the earliest possible planting dates, mid-April to mid-May. Practically speaking, the mid-April date is not very feasible in Ames, Iowa because of poor soil workability and low soil temperatures, which result in poor stands (this is not simulated in CROPGRO). With mid-April plantings, there were three occurrences of subzero (Celsius) temperatures after planting that would not have killed the shoot because emergence required 2–3 weeks. Later plantings were progressively lower yielding. Killing freeze occurrences (below −2.2°C) were predicted in two of 10, four of 10, and nine of 10 years for planting on days 168, 183, and 198 (dates from mid-June to mid-July) in Iowa. As modeled, foliage is killed (if at or below −2.2°C) and the crop progresses to maturity without photosynthetic input except for mobilized protein and carbohydrate. These responses to planting date mostly mimic those reported in the literature for soybeans grown at the three locations.

D. In-Season Growth Responses to Water Deficit

Water deficit decreases canopy assimilation, reduces leaf area expansion, and enhances leaf area abscision. These features are illustrated in a simulation of 'Cobb' soybeans planted June 26, 1981 at Gainesville, Florida and grown under full irrigation compared to a treatment that endured two drought periods during vegetative growth (Fig. 13). During the two periods of soil water deficit, there was a complete cessation in LAI increase and minor leaf abscision (Fig. 13a). Leaf area growth was prolonged after the water deficit was relieved, resulting in only slightly lower LAI values during seed fill. Biomass accumulation was slowed during each drought period and continued to lag the fully irrigated treatment (Fig. 13b). Onset of seed growth was slightly delayed for the plants suffering soil water deficits during vegetative growth, but final seed yield was reduced less than 10% because the crop received full irrigation during seed fill and the LAI was nearly adequate. Droughts during reproductive growth typically decrease seed yield more, as we have seen with data from other years.

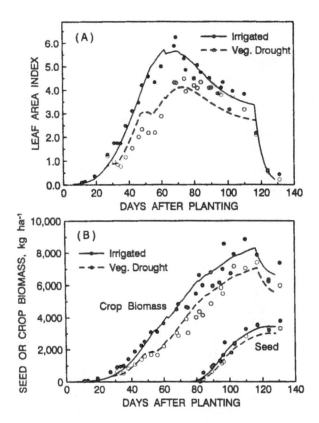

FIGURE 13 (a) Leaf area index and (b) total crop mass and seed mass of 'Cobb' soybeans (MG 8) planted June 26, 1981 at Gainesville, Florida and grown under full irrigation compared to a treatment exposed to two drought periods during vegetative growth. Lines are simulations, and symbols indicate field-observed values.

E. Final Seed Yield Response to Rainfall and Total Evapotranspiration

The Environmental Modifications section of the crop management file (File X) was used to vary rainfall from 10% to 150% of the actual rainfall for 10 weather years, 1978–1987, at Gainesville for 'Bragg' soybeans planted on day 125. Simulations were initialized with the soil profile containing 50% available soil water to 180 cm. Soybean seed yield increased with increasing rainfall, gradually achieving a plateau above 900 mm (Fig. 14a). At normal rainfall (the 100% case),

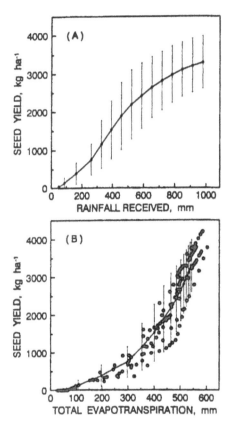

FIGURE 14   Predicted seed yield of 'Bragg' soybeans (MG 7) in response to (a) rainfall received and (b) total evapotranspiration. Simulated with 10 weather years (1978–1987) at Gainesville by modifying rainfall in steps of 10% from 10% to 150% of actual. Crop was planted on day 125, starting with 50% of available soil water in the soil profile to 180 cm.

the vertical bars represent plus or minus one standard deviation (± 1 SD) in yield associated with year-to-year variation in rainfall. Seed yield plotted against total E_t illustrates initial curvilinear increase, with a transition toward a linear response above 300 mm (Fig. 14b). In the curvilinear part of the relationship (at low rainfall), soil evaporation is initially a large fraction of total E_t. As rainfall increases, soil evaporation becomes a smaller fraction of E_t because a larger crop canopy is produced and shades the soil and more of the water use goes to transpiration in the linear phase. The scatter of points illustrates that different seed yields are possible with the same seasonal E_t owing to variation in drought patterns.

F. Response to Climatic Temperature Change at Two Locations

Variation in temperature considerably influences yield potential across locations, and year-to-year variation influences yield within a location. Furthermore, global warming of 2–4°C has been predicted as a consequence of a doubling in "greenhouse" gases (primarily carbon dioxide increase from fossil fuel burning). For this reason, we are interested not only in overall temperature effects on yield but also in the effects of temperature variation at a given location (Jones et al., 1996). Figure 15 illustrates predicted seed yield response to decrease or increase in temperature for MG 3 soybeans grown at Ames, Iowa and MG 6 soybeans grown at Gainesville, Florida under rain-fed conditions using 6 years of historical weather at each site. The response to temperature change differed across sites and depended on the site mean seasonal temperature. In Iowa where seasonal mean temperature was 21.9°C, seed yield was decreased by 2 or 4°C tem-

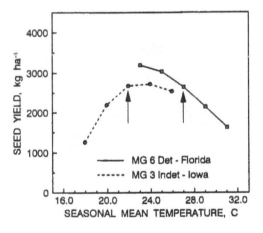

FIGURE 15 Predicted seed yield response to decrease or increase in temperature for maturity group 3 soybeans grown at Ames, Iowa or for MG 6 soybeans grown at Gainesville, Florida under rain-fed conditions in 91 cm row spacing at 30 plants/m², using 6 years of historical weather (1984–1989) at each site. Mean seasonal temperature over the soybean growth cycle for the 6 years was 21.9 and 27.0°C at Ames and Gainesville, respectively. [From Boote et al. (1996).]

perature decrease because temperatures became suboptimal for growth, photosynthesis, and seed fill in addition to extending the life cycle into freeze hazards. A 2°C increase had no effect, and a 4°C increase caused only a small yield decrease in Iowa. By contrast, in Florida where seasonal mean temperature was much warmer (27.0°C), yield was increased by 2 or 4°C decreases in temperature, and yield was decreased if temperature was increased. One implication of Fig. 15 is that a global temperature increase of 2–4°C would be expected to decrease soybean yields in the southern states but with little effect in the upper midwestern states. From Fig. 15, it appears that 22–24°C is the optimum mean seasonal temperature for seed yield. This optimum for seed yield is the integrated result of all model processes and temperature influences on processes of vegetative development, reproductive development, photosynthesis, respiration, N_2 fixation, vegetative growth, and seed growth processes. CROP-GRO's predicted soybean yield response to temperature is fairly similar to that predicted by SOYGRO V5.42 (Boote et al., 1989a), in part because many of the temperature relationships are similar. The decline in seed yield at higher than optimum temperature is consistent with recent experimental data (Pan et al., 1994).

G. Model Applications in Crop Production Management

The CROPGRO model and its precursor models have been used considerably in the research mode to synthesize understanding of growth physiology, to hypothesize genetic improvement, to devise production strategies to optimize yield relative to inputs of water and other natural resources, and to develop policy recommendations relative to global climate change (Boote et al., 1996). Possibly the greatest future potential (and challenge) is to operationalize the model to be used as a decision support tool (DSS) for production agriculture (Hoogenboom et al., 1994a). The goal is to help producers make preseason and in-season decisions related to cultivar choice, planting date, row spacing, fertilization, irrigation, crop rotation, land area devoted to each crop enterprise, replant decisions, and minimization of water, soil, and agrochemical losses. We are addressing this DSS use of the CROPGRO model in a large interdisciplinary project in soybean-producing states of the United States. This project is also dealing with problems of data availability (historical weather records, current year's weather, soils information, cultivar traits, and prior field history). Site-specific production recommendations with the model will potentially use geographic information systems and global positioning sensors as producers develop maps of grain yield production with yield monitors on combines and then plan their future production management decisions. We think crop models can be an important interface in such site-specific recommendations. The CROPGRO model has already been coupled to such GIS packages.

VII. CONCLUSIONS

The CROPGRO model is a process-based model for grain legumes that includes leaf-level photosynthesis, hedgerow canopy light interception, soil N balance, N uptake, N_2 fixation, soil water balance, evapotranspiration, respiration, leaf area growth, pod and seed addition, growth of component parts, root growth, senescence, N mobilization, carbohydrate dynamics, and crop development processes. The code is generic and modular and will simulate three legumes (soybeans, peanuts, and dry beans), using species characteristics in a read-in species file. Model structure, processes, relationships to climatic factors, and example simulations have been described. Most of the example simulations were for soybeans as we have done the most model testing for soybeans (Boote et al., 1997b). Nevertheless, considerable model evaluations (with PNUTGRO and BEANGRO versions) were done previously on peanuts (Boote et al., 1991; Singh et al., 1994a, 1994b) and dry beans (Hoogenboom et al., 1994c).

CROPGRO is released as part of the DSSAT V3.0 Decision Support software and benefits from using standard input/output files for weather, soils, and management data. The DSSAT V3.0 software, including CROPGRO and the CERES models, can be purchased from IBSNAT, University of Hawaii (Uehara et al., 1994).

REFERENCES

Baker, J. T., L. H. Allen, Jr., K. J. Boote, P. H. Jones, and J. W. Jones. 1989. Response of soybean to air temperature and CO_2 concentration. *Crop Sci.* 29:98–105.

Batchelor, W. D., J. W. Jones, K. J. Boote, and H. O. Pinnschmidt. 1993. Extending the use of crop models to study pest damage. *Trans. ASAE 36*: 551–558.

Boote, K. J., and N. B. Pickering, 1994. Modeling photosynthesis of row crop canopies. *HortScience 29*:1423–1434.

Boote, K. J., and M. Tollenaar. 1994. Modeling genetic yield potential. In: *Physiology and Determination of Crop Yield.* K. J. Boote, J. M. Bennett, T. R.

Sinclair, and G. M. Paulsen (Eds.). Madison, WI: ASA-CSSA-SSSA, pp. 533–565.

Boote, K. J., J. W. Jones, and J. M. Bennett. 1984. Factors influencing crop canopy CO_2 assimilation of soybean. In: Proceedings, World Soybean Research Conference III. Boulder, CO: Westview Press, pp. 780–788.

Boote, K. J., J. W. Jones, J. W. Mishoe, and G. G. Wilkerson. 1986. Modeling growth and yield of groundnut. In: *Agrometeorology of Groundnut*. Proc. Int. Symp. Aug. 21–26, 1985, ICRISAT Sahelian Center, Niamey, Niger. Patancheru, India: ICRISAT, pp. 243–254.

Boote, K. J., J. W. Jones, G. Hoogenboom, G. G. Wilkerson, and S. S. Jagtap. 1987. PNUTGRO V1.0, Peanut Crop Growth Simulation Model, User's Guide. Florida Agric. Exp. Sta., Journal No. 8420. Gainesville, FL: University of Florida.

Boote, K. J., J. W. Jones, and G. Hoogenboom. 1989a. Simulating growth and yield response of soybean to temperature and photoperiod. Proc. World Soybean Research Conf. IV, Mar. 5–9, 1989, Buenos Aires, Argentina, pp. 273–278.

Boote, K. J., J. W. Jones, G. Hoogenboom, G. G. Wilkerson, and S. S. Jagtap. 1989b. PNUTGRO V1.02, Peanut Crop Growth Simulation Model, User's Guide. Florida Agric. Exp. Sta., Gainesville, FL: University of Florida.

Boote, K. J., J. W. Jones, and P. Singh. 1991. Modeling growth and yield of groundnut—State of the art. In: *Groundnut—A Global Perspective*. Proc. Int. Workshop, Nov. 25–29, 1991, Patancheru, India: ICRISAT, pp. 331–343.

Boote, K. J., W. D. Batchelor, J. W. Jones, H. Pinnschmidt, and G. Bourgeois. 1993. Pest damage relations at the field level. In: *Systems Approaches for Agricultural Development*. F. W. T. Penning de Vries et al. (Eds.). Dordrecht, The Netherlands: Kluwer, pp. 277–296.

Boote, K. J., M. B. Kirkham, L. H. Allen, Jr., and J. T. Baker. 1994. Effects of temperature, light, and elevated CO_2 on assimilate allocation. (Abstract.) *Agron. Abstr. 1994*:150.

Boote, K. J., J. W. Jones, and N. B. Pickering. 1996. Potential uses and limitations of crop models. *Agron. J. 88*:704–716.

Boote, K. J., J. W. Jones, G. Hoogenboom, and N. B. Pickering. 1997a. CROP-GRO model for grain legumes. In: *A Systems Approach to Research and Decision Making*. G. Y. Tsuji, G. Hoogenboom, and P. K. Thornton (Eds.). Dordrecht, The Netherlands, Kluwer, Chapter 6.

Boote, K. J., J. W. Jones, G. Hoogenboom, and G. G. Wilkerson. 1997b. Evaluation of the CROPGRO-soybean model over a wide range of experiments. In: *Systems Approaches for Sustainable Agricultural Development: Applications of Systems Approaches at the Field Level*. Volume 2. M. J. Kropff et al. (Eds.). Dordrecht, The Netherlands: Kluwer, pp. 113–133.

Campbell, I. S. 1980. Growth and development of Florunner peanuts as affected by temperature. Ph.D. Dissertation. Gainesville, FL: University of Florida.

Cox, F. R. 1979. Effect of temperature treatment on peanut vegetative and fruit growth. *Peanut Sci. 6*:14–17.

de Wit, C. T. 1965. Photosynthesis of leaf canopies. Verslagen van landbouw kundige Onderzoekingen (Agricultural Research Report) 663. Wageningen, The Netherlands: Pudoc.

de Wit, C. T., J. Goudriann, H. H. van Laar, F. W. T. Penning de Vries, R. Rabbinge, H. van Keulen, W. Louwerse, L. Sibma, and C. de Jonge. 1978. *Simulation of Assimilation, Respiration, and Transpiration of Crops.* Simulation Monographs. Wageningen, The Netherlands: Pudoc.

Duncan, W. G., R. S. Loomis, W. A. Williams, and R. Hanau. 1967. A model for simulating photosynthesis in plant communities. *Hilgardia 38*:181–205.

Egli, D. B., and I. F. Wardlaw. 1980. Temperature response of seed growth characteristics of soybeans. *Agron. J. 72*:560–564.

Erbs, D. G., S. A. Klein, and J. A. Duffie. 1982. Estimation of the diffuse radiation fraction for hourly, daily and monthly-average global radiation. *Solar Energy 28*:293–302.

Farquhar, G. D., and S. von Caemmerer. 1982. Modelling of photosynthetic response to environment. In: *Encyclopedia of Plant Physiology.* Vol. 12B. *Physiological Plant Ecology II.* O. L. Lange, P. S. Nobel, C. B. Osmond, and H. Ziegler (Eds.). Berlin: Springer-Verlag, pp. 549–587.

Godwin, D. C., and C. A. Jones. 1991. Nitrogen dynamics in soil–plant systems. In: *Modeling Soil and Plant Systems.* J. Hanks and J. T. Ritchie (Eds.). ASA Monograph 31. Madison, WI: American Society of Agronomy, pp. 287–321.

Godwin, D. C., J. T. Ritchie, U. Singh, and L. Hunt. 1989. *A User's Guide to CERES-Wheat V2.10.* Muscle Shoals, AL: International Fertilizer Development Center.

Grimm, S. S., J. W. Jones, K. J. Boote, and J. D. Hesketh. 1993. Parameter estimation for predicting flowering date of soybean cultivars. *Crop Sci. 33*: 137–144.

Grimm, S. S., J. W. Jones, K. J. Boote, and D. C. Herzog. 1994. Modeling the occurrence of reproductive stages after flowering for four soybean cultivars. *Agron. J. 86*:31–38.

Hammer, G. L., T. R. Sinclair, K. J. Boote, G. C. Wright, H. Meinke, and M. J. Bell. 1995. A peanut simulation model. I. Model development and testing. *Agron. J. 87*:1085–1093.

Harley, P. C., J. A. Weber, and D. M. Gates. 1985. Interactive effects of light, leaf temperature, [CO_2] and [O_2] on photosynthesis in soybean. *Planta 165*: 249–263.

Hesketh, J. D., K. L. Myhre, and C. R. Willey. 1973. Temperature control of time intervals between vegetative and reproductive events in soybeans. *Crop Sci. 13*:250–254.

Hofstra, G., and J. D. Hesketh. 1975. The effects of temperature and CO_2 enrichment on photosynthesis in soybean. In: *Environmental and Biological Control of Photosynthesis.* R. Marcelle (Ed.). The Hague: Dr. W. Junk, b.v. Publishers, pp. 71–80.

Hoogenboom, G., J. W. White, J. W. Jones, and K. J. Boote. 1990. BEANGRO V1.0: Dry Bean Crop Growth Simulation Model. User's Guide. Florida

Agric. Exp. Sta., Journal No. N-00379. Gainesville, FL: University of Florida.

Hoogenboom, G., J. W. Jones, and K. J. Boote. 1991. Predicting growth and development of grain legumes with a generic model. ASAE Paper No. 91-4501. St. Joseph, MI: ASAE.

Hoogenboom, G., J. W. Jones, and K. J. Boote. 1992. Modeling growth, development, and yield of grain legumes using SOYGRO, PNUTGRO, and BEANGRO: A review. *Trans. ASAE* 35:2043–2056.

Hoogenboom, G., J. W. Jones, K. J. Boote, W. T. Bowen, N. B. Pickering, and W. D. Batchelor. 1993. Advancement in modeling grain legume crops. ASAE Paper No. 93-4511. St. Joseph, MI: ASAE.

Hoogenboom, G., J. W. Jones, L. A. Hunt, P. K. Thornton, and G. Y. Tsuji. 1994a. An integrated decision support system for crop model applications. ASAE Paper No. 94-3025. St. Joseph, MI: ASAE.

Hoogenboom, G., J. W. Jones, P. W. Wilkens, W. D. Batchelor, W. T. Bowen, L. A. Hunt, N. B. Pickering, U. Singh, D. C. Godwin, B. Baer, K. J. Boote, J. T. Ritchie, and J. W. White. 1994b. Crop models. In: *DSSAT Version 3*. Vol. 2. G. Y. Tsuji, G. Uehara, and S. Balas (Eds.). Honolulu, HI: University of Hawaii, pp. 95–244.

Hoogenboom, G., J. W. White, J. W. Jones, and K. J. Boote. 1994c. BEANGRO, a process-oriented dry bean model with a versatile user interface. *Agron. J.* 86:182–190.

Hume, D. J., and A. K. H. Jackson. 1981. Pod formation in soybean at low temperatures. *Crop Sci.* 21:933–937.

IBSNAT. 1989. Decision Support System for Agrotechnology Transfer Version 2.1 (DSSAT V2.1). International Benchmark Sites Network for Agrotechnology Transfer Project. Honolulu, HI: Dept. of Agronomy and Soil Sci., College of Tropical Agriculture and Human Resources, University of Hawaii.

Jeffers, D. L., and R. M. Shibles. 1969. Some effects of leaf area, solar radiation, air temperature, and variety on net photosynthesis in field-grown soybeans. *Crop Sci.* 9:762–764.

Jensen, M. E., R. D. Burman, and R. G. Allen (Eds.). 1990. *Evapotranspiration and Irrigation Water Requirements: A Manual.* New York: Am. Soc. of Civil Engineers.

Jones, J. W., and K. J. Boote. 1987. Simulation Models for Soybeans and Other Crops. Concepts of Crop Systems. Tech. Bull. 106. Taipei City, Taiwan: Food and Fertilizer Technology Center, pp. 1–7.

Jones, J. W., K. J. Boote, S. S. Jagtap, G. Hoogenboom, and G. G. Wilkerson. 1987. SOYGRO V5.4, Soybean Crop Growth Model, User's Guide. Florida Agric. Exp. Sta., Journal No. 8304. Gainesville, FL: University of Florida.

Jones, J. W., K. J. Boote, G. Hoogenboom, S. S. Jagtap, and G. G. Wilkerson. 1989. SOYGRO V5.42, Soybean Crop Growth Simulation Model. User's Guide. Florida Agric. Exp. Sta., Journal No. 8304. Gainesville, FL: University of Florida.

Jones, J. W., N. B. Pickering, C. Rosenzweig, and K. J. Boote. 1996. Simulated impacts of global climate change on crops. In: *Climate Change and Rice.* S. Peng, K. T. Ingram, H.-U. Neue, and L. H. Ziska (Eds.). New York: Springer-Verlag, pp. 218–231.

Keulen, H. van. 1975. *Simulation of Water Use and Herbage Growth in Arid Regions.* Simulation Monographs. Wageningen, The Netherlands: Pudoc.

Kimball, B. A., and L. A. Bellamy. 1986. Generation of diurnal solar radiation, temperature, and humidity patterns. *Energy Agric.* 5:185–197.

Lawn, R. J., and D. J. Hume. 1985. Response of tropical and temperate soybean genotypes to temperature during early reproductive growth. *Crop Sci.* 25:137–142.

McCree, K. J. 1974. Equations for the rate of dark respiration of white clover and grain sorghum as functions of dry weight, photosynthetic rate, and temperature. *Crop Sci.* 14:509–514.

Pan, D., K. J. Boote, J. T. Baker, L. H. Allen, Jr., and N. B. Pickering. 1994. Effects of elevated temperature and CO_2 on soybean growth, yield, and photosynthesis. (Abstract.) *Agron. Abstr. 1994*:150.

Parton, W. J., and J. A. Logan. 1981. A model for diurnal variation in soil and air temperature. *Agric. For. Meteorol.* 23:205–216.

Penning de Vries, F. W. T., and H. H. van Laar. 1982. Simulation of growth processes and the model BACROS. In: *Simulation of Plant Growth and Crop Production.* F. W. T. Penning de Vries and H. H. van Laar (Eds.). Wageningen, The Netherlands: Pudoc, pp. 114–135.

Penning de Vries, F. W. T., A. H. M. Brunsting, and H. H. van Laar. 1974. Products, requirements and efficiency of biosynthesis: A quantitative approach. *J. Theor. Biol.* 45:339–377.

Pickering, N. B., J. W. Jones, and K. J. Boote. 1990. A moisture- and CO_2-sensitive model of evapotranspiration and photosynthesis. ASAE Paper No. 90-2519. St. Joseph, MI: ASAE.

Pickering, N. B., J. W. Jones, K. J. Boote, G. H. Hoogenboom, L. H. Allen, Jr., and J. T. Baker. 1993. Modeling soybean growth under climate change conditions. ASAE Paper No. 93-4510. St. Joseph, MI: ASAE.

Pickering, N. B., J. W. Jones, and K. J. Boote. 1995. Adapting SOYGRO V5.42 for prediction under climate change conditions. In: *Climate Change and Agriculture: Analysis of Potential International Impacts.* C. Rosenzweig, J. W. Jones, and L. H. Allen, Jr. (Eds.). ASA Spec. Pub. No. 59. Madison, WI: ASA-CSSA-SSSA, pp. 77–98.

Priestley, C. H. B., and R. J. Taylor. 1972. On the assessment of surface heat flux and evaporation using large-scale parameters. *Mon. Weather Rev.* 100: 81–92.

Ritchie, J. T. 1985. A user-oriented model of the soil water balance in wheat. In: *Wheat Growth and Modeling.* E. Fry and T. K. Atkin (Eds.). NATO-ASI Ser. New York: Plenum, pp. 293–305.

Scholberg, J. M. S., B. L. McNeal, J. W. Jones, K. J. Boote, and C. D. Stanley. 1995. Calibration of a generic crop-growth model (CROPGRO) for field-grown tomato. (Abstract.) *Agron. Abstr. 1995*:17.

Scholberg, J. M. S., K. J. Boote, J. W. Jones, and B. L. McNeal. 1997. Adaptation of the CROPGRO model to simulate the growth of field-grown tomato. In: *Systems Approaches for Sustainable Agricultural Development: Applications of Systems Approaches at the Field Level*. M. J. Kropff et al. (Eds.). Dordrecht, The Netherlands: Kluwer, pp. 135–151

Singh, P., K. J. Boote, A. Yogeswara Rao, M. R. Iruthayaraj, A. M. Sheikh, S. S. Hundal, R. S. Narang, and Phool Singh. 1994a. Evaluation of the groundnut model PNUTGRO for crop response to water availability, sowing dates, and seasons. *Field Crops Res. 39*:147–162.

Singh, P., K. J. Boote, and S. M. Virmani. 1994b. Evaluation of the groundnut model PNUTGRO for crop response to plant population and row spacing. *Field Crops Res. 39*:163–170.

Spitters, C. J. T. 1986. Separating the diffuse and direct component of global radiation and its implication for modeling canopy photosynthesis. Part II. Calculation of canopy photosynthesis. *Agric. For. Meteorol. 38*:231–242.

Spitters, C. J. T., H. A. J. M. Toussaint, and J. Goudriaan. 1986. Separating the diffuse and direct components of global radiation and its implications for model canopy photosynthesis. I. Components of incoming radiation. *Agric. For. Meteorol. 38*:217–229.

Thomas, J. F., and C. D. Raper, Jr. 1977. Morphological response of soybeans as governed by photoperiod, temperature, and age at treatment. *Bot. Gaz. 138*:321–328.

Thomas, J. F., and D. D. Raper, Jr. 1978. Effect of day and night temperature during floral induction on morphology of soybean. *Agron. J. 70*:893–898.

Uehara, G., G. Y. Tsuji, and S. Balas (Eds.). 1994. *DSSAT Version 3*. Vols. 1–3. Honolulu, HI: University of Hawaii.

Wilkerson, G. G., J. W. Jones, K. J. Boote, K. T. Ingram, and J. W. Mishoe. 1983. Modeling soybean growth for crop management. *Trans. ASAE 26*: 63–73.

Wilkerson, G. G., J. W. Jones, K. J. Boote, and J. W. Mishoe. 1985. SOYGRO V5.0: Soybean Crop Growth and Yield Model. Technical documentation. Gainesville, FL: Agricultural Engineering Department, University of Florida.

Index